Bruchids and Legumes: Economics, Ecology and Coevolution

SERIES ENTOMOLOGICA

EDITOR

K. A. SPENCER

VOLUME 46

Bruchids and Legumes: Economics, Ecology and Coevolution

Proceedings of the Second International Symposium on
Bruchids and Legumes (ISBL-2) held at Okayama (Japan),
September 6–9, 1989

Edited by

K. Fujii, A. M. R. Gatehouse, C. D. Johnson,
R. Mitchel and T. Yoshida

KLUWER ACADEMIC PUBLISHERS
DORDRECHT / BOSTON / LONDON

Library of Congress Cataloging in Publication Data

International Symposium on Bruchids and Legumes (2nd : 1989: Okayama
-shi, Japan)
    Bruchids and legumes : economics, ecology, and coevolution :
proceedings of the Second International Symposium on Bruchids and
Legumes (ISBL-2) held at Okayama, Japan, September 6-9, 1989 /
edited by K. Fujii ... [et al.].
        p.   cm. -- (Series entomologica ; 46)
    ISBN 0-7923-0701-1 (alk. paper)
    1. Legumes--Diseases and pests--Congresses. 2. Bruchidae--Host
plants--Congresses. 3. Bruchidae--Control--Congresses.
4. Bruchidae--Ecology--Congresses. 5. Coevolution--Congresses.
I. Fujii, Kōichi, 1942-   . II. Title. III. Series: Series
entomologica ; v. 46.
SB608.L4I55 1989
633.3'0497648--dc20   90-32079

ISBN 0-7923-0701-1

Published by Kluwer Academic Publishers,
P.O. Box 17, 3300 AA Dordrecht, The Netherlands.

Kluwer Academic Publishers incorporates
the publishing programmes of
D. Reidel, Martinus Nijhoff, Dr W. Junk and MTP Press.

Sold and distributed in the U.S.A. and Canada
by Kluwer Academic Publishers,
101 Philip Drive, Norwell, MA 02061, U.S.A.

In all other countries, sold and distributed
by Kluwer Academic Publishers Group,
P.O. Box 322, 3300 AH Dordrecht, The Netherlands.

Printed on acid-free paper

Printed in the Netherlands

CONTENTS

Part 1.   Biology and Control of Bruchid

Biology:

Control:

vi

Part 2. Life Cycle and Coevolution of Bruchid and Legume

Taxonomic Feature and Life Cycle of Bruchidae:

Coevolution of Bruchid and Legume:

Breeding of Bruchid-resistant Legumes:

Part 3.   Laboratory Population Ecology of Bruchids and their Parasites

Bruchid Life History, Ecology, and Evolution:

Behavioral Interaction:

Population Dynamics:

Preface

In 1980, the International Symposium on the Ecology of Bruchids Attacking Legumes (Pulses), organized by Dr. Labeyrie, was held at Tours, France. Since then, there has been tremendous progress in the area of Bruchid and Legume research. At the same time, as we face the problems of world-wide population explosion and food shortage, the importance of legumes as the world's major protein source is rapidly increasing, especially in tropical regions. Thus, it seemed appropriate to hold the Second Symposium in order to review the recent progress in the control of Bruchids and in the biology and ecology of Bruchids and legumes. This is an important part of the search for ways to integrate these fields with a common perspective.

The Second International Symposium on Bruchids and Legumes (ISBL-II) was held in September 6-9, 1989 at Okayama, Japan under the joint auspices of the Japanese Society of Applied Entomology and Zoology and of the Foundation for Advancement of International Science. Significant contributions have originated in Japan on the study of Bruchid and legumes. Most notably, the study on population ecology by Professor S. Utida with *Callosobruchus chinensis* paved the way to modern experimental population ecology, and the study on Bruchid resistance by some legumes by Professor S. Ishii was the starting point of recent research on the Bruchid resistance of legumes. Thus it was very appropriate to hold the Second Symposium in Japan. The Symposium was also held to honor Prof. T. Yoshida, who contributed so much to field and laboratory studies of Bruchids, on the occasion of his retirement from Okayama University.

By the generous support from many organizations, many scientists from developing countries could participate in the Symposium. The Symposium had a total participants of 140 from 23 countries. Thanks to the active discussions among participants, the Symposium achieved important and memorable results. This volume is based on the papers presented at the Special Sessions of the Symposium. The contents of the articles were further edited and many comments and discussions during the Symposium have been incorporated.

The Symposium was cosponsored by 13 organizations including Science Council of Japan. More than 100 organizations and companies contributed financial support to the Symposium. The Organizing Committee and editorial committee of the Symposium greatly acknowledge this support, especially by the Commemorative Association for the Japan World Exposition (1970), Japan Beans and Peas Foundation, Okayama Prefecture, and Okayama City.

The administrative office for the Symposium, headed by Professor F. Nakasuji, and the Fund Raising Committee headed by Professor I. Yamamoto should be given special thanks by all the participants for the superlative jobs done for the Symposium. Finally, I thank Prof. L. B. Slobodkin, Dr. M. L. Taper, Dr. I. Sano-Fujii, Mr. Y. Toquenaga and Y. Moriguchi for their assistance to the Editorial Committee for this volume.

December 1989

Koichi Fujii
Editor in Chief

HISTORICAL REVIEW OF BRUCHID STUDIES IN JAPAN

TOSHIHARU YOSHIDA
Laboratory of Applied Entomology, Faculty of Agriculture
Okayama University, Tsushima, Okayama 700, Japan

ABSTRACT. In the Meiji era we had two serious Bruchid pests, *Callosobruchus chinensis* and *Bruchus pisorum* in Japan. In 1926 *B. rufimanus* was found in Kyushu. Urgent countermeasures were taken to meet the new pest: biology of and control measures against the weevil were studied extensively especially by Kamito. In 1937 the basis of classification of Bruchids in Japan was established by Chujo. Recently the introduction of *C. maculatus* was noticed. During and after World War II Ishii started his pioneering study on biochemical resistance to Bruchid attack in legume seeds. Also, Utida was carrying on his continuing study of experimental populations of *C. chinensis*. The range of his study was enlarged from density effect to interspecific competition, host-parasite relationship and phase dimorphism in *Callosobruchus* spp. The new pheromones of "oviposition marker" and "erectin" were identified and synthesized by Yamamoto and his school in the bean weevils. Some authors took a growing interest in the life history strategy of Bruchids, coevolution between bean and weevil, and evolution or domestication of the bean weevils. Dealing with the increase of imported beans from foreign countries into Japan, the techniques of bean fumigation with methyl bromide and phosphine for quarantine fumigation was developed. In the laboratories of Universities and the private laboratories of agricultural Chemicals Companies many investigations were carried on using the bean weevil as an experimental insect to develop new chemicals without environmental disruption.

1. Introduction

A brief review of the research on bruchids in Japan is presented below as an introduction to the Symposium. There were too many papers to review completely so that it was necessary to omit some contributions.

2. The Edo Period

Many books of agriculture and natural history were published in the Edo period (1603-1867). Kurimoto (1805) published "Risshi Senchufu" (1000 Insects Picture Book) in which he described that the azuki bean weevil emerged leaving a hole in the azuki beans and the consumed beans were in moist condition. He also stated that the weevil was winged and could fly (Yasue personal communication).

3. The Pea Weevil

We had two serious bruchid pests in the Meiji period (1868-1912), the azuki bean weevil, *Callosobruchus chinensis* (L.), and the pea weevil, *Bruchus pisorum* (L.).

1

K. Fujii et al. (eds.), Bruchids and Legumes: Economics, Ecology and Coevolution, 1–24.
© 1990 *Kluwer Academic Publishers. Printed in the Netherlands.*

The pea weevil was imported together with pea seed (*Pisum sativum* L.) from U.S.A. about 1888. At the last years of Meiji it became difficult to obtain sound seed peas, and the cultivation of peas declined markedly. The weevil invaded Hokkaido in 1917. After that the peas had to be harvested at an immature stage while the weevils were still at an early developmental stage in the pea. The weevil caused an annual loss of about 5 per cent on the average all over the country. Okamoto (1944) studied the damage effect on germination and growth of peas by the weevil. Tsuchiyama, Oyama and Nagaoka (1960) studied the biology of the weevil and analyzed the difference in damage among varieties of peas and found a high correlation between damage and thickness of the leaf and stem. General comments on the weevil are found in Tsutsui (1955) and Kamito (1952) among others.

4. The Broad Bean Weevil

The broad bean weevil, *Bruchus rufimanus* Boh., may have been introduced in Japan together with seed of a broad bean, *Vicia faba* L., from England or other countries in Europe about 1921. At first, people confused this species with the pea weevil. On 1926 the weevil was found and identified definitely in Kumamoto. The weevil inflicted serious damage on broad bean cultivation, therefore the government and the people took strong measures to meet the situation. The weevil then spread over Kyusyu, Shikoku and Kinki district and later all over the country (Kinoshita and Sakai, 1930; Mishima, 1931; Kamito, 1937).

Takahashi (1929), Kamito and Sakai (1930, 1931), Kinoshita and Sakai (1930), and Maeda (1931) studied the biology of the weevil and Kamito, Sakai and Obi (1977) summarized the results of their continuing investigations on the weevil. Maehara (1954) examined the number of entrance holes per bean and found that the average number of holes per bean decreased with delay of the planting date. Ishikura, Ozeki and Watanabe (1962a) studied emergence and ovipositional activity of the weevil; the adult weevil emerged from the infested bean from late July till middle October, the peak of emergence was in September. Most of the adult weevils hibernated outdoors, and the emerged adults began to gather in the broad bean field from early April. The weevils began to oviposit soon after the setting of young pods in the middle of April and oviposition ended in the latter half of middle May or at the beginning of late May.

Okamoto (1951) estimated the weight loss of the bean caused by the weevil. For control of the weevil Mishima (1931), Onoue and Murakami (1933), Maeda (1934), and Onoue (1935) tested the effect of fumigants on the weevil. Ishikura and Ozaki (1951) tried to control the oviposition of the weevil by spraying BHC. Suenaga, Matsuo and Sakai (1951) and Ishikura and Ozaki (1962b) carried out studies on varietal difference in damage caused by the broad bean weevil. The number of eggs of the weevil deposited on the bean differed among varieties, but it varied also with year, sowing time and observation time. Large number of eggs were deposited on Shimizu variety, but the number of eggs decreased substantially with the delay of sowing time. However, this trend was much less conspicuous among the pea varieties when the infestation was light. The number of weevils that emerged differed largely with variety. The proportion of emergence and the ratio of the number of adults that emerged to the number of larvae in per bean, differed also largely with variety. In general, the varieties of small sized beans showed a low proportion of emergence: the small bean might not have enough resource for the larval development. Thoyama (1952) tried to develop an air-tight storage container using an Erlenmeyer's flask. Yamada (1944) examined the resistance of the weevil to hot water. All the weevils were killed after 8 minutes at 64°C. The germination of the bean was not affected by this treatment.

General comments on the broad bean weevil are also found in Kamito (1952), Tsutsui (1955), and Kamito, Sakai and Obi (1977).

5. Observation on Biology of Bruchids

In the 1930's the biology of the bean weevils having a possibility of invading Japan was examined systematically by the plant quarantine officers (Miyake, 1938; Miyake and Kodera, 1939; Kuroda, 1939; among others).

6. Classification

Chujo (1937) published the first book on the classification of Bruchids in Japan. Dr. Morimoto (in this volume) presents his subsequent contribution to the taxonomy of Bruchids.

Ishii and Nagasawa (1942) described a new chalcidoid, *Neocatolaccus mamezophagus* Ishii et Nagasawa, reared from *C. chinensis*, although later this species was identified correctly as *Anisopteromalus calandrae* (Howard) by Tachikawa(1966).

7. Morphology and Physiology

Ishikura (1939b) measured the length of adult elytra of *C. chinensis*, and proved the relationship between the length and the conditions of temperature and moisture at which the weevil was kept in the developmental stages. Nagasawa (1952) and Ishikawa, Miyamoto and Matsuzawa (1957) reported the relationship between parental density and the length and width of elytra of progeny for the azuki bean weevil. Kishimoto (1953) evaluated the effect of temperature and moisture on the morphological and physiological characters of the azuki bean weevil. Utida (1966) measured the water content of adult body in *B. rufimanus, C. chinensis, C. maculatus* (F.), *C. analis* (F.), *C. phaseoli* (Gyllenhal), *C. rhodesianus* (Pic)(?), and *Zabrotes subfasciatus* Boheman and discussed the relationship between the water content and the body weight, adult age, and time after emergence. Watanabe and Sugimoto (1988) studied geographic variation in male antenna of the azuki bean weevil.

Ishii (1952) and Sakagami (1957) amputated the antenna of the azuki bean weevil and tried to evaluate the effect of antennectomization upon the oviposition and reproductivity of the weevil, but could not demonstrate any effect. Nakamura (1971) repeated the experiment with *C. chinensis* and *C. rhodesianus*(?), and could find only a minute effect on the oviposition rates of the latter species.

Yoshida (1978) studied a sex pheromone in the parasitoid wasp, *Anisopteromalus calandrae*.

8. Chromosome and Mutation

Takenouchi (1955) presented a note on the chromosomes in three species of Bruchidae, and Takenouchi (1971) made a note on the chromosomes in male of five strains of *C. chinensis*. Kashiwagi and Utida (1972) described a new mutant in *C. chinensis*.

9. Variation

Kisimoto (1953) analyzed the individual variation of several morphological and physiological characters, the length of elytron and the width of thorax of adult, mortality, duration of development, adult longevity and sex ratio, of *C. chinensis* under different physical environmental conditions and population density. Nakamura (1969b,c) investigated the geographic variations of sensitivities to high density and other ecological characters such as duration of developmental stage and fecundity, among local populations of *C. chinensis*. Yoshida (1980) made a comprehensive review of studies on variation in stored-product insects and discussed its significance to plant quarantine.

Yoshida (1958) transferred azuki bean weevils that had been reared continuously on azuki beans to soy beans and traced the changes of population parameters produced by growth of successive generations on soy beans. The rate of increase and the number of progeny per female increased to an asymptote after a number of generations on soy beans.

10. Tolerance for Temperature and Moisture

Ouchi (1936, 1937) traced the effect of temperature and moisture on oviposition of *C. chinensis*. Ishikura (1939a) studied the effect of temperature and moisture on the oviposition of the azuki bean weevil. Also Ishikura (1940) analyzed the effect of temperature and moisture on duration of the egg stage. Ishikura (1941) investigated the development of all stages at various condition of temperature and moisture and found that development did not occur at 10.8-9°C and that developmental velocity decline with decrease of moisture (water content of bean). Utida (1971) measured the influence of temperature on the fecundity, mortality, development and adult longevity of *C. chinensis*, *C. maculatus*, *C. rhodesianus*, *C. phaseoli*, and *Z. subfasciatus*, and also made a comparison of the developmental limiting temperatures and accumulated temperatures for development among the species.

Umeya, Kato and Sekiguchi (1970) studied tolerance to low temperature of the five species of bean weevil and speculated that *C. maculatus* and *C. analis* can overwinter outdoors in some areas in the south of Japan and a slight chance for overwintering might exist for *C. rhodesianus*, while such possibility was totally precluded for *Z. subfasciatus*. With the exception of *C. rhodesianus*, however, these species had expanded their distribution toward the north of the Northern Hemisphere much beyond the limit of their low temperature tolerance as obtained in the study. The reasons for this were considered.

Kiyoku (1960) carried on a long term study on the lethal action of high temperature on *C. chinensis* and the results of his study were summarized in his thesis for a doctor's degree (Kiyoku, 1961). Mori (1944) investigated vacuum heat measure for insect control using the azuki bean weevil. Yoshida and Suzuki (1953) immersed azuki beans infested by the azuki bean weevil into hot water and examined the susceptibility of the weevil to hot water. Yoshida and Gichuki (1983) investigated use of solar heat to control the bean weevils.

11. Ecology

11.1. ADULT LONGEVITY

Hirano and Umeya (1953) studied the effect of temperature on adult longevity of the azuki bean weevil and found that the longevity extends to about 100 days at a temperature of 10°C and the fat in the body decreases with time: the decreasing of fat in the body of weevils strongly affects their longevity. Umeya and Shimizu (1968) analyzed the effect of feeding on longevity and oviposition of three species of bean weevil, *C. chinensis*, *C. maculatus*, and *Z. subfasciatus*, and found that the longevity and the number of eggs deposited increased with feeding. Yoshida, Igarashi and Shinoda (1986) studied adult longevity of the azuki bean weevil under favorable laboratory conditions: moderately low temperature (18°C), high relative humidity (75% r.h.) and a food supply and recorded 120 days of male average longevity (maximum 240 days) and 223 days of female average longevity (maximum 408 days).

11.2. MATING AND OVIPOSITION BEHAVIOR

Nakamura (1968) compared the mating behavior and spatial egg distribution pattern between two Bruchid species, *C. chinensis* and *C. rhodesianus*, changing female density and the spatial distribution pattern of beans (scattered evenly or in clumps), and found that *C. chinensis* had a tendency to distribute their eggs evenly on beans regardless of the density and distribution pattern of the beans, although *C. rhodesianus* changed the total number of eggs deposited and the pattern of egg distribution according to density and distribution pattern of beans. Nakamura (1969a) also observed the mating behavior of the two species. When a female and a male of *C. chinensis* were introduced the male first moved around rather sluggishly, but suddenly he began to move quickly with rapid movement of his antennae. Such change of behavior, called "activation", could not be seen when a male was alone. In *C. rhodesianus* no behavioral change was observed after first walking around.

Tanaka et al. (1980, 1981, 1982) Yamamoto et al. (1980), Mori et al. (1983), and Yamamoto (1986) found erectin, the substance introducing male mating behavior. Honda (1982) commented on the results of these studies.

11.3. DISTRIBUTION OF EGGS ON BEANS

Utida (1943a), Ueno (1955) Yoshida (1961) Umeya (1966a), Nakamura (1968), and Umeya and Kato (1970) made clear that weevils depositing their eggs avoided the beans already infested by weevils and preferred fresh beans. Yoshida (1961), Ohshima, Honda and Yamamoto (1973), Ohshima (1975), Honda, Oshima and Yamamoto (1976), Yamamoto (1976) and Sakai et al. (1985) found that the rather uniformly egg laying among the beans was due to the fact that the weevils conditioned the beans while creeping and egg-laying and preferred fresh or less-conditioned beans for oviposition. One of the conditioning factors was the deposited eggs. Another one was a mixture of lipids such as fatty acids, hydrocarbons and triglycerides that was named "oviposition marker". Honda (1979) commented on the results of these studies.

11.4. DISPERSAL AND MIGRATION

Watanabe, Utida and Yoshida (1952) studied experimentally the dispersal process of the adults of *C. chinensis* and traced mathematically the change of distribution pattern in the dispersal process. Umeya (1966b) investigated the escaping behaviors of *C. chinensis, C. maculatus* and *Z. subfasciatus* by caging single or mixed species population of adults in a semi-open petri dishes and discussed the significance of putting beans or another species' adults in the dish to the escaping behavior. Shinoda and Yoshida (1984) made clear the significance of supplying food to emigration from storage beans. Watanabe (1984) observed the effect of seed size on female crawling of *C. chinensis* and *C. maculatus* into pile of seeds placed in glass tubes. Next, Watanabe (1985a) proved the relationship between female body size and degree of creeping into a pile of stored seeds. Furthermore, Watanabe (1986c) analyzed the effect of adult body size on escaping behavior from the bottom of piled seeds.

Kiritani (1968) discussed the relation between import of grains and invasion of insect pests in reference to plant quarantine.

11.5. LIFE HISTORY OF *C. CHINENSIS*

In recent years Yoshida and his coworkers made clear the natural life history of the azuki bean weevil (Yoshida, Shinoda and Okamoto, 1984, Shinoda and Yoshida, 1985, 1987, Yoshida, Igarashi and Shinoda, 1986, Shinoda, 1989, Yoshida, 1989). It had been believed that there was a compulsory movement of weevils to and from the bean field and storage sites every year. In 1985 a huge weevil population was found on wild legume plants, *Vigna angularis* var. *nipponensis* (Ohwi) Ohwi & Ohashi and *Dunbaria villosa* (Thunb.). After that it was proved that these weevils in the wild population overwintered on the wild beans and the adult weevils emerging from the wild beans prolonged their life by taking some foods and water in the field. In autumn they visited the legumes, waiting for the flowering, and fruiting and then laid eggs on the pods. There was one generation a year for the wild population of the azuki bean weevil.

12. Invasion of Bruchids in Japan in the Post-World War II Period

Umeya (1968) reviewed historically the invasion of pests in Japan. Takara and Azuma (1971) recorded the invasion of *Acanthoscelides obtectus* (Say) in Okinawa on 1951. Nagayasu and Matsushita (1981) carried on faunal survey of bean weevils in bean processing mills and confirmed the establishment of *C. maculatus* in western part of Japan.

13. Experimental Population Ecology

13.1. STUDY ON EXPERIMENTAL POPULATION OF THE AZUKI BEAN WEEVIL

Utida started his studies on the experimental population of *C. chinensis* during the last War, ranking among research pioneers of population ecology in the world (Utida , 1941a to f, 1942a,b, 1943a,b,c, 1947b, 1949b, 1952a, 1956e, 1957a, 1959b,c, 1967a,b, 1971, 1972b, Utida and Kakemi, 1959). He analyzed experimentally the mechanism of population growth and found the role of the population density as a factor producing the fluctuation of population density, that is "density effect". A part of his findings was summarized in Fig. 2 (Utida, 1941d) that showed the relations of parent density to mortalities in egg and preimaginal stages and per cent reduction in fecundity, and in Fig. 1 (Utida, 1941d) that showed the relations of density of fecundated females to its reproductive rate and mortality of eggs.

Many studies were carried on by members of his school on the density effect. Fujita and Utida (1952) advanced a mathematical theory on the growing process of population influenced by the density. Fujita (1953) developed a mathematical theory that described the process of oviposition in the population and discussed the type of density effect represented by the equation with respect to the parameters involved. Yoshihara (1956) examined mathematically the reproduction curve (refer to Takahashi, 1976, too).

Ishida (1952) analyzed the effect of available space for oviposition on fecundity, fertility and longevity of adults. Nagasawa (1952) and Ishikawa, Miyamoto and Matsuzawa (1957) analyzed the effect of larval density on the development and survival rate in immature stage and the size and longevity of emerged adults. Nishigaki (1963) observed the effect of low density on mating chance and fecundity of the weevil. Nakamura (1967) compared different effects of density between *C. chinensis* and *C. rhodesianus*. Nakamura (1969) also compared the different effects of density on progeny population among different local strains of *C. chinensis*. Nagasawa (1952) made clear the relationship between the parental density and the length and width of elytra of progeny.

Nakamura (1962) studied the problem of density by removing and adding the azuki bean weevils in the experimental population. Also Kiyoku (1966, 1982) approached to the problem experimentally.

13.2. DAILY EMERGENCE RATES

Yoshida (1952) analyzed daily oviposition rates and emergence-rates experimentally using *C. chinensis* and *C. maculatus*. The eggs deposited in each successive day were reared separately and the daily emergence rates were obtained in each culture. By summing up these data the total daily emergence rates were calculated. The mortality in immature stage increased largely with the time of oviposition by different ages of the parent weevils. The duration of the immature stage was longer for the progenies produced by the older parents (age effect). Utida (1959b) described the daily emergence rates as normal distribution by transforming reciprocally the time scale and analyzed the effect of density and moisture on the emergence rates. Murai and Fujii (1970) analyzed the effect of interspecific competition on the emergence rates. Fujii (1975) constructed a general simulation model for the fate of a cohort from eggs to adult emergences. Nagasawa and Yamada (1988) described emergence rates as a mathematical regression model.

13.3. INTERSPECIFIC COMPETITION

Utida's study on population ecology expanded to mixed species population behavior and interspecific competition (Utida, 1953a). Yoshida (1957, 1960a,b, 1961) carried on a study of the interspecific competition between *C. chinensis* and *C. maculatus* and a main part of the results of his study was presented in Yoshida (1966). He emphasized the finding that the early occupation of the environment by a species brought about a considerable advantage in competition with the other species: this advantage was called the "preoccupation effect" (Yoshida, 1967, 1976). Fujii (1965, 1967, 1968, 1969, 1970) also studied competition between these species. The outcome of competition between the two species was reversed under air-tight condition. His study continued to the experiments using different strains of the two species. Shimada (1985) supplemented these studies.

13.4. HOST PARASITE RELATIONSHIP

The other expansion of Utida's study on population ecology was a series of studies on the host parasite interaction in the experimental population of the azuki bean weevil (Utida, 1943d,e, 1944, 1948a,b, 1951a,b, 1952b, 1953b,c,d, 1955a,b, 1956a,b,c, 1957a). The results of the famous and elegant studies (Figures 1 and 2 in Utida (1957b), showing oscillation in host-parasite system) are cited in many books; for example, Boughey, A. S. (1968) Ecology of Populations; Sladen, B. K. and Bang, F. B. (eds.) (1969) Biology of Populations; Krebs, C. J. (1972) Ecology; Ricklefs R. E. (1973) Ecology; Colinvaux, P. A. (1973) Introduction to Ecology; Pielou, E. C. (1974) Population and Community Ecology; and Begon, M. and Mortimer, M. (1981) Population Ecology, among others. Watanabe (1950), Nakamura (1963a,b), and Shimada and Fujii (1985) further contributed on the analysis of host-parasite system.

13.5. COMMUNITY DIVERSITY AND STABILITY

Fujii (1983), Shimada (1984, 1985), Shimada and Fujii (1985a,b), and Lai and Yoshida (1989) approached experimentally the problem of community diversity and stability.

14. Phase Dimorphism

Utida (1954) found the dimorphic forms, the normal and flight, in laboratory population of *C. maculatus* and listed the differences in morphological, physiological and behavioral characters between the two forms. As the results of his studies, the mechanism of emergence of the flight form was analyzed completely (Utida, 1956d, 1967, 1968, 1969, 1970, 1972a, 1974, 1981; Utida and Takahashi, 1958). Sano-Fujii (1967, 1980) added new information to the problem. Nakamura (1966) found the similar dimorphic forms in *C. chinensis* too.

15. Origin and Evolution of Domestication of Bruchid Pests

Kiritani (1955, 1956a,b) discussed the process of domestication by which Bruchids evolved into pests. Yoshida (1958b) reviewed comprehensively the literatures on the origin and evolution or domestication of stored-product insects and stimulated attention to the subject. Later, Kiritani (1959, 1961a,b, 1964, 1968) expanded his ideas on the subject. He suggested the domesticating process of the bean weevils from the univoltine species, that were probably monophagus and needed taking food in adult stage, to the polyvoltine species that were polyphagus, could breed on dried beans, needed not take food in the adult stage and sometimes even lost their flying ability. Kiritani (1955) and Utida (1967) discussed the domestication of the weevil in reference to phase dimorphism. Yoshida (1982a,b, 1983, 1984) recognized three stages in the domesticating process; the first age when cereal food was stored as grain, the second age when a large amount of flour was stored, and the third age when processed food was stored. In conclusion he emphasized that in all these examples all species becoming a serious pest were already preadapted to the new condition and in the new prevailing condition they were selected over the others. Watanabe (1985c, 1986d), Watanabe and Sugimoto (1988) and Umeya (1987) made additional studies of the problem of Bruchid domestication.

16. Control Measures

16.1. BIOLOGICAL CONTROL

Ishii and Mizutani (1933), Ishii (1934), and Kariya and Kurozawa (1939) explored the possibility of introducing parasites from foreign countries for control of Bruchid pests.

16.2. BRUCHID-RESISTANT LEGUMES

Ishii (1952) has studied host preference of the azuki bean weevil since 1937. The problems were divided into two items: the first was the mechanism of host selections, the second was to ascertain why the larvae did not develop in the kidney bean, *Phaseolus vulgaris* L. He reared the weevils with eight kinds of beans and peas and these were divided into the following three groups: (A) favorable; *Vigna angularis* (Willd.) Ohwi and Ohashi and others, (B) not so favorable; *Vicia faba* L. and others, (C) unfavorable; *P. vulgaris* and others. The weevils could penetrate the unfavorable beans but could not develop and died in the first instar. The sensory organ concerned in oviposition response was traced. The female antennae were not used for the tactile sense of oviposition. The smoothness of the surfaces of the beans seemed to be an important factor to oviposition response. The numbers of eggs laid on large glass beads were much fewer than those on the small beads. So, the sizes and curvatures of the beans seemed to be correlated with the oviposition response of the weevils.

The physical properties, size, hardness, specific gravity and tissue were not concerned in the development of the larvae, but the chemical components of the beans might be important factors checking larval development. The beans were ground into flour and water added to the flour and pills were made from it. The pills were covered with collodion and these were named "synthetic beans". The test of breeding the weevil on these synthetic beans was carried out. Some unknown substances contained in the bean possibly hindered the larvae from development. The chemical organic compounds contained in the cotyledons might be important factors checking larval development. The growth hindering substances contained in the beans should be stable at 100°C. The growth hindering substance might be not a single compound. That compound alone might not be so poisonous for the larvae, but when several compounds were present together, the development of the larvae might be hindered seriously. The pentoses contained in the beans might be a substance hindered the larvae from the growth. The reason why the larvae could not develop in *P. vulgaris* was not the difference of proteins which were contained in the two kinds of beans.

Umeya and Imai (1965) tried to determine the part of the *P. vulgaris* plant that was responsible for synthesis of this inhibiting critical substances for the growth of the azuki bean weevil. Grafting was introduced and the resulting seeds were submitted for feeding to the weevils. The kidney bean plant was used as a stock and the azuki plant as a scion. When the two species of the bean weevils, the azuki bean weevil and the Mexican bean weevil (*Z. subfaciatus*), were fed on the beans harvested from these grafted plants, the latter species is able to develop on the kidney bean. As the results of these experiments it was concluded that the substance in the kidney beans, which inhibited development of the azuki bean weevil, was derived mainly from the metabolic activity of the kidney bean leaves. The different chemical composition of the two species of beans which caused the different growth rate of the Mexican bean weevil was also ascribed to the function of leaves. The root system seemed to be of little importance for the synthesis of these critical substances.

Ishimoto and Kitamura (1989) evaluated the inhibitory effects of an α-amylase inhibitor from the kidney beans. Fujii and Miyazaki (1987) studied the infestation resistance of wild legumes (*Vigna sublobata*) to the azuki bean weevil.

16.3. STERILIZATION

16.3.1. *Sterilization with Gamma Radiation*

The effect of gamma radiation on sterility, survival and reproduction of *C. chinensis* was investigated by Kiyoku and Tsukuda (1968a,b), Kumagai (1969), Kiyoku and Fukushima (1977) and Hussain and Imura (1989). The results of studies on sterilization of the weevil with gamma-radiation by Kiyoku and his coworkers were summarized in a booklet, "Studies on Sterile Male Technique" (Kiyoku, 1980).

16.3.2. *Sterilization with Chemosterilants*

Starting with the study of Shinohara and Nagasawa (1963), studies on sterilizing effect of apholate and metepa on *C. chinensis* was carried on by Nagasawa and Shinohara (1964a,b, 1965a,b, 1967), Nagasawa, Shinohara and Shiba (1965, 1966, 1967), Nagasawa (1966, 1968), Nakayama and Nagasawa (1966), Nagasawa and Nakayama (1968a,b), Nagasawa et al. (1974), Kazano et al. (1975), and Nagasawa et al. (1980). Nagasawa (1969) explained the problem in a nontechnical way.

16.4. BIOASSAY OF CHEMICALS

16.4.1. *Chemicals*

In the postwar period many new chemicals were developed for control of insect pests using the bean weevils as a experimental test insect in the laboratories of Universities and Chemical Companies.

Utida (1946) observed the toxic effect of mercury vapor to eggs of *C. chinensis*. Harukawa and Tokunaga (1948), Harukawa (1951), Sato, Higuchi and Suwanai (1973), and Sato and Suwanai (1973, 1975) carried on the study of fumigants using the azuki bean weevil. Sato, Higuchi and Suwanai (1973) and Sato and Suwanai (1973, 1975) analyzed the toxic effect of hydrogen phosphide. Sato (1983) studied methyl bromide vapor.

Utida (1947) and Nagasawa (1947, 1950a,b, 1954a,b), Nagasawa and Yoshinobu (1951), and Nagasawa and Arakawa (1952) evaluated the inert mineral dusts (silica) for control of insects.

Kono (1952), Ueki (1952) and Ishikura and Ozaki (1953) studied the toxicity of BHC for the azuki bean weevil. Ohota and Ikeda (1957b) examined DDT and BHC. Hatai and Kimura (1956) studied the action of DDT and Uchida et al. (1974) studied its analogs.

Ohta and Ikeda (1957a) investigated the insecticidal effect of brominated camphor and Kazano (1983), Kazano et al. (1968), and Kazano, Asakawa and Fukunaga (1975) of carbamate compounds.

Many other authors carried on the study of new chemicals: Ohota (1961), Sakai et al. (1964), Kato (1967), Kato, Sato and Saki (1967), Miyakado et al. (1979), Yajima and Munakata (1979), Nagasawa et al. (1980), Ueji and Tomizawa (1984, 1986), Miyakado et al. (1985 a,b), and Tomizawa, Ueji and Yoshida (1988).

The equality of test insects was analyzed in reference to rearing density (Ishikura, Ozaki, 1953) and to age of adult (Gotoh, 1955, Ishikura, Ozaki, 1955). Testing technique and equipment for new chemicals was developed by Hatai and Kimura (1956) Sugimoto (1963) and Nishiuchi and Sugimoto (1971).

16.4.2. *Miscellaneous*

Fukami, Nakatsugawa and Narahashi (1959) studied the toxicity of rotenone.

Sherman and Hayakawa (1961) studied carbon dioxide as an anesthetizing agent for the azuki bean weevil. Ohguchi et al. (1983) investigated lethal effect of oxygen absorber.

Ando et al. (1971) studied new antibiotic for the azuki bean weevil. Suwanai (1959, 1964) studied insect integument permeability of insecticides.

Kiyoku and Tsukuda (1963, 1964) and Kiyoku (1964) studied the effect of insecticide upon the treated weevil population and the results of their study were summarized in Kiyoku (1982).

17. On References

Hirano (1959a, b) lists almost all the literatures on Bruchids published from 1899 to about 1959 in Japan. Many miscellaneous short notes listed in the list are not referred to in the present paper.

REFERENCES

(Ja.) at the end of the reference indicates the article in Japanese, and (Ja. en) indicates the article in Japanese with English summary.

Ando, K., Oishi, H. , Hirano, S., Okutomi, T., Suzuki, K., Okazaki, H., Sawada, M., and Sagawa, T. (1971) Tetranactin, a new miticidal antibiotic. I. Isolation, characterization and properties of tetranactin, J. Antibiotics 24, 347-352.

Chujo, M. (1937) Bruchidae. Classification of Japanese Animal X, Sanseido, Tokyo. (Ja.)

Fujii, K. (1965) Studies on interspecies competition between the azuki bean weevil and the southern cowpea weevil. I. The reversal in competition result, Res. Popul. Ecol. 7, 43-51.

Fujii, K. (1967) Ditto. II. Competition under different environmental conditions, Res. Popul. Ecol. 9, 192-200.

Fujii, K. (1968) Ditto. III. Some characteristics of strains of two species, Res. Popul. Ecol. 10, 87-98.

Fujii, K. (1969) Ditto. IV. Competition between strains, Res. Popul. Ecol. 11, 84-91.

Fujii, K. (1970) Ditto. V. The role of adult behavior in competition, Res. Popul. Ecol. 12, 233-242.

Fujii, K. (1975) A general simulation model for laboratory insect populations I. From cohort of eggs to adult emergences, Res. Popul. Ecol. 17, 85-133.

Fujii, K. (1983) Resource dependent stability in an experimental laboratory resource-herbivore system, Res. Popul. Ecol. Suppl. 3, 15-26.

Fujii, K. and Miyazaki, S. (1987) Infestation resistance of wild legumes (*Vigna sublobata*) to azuki bean weevil, *Callosobruchus chinensis* (L.) (Coleoptera: Bruchidae) and its relationship with cytogenetic classification, Appl. Ent. Zool. 22, 229-230.

Fujita, H. (1953) Factors affecting the type of population density effect upon average rate of oviposition, Res. Popul. Ecol. 2, 1-7. (Ja. en.)

Fujita, H. and Utida, S. (1952) The effect of population density on the growth of an animal population, Res. Popul. Ecol. 1, 1-14. (Ja. en.)

Fukami, J., Nakatsugawa, T., and Narahashi, T. (1959) The relation between chemical structure and toxicity in rotenone derivatives, Jpn. J. appl. Ent. Zool. 3, 259-265. (Ja. en.)

Gotoh, A. (1955) Relationship between the age and the susceptibility to BHC of adults of rice weevil and azuki bean weevil, Botyu-Kagaku 20, 126-132. (Ja. en.)

12

Harukawa, C. and Tokunaga, M. (1948) On the toxicity and toxic action of poisonous gases (I), Matsumushi 3, 1-10, 33-39. (Ja.)

Harukawa, C., Utida, S., The late Nishikawa, Y., Kiyoku, M., Kondo, T., Yosida, M., and Suzuki, N. (1951) Insecticidal action of volatile compounds. Toxicity and action of gaseous insecticides (3rd Report), Botyu-Kagaku 16, 193-212. (Ja. en.)

Hatai, N. and Kimura, N. (1956) The influence of the size of spray particle of DDT emulsion on the mortality of azuki-bean weevil, Bull. Nat. Inst. Agr. Sci. Jpn. Ser. C. 6, 17-24. (Ja. en.)

Hirano, C. and Umeya, K. (1953) On the relation of the duration of adult life of the azuki bean weevil, *Callosobruchus chinensis* L. and the quantity of the decreasing of fat content in the body, Oyo-Kontyu 9, 111-114. (Ja, en,)

Hirano, I. (1959a) The list of Japanese references on insects. No. 203. Bruchidae (A). A General and a Miscellaneous, Osaka Pl. Prot. 7, 317-334. (Ja.)

Hirano, I. (1959b) The list of Japanese references on insects. No. 204. *Bruchus pisorum* Linnaeus, *B. rufimanus* Boheman, Osaka Pl. Prot. 7, 335-344. (Ja.)

Hussain, T. and Imura, O. (1989) Effects of gamma radiation on survival and reproduction of *Callosobruchus chinensis* (L.) (Coleoptera: Bruchidae), Appl. Ent. Zool. 24, 273-280.

Honda, H. (1979) Oviposition-regulating material in *Callosobruchus chinensis*, in T. Hidaka et al. (eds.), Physiology and Chemistry of Insect, Kitami-Shobo, Tokyo, pp. 165-178.

Honda, H. (1982) Eco-chemical materials of stored product insect pests, Chem. Bio. 20,185-191

Honda, H., Oshima, K., and Yamamoto, I. (1976) Oviposition marker of azuki bean weevil, *Callosobruchus chinensis* L., Proc. Joint US-Jpn. Sem. Stored Prod. Insect. Manhattan, Kansas, 116-128.

Ishida, H. (1952) Studies on the density effect and the extent of available space in the experimental population of the azuki bean weevil, Res. Popul. Ecol. 1, 25-35. (Ja. en.)

Ishii, S. (1952) Studies on the host preference of the cowpea weevil (*Callosobruchus chinensis* L.), Bull. Nat. Inst. Agr. Sci. (Japan) Ser. C, 1, 185-256. (Ja. en.)

Ishii, S. (1970) Introduction to Entomology, Iwanami Pub. Tokyo. (Ja.)

Ishii, T. (1934) Introduction of natural enemy, Kontyu 8, 313-314. (Ja.)

Ishii, T. and Mizutani, Y. (1933) Use of egg parasite of Bruchids that was imported from Hawai, J. Appl. Zool. 5, 133-134. (Ja.)

Ishii, T. and Nagasawa, S. (1942) A new chalcidoid reared from *Callosobruchus chinensis* (L.), Kontyu 16, 67-68.

Ishikawa, R., Miyamoto, Y. and Matsuzawa, H. (1957) Effects of the larval density of the azuki bean weevil on some adult characters, Botyu-Kagaku 22, 182-185. (Ja. en.)

Ishikura, H. (1939a) Effect of temperature and moisture on oviposition of the azuki bean weevil, *Callosobruchus chinensis* (L.), J. Appl. Zool. 11, 41-52. (Ja.)

Ishikura, H. (1939b) On the relative size of the bean weevil, *Bruchus chinensis* L., Kontyu 13, 222-224. (Ja.)

Ishikura, H. (1940) Effect of temperature and moisture on egg development of the azuki bean weevil, *Callosobruchus chinensis* (L.), J. Appl. Zool. 11, 218-229. (Ja.)

Ishikura, H. (1941) Effect of temperature and moisture on all developmental period of the azuki bean weevil, *Callosobruchus chinensis* (L.), J. Appl. Zool. 13, 118-131. (Ja.)

Ishikura, H. and Ozaki, K. (1951) On the action of BHC sprayed and dusted for preventing oviposition of broad bean weevil, Oyo-Kontyu 7, 35-39. (Ja. en.)

Ishikura, H. and Ozaki, K. (1953) On the difference in the resistance to BHC of azuki bean weevils reared under different densities, Botyu-Kagaku 18, 85-89. (Ja. en.)

Ishikura, H. and Ozaki, K. (1955) On the relationship of the order of development of some insects to their resistance to insecticides, Bull. Nat. Inst. Agr. Sci. Jpn. Ser. C. 5, 81-98. (Ja. en.)

Ishikura, H., Ozaki, K., and Watanabe, Y. (1962a) Studies on broad bean weevil. (Pt. 1) Emergence and oviposition activity, Bull. Shikoku Agr. Exp. Sta. 7,197-216. (Ja. en.)

Ishikura, H., Ozaki, K. and Watanabe, Y. (1962b) Ditto. (Pt. 2) Varietal differences of damage caused by the broad bean weevil, Bull. Shikoku Agr. Exp. Sta. 7, 217-237. (Ja. en.)

Ishimoto, M. and Kitamura, K. (1989) Inhibitory effects of an α-amylase inhibitor from the kidney bean, *Phaseolus vulgaris* (L.) on three species of Bruchids (Coleoptera: Bruchidae), Appl. Ent. Zool. 24, 281-286.

Jpn. Fumi. Tech. Associ. (1978) Insect Pests of Pulse and Its Control. Plant Quarantine of Imported Pulse, Jpn. Fumi. Tech. Associ. Tokyo. (Ja.)

Jpn. Fumi. Tech. Associ. (1982) Study on the actual condition of insect pest damage and the development of insect control measure in pulse storage warehouse at seaport, Jpn. Fumi. Tech.Associ. Tokyo. (Ja.)

Kamito, A. (1937) Process of spread of the serious exotic pests in Japan, Kontyu 9, 29-34. (Ja.)

Kamito, A. (1952) Bruchids, in H. Yuasa and A. Kawada (eds.), A New View of Agricultural Insect Pests, Asakura Pub. Tokyo. (Ja.)

Kamito, A., Sakai, K., and Obi, A. (1977) Studies on broad bean weevils, *Bruchus rufimanus* Boheman with reference to its bionomics and control, Nihon Tokushu Noyaku Seizo K.K. Tokyo, pp. 75. (Ja, en,)

Kariya, K. and Kurozawa, M. (1939) Investigation on the use of natural enemy to control of the azuki bean weevil, Data Agr. Improve. 137, 89-90. (Ja.)

Kashiwagi, M. and Utida, S. (1972) A new mutant in *Callosobruchus chinensis* L. (Coleoptera: Bruchidae), Appl. Ent. Zool. 7, 95-96.

Kato, M. (1967) The conversion of an insecticidal compound, 1,3-dithiocyanato-2-N,N-dimethylaminopropane, to nereistoxin, Botyu-Kagaku 32, 70-79. (Ja. en.)

Kato, M., Sato, Y., and Saki, M. (1967) Foliage spray treatment of cartap mixed with DCPA for simultaneous control of rice stem borer and barnyard grass, Jpn. J. appl. Ent. Zool. 11, 135-139. (Ja. en.)

Kazano, H. (1983) Studies on insecticidal characteristics of carbamate compounds and their metabolism in insects, soils and model ecosystem, Bull. Nat. Inst. Agr. Sci. Ser. C. 37, 31-89. (Ja. en.)

Kazano, H., Asakawa, M., Tanaka, T., and Fukunaga, K. (1968) Studies on carbamate insecticides. I. Insecticidal activities of substituted phenyl N-methyl- and N,N-dimethylcarbamates to several species of insects, Jpn. J. appl. Ent. Zool. 12, 202-210. (Ja. en.)

Kazano, H, Asakawa, M., and Fukunaga, K. (1975) Evaluation methods of insect-sterilizing compounds and search for new chemicals, Bull. Nat. Inst. Agr. Sci. C., 29, 1-43. (Ja. en.)

Kinoshita, S. and Sakai, K. (1930) Survey of distribution of and damage from the broad bean weevil in Tokyo and nearby prefectures, Kontyu 4, 271-276. (Ja.)

Kiritani, K. (1955) Phase variation found in stored-product insects and their domestication, Biol. Sci. (Separate Vol., Interrelations Between Organisms and Their Environments), 55-58. (Ja.)

Kiritani, K. (1956a) Ecology of Bruchids and their domestication. I, New Entomol. 9(5), 2-7. (Ja.)

Kiritani, K. (1956b) Ecology of Bruchids and their domestication. II, New Entomol. 9(6), 7-11. (Ja.)

Kiritani, K. (1959) Problems in the stored product entomology, Osaka Pl. Prot. 7, 1-44. (Ja.)

Kiritani, K. (1961a) Origin of household insect pests and process of their domestication, Pl. Prot. 16, 295-300. (Ja.)

Kiritani, K. (1961b) The origin of the household pests, and the process of adaptation to indoor conditions, Ecol. Insect 9, 22-40. (Ja.)

Kiritani, K. (1964) The origin of the vermin, Asahi-Kagaku 24(7), 23-27. (Ja.)

Kiritani, K. (1968) Transport of grain and dispersal of insect pests, Pl. Prot. 22, 204-209. (Ja.)

Kiritani, K. and Utida, S. (1956) The stored products pests, their distribution and establishment, Osaka Pl. Prot. 5, 217-222, 267-340. (Ja.)

Kisimoto, R. (1953) On the individual variation of several morphological and physiological characters under different environmental conditions, Res. Popul. Ecol. 2, 65-78. (Ja. en.)

Kiyoku, M. (1960) Experimental studies on the influence of an abnormally high temperature upon some biological characters in insects survived the heat-treatment and those in their offsprings, Sci. Rep. Fac. Agr. Okayama Univ. 16, 25-32. (Ja. en.)

14

Kiyoku M. (1961) The lethal action of high temperature upon insects and the resistance of insects to heat, Dr. Thesis. Kyoto Univ., Kyoto. pp.172. (Ja. en.)

Kiyoku, M. (1964) Growth form of the experimental population of Azuki-bean weevils under an insecticide treatment, Sci. Rep. Fac. Agr. Okayama Univ. 23, 1-6. (Ja. en.)

Kiyoku, M. (1966) Self-regulation in the experimental population of azuki-bean weevil, *Callosobruchus chinensis*, Sci. Rep. Fac. Agr. Okayama Univ. 27, 9-15. (Ja. en.)

Kiyoku, M. (1978) Studies of the ecology of insects sterilized artificially (gamma radiation). XII. Ecological considerations on the gamma rays-sterile insects based upon the results obtained from serial studies, Sci. Rep. Fac. Agr. Okayama Univ. 51, 1-16. (Ja. en.)

Kiyoku, M. (1980) Studies on sterile male technique, Private Printing. (Ja.)

Kiyoku, M. (1982) Special mention of the azuki bean weevil, *Callosobruchus chinensis* (L.), Private Printing. (Ja.)

Kiyoku, M. and Tamaki, H. (1959) Modification in biotic potential of insects surviving exposure to an insecticide, Sci. Rep. Fac. Agr. Okayama Univ. 14, 1-5. (Ja. en.)

Kiyoku, M. and Tsukuda, R. (1963) Influence of insecticide (malathion) on the population trend of Azuki-bean weevil, *Callosobruchus chinensis*, Sci. Rep. Fac. Agr. Okayama Univ. 22, 1-7. (Ja. en.)

Kiyoku, M. and Tsukuda, R. (1964) Considerations on the adaptive changes induced by the insecticidal selection in experimental populations of *Callosobruchus chinensis*, Sci. Rep. Fac. Agr. Okayama Univ. 24, 9-18. (Ja. en.)

Kiyoku, M. and Tsukuda, R. (1968a) Effect of ^{137}Cs on the sterilities of *Callosobruchus chinensis* (L.) and *Spodoptera litura* (F.). Studies on the ecology of insects sterilized artificially (gamma radiation). I., Odokon-Chugoku 10, 10-12. (Ja.)

Kiyoku, M. and Tsukuda, R. (l968b) Effect of ^{137}Cs on the sterilities of *Callosobruchus chinensis* (L.). Ditto. II, Sci. Rep. Fac. Agr. Okayama Univ. 32, 15-23. (Ja. en.)

Kiyoku, M. and Fukushima, H. (1977) Effects of ^{137}Cs gamma radiation on the inherited sterility of *Callosobruchus chinensis* (L.), Odokon-Chugoku 19, 18-25. (Ja.)

Kono, T. (1952) On the relative resistance of several species of insect pest of stored products to the gaseous gamma-BHC, Botyu-Kagaku 25, 153-156. (Ja. en.)

Kumagai, M. (1969) Effect of irradiation on the eggs of Azuki-bean weevil, *Callosobruchus chinensis* L. (Coleoptera: Bruchidae), Appl. Ent. Zool. 4, 9-15.

Kurimoto, T. (1805) Risshi Senchufu (An illustrated book of the 1000 Japanese insects), (Three Volumes), Yamamoto Youdou, Edo (Tokyo). (Ja.)

Kuroda, M. (1939) Investigation on the common bean weevil, Data Agr. Improve. 137, 102-106. (Ja.)

Lai, C. H. and Yoshida, T. (1989) Experimental study on the diversity and stability in the laboratory ecosystem. 1. Persistency of community, Chinese J. Entomol. 9, 169-187. (Chinese with English summary)

Maebara, H. (1954) On the number of entrance holes on the broad bean bored by the broad bean weevil, Bull. Fac. Agr. Kagoshima Univ. 3, 28-33. (Ja. en.)

Maeda, A. (1931) Study on morphology and ecology of the broad bean weevil, *Bruchus rufimanus* Boheman, Agr. Hort. 6, 1757-1768. (Ja.)

Maehara, H. (1954) On the number of emergence holes on the broad bean bored by the broad bean weevil, Bull. Fac. Agr. Kagoshima Univ. 3:28-33. (Ja. en.)

Mishima, Y. (1931) Dispersion process of and control exercised over the broad bean weevil, J. Disease Insect Pest 18, 88-93, 168-175. (Ja.)

Miyakado, M., Nakayama, I., Inoue, A., Hatakoshi, M., and Ohno, N. (1985a) Chemistry and insecticidal activities of piperaceae amides and their synthetic analogs, J. Pesticide Sci. 10, 11-17.

Miyakado, M., Nakayama, I., Inoue, A., Hatakoshi, M., and Ohno, N. (1985b) Insecticidal activities of phenoxy analogues of dihydropipercide, J. Pesticide Sci. 10, 25-30.

Miyakado, M., Nakayama, I., Yoshioka,H., and Nakatani, N. (1979, recd. 1981) The piperaceae amids I: structure of pipercide, a new insecticidal amide from *Piper nigrum* L., Agr. Biol. Chem. 43,

1609-1611.

Miyake, T. (1938) Bionomics of the bean weevils or Bruchids (Report I), with notes on structural characters of the important species of beetles found at the time when plants were quarantined, Oyo-Kontyu 1, 61-70. (Ja.)

Miyake, T. (1950) Ditto (Report II), Hiroshima Agr. Spec. Rep. 3, 8-16. (Ja.)

Miyake, T. and Kodera, S. (1939) Study on ecology of Bruchids, Data Agr. Improve.137, 154-155. (Ja.)

Mori. H. (1944) Study on vacuum heat method for insect control. (I), J. Appl. Zool. 15, 86-101. (Ja.)

Mori, K., Ito, T., Tanaka,K. Honda, H. and Yamamoto, I. (1983) Synthesis and biological activity of optically active forms of (E)-3,7-dimethyl-2-octen-1,8-dioic acid (callosobruchisic acid), Tetrahedron 39, 2303-2306.

Murai, M. and Fujii, K. (1970) Examination of the influences of density pressure on the pattern of adult emergence with reference to the azuki bean weevil, *Callosobruchus chinensis*, Res. Popul. Ecol. 12, 219-232.

Nagasawa, S. (1947) Lethal effect of the diatomaceous earth against the azuki bean weevil, *Callosobruchus chinensis* (L.), especially on the problem of the relation of this lethal effect to the moisture, Botyu-Kagaku 7, 8, 9, 38-44. (Ja. en.)

Nagasawa, S. (1950a) On the lethal effect of the powder of silicon carbide to the adult of the azuki bean weevil (*Callosobruchus chinensis* L.), with special reference to the relation between the lethal effect and the particle size. (Preliminary Report). Studies on the lethal effect of so-called "inert" pulverized dusts to insects. II, Botyu-Kagaku 15, 79-85. (Ja. en.)

Nagasawa, S. (1950b) On the lethal effect of the powder of "Yamagata-bentonite" to the adult of the azuki bean weevil (*Callosobruchus chinensis* L.), with special reference to the relation between the lethal effect and the particle size. Ditto. III., Botyu-Kagaku 15, 178-180. (Ja. en.)

Nagasawa, S. (1952) On the relation between the population densities of the azuki bean weevil (*Callosobruchus chinensis* L.) and the length and the width of elytra of its progeny, Res. Popul. Ecol. 1, 136-142. (Ja. en.)

Nagasawa, S. (1954a) On the lethal effect of silicon carbide powder in various moisture contents to adults of the azuki bean weevil, *Callosobruchus chinensis* L. Studies on the lethal effect of so-called "inert" pulverized dust to insects. V, Botyu-Kagaku 19, 100-102. (Ja. en.)

Nagasawa, S. (1954b) Lethal effect of some inert pulverized dusts to adults of the azuki bean weevil, *Callosobruchus chinensis* L., under different relative humidities. Studies on the lethal effect of so-called "inert" pulverized dust to insects. VI, Botyu-Kagaku 19, 127-139. (Ja. en.)

Nagasawa, S. (1969) Insect pest control by chemosterilant, Iden 23, 46-52. (Ja.)

Nagasawa, S. and Arakawa, M. (1952) On the relation between the particle shape of powder of calcium carbonate and the lethal effect to adults of the azuki bean weevil (*Callosobruchus chinensis* L.), and the difference of knock down effect between DDT powders prepared with these calcium carbonates to adults of common housefly (*Musca domestica* L.). Studies on the biological assay of insecticides. XIX, Botyu-Kagaku 17, 14-19. (Ja. en.)

Nagasawa, S. and Nakayama, I. (1968a) Presumption of dosage-response curve obtained by the treatment of chemosterilant for both sexes of the azuki bean weevil, *Callosobruchus chinensis* L., Botyu-Kagaku 33, 146-152. (Ja. en.)

Nagasawa, S. and Nakayama, I. (1968b) Joint sterilizing effect of a mixture of hempa and thiohempa, and of hempa and N,N,N',N'-tetramethyl-P-morpholinophosphinic diamide on the azuki bean weevil, *Callosobruchus chinensis* L. Studies on the chemosterilants of insects XIII, Jpn. J. appl. Ent. Zool. 12, 194-201. (Ja. en.)

Nagasawa, S., Nakayama, I., Borkovec, A. B., Terry, P. H., and DeMilo, A. B. (1974) Screening of phosphorus amide and s-triazine chemosterilants in house flies, fruit flies, and azuki bean weevils, Botyu-Kagaku 39, 105-109. (Ja. en.)

Nagasawa, S. and Shinohara, H. (1964a) Sterilizing effect of metepa on the azuki bean weevil, *Callosobruchus chinensis* L., with special reference to the hatching of the eggs deposited by treated weevils. Studies on the chemosterilants of insects. II, Jpn. J. appl. Ent. Zool. 8, 123-128. (Ja. en.)

Nagasawa, S. and Shinohara, H. (1964b) Sterilizing effect of apholate on the azuki bean weevil, *Callosobruchus chinensis* L., with special reference to the hatching of the eggs deposited by treated weevils, Jpn. J. appl. Ent. Zool. 8, 272-276. (Ja. en.)

Nagasawa, S. and Shinohara, H. (1965a) Joint sterilizing of a mixture of apholate and metepa on the azuki bean weevil, *Callosobruchus chinensis* L., with special reference to the hatchability of eggs deposited by treated weevils, Jpn. J. appl. Ent. Zool. 9, 162-165. (Ja. en.)

Nagasawa, S. and Shinohara, H. (1965b) Mating competition between aholate-sterilized and normal males of the azuki bean weevil, *Callosobruchus chinensis* L., Jpn. J. appl. Ent. Zool. 9, 271-274. (Ja. en.)

Nagasawa, S. and Shinohara, H. (1967) Joint sterilizing effect of a mixture of apholate and hempa on the azuki bean weevil, *Callosobruchus chinensis* L. Studies on the chemosterilants of insects. VIII, Botyu-Kagaku 32, 39-43. (Ja. en.)

Nagasawa, S., Shinohara, H. and Shiba, M. (1965) Sterilizing effect of Dowco-186 on the Azuki bean weevil, *Callosobruchus chinensis* L., with special reference to the hatchability of the eggs deposited by treated weevils. Ditto. VI. , Botyu-Kagaku 30, 91-95. (Ja. en.)

Nagasawa, S., Shinohara, H. and Shiba, M. (1966) Differential susceptibilities in sexes of the azuki bean weevil *Callosobruchus chinensis* L., to the sterilizing effect of hempa. Ditto. IX, Botyu-Kagaku 31, 108-113. (Ja. en.)

Nagasawa, S., Shinohara, H. and Shiba, M. (1967) Differential susceptibilities in sexes of *Callosobruchus chinensis* L. (Coleoptera, Bruchidae) to the sterilizing effects of triphenyltin hydroxide, J. stored Prod. Res. 3, 177-184.

Nagasawa, S., Ushirokita, M, Borkovec, A. B. and DeMiro, A. B. (1980) Effects of a substituted 1,3-benzodioxole on reproduction of male and female *Callosobruchus chinensis* L. (Coleoptera: Bruchidae), Appl. Ent. Zool. 15, 494-495.

Nagasawa, S. and Yamada, K. (1988) An asymptotic regression curve for describing the emergence pattern of the azuki bean weevil, *Callosobruchus chinensis* L. (Coleoptera: Bruchidae), Appl. Ent. Zool. 23, 181-185.

Nagasawa, S. and Yoshinobu, M. (1951) On the lethal effect of the powder of "volcay bentonite" and "panther creek bentonite" to the azuki bean weevil (*Callosobruchus chinensis* L.). Studies on the lethal effect of so-called "inert" pulverized dusts to insects. IV, Botyu-Kagaku 16, 35-40. (Ja. en.)

Nagayasu, M. and Matsushita, K. (1981) Faunal survey of bean weevils in bean processing mills, Res. Bull. Pl. Prot. Japan 17, 87-91. (Ja. en.)

Nakagita, K. (1953) Relationship between female longevity and fecundity, J. Appl. Zool. 17, 191-198. (Japanese with German summary)

Nakamura, H. (1962) The experimental demonstration of population balance in the azuki bean weevil, by removing and adding, Jpn. J. Ecol. 12, 141-146. (Ja. en.)

Nakamura, H. (1963a) Population balance in the host-parasite interacting system demonstrated by partial removal of population, Jpn. J. Ecol. 13, 59-57. (Ja. en.)

Nakamura, H. (1963b) Population balance in the host-parasite interacting system demonstrated by adding host or parasite individuals, Jpn. J. Ecol. 13, 167-172. (Ja. en.)

Nakamura, H. (1966) The active-types observed in the adult of *Callosobruchus chinensis* L., Jpn. J. Ecol. 16, 236-241. (Ja. en.)

Nakamura, H. (1967) Comparative study of adaptability to the density in two species of *Callosobruchus*, Jpn. J. Ecol. 17, 57-63. (Ja. en.)

Nakamura, H. (1968) A comparative study on the ovipositional behavior of two species of *Callosobruchus* (Coleoptera: Bruchidae), Jpn. J. Ecol. 18,192-197. (Ja. en.)

Nakamura, H. (1969a) Comparative studies on the mating behaviour of two species of *Callosobruchus* (Coleoptera: Bruchidae), Jpn. J. Ecol. 19, 20-26. (Ja. en.)

Nakamura, H. (1969b) The effect of density on progeny population in *Callosobruchus chinensis* L.

from different localities, Jpn. J. Ecol. 19, 92-96. (Ja. en.)

Nakamura, H. (1969c) Geographic variation of the ecological characters in *Callosobruchus chinensis* L., Jpn. J. Ecol. 19, 127-131. (Ja. en.)

Nakamura, H. (1971) Effect of the amputation of female's antennae on the oviposition in *Callosobruchus chinensis* and *C.* sp. (*C. rhodesianus*?), Jpn. J. Ecol. 21, 167-169. (Ja.)

Nakayama, I. (1977) Histopathological observation of chemosterilization effect of metepa and hempa on female adults of the azuki bean weevil, *Callosobruchus chinensis* L., Botyu-Kagaku 42, 92-96. (Ja. en.)

Nakayama, I. and Nagasawa, S. (1966) Histopathological observation of chemosterilizing effect of metepa on male adults of the azuki bean weevil, *Callosobruchus chinensis* L. Studies on the chemosterilants of insects. XII, Jpn. J. appl. Ent. Zool. 10, 192-196. (Ja. en.)

Nishigaki, J. (1963) The effect of low population density on the mating chance and the fecundity of the azuki bean weevil, *Callosobruchus chinensis* L., Jpn. J. Ecol. 13, 178-184. (Ja. en.)

Nishiuchi, Y. and Sugimoto, A. (1971) On a simplified testing method for screening soil insecticides by the use of Azuki-bean weevils, Bull. Agr. Chem. Insp. Sta. 11, 108-111. (Ja. en.)

Ohguchi, Y., Suzuki, H., Tatsuki, S. and Fukami, J. (1983) Lethal effect of oxygen absorber (Ageless[R]) on several stored grain and clothes pest insects, Jpn. J. Appl. Ent. Zool. 27, 270- 275. (Ja. en.)

Ohta, K. (1961) On the insecticidal and repellent activity of the household insecticides, Botyu-Kagaku 26, 66-69. (Ja. en.)

Ohta, K. and Ikeda, Y. (1957a) On the insecticidal effect of brominated camphor, Botyu-Kagaku 22, 219-223. (Ja. en.)

Ohota, K. and Ikeda, Y. (1957b) On the formation of eutectic mixture with DDT and gamma-BHC, Botyu-Kagaku 22, 318-323.(Ja. en.)

Okamoto, D. (1944) Damage to pea by the pea weevil and its effect on germination and growth of pea, Bull. Korea Agr. 18, 33-44. (Ja.)

Okamoto, D. (1951) The data concerning loss assessment from insects, Agr. Hort. 26, 324-330. (Ja.)

Okuni, T. (1924) On the grain-pest in Formosa, Rep. Gov. Res. Inst., Formosa, Dept. Agr. 9, 1-166. (Ja.)

Onoue, T. (1935) On fumigation of broad beans, J. Appl. Zool. 7, 121-122. (Ja.)

Onoue T. and Murakami, Y. (1933) On the lethal time of several species of Bruchic by fumigation, J. Appl. Zool. 5, 143. (Ja.)

Oshima, K. (1975) Why do the azuki bean weevils lay their eggs avoiding the already egg-laid beans?, Pl. Prot. 29, 61-63. (Ja.)

Oshima, K., Honda, H. and Yamamoto, I. (1973) Isolation of an oviposition maker from azuki bean weevil, *Callosobruchus chinensis* (L.), Agr. Biol. Chem. 37, 2679-2680.

Ouchi, M. (1936) Effects of moisture and temperature on the oviposition of *Callosobruchus chinensis* (L.), J. Appl. Zool. 8, 308-314. (Ja.)

Ouchi, M. (1937) Effects of humidity and temperature on the oviposition of *Bruchus chinensis* Linn., Kontyu 11, 196-197. (Ja.)

Sakagami, S. (1957) Effect of antennectomization upon the reproductivity of the azuki bean weevil, *Callosobruchus chinensis* L., Botyu-Kagaku 22, 10-12. (Ja, en,)

Sakai, A., Honda, H., Oshima, K. and Yamamoto, I. (1985) Oviposition marking pheromone of two bean weevils, *Callosobruchus chinensis* and *Callosobruchus maculatus*, J. Pesticide Sci. 11, 163-168.

Sakai, S., Gohda, M., Yonebayashi, H., and Matsuishi, K. (1964) A case on the dynamics control of pine tree boring insects, Botyu-Kagaku 29, 61-68. (Ja. en.)

Sano-Fujii, I. (1967) Density effect and environmental temperature as the factors producing the active form of *Callosobruchus maculatus* (F.) (Coleoptera, Bruchidae), J. Stored Prod. Res. 2, 187-195.

Sano-Fujii, I. (1980) Effect of parental age and developmental rate on the production of active form of *Callosobruchus maculatus* (F.) (Coleoptera: Bruchidae), Mech. Ageing Development 10, 283-

293.

Sato, K. (1983) Movement of methyl bromide vapour through a concrete plate, Proc. 10th Intern. Cong. Pl. Prot. 1983, Brighton, England, Vol. 2, 589.

Sato, K., Higuchi, Y., and Suwanai, M. (1973) Studies on the characteristics of action of fumigants. I. The fifty per cent knock down dose of hydrogen phosphide to the azuki bean weevil, *Callosobruchus chinensis* L., calculated from the uptake amounts of oxygen by the weevil, Botyu-Kagaku 38, 22-25. (Ja. en.)

Sato, K. and Suwanai, M. (1973) Ditto. II. Entrance of hydrogen phosphide into weevil body under conditions of the failure to respire for the weevil, Botyu-Kagaku 38, 213-216. (Ja. en.)

Sato, K. and Suwanai, M. (1975) Ditto. III. Emergence of the azuki bean weevil, *Callosobruchus chinensis* L., from azuki beans fumigated with hydrogen phosphide at some developmental stage of ones, Botyu-Kagaku 40, 85-89. (Ja. en.)

Sherman, M. and Hayakawa, M. (1961) Carbon dioxide as an anesthetizing agent for the flesh fly, *Sarophaga peregrina* Robineau-Desvoidy, and the Azuki-bean weevil, *Callosobruchus chinensis* L., Jpn. J. appl. Ent. Zool. 5, 151-153

Shimada, M. (1984) Niche modification process and stability of competitive systems, Dr. Dissertation. University of Tsukuba, pp. 146.

Shimada, M. (1985) Niche modification and stability of competitive systems. II. Persistence of interspecific competitive systems with parasitic wasps, Res. Popul.Ecol. 27, 203-216.

Shimada, M. and Fujii, K. (1985a) Ditto. I. Niche modification process, Res. Popul. Ecol. 27, 185-201.

Shimada, M. and Fujii, K. (1985b) Ditto. III. Simulation model analysis, Res. Popul. Ecol. 27, 217-230.

Shinoda, K. (1989) Study on the life history of *C. chinensis* in the field, Dr. Dissertation, Okayama University, pp. 116. (Ja)

Shinoda, K. and Yoshida, T. (1984) Relationship between adult feeding and emigration from beans of azuki bean weevil, *Callosobruchus chinensis* Linn? (Coleoptera: Bruchidae), Appl. Ent. Zool. 19, 202-211.

Shinoda, K. and Yoshida, T. (1985) Field biology of the azuki bean weevil, *Callosobruchus chinensis* (L.) (Coleoptera: Bruchidae). I. Seasonal prevalence and assessment of field infestation of Akiazuki, autumn variety of *Phaseolus angularis* W., Jpn. J. Appl. Ent. Zool. 29, 14-20. (Ja. en.)

Shinoda, K. and Yoshida, T. (1987) Effect of fungal feeding on longevity and fecundity of the azuki bean weevil, *Callosobruchus chinensis* (L.) (Coleoptera: Bruchidae), in the azuki bean field, Appl. Ent. Zool. 22, 465-473.

Shinohara, H. and Nagasawa, S. (1963) Sterilizing effect of apholate and metepa on adults of the azuki bean weevil, *Callosobruchus chinensis* L. Studies on the chemosterilants of insects. I, Ent. exp. appl. 6, 263-267.

Sugimoto, A. (1963) On the method of bioassay of insecticide residues using adzuki bean weevil, Jpn. J. appl. Ent. Zool. 7, 20-25. (Ja. en.)

Suenaga, H., Matsuo, K., and Sakai, H. (1951) Varietal difference of damage caused by the broad bean weevil, Bull. Kyushu Agr. Exp. Sta. 1, 82. (Ja.)

Suwanai, M. (1959) Absorption of parathion in insect integument and its osmosis in the body system, Proc. 3rd Symp., Jpn. Soc. Appl. Ent. Zool. 41-43. (Ja.)

Suwanai, M. (1964) Insect integument permeability of insecticides. Laying stress on the action of parathion on the azuki bean weevil, Pl. Prot. 18, 485-489. (Ja.)

Tachikawa, T. (1966) On the identity of *Neocatolaccus mamezophagus* Ishii et Nagasawa (Hymenoptera: Pteromalidae), Jpn. J. Appl. Ent. Zool. 10, 99. (Ja.)

Takahashi, S. (1929) On the infestation of newly introduced insect pest, the broad bean weevil, Konchu-Sekai 33, 219-222. (Ja.)

Takahashi, F. (1976) Reproduction curve in the experimental population of *Callosobruchus chinensis*. Physiol. Ecol. Jpn. 17, 495-501.

Takara, T. and Azuma, S. (1971) Alien insect pests in Okinawa, Pl. Prot. 25, 449-452. (Ja.)

Takasugi, K. (1924) Research on the azuki bean weevil, *Callosobruchus chinensis* (L.). 2nd Report of the Investigation on the Control of Stored-Product Insect Pests, Plant Quarantine Station. pp. 12 (Ja.)

Takenouchi, Y. (1955) A short note on the chromosomes in three species of the Bruchidae (Coleoptera), Jpn. J. Genet. 30, 7-9.

Takenouchi, Y. (1971a) Chromosomes in males of five strains of *Callosobruchus maculatus* (Coleoptera: Bruchidae), Can. J. Genet. Cytol. 13, 708-713.

Takenouchi, Y. (1971b) A further study on the chromosomes of *Callosobruchus chinensis* L. (Coleoptera: Bruchidae), Kontyu 39, 332-337.

Takeuchi, H., Watanabe, N. and Sonda, M. (1977) Loss in weight of adzuki-bean caused by a Bruchid beetle, *Callosobruchus analis* Fabricius, Res. Rep. Pl. Prot. Sta. 14, 60-63. (Ja. en.)

Tanaka, K. Ohsawa, K., Honda, H. and Yamamoto, I. (1981) Copulation release pheromone, erectin, from the azuki bean weevil (*Callosobruchus chinensis* L.), J. Pesticide Sci. 7, 75-82.

Tanaka, K., Oshima, K., Honda, H., and Yamamoto, I. (1982) Synthesis of erectin, a copulation release pheromone of the azuki bean weevil, *Callosobruchus chinensis* L., J. Pesticide Sci. 7, 535-537.

Thoyama, Y. (1952) Study on control of the broad bean weevil by storage management of seed, Agr. Hort. 27, 748. (Ja.)

Tomizawa, C., Ueji, M., and Yoshida, T. (1988) Insecticidal activity of phosphoramidothioates and bioactivation of their oxon homologs, J. Pesticide Sci. 13, 449-454.

Tsuchiyama, T., Oyama, M., and Nagaoka, N. (1960) Analytical studies on the damage caused by the pea weevil, *Bruchus pisorum* L., Bull. Shikoku Agr. Exp. Sta. 5, 359-380. (Ja. en.)

Tsutsui, K. (1955) Insect pests of pulse. Bruchids, Agr. Chem. 2(4), 21-26. (Ja.)

Uchida, M., Naka, H. Irie, Y., Fujita, T., and Nakajima, M. (1974) Insecticidal and neuroexciting actions of DDT analogs, Pesticide Biochem. Physiol. 4, 451-455.

Ueji, M. and Tomizawa, C. (1984) Bioactivation of n-alkyl substituted phosphoramidothioate insecticides, J. Pesticide Sci. 9, 675-680.

Ueji, M. and Tomizawa, C. (1986) Insect toxicity and anti-acetylcholinesterase activity of chiral isomers of isofenphos and its oxon, J. Pesticide Sci. 11, 447-451.

Ueki, K. (1952) The control of azuki bean weevil (*Callosobruchus chinensis* L.) by benzene hexachloride, Botyu-Kagaku 17, 103-106. (Ja. en.)

Ueno, H. (1955) On the oviposic behaviour of the azuki bean weevil, OYO-KONTYU 10, 196-200. (Ja. en.)

Umeya K. (1966a) Studies on the comparative ecology of bean weevils. I. On the egg distribution and the oviposition behaviors of three species of bean weevils infesting azuki bean, Res. Bull. Pl. Prot. Serv. Jpn. 3, 1-11. (Ja. en.)

Umeya, K. (1966b) Ditto. II. An interspecific comparison of the escaping behaviors of adult and reproduction ratio in a semi-open cage, Res. Bull. Pl. Prot. Serv. Jpn. 4, 1-15. (Ja. en.)

Umeya, K. (1968) Historical review of the invasion of insect pests in Japan, Pl. Prot. 22, 3-8. (Ja.)

Umeya, K. (1987) Biology of Bruchids, Tukizi-Shokan, Tokyo. (Ja.)

Umeya, K. and Imai, E. (1965) Growth of the azuki bean weevil (*Callosobruchus chinensis* L.) and the Mexican bean weevil (*Zabrotes subfasciatus* Boh.) on bean of grafted *Phaseolus* plants, Jpn. Appl. Ent. Zool. 9(3), 238-246.

Umeya, K. and Kato, T. (1970) Studies of the comparative ecology of bean weevils. V. Distribution of eggs and larvae of *Acanthoscelides obtectus* in relation to its oviposition and boring behavior, Res. Popu. Ecol. 12, 35-50.

Umeya, K., Kato, T. and Kocha, T. (1975) Ditto. VI. Intraspecific larval cometition in *Callosobruchus analis* (F.), Jpn. Appl. Zool. Ent. 19, 47-53. (Ja. en.)

Umeya, K., Kato, T. and Sekiguchi, Y. (1970) Ditto. IV. Tolerance to low temperature (5°C) of the five species of bean weevil infesting azuki bean, Res. Bull. Pl. Prot. Serv. Jpn. 8, 39-48

Umeya, K. and Shimizu, (1968) Ditto. III. Effect of feeding on the life span and oviposition of the adult of three species of bean weevils, Res. Bull. Pl. Prot. Serv. Jpn. 5, 39-49 (Ja. en.)

Utida, S. (1941a) On the growth of the larvae of the azuki bean weevil, *Callosobruchus chinensis* (L.), Pl. Zool. 9, 322-328. (Ja. en.)

Utida, S. (1941b) Studies on experimental population of the azuki bean weevil, *Callosobruchus chinensis* (L.) I. The effect of population density on the progeny population, Mem. Coll. Agr. Kyoto Imp. Univ. 48,1-30.

Utida, S. (1941c) Ditto. II. The effect of population density on progeny populations under different conditions of atmospheric moisture, Mem. Coll. Agr. Kyoto Imp. Univ. 49, 1-20.

Utida, S. (1941d) Ditto. III. The effect of population density upon the mortalities of different stages of life cycle, Mem. Coll. Agr. Kyoto Imp. Univ. 49, 21-42.

Utida, S. (1941e) Ditto. IV. Analysis of density effect with respect to fecundity and fertility of eggs, Mem. Coll. Agr. Kyoto Imp. Univ. 51, 1-26.

Utida, S. (1941f) Ditto. V. Trend of population density at the equilibrium position, Mem. Coll. Agr. Kyoto Imp. Univ. 51, 27-34.

Utida, S. (1942a) Ditto. VI. The relations between the size of environment and the rate of population growth, Mem. Coll. Agr. Kyoto Imp. Univ. 53, 1-18.

Utida, S. (1942b) Ditto. VII. Analysis of the density effect in the preimaginal stage, Mem. Coll. Agr. Kyoto Imp. Univ. 53, 19-31.

Utida, S. (1943a) Ditto. VIII. Statistical analysis of the frequency distribution of the emerging weevils on beans, Mem. Coll. Agr. Kyoto Imp. Univ. 54, 1-22.

Utida, S. (1943b) Ditto. IX. General consideration and summary of the serial reports from I to VII, Mem. Coll. Agr. Kyoto Imp. Univ. 54, 23-40.

Utida, S. (1943c) Relationship between population density and space in *Callosobruchus chinensis*, J. Appl. Zool. 14, 232-235. (Ja.)

Utida, S. (1943d) Host parasite interaction in the experimental population of the azuki bean weevil, *Callosobruchus chinensis* (L.). I. The effect of density of parasite population on the growth of the host population and also of the parasite population (1), Oyo-Kontyu 4, 117-128. (Ja. en.)

Utida, S. (1943e) Ditto. III. The effect of host density on the growth of host and parasite population, Ecol. Rev. (Sendai) 9, 40-54. (Ja. en.)

Utida, S. (1944) Ditto. II. The effect of density of parasite population on the growth of the host population and also of the parasite population (2), J. Appl. Zool. 15, 1-18. (Ja.)

Utida, S. (1946) Toxic effect of mercury vapour to insect eggs, Matsumusi 1, 20-23. (Ja.)

Utida, S. (1947a) Lethal effect of inert materials on insect, Botyu-Kagaku 7, 8, 9, 49-52. (Ja.)

Utida, S. (1947b) The effect of different sex-ratios upon the growth of population. Studies on experimental population of the azuki bean weevil, *Callosobruchus chinensis* (L.). X, Physiol. Ecol. 1, 67-78. (Ja. en.)

Utida, S. (1848a) Host-parasite interaction in the experimental population of the azuki bean weevil, *Callosobruchus chinensis* (l.). IV. The effect of host density on the growth of host and parasite populations. (2), Oyo-Kontyu 4, 164-174. (Ja. en.)

Utida, S. (1948b) Population fluctuations caused by host-parasite interaction. Host-parasite interaction in the experimental population of the azuki bean weevil, *Callosobruchus chinensis* (L.) V, Physiol. Ecol. 2, 1-11. (Ja. en.)

Utida, S. (1948c) Sex-ratios at the equilibrium position of population. Ditto. VIII, Seibutu 3, 218-221. (Ja. en.)

Utida, S. (1949a) On the equilibrium state of population. Ditto, VII, Physiol. Ecol. 3, 89-96. (Ja. en.)

Utida, S. (1949b) Generation overlapping and the effect of population density, Oyo-Kontyu 5, 55-61. (Ja.)

Utida, S. (1950a) On the equilibrium state of the interacting population of an insect and its parasite, Ecol. 31, 165-175.

Utida, S. (1950b) The experimental population of insects, in Soc. Res. Theor. Biol., Demo. Sci. Asso. (ed.), Biotic Population and Environment, Iwanami Pub., Tokyo, pp. 10-23. (Ja.)

Utida, S. (1950c) Ecology of insect population, in Recent Biology, Baifukan, Tokyo, pp. 26-67. (Ja.)

Utida, S. (1951a) Role of parasite in determining equilibrium state of interacting population of a host and its parasite. Host-parasite interaction in the experimental population of the azuki bean weevil, *Callosobruchus chinensis*. X, Oyo-Kontyu 7, 1-7. (Ja. en.)

Utida, S. (1951b) Population fluctuations caused by host-parasite interaction. 2nd Report. Host-parasite interaction in the experimental population of the azuki bean weevil, *Callosobruchus chinensis*. IX, J. Appl. Zool. 16, 111-118. (Ja. en.)

Utida, S. (1952a) Space of iso-effect of density, Res. Popul. Ecol. 1, 119-121. (Ja, en,)

Utida, S. (1952b) Density effects observed in the interacting populations of a host and its two parasites. Experimental studies on synparasitism. 1st Report, Oyo-Kontyu 8, 1-7 (Ja. en.)

Utida, S. (1953a) Interspecific competition between two species of bean weevil, Ecol. 34, 301-307.

Utida, S. (1953b) Population fluctuation in the system of host-parasite interaction, Res. Popul. Ecol. 2, 22-46. (Ja. en.)

Utida, S. (1953c) Fluctuations in the interacting populations of host and its parasite in relation to the biotic potential of host species, Oyo-Kontyu 9, 59-62. (Ja. en.)

Utida, S. (1953d) Effect of host density upon the population growth of interacting two species of parasite. Experimental studies on synparasitism. Second Report, Oyo-Kontyu 9, 102-107. (Ja. en.)

Utida, S. (1954) "Phase" dimorphism observed in the laboratory population of the cowpea weevil, *Callosobruchus quadrimaculatus*, J. Appl. Zool. 18, 161-168. (Ja. en.)

Utida, S. (1955a) Population fluctuation in the system of host-parasite action, Mem. Coll. Agr. Kyoto Univ. 71, 1-34.

Utida, S. (1955b) Fluctuations in the interacting populations of host and parasite in relation to the biotic potential of the host, Ecol. 36, 202-206.

Utida, S. (1956a) Ecological studies on the experimental populations of some insects, in T. Komai and K. Saka (eds.), Syudan Idengaku (Population Genetics), Baifukan, Tokyo, pp. 121-142. (Ja. en.)

Utida, S. (1956b) Population fluctuation in the system of host-parasite interaction in different size of their environment, Res. Popul. Ecol. 3, 45-51. (Ja. en.)

Utida, S. (1956c) Long-term fluctuation of population in the system of host-parasite interaction, Res. Popul. Ecol. 3, 52-59. (Ja. en.)

Utida, S. (1956d) "Phase" dimorphism observed in the laboratory population of the cowpea weevil, *Callosobruchus quadrimaculatus*. 2nd Report, Res. Popul. Ecol. 3, 93-104. (Ja. en.)

Utida, S. (1956e) Effect of population density upon reproductive rate in relation to adult longevity, Jpn. J. Ecol. 5, 137-140. (Ja. en.)

Utida, S. (1957a) Population fluctuation, an experimental and theoretical approach, Cold Spring Harbor Symp. Quant. Biol. 22, 139-151.

Utida, S. (1957b) Cyclic fluctuations of population density intrinsic to the host-parasite system, Ecol. 38, 442-449.

Utida, S. (1957c) Outbreak of insect pests and its mechanism, Pl. Prot. 11, 55-59. (Ja.)

Utida, S. (1959a) The effect of population density on progeny populations observed in the different strains of the azuki bean weevil, Kontyu 27, 41-46.

Utida, S. (1959b) Sequential frequency of the emergence of adult insect, in relation to the change of environmental condition, Jpn. J. Ecol. 9, 139-143. (Ja. en.)

Utida, S. (1959c) The effect of population density on progeny populations observed in the different strains of the azuki bean weevil II, Jpn. J. Ecol. 9, 172-178.

Utida, S. (1965) "Phase" dimorphism observed in the laboratory population of the cowpea weevil, *Callosobruchus maculatus*. IV. The mechanism of induction of the flight form, Jpn. J. Ecol. 15, 193-199. (Ja. en.)

Utida, S. (1966) Water content of body in several kinds of the bean weevil, Jpn. J. Appl. Ent. Zool. 10, 39-43. (Ja. en.)

Utida, S. (1967) Polymorphism in *Callosobruchus maculatus* and domestication of stored product insect pests, Pl. Prot. 21, 243-248. (Ja.)

22

Utida, S. (1967a) Damped oscillation of population density at equilibrium, Res. Popul. Ecol. 9, 1-9.

Utida, S. (1967b) Collective oviposition and larval aggregation in *Zabrotes subfasciatus* (Boh.) (Coleoptera, Bruchidae), J. stored Prod. Res. 2, 315-322.

Utida, S. (1968) The influence of the parental condition on the production of flight form in the population of *Callosobruchus maculatus*, Jpn. J. Ecol. 18, 246-249. (Ja. en.)

Utida, S. (1969) Photoperiod as a factor inducing the flight form in the population of southern cowpea weevil, *Callosobruchus maculatus*, Jpn. J. Appl. Ent. Zool. 13, 129-134. (Ja. en.)

Utida, S. (1970) Secular change of percent emergence of the flight form in the population of the cowpea weevil, *Callosobruchus maculatus*, Jpn. J. Appl. Ent. Zool. 14, 71-78. (Ja. en.)

Utida, S. (1971) Influence of temperature on the number of eggs, mortality and development of several species of Bruchid infesting stored beans, Jpn. J. Appl. Ent. Zool. 15, 23-30. (Ja. en.)

Utida, S. (1972a) Density dependent polymorphism in the adult of *Callosobruchus maculatus* (Coleoptera, Bruchidae), J. stored Prod. Res. 8, 111-126

Utida, S. (1972b) Animal Demography. Considering the Ecology of Over- and Under-population, Japan Broadcasting Corporation (NHK) Pub. Tokyo. (Ja.)

Utida, S. (1974) Polymorphism in the adult of *Callosobruchus maculatus* - A possible process of evolution to stored product pest -, Proc. First Intern. Wkng. Conf. Stored-Prod. Entom. Savannah, Georgia USA, 686-691.

Utida, S. (1981) Polymorphism and phase dimorphism in *Callosobruchus*, in Series Entomologica, Vol. 19: ed by V. Labeyrie (ed.), The ecology of Bruchids attacking legumes (pulses), Dr W. Junk Publishers, The Hague, pp. 143-147.

Utida, S. and Kakemi, H. (1959) Growth and duration of larval instars of the cowpea weevil, *Callosobruchus quadrimaculatus*, Jpn. J. Appl. Ent. Zool. 3, 29-33. (Ja. en.)

Utida, S. and Nagasawa, S. (1949) On the developmental period and that of adult life of *Neocatolaccus mamezophagus*, a pteromalid parasite of the azuki bean weevil, KONTYU 17, 1-21. (Ja.)

Utida, S. and Takahashi, F. (1958) "Phase" dimorphism observed in the laboratory population of the cowpea weevil, *Callosobruchus quadrimaculatus*. III. Chemical differences of body constituents between two phases, Jpn. J. Appl. Ent. Zool. 2, 33-37. (Ja. en.)

Watanabe, N. (1984) Comparative studies on ecology of the azuki bean weevil, *Callosobruchus chinensis* (L.), and the southern cowpea weevil, *C. maculatus* (F.) (Coleoptera: Bruchidae). I. Effect of seed size on creeping of females into the pile, Jpn. J. Appl. Ent. Zool. 28, 223-228. (Ja. en.)

Watanabe, N. (1985a) Ditto. II. Relation between female body size and degree of creeping into the stored pile of seeds, Jpn. J. Appl. Ent. Zool. 29, 107-112. (Ja. en.)

Watanabe, N. (1985b) Ditto. III. Oviposition preference for some legume seeds, House Household Insect Pests, 25/26, 78-85. (Ja. en.)

Watanabe, N. (1985c) Establishment of indoor life in insects, Insectarium 22, 184-190. (Ja.)

Watanabe, N. (1985d) Oviposition habit of *Sulcobruchus sauteri* (Pic) and its significance in speculation on the pre-agricultural life of seed beetles attacking stored pulses (Coleoptera:Bruchidae), Kontyu 53, 391-397.

Watanabe, N. (1986a) Comparative studies on *Callosobruchus chinensis* (L.) and *C. maculatus* (F.) with special reference to their adaptation to cultivated and stored pulses, Thesis of Kyoto University. pp. 114. (Ja.)

Watanabe, N. (1986b) Quarantine significance against expansion of insect pests caused by agricultural diffusion of their host plants; with special reference to the seed beetles, Res. Bull. Pl. Prot. Japan 22, 1-9. (Ja. en.)

Watanabe, N. (1986c) Comparative studies of the azuki bean weevil, *Callosobruchus chinensis* (L.), and the southern cowpea weevil, *C. maculatus* (F.) (Coleoptera: Bruchidae). IV. Effect of adult body size on easiness of escaping from the bottom of piled seeds, Jpn. J. Appl. Ent. Zool. 30, 64-66. (Ja.)

Watanabe, N. (1986d) Bruchids, in K. Kiritani (ed.), Insect in Japan, Ecology of Invasion and Disturbance, Tokai Univ. Pub. Tokyo, pp. 52-60. (Ja.)

Watanabe, N. and Sugimoto, S. (1988) Geographic variation in male antenna of the azuki bean weevil, *Callosobruchus chinensis* (L.) (Coleoptera: Bruchidae), Appl. Ent. Zool. 23, 282-290.

Watanabe, S. (1950) Interaction between a host and its parasite, Botyu-Kagaku 15, 73-79. (Ja. en.)

Watanabe, S., Utida, S. and Yoshida, T. (1952) Dispersion of insect and change of distribution type in its process, Res. Popul. Ecol. 1, 94-108. (Ja. en.)

Yajima, T. and Munakata, K. (1979, recd. 1981) Phloroglucinol-type furocoumarins, a group of potent naturally-occurring insect antifeedants, Agr.Biol. Chem. 43, 1701-1706.

Yamada, S. (1944) On the use of hot water for control of the broad bean weevil, New Agr. 46 (6), 16-18. (Ja.)

Yamamoto, I. (1976) Approaches to insect control based on chemical ecology - Case studies, in F. Coulston and F. Korte (eds.), Environmental Quality Safety 5, pp. 73-77.

Yamamoto, I. (1986) New insect pheromone, erectin-like substances, in E. Donahue and S. Navarr (eds.), Proc. 4th Intern. Work. Conf. Stored-Product Prot. Tel Aviv, Israel, pp. 418-424.

Yamamoto, I., Honda, H. and Ohsawa, K. (1980) Chemical ecology of pest insects, House Household Insect Pests 7, 8, 8-15. (Ja. en.)

Yoshida, M. and Suzuki, Y. (1953) Immersion into hot water as a controlling measure of the azuki bean weevil, *Callosobruchus chinensis*, Botyu-Kagaku 18, 109-117. (Ja. en.)

Yoshida, S. (1978) Behaviour of males in relation to the female sex pheromone in the parasitoid wasp, *Anisopteromalus calandrae* (Hymenoptera: Pteromalidae), Ent. Exp. Appl. 23, 152-162.

Yoshida, T. (1952) Experimental analysis of emergency curve, Res. Popul. Ecol. 1, 152-165. (Ja. en.)

Yoshida, T. (1957) Experimental studies on the interspecific competition among the granary beetles. No.4 Competition between the azuki bean weevil, *Callosobruchus chinensis* and the southern cowpea weevil, *C. quadrimaculatus*, Mem. Fac. Lib. Arts Educ., Miyazaki Univ. 1, 55-80. (Ja. en.)

Yoshida, T. (1958a) Increase of the reproductivity of the azuki ban weevil, *Callosobruchus chinensis* (L.) in the soy bean, Jpn. J. Ecol. 8, 171-176. (Ja. en.)

Yoshida, T. (1958b) Origin and domestication of stored product insects, Biol. Sci. 10, 60-68. (Ja.)

Yoshida, T. (1959) Local distribution of the eggs of the pea weevil, *Bruchus pisorum* L., Mem. Fac. Lib. Arts Educ., Nat. Sci., Miyazaki Univ. 6, 11-21. (Ja. en.)

Yoshida, T. (1960a) Adult longevity under the condition of interspecific competition, Mem. Fac. Lib. Arts Educ., Miyazaki Univ. 9, 463-472.

Yoshida, T. (1960b) Effect of interspecific competition between two species of bean weevil upon fecundity and fertility, Mem. Fac. Lib. Arts Educ., Nat. Sci. Miyazaki Univ. 10, 17-31.

Yoshida, T. (1961) Oviposition behaviours of two species of bean weevils and interspecific competition between them, Mem. Fac. Lib. Arts Educ., Nat. Sci., Miyazaki Univ. 11, 41-65. (Ja. en.)

Yoshida, T. (1966) Studies on the interspecific competition between bean weevils, Mem. Fac. Lib. Arts Educ., Miyazaki Univ. 20, 59-98.

Yoshida, T, (1967) Struggle for existence, Kokin Shoin, Tokyo. (Ja.)

Yoshida, T. (1976) Interspecific competition among the stored-product insects, Proc. Joint US-Jpn. Sem. Stored Prod. Insect., Manhattan, Kansas, pp. 1-19.

Yoshida, T. (1980) Plant quarantine and intra-specific variation in insect pests: On strain of stored-product insects, Pl. Prot. 34, 35-41. (Ja.)

Yoshida, T. (1982a) Historical change in the status of stored-product insects, in T. Okutani (ed.), Cities and Insect, pp. 35-45. (Ja.)

Yoshida, T. (1982b) History of stored-product insect pests infestation, Insect and Nature 17, 8-12. (Ja.)

Yoshida, T, (1983) Historical change in the status of stored product insect pests especially in Japan, Proc. Third Intern. Wkng. Conf. Stored-Prod. Entomol., Manhattan, Kansas USA, 655-668.

Yoshida, T. (1984) Colonization by forest insects of stored barley or wheat in ancient times, in the Near and Middle East, Proc. Sixth Intern. Biodeterio. Symp., Washington D.C., USA, (Biodeterioration VI), 660-663.

Yoshida, T. (1989) Life history of *Callosobruchus chinensis* (Coleoptera: Bruchidae). One generation a year, Proc. 1st Asia-Pasif. Conf. Entomol. (in press)

Yoshida, T. and Gichuki, E. M. (1983) Use of solar heat to control stored-product insect pests: bean depth, Sci. Rep. Fac. Agr., Okayama Univ. 61, 5-8.

Yoshida, T., Igarashi, H. and Shinoda, K. (1987) Life history of *Callosobruchus chinensis* (L.) (Coleoptera, Bruchidae), Proc. 4th Intern. Work. Conf. Stored-Prod. Prot., Tel Aviv, Israel, 471-477.

Yoshida, T., Shinoda, K., and Okamoto, T. (1984) Life history and new wild host legume of *Callosobruchus chinensis* (L.), Tribolium Inf. Bull. 24, 144.

Yoshihara, T. (1956) On some theoretical considerations of the reproduction curve, Res. Popul. Ecol. 3, 1-7. (Ja. en.)

BIONOMICS OF *BRUCHIDIUS INCARNATUS* BOH. IN EGYPT

MONIR. M. METWALLY
Dept. of Plant Protection,
Faculty of Agriculture,
Al-Azhar University, Cairo, Egypt

ABSTRACT. Bionomics of *B. incarnatus* Boh. were studied under natural and controlled conditions. Seven generations per year were bred on broad beans, eight on common peas, and nine on lentils. The sex ratio was nearly 1:0.8 and the number of males exceeded that of the females by about 18%.

The pre-oviposition period from March till October, ranged between 1 and 1.03 days under natural conditions and less than one and one half days under controlled conditions. The oviposition period varied greatly, from 4.57 ± 0.26 days during the period from July-August and 10.33 ± 1.60 days during the period from Nov.-February. Under controlled conditions, the temperature and the relative humidity, either separately or combined, had highly significant effects on the oviposition period. The mean number of eggs laid per female during winter was less than 22 eggs, while this number was no less than 57 eggs during any other period of the year. In the period from March-April it may reach 75.6 ± 6.88 eggs. Under controlled conditions (28°C and 60% R.H.), the number of eggs laid per female reached its maximum (mean 77.10 eggs). The post-oviposition period was longer during winter. Under controlled conditions, this period did not exceed 3.8 days. The incubation period lasted six days during July-August and 21 days (Nov.-Feb.). It was 26.11 ± 0.01 days (20°C and 20% R.H.) and only 6.0 days under 32°C at 40% R.H. The larval period lasted 13.8 days (July-August) and 31.17 days (Nov.-Feb.). The shortest larval period occurred under 28°C at 20, 40 and 60% R.H. Under natural conditions, the pupal stage lasted for only 3.12 ± 0.07 in July-August and 11.85 days in winter.

The adult male (under natural conditions) could survive for more than 24 days in winter and 10.10 days in summer, while this period was 28 and 9.62 days, respectively for the female.

1. Introduction

Broad bean (*Vicia faba*) is one of the most important economic crops in Egypt (Kamel 1981) and bean seeds are the principal leguminous food for human and animal consumption in many parts of the world. According to statistical records of the Central Agency of Public Mobilization and Statistics (1987), an area of 324,000 fedans had been planted with broad bean which yielded 499,000 tons.

Although broad beans are considered the most favored food of *Bruchidius incarnatus* Boh., other leguminous crops such as peas and beans are liable to its attack. Serious damage is caused to stored dry beans on which this pest reproduces. De Luca (1962) found that *B. incarnatus* could infest seeds of different plants such as flax, chick peas (*Cicer arietinum*), *Dolichos lablab* and peas.

In spite of its great economic importance, *B. incarnatus* has been studied by relatively few workers, e.g., Winkler (1927), Zacher (1930), De Luca (1962) and El-Gendy (1984). Although this insect is an important pest in Egypt, little work has been done on it as shown by the scanty literature. The literature of other stored grain insects, especially

K. Fujii et al. (eds.), Bruchids and Legumes: Economics, Ecology and Coevolution, 25–36.
© 1990 *Kluwer Academic Publishers. Printed in the Netherlands.*

those infesting leguminous seeds, on the contrary, is voluminous (e.g., Schoof, 1941; El-Sawaf, 1956, 1958; Abdel-Salam et al., 1984; Abdul-Rassoul et al., 1986).

The present work deals with the biology and behavior of this important pest.

2.   Materials and Methods

Studies on *B. incarnatus* Boh. were conducted under natural and controlled laboratory conditions. Under natural conditions, adult beetles were collected from various infested seeds in various regions of the country. Jars (2 liters) containing 1/2 kg of various sterilized seeds (e.g., broad beans, cowpeas, lentils, common peas, and French beans) were infested by the beetles. In order to avoid crowding (which is toxic to the beetles (El-Sawaf, 1956)) in the original media, more beans were added from time to time.

Jars were covered with muslin held tightly by rubber bands and all cultures were kept in an incubator at a temperature of 28°C and 70% R.H. Certain kinds of sterilized seeds were infested by newly emerged male and female beetles selected at 10 am. daily. Seed sterilization was made at a temperature of 70-80°C for five hours.

Relative humidities were produced by using sulfuric acid solutions (Buxton 1931; Solomon 1951; Hafez and Chapman 1962). Desiccators with well-fitting and slightly greased covers were used. The solutions needed for any desired relative humidity were put into the desiccators at least a day before starting the experiment and the solutions were also renewed weekly to assure desired humidities.

Counting the eggs and the methods of differentiating various insect instars will be illustrated in their proper places in the text. Temperatures and relative humidities were recorded daily.

Under controlled conditions four constant temperatures were used viz. 20, 24, 28 and 32°C. and four relative humidities viz. 20, 40, 60 and 80%.

Adult insects were taken from stock cultures of infested broad beans and the seeds in which new eggs were deposited were collected directly and kept in glass tubes (1x2 inches) covered with muslin and incubated. Newly hatched larvae were isolated in separate tubes covered with muslin and fixed tightly with rubber bands. Fifty pairs of male and female adults were used in each experiment. All observations were recorded daily and as was the duration of the different developmental stages. When studying the pre-oviposition period, the oviposition period and the number of eggs laid per female, pairs of newly emerged adults were isolated in 1x2 inch tubes.

3.   Results and Discussion

3.1. ANNUAL GENERATIONS

Data obtained from ten replicates under laboratory conditions revealed the following (Fig. 1).

3.1.1. *Broad beans (Vicia faba)*.   Two generations were obtained during the spring season. The first, which started late in the previous winter, lasted for a period of 73 days. The second lasted for 33 days under an average temperature of 22.7 and 27.8°C, respectively. In the summer, three generations were recorded. The first lasted for 33 days, the second 26 days and the third 26 days under average temperatures of 29.6, 33.1 and 33.3°C, respectively. In autumn and winter, two generations occurred. The first lasted for 27 days with an average of 31°C. The second started about mid October

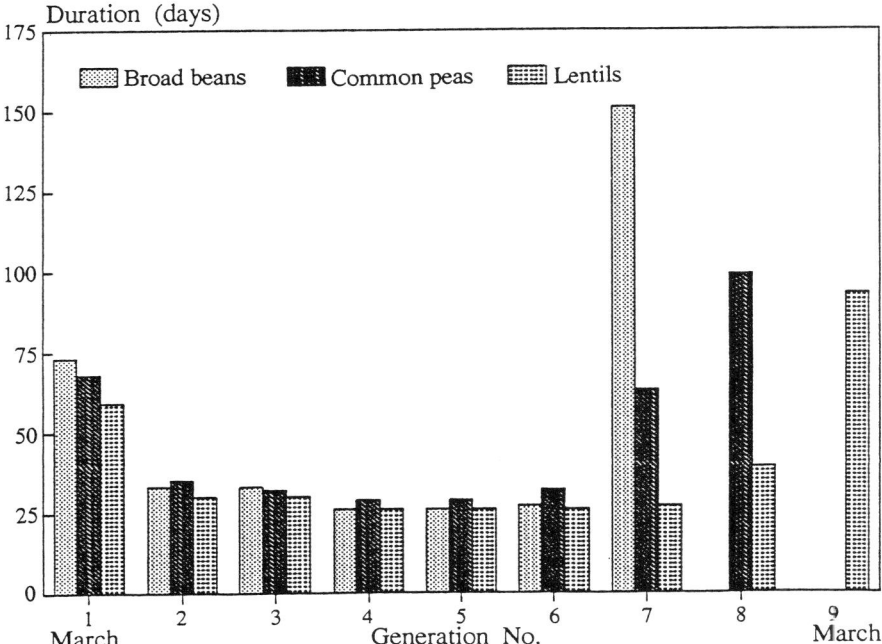

Figure 1. Annual successive generation periods on various legumes.

when the temperature started to decrease causing the elongation of the immature periods inside the seeds so that this generation extended during the whole winter (about 151 days) under an average temperature of 20.5°C. The previous results indicated that if *B. incarnatus* is bred on broad beans, it will have about seven generations per year. The correlation coefficient between the average temperature and generation period resulted in r = 0.809 (with F = 7.50). This means that as the average temperature increased the generation period decreased. An increase in the average temperature by 1°C resulted in the decrease of the duration of the generation period by 7.50 days.

3.1.2. *Common peas (Pisum sativum).* The number of annual generations was eight (Fig. 1) in 287 days. In the spring period, owing to the low average temperature (22.1°C) during the first half of this period, the length of the generation was 68 days. During late spring, the generation period was rather short, only 35 days (27.8°C). In summer, owing to the homogeneity of the average temperature, the generation periods were almost equal, resulting in 32, 29, and 29 days under average temperatures of 29.6, 33.1 and 33.3°C, respectively. In autumn, two generations obtained (32 and 63 days) under average temperature of 31 and 24.3°C, respectively. In winter, one generation took place lasting for 99 days under 19.6°C. Statistical analysis showed that an increase in the average temperature by 1°C resulted in the decrease of the generation period by 4.73 days when using common peas as a diet.

3.1.3. *Lentils (Lens culinaris).* Nine generations occurred per year (Fig. 1). In spring, two generations occurred lasting 59 and 30 days under average temperatures of 22.1 and 27.8°C. In summer, the three generations obtained lasted 30 (29.6°C), 26 (33.1°C), and 26 (33.3°C) days.

Also three generations occurred in autumn which lasted 26 (31°C), 27 (30.8°C) and 39 (26.4°C) days. In winter, only one generation occurred, lasting 93 days under an average temperature of 19.6°C. By using lentils as a diet, an increase in the average temperature by 1°C. resulted in the statistically significant decrease of the generation period by 4.43 days.

3.1.4. *Cowpeas (Vigna unguiculata)*. Through a period from April-June, the insect was unable to complete its life cycle. Adults did lay eggs on cowpea seeds. Although the eggs hatched and the larvae bored inside the seeds, they failed to continue their life cycles. Cowpea seeds are more suitable as a host for *C. maculatus* (Dick and Credland, 1986).

3.1.5. *French beans (Phaseolus vulgaris)*. The insect laid eggs but the newly hatched larvae were unable to bore inside the seed. Thus, French beans are considered as resistant to infestation by this insect (Applebaum and Guez, 1972).

When data concerning emerging adults are compared, lentil seeds were the most suitable diet and habitat for this insect. Larson (1927) dealing with *B. quadrimaculatus* (*Callosobruchus maculatus*), El-Sawaf (1956 and 1958), Metwelly et al. (1984), and Pajni (1987), dealing with *Callosobruchus maculatus* and *C. chinensis*, found that such insects showed a predilection to certain seeds and refused to oviposit or complete larval development in other kinds of seeds. This may due to the physical and chemical (protein and reducing sugar) properties of these kinds of seeds.

3.2. EFFECT OF KINDS OF LEGUMES ON SEX RATIOS

When the larval diet was broad beans, out of 1302 emerging beetles, 696 were males and 606 were females with a ratio 1:0.871. But when the diet was common peas among 580 adults, 390 were males and 290 females, a sex ratio of 1:0.746. On the other hand, when lentils were used, among 2123 beetles, 1136 were males and 987 females (1:0.869). In all diets the number of males exceeded the number of females by about 18%.

3.3. PRE-OVIPOSITION PERIOD

3.3.1. *Natural conditions*. Under natural conditions, the mean pre-oviposition period (Fig. 2) was nearly equal from March till October, ranging between 1 and 1.03 days, but in the cooler months of the year (November-February) the pre-oviposition period was 8.4 days.

3.3.2. *Controlled conditions*. On the other hand, under controlled conditions, this period varied from one day (at 32°C and 40% R.H.) to 1.42 ± 0.97 days (at 20°C and 20% R.H.). The general trend was that at various relative humidities the pre-oviposition period decreased as the temperature increased.

Data obtained from analysis of variance revealed that both the relative humidity and temperature and the interaction between those two factors had a significant effect on the pre-oviposition period.

3.4. OVIPOSITION PERIOD

3.4.1. *Natural conditions*. Under natural conditions during April and March when the average temperature reaches 23.5°C, the oviposition period (Fig. 2) was 7.65 ± 0.61 days while it was 5.3 ± 0.25 days in May and June (28.65°C). In July and August it

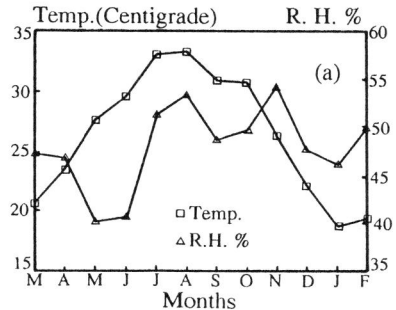

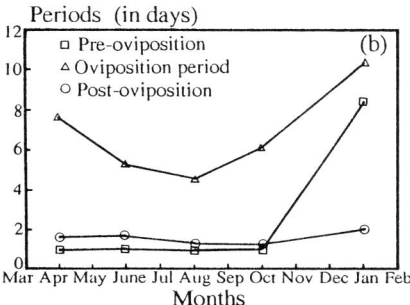

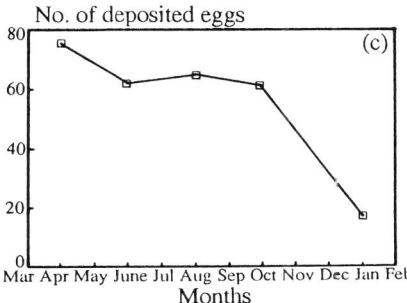

Figure 2. The preoviposition, oviposition and postoviposition periods in days during the different months of the year (b), mean number of eggs laid per female (c), and their relation with temperature and R.H. % (a).

decreased to 4.57 ± 0.26 days (33.20°C). In September and October it was 6.43 ± 0.49 days at 30.9°C average temperature. It reached 10.33 ± 1.66 days during November to February (21.47°C average temperature).

Data obtained from statistical analysis indicate that temperature had a highly significant effect on the oviposition period. As the average temperature increases, the oviposition period decreases. An increase in the average temperature by 1°C results in the decrease of oviposition period by 0.45 days and vice versa.

3.4.2. *Controlled conditions.* Under controlled conditions (Table 1) I compared the effect of various temperatures under constant relative humidities. I found that there were significant differences between the mean oviposition period at 20°C and those at

Table 1. Effect of temperature and relative humidity on the mean oviposition periods in days.

Temp. (°C)	% R. H.			
	20	40	60	80
20	8.77 ± 0.49	8.42 ± 0.34	9.42 ± 0.34	8.08 ± 0.37
24	4.36 ± 0.20	2.74 ± 0.14	3.16 ± 0.14	3.38 ± 0.14
28	4.88 ± 0.14	5.18 ± 0.20	6.50 ± 0.20	5.72 ± 0.20
32	2.68 ± 0.13	4.08 ± 0.20	4.14 ± 0.14	2.62 ± 0.12

24°C, 28°C and 32°C, with a difference equal to 5.08, 2.92 and 4.86 days, respectively. There are significant differences equal to 2.16 and 1.94 days between means of oviposition period at 28°C and both at 24°C and 32°C. But there was no significant difference between means at 32°C and 24°C. When comparing the effects of various humidities on the oviposition period under constant temperature, there was a significant difference between the mean oviposition period at 60% R.H. and at 20, 40 and 80% R.H., a difference equal to 0.88, 0.86, and 1.01 days, respectively.

3.5. NUMBER OF EGGS LAID

Eggs were oviposited singly and indiscriminately on the seed surface. Unfertilized eggs were similar in size and color to fertilized ones but did not hatch. The gradual development of embryos differentiates the latter from the former.

3.5.1. *Natural conditions.* Under natural conditions, the mean number of eggs laid per female (Fig. 2) was 75.6 ± 6.89 (April-March) 62.3 ± 3.0 (May-June), 65.25 ± 3.39 (Nov.-Feb.). The highest number of eggs was during the winter period. The average number of eggs in all the periods of the year other than winter did not decrease very much. This means that the potential ability of egg laying by the insect is high during the whole year with the exception of the winter period.

The correlation between the average temperature and number of eggs laid per female indicates that an increase in the average temperature by 1°C results in the increase of the number of eggs by 4.03 and vice versa.

3.5.2. *Controlled conditions* (Table 2). Under controlled conditions the mean number of eggs laid per female at 20°C ranged between 32.75 ± 3.28 and 55.84 ± 1.75 eggs at 20 and 60% R.H., the mean at 24°C ranged between 25.10 ± 1.59 and 42.00 ± 2.35 eggs at 40 and 20% R.H., respectively. The suitable relative humidity for laying eggs at 24°C was 20%, but when the temperature was 28 or 32°C, the mean number of eggs was highestat 60% R.H. Under 32°C the mean number of eggs laid ranged between 34.10 ± 2.15 and 74.21 ± 2.77 eggs at 20 and 60% R.H., respectively. The latter was the most favorable relative humidity for laying eggs at 28 and 32°C temperature.

When comparing the effect of various temperatures on the number of eggs laid under constant relative humidities, it was found that there were significant differences between the mean number of eggs laid under different temperatures. When comparing the effect of various humidities it was found that the difference between the mean number of eggs laid at 60% and at 20, 40 and 80% R.H., is 14.66, 14.58 and 12.45 eggs, respectively.

Table 2. Effect of temperature and relative humidity on the mean number of eggs laid per female.

Temp. (°C)	% R. H.			
	20	40	60	80
20	32.75 ± 3.28	45.02 ± 1.97	55.84 ± 1.75	42.72 ± 2.02
24	42.00 ± 2.35	25.10 ± 1.59	30.70 ± 2.32	29.34 ± 2.28
28	61.62 ± 2.01	64.88 ± 2.67	72.10 ± 3.61	60.72 ± 3.35
32	34.10 ± 2.15	40.38 ± 3.60	74.21 ± 2.77	46.38 ± 2.21

N.B. Number of fertilized females used in each case was 50.

Statistically these differences are significant. Comparing the number of eggs laid at 20, 40 and 80% R.H., it could be concluded that there were no significant differences between them. My conclusions are that the number of eggs laid by fertilized females of *B. incarnatus* varies greatly according to the prevailing temperatures and relative humidities.

3.6. POST-OVIPOSITION PERIOD

3.6.1. *Natural conditions.* Under natural conditions, the post-oviposition period (Fig. 2) increased as temperature decreased. It was 1.24 ± 0.12 days under an average temperature of 30.9°C during September-October but it was prolonged to 2.0 ± 0.19 days under temperature of 21.47°C during Nov.-February. Data obtained from the analysis of variance indicate that the temperature had a significant effect on the post-oviposition period.

3.6.2. *Controlled conditions.* Under controlled conditions, the post-oviposition period at 20°C ranged between 2.28 ± 0.19 and 3.43 ± 0.28 days at 20 and 40% R.H, respectively. Under 24°C, it ranged between 1.32 ± 0.07 and 1.80 ± 0.11 days at 80 and 40% R.H., respectively. Under 28°C it ranged between 1.32 ± 0.14 and 1.74 ± 0.14 days at 60 and 20% R.H., respectively. Under 32°C, it ranged between 1.20 ± 0.06 and 1.92 ± 0.20 days at 80 and 40% R.H., respectively.

Statistical analysis revealed a highly significant effect when comparing the effect of various temperatures on mean post-oviposition period under constant relative humidities. It was found that under 20°C the mean post-oviposition periods were 1.32, 1.36 and 1.45 days longer than under 24, 28 and 32°C, respectively. There were no significant differences between means at 24, 28 and 32°C. When comparing the effect of various humidities, it appeared that there were significant differences at 20% R.H. and at 40, 60 and 80% R.H., resulting in 0.74, 0.59 and 0.69 days, respectively. No significant differences occurred between means at 40, 60 and 90% R.H.

3.7. THE INCUBATION PERIOD AND HATCHABILITY

3.7.1. *Natural conditions.* Under natural conditions, the incubation period (Fig. 3) was 15.8 ± 1.22 days in April-March, 8.06 ± 0.26 days in May-June, 6.5 ± 0.09 in July-August, 14.17 ± 4.7 days in September-October and 20.8 ± 0.94 days during November-February. The percentages of hatchability were 60.85 ± 3.69, 56.71 ± 2.49, 65.11 ± 3.33, 74.77 ± 6.41 and 32.6 ± 7.07%, respectively under average temperatures of 23.5, 28.65, 33.20, 30.9 and 21.47°C.

Statistical analysis of data revealed that the temperature had a highly significant effect. There is an inverse correlation between the temperature and incubation period and proportional correlation between the temperature and hatchability percentage. Hatchability reached its utmost percentage during autumn and its minimal percentage during winter.

3.7.2. *Controlled conditions.* Under controlled conditions, the mean incubation period under 20°C, ranged between 21.11 ± 0.003 and 26.11 ± 0.001 days at 80 and 20% R.H., respectively. Under 24°C it ranged between 10 ± 0.0 and 12.03 ± 0.007 days at 20 and 80% R.H., under 28°C, it ranged between 7.0 ± 0.0 and 10.01 ± 0.005 days at 80 and 40% R.H., and under 32°C, it was between 6 ± 0.0 and 7 ± 0.0 days at 40 and 80% R.H., respectively.

32

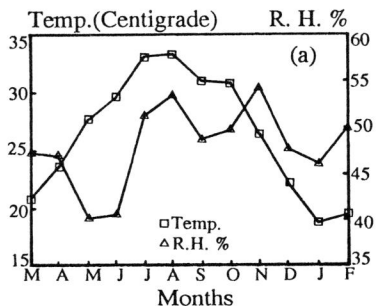

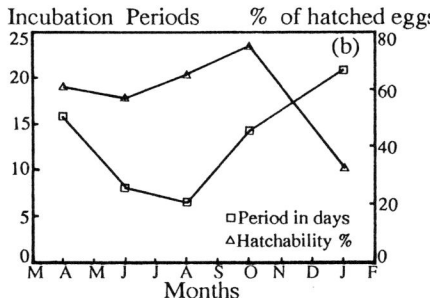
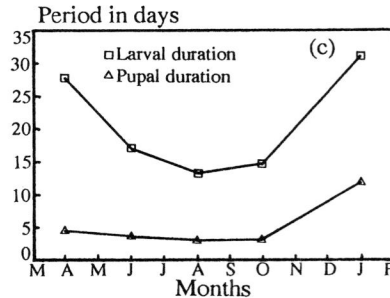

Figure 3. Incubation period and hatchability % (b), larval and pupal periods (c), and how affected by temperature and R.H. % during one year (a).

Comparison of the effect of various temperatures showed that under 20°C the incubation periods resulted in 11.97, 14.22 and 16.21 days longer than under 24, 28 and 32°C, showing a statistically significant difference. No significant differences occurred under 24 and 28°C or between 28 and 32°C.

3.7.3. *Hatchability.* The mean percentage of hatchability under 20°C ranged between 40.31 ± 7.47 and 59.96 ± 1.39% at 20 and 60% R.H., respectively. Under 24°C, it ranged between 66.96 ± 2.16 and 78.47 ± 2.73% at 20 and 40% R.H.. Under 28°C, it ranged between 56.72 ± 2.50 and 70.26 ± 2.39% at 80 and 40% R.H., respectively. Under 32°C, it ranged between 61.46 ± 1.88 and 76.06 ± 2.78% at 60 and 80% R.H.

As a whole the optimum temperature for hatchability was 24°C at 40 or 80% R.H. Statistical analysis indicated only a significant effect of temperature while there were no significant differences at 40, 60 and 80% R.H.

3.8. DURATION OF THE LARVAL STAGES

3.8.1. *Natural conditions.* Under natural conditions, the shortest period for the larval development was 13.9 ± 0.06 days at 33.20°C during July-August while the longest period was 31.0 ± 0.71 days at 21.47°C. during Nov.-Feb. (Fig. 3). The larval duration at different periods of the year was 31 ± 0.71, 27.72 ± 1.76, 16.96 ± 0.12, 14.71 ± 0.10 and 13.19 ± 0.6 days at 21.47, 23.50, 28.65, 30.90 and 33.20°C, respectively. Statistical analysis revealed a highly significant effect of temperature. It was found that an increase in the average temperature by 1°C, results in the decrease of larval duration by 1.6 days and vice versa.

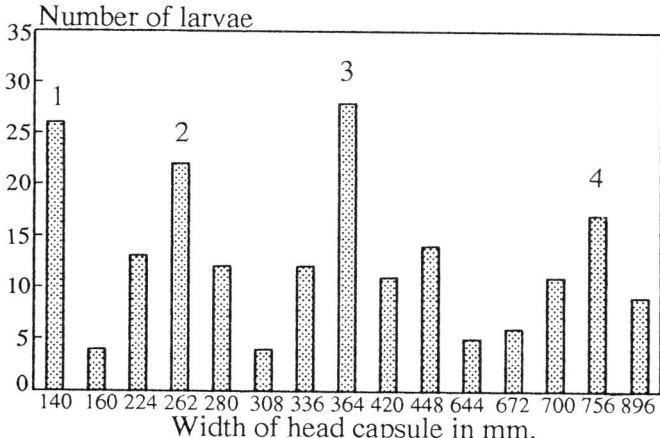

Figure 4. Histograms showing the measurements of the larval head capsules denoting the four larval instars.

3.8.2. *Head capsule.* Head capsule measurements taken until larvae reached pupation revealed four larval instars (Fig. 4).

3.8.3. *Controlled conditions.* Under controlled conditions (Table 3) the larval duration at 24°C ranged between 36.30 ± 0.49 and 61.39 ± 0.2 days at 80 and 40% R.H., respectively. Under 28°C it ranged between 20.08 ± 0.16 and 26.27 ± 0.23 days at 40 and 80% R.H., respectively. Under 32°C, it ranged between 23.53 ± 0.81 and 27.27 ± 0.12 days at 20 and 40% R.H.

Statistical analysis showed significant differences of both temperature and relative humidity alone. Comparison of various temperatures revealed significant differences of 3.07, 18.4 and 4.43 days, respectively between the mean larval periods at 40% R.H. and each of them at 20, 60 and 80% R.H. Also, significant differences occurred resulting in 1.23 and 2.59 days longer at 60% R.H. than at 20 and 80% R.H.

Table 3. Effect of temperature and relative humidity on larval duration.

Temp. (°C)	% R. H.			
	20	40	60	80
24	46.14 ± 0.34 (65)	61.39 ± 0.22 (43)	54.06 ± 0.35 (46)	36.30 ± 0.49 (50)
28	21.83 ± 0.11 (100)	20.08 ± 0.16 (67)	25.66 ± 0.06 (97)	26.27 ± 0.23 (112)
32	23.53 ± 0.81 (60)	27.27 ± 0.12 (83)	26.31 ± 0.17 (144)	23.71 ± 0.88 (58)

N.B. Number between brackets indicate the number of individuals experimented on.

The optimum temperature for larval growth seems to be 28°C at 40%. Also it was clear that the larval period was longer under 24°C at 40 and 60% R.H., than at 20 and 80% R.H. These two prolonged periods might be due to the previous environmental conditions surrounding the larva before carrying out the experiment. The shorter periods at 20 and 80% R.H. took place during summer. The larvae were inclined normally to hibernate during winter.

3.9. DURATION OF PUPAL STAGE

3.9.1. *Natural conditions.* Under natural laboratory conditions, the longevity of the pupal stage decreased with the increase of temperature. At an average temperature of 33.20°C in July-August it lasted for only 3.13 ± 0.07 days, while in winter, it was prolonged to 11.85 ± 0.51 days. Under an average temperature of 21.47°C, in winter, the longevity was approximately three times as long as its duration in other periods of the year (Fig. 3). Analysis of variance revealed a highly significant effect of the temperature.

3.9.2. *Controlled conditions.* Under controlled conditions at 24°C, it ranged between 6.05 ± 0.21 and 17.23 ± 0.15 days at 20 and 40% R.H., while under 28°C, it ranged between 3.20 ± 0.11 and 5.11 ± 0.12 days at 20 and 60% R.H., respectively. Under 32°C it ranged between 3.21 ± 0.09 and 4.88 ± 0.10 days at 60 and 80% R.H.

Statistical analysis indicated that the relative humidity, temperature and their interaction had highly significant effects. Comparison between the various temperatures revealed that under 24°C the pupal periods were 6.57 and 6.54 days longer than under 28 and 32°C. No significant differences occurred between 32 and 28°C. Comparison between the various percentages of humidity revealed significant differences between 40% R.H and each of 20, 60 and 80% R.H., but no significant difference were found between 80 and 60% R.H.

3.10. LONGEVITY OF ADULT STAGE

3.10.1. *Natural conditions.* The mean longevity of the male was 17.69 ± 0.42 days during the period of March-April, whereas it was 12.10 ± 0.03, 10.39 ± 0.29, 13.36 ± 1.39 and 22.22 ± 2.1 days during the other four successive periods, respectively (Fig. 5).

In the female it was 19.42 ± 0.56 days during March-April and 11.02 ± 0.27, 9.82 ± 0.22, 10.61 ± 0.59 and 24.59 ± 2.69 days during the other four successive periods (Fig. 5). Analysis of variance indicated a highly significant effect of temperature on both sexes. The female lives longer than the male in winter and spring periods. This phenomenon is reversed during other periods. However, the difference in longevity between sexes does not exceed approximately three days.

3.10.2. *Controlled conditions.* In the males at 24°C, it ranged between 12.97 ± 0.28 and 15.00 ± 0.15 days under 80 and 40% R.H., while under 28°C, at ranged between 11.83 ± 0.19 and 14.44 ± 0.22 days at 20 and 60% R.H. Under 32°C it ranged between 9.58 ± 0.13 and 12.83 ± 0.34 days at 20 and 40% R.H., respectively.

Analysis of variance revealed a significant effect of the two factors. Comparison of the various temperatures indicated that under 28°C the male longevity was 2.37 and 3.83 days longer than under 32 and 24°C. Significant differences occurred also between 32 and 24°C, resulting in 1.49 days difference.

If we compare the effect of humidity, it was 0.88 and 1.25 days longer at 40% R.H. than those at 20 and 80% R.H., showing significant differences which were obtained

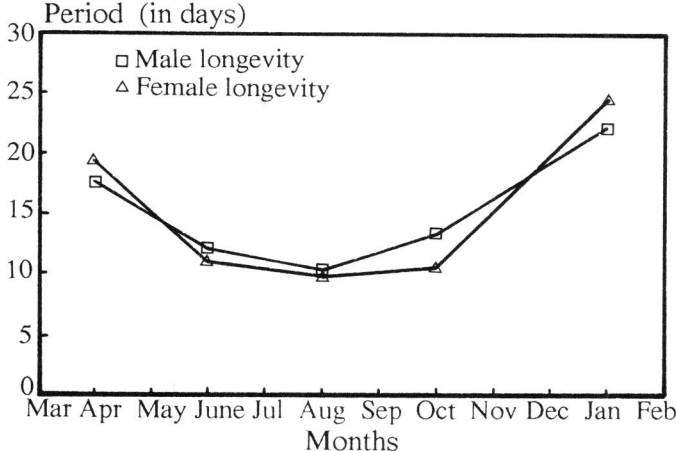

Figure 5. Longevity of adult stage during the different periods of the year under normal conditions (temp. and R.H. %).

between 60% R.H. and 20 and 80% R.H., but there was no significant difference between 20 and 80% R.H.

The female longevity under 24°C ranged between 13.50 ± 0.27 and 15.80 ± 0.10 days at 20 and 40% R.H., while under 28°C, it ranged between 11.42 ± 0.17 and 33.44 ± 0.35 days at 20 and 60%, respectively. Under 32°C, it ranged between 9.81 ± 0.11 and 11.84 ± 0.29 days at 20 and 40% R.H. Statistical analysis indicated that either temperature or relative humidity alone or combined, has a significant effect.

REFERENCES

Abdel-Salam, A. L., Metwally, M. M., and Ahmed, S. G. (1984) The susceptibility of different pulse grains to bruchid infestation. II. Effect of biological and ecological factors, 9th. Intern. Cong. Stat. Comp. Sc. March, 31 - April, 10.

Abdul-Rassoul, M. E., Othman, N. Y., and Daweh, H. A. (1986) Observation on the biology, host plants and distribution of Iraqi Bruchidae (Insecta, coleoptera), J. Biol. Sc. Res. Iraq 17, 207-222.

Applebaum, S. W. and Guez, M. (1972) Comparative resistance of *Phaseolus vulgaris* beans to *C. chinensis* and *Acanthoscelids obtectus* (Col. Bruchidae). The differential digestion of soluble heteropolysaccharide, Entomol. Exp. Appl. 15, 203-207.

Buxton, P.A. (1931) The measurement and control of atmospheric humidity in relation to entomological problem, Bull. Entomol. Res. 22, 431-447.

De Luca, I. (1962) Au sujet de *Brnchidius algiricus* all. et *B. incarnatus* Boh., Rev. App. Entomol. 50, 24

Dick, K. M. and Credland, P. F. (1986) Variation in the response of *Callosobruchus maculatus* (f) to a resistant variety of cowpea, J. stored Prod. Res. 22, 43-48.

El-Gendy, H. O. (1984) Biological studies on *Bruchidius incarnatus* Boh. and its control, M. Sc. Thesis, Fac. of Agric, Al-Azhar Univ., Cairo.

El-Sawaf, S. K. (1956) A contribution to the host-selection principle as applied to *Bruchus* (*Callosobruchus*) *maculatus* F., Bull. Soc. Entomol. Egypt 38, 297-203.

El-Sawaf, S. K. (1958) Some factors affecting the longivety, oviposition and rate of development in the southern cowpea *C. maculatus* F., Bull. Soc. Entomol. Egypt 40, 29-95.

Hafez, M. A. and Chapman, C. (1962) A simple apparatus for the study of small insects under controlled conditions of humidity and temperatures, Issue of laboratory practice.

Kamel, A. H. (1981) Faba bean pests in Egypt, in Proc. of the Faba Bean Conference, March, 7-11, Cairo, Egypt, pp. 271-285.

Larson, A. O. (1927) The host selection principle as applied to *B. quadrimaculatus* Fab., Ann. Entomol. Soc. Amer. 20, 37-77.

Metwally, M. M., Abdel-Salam, A. L., and Ahmed, S. E. (1984) The susceptibility of different pulse grains to bruchid infestation III. Effect of physical and chemical factors, 9th Intern. Cong. Stat. Comp. Sci., March, 31-April, 10, Cairo, pp. 417-428.

Pajni, H. R. (1987) Some aspects of biosystematics of Bruchdae (Coleoptera), Proc. Indian Acad. Sci., Animal Sci. 96, 509-515.

Report of Central Agency of Public Mobilization and Statistics (1987) Cairo-Egypt (CAPMS).

Schoof, H. F. (1941) The effects of various relative humidities on the life process of the southern cowpea weevil, *Callosobruchus maculatus* Fab., Rev. App. Entomol. 29, 566.

Solomon, M. E. (1951) Control of humidity with potassium hydroxide, sulfuric acid and other solutions, Bull. Entomol. Res. 42, 543-544.

Winkler, A. (1927) Catalogue coleopterorum regions paleareticae (*B. incarnatus*), Wien. 18, 1365.

Zacher, F. (1930) Unteruchunger zur morphologie und biologie der samenkafer (Bruchidae Laridae), Arb. Biol. Reichsant. Land. 18, 233-384.

REGULATION OF AUTONOMIC PHYSIOLOGICAL FUNCTIONS DURING REPRODUCTIVE DIAPAUSE OF *BRUCHUS AFFINIS*

M. -S. COQUILLAUD[1], K. SLÁMA[2], AND V. LABEYRIE[-]
[1] *I.B.E.A.S., Université de Pau et des pays de l'Adour,*
Campus Universitaire, 64000 Pau, France
[2] *Insect Chemical Ecology Unit, UOCHAB,*
Czechoslovak Academy of Sciences, U Šalamounky 41,
15800 Praha 5 Czechoslovakia

ABSTRACT. Physiological mechanisms controlling diapause in *B. affinis* were investigated using several electrophysiological methods. The methods included: direct hydraulic recording of haemocoelic pulsations; indirect recording of the pulsations from the body surface; inductance recording with unrestrained insects, and; ultra-microrespirographic scanning of O_2 consumption and CO_2 output. The diapausing beetles have shown almost continuous pulsations in haemocoelic pressure. The frequency and amplitude of the pulsations undergo specific variations in connection with changes of environmental or internal physiological factors. The mechanism is controlled by an autonomic nervous system (coelopulse) with the center located in the thoracic ganglia. The system operates on a similar basis to an electronic microprocessor: it takes the signals from multiple sensors and transforms them into the optimum and most economic regimen of the pulsations. The respirographic data provide direct evidence that the coelopulse system also regulates the saturation of haemolymph and tissues with the metabolic CO_2. The system liberates gaseous CO_2 in 30 - 50 nl bursts once in 2 - 5 min., which is so far unknown but a very efficient method of water retention.

1. Introduction

Growth and development of certain species of Bruchids proceed in relatively short periods, which are precisely synchronized with the periods of growth and flowering in the host plant. This has been well studied and described especially in *Bruchus affinis* (Bashar et al., 1987) or *Bruchidius atrolineatus* (Huignard et al., 1987). By contrast, the newly emerged adults of a new generation enter several months of adult diapause. They do not move, feed or reproduce and their metabolic functions become reduced to the minimum for maintenance. During diapause the beetles are quite resistant to a number of unfavourable environmental factors, e.g., starvation, desiccation, or freezing. It is also generally known that diapausing adults are resistant to most of the common insecticides. However, in spite of all these exceptional features, we have no reliable physiological explanation for the regulatory mechanisms of very low, diapause metabolism. In fact, it is a physiological miracle that a delicate system like an adult beetle with enormously developed flight musculature can freely breathe and survive for more than just a few minutes.

In immature developmental stages the regulation of some homeostatic functions is achieved by a parasympathetic-like, autonomic nervous system, which is centered in the thoracic ganglia of the ventral nerve cord (Sláma, 1984a). The system regulates impor-

K. Fujii et al. (eds.), Bruchids and Legumes: Economics, Ecology and Coevolution, 37–44.

tant physiological functions like, for instance, water balance and isoosmosis, gaseous exchange through the spiracles or haemolymph circulation. The regulation is by rhythmical pulsations in haemocoelic pressure. The methods based on monitoring the functions of the autonomic nervous system have been already used for pharmacodynamic studies in the action of insect hormones, neuroactive chemicals, insecticides, or insect pathogens (for references see Sláma, 1989). More recently, the autonomic nervous system has been found responsible for regulation of life in various diapausing or actively reproducing adult insects. It has been proposed that we call this system the coelopulse system, from the Greek word *koiloma* or Latin *coelom* for body cavity and *pulsus* for beating or striking (see Sláma, 1989).

Here we have continued our studies on hormonal control of diapause in *B. affinis* (Coquillaud and Sláma, 1989) by electrophysiological investigations connected with the coelopulse system. Due to the small size of these beetles, we were obliged to use very sensitive electronic methods, some of which were just developed and used for the first time. Due to the limited space here we have briefly described only a few selected topics to establish a pharmacological basis for future studies in Bruchids.

2. Materials and Methods

The diapausing adults of *B. affinis* were collected in the vicinity of Pau, Southern France. They were stored in refrigerators at + 5°C in complete darkness. Once in 2 - 3 weeks they were brought to room temperature and allowed to drink water. Prior to installation of the measuring devices, the beetles (mostly diapausing females) were immobilized by submersion in water or insect Ringer for about 15 min. The attachments of the sensors, filaments, the fixation of the hind legs or decapitation were made by means of quickly polymerizing cyanoacryllic glues.

Recordings of the extracardiac pulsations in haemocoelic pressure were made as in previously described methods (Sláma, 1984a). The respirometric data were obtained by the method of ultra-microrespirometric scanning (originally devised for measuring respiration of single cells or small tissue samples) (see Sláma, 1984b). The system of electronic transducers, amplifiers and recorders was the same as has been described by Sláma (1988).

3. Results and Discussion

3.1. RECORDING FROM THE HAEMOCOEL

The adults of *B. affinis* are very inconvenient insects for recording functions of the coelopulse system. They hear, look and feel, and are easily disturbed by almost anything that passes around the laboratory. They always try vigorously to escape from the delicate instruments and the most discouraging thing is their instinct to brush and wipe off anything that touches their body with their long hind legs. Nevertheless, we managed a few successful recordings of haemocoelic pressure with direct hydraulic transducers. Fig. 1 shows that there are very frequent extracardiac pulsations of rather small, infrasonic frequencies (7 - 10 pulses per min. at 25°C) and of very small amplitude of just a few Pa (note that 1 mm hydrostatic pressure is approx. 10 Pa). Similarly, as in most other insect species, the baseline haemocoelic pressure of *B. affinis* shows also subatmospheric values. The direct hydraulic measurements appeared quite inconvenient for the routine work, because the thin connecting needle of the transducer was often blocked by the coagulated haemolymph.

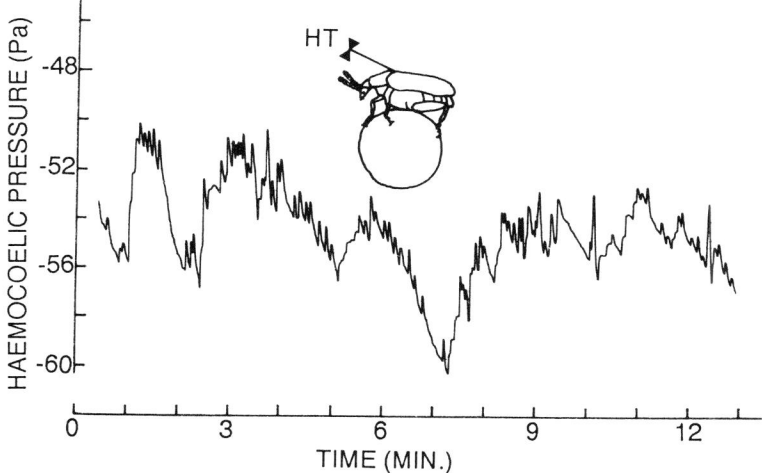

Figure 1. Extracardiac pulsations in haemocoelic pressure in the diapausing female of *B. affinis*, recorded by means of the direct hydraulic transducer inserted through the mesonotum (HT), 25°C.

3.2. INDIRECT RECORDING WITH UNRESTRAINED INSECTS

It was found previously that mechanical changes in insect haemocoel are always compensated by slight movements of some elastic membranes or segments. This principle has been used earlier for development of some indirect methods for recording the haemocoelic pulsations from the body surface (Sláma, 1984a). The adults of *B. affinis* are heavily sclerotised. Their abdominal segments contain elastic intersegmental membranes which allow telescopic contractions or elongations of the abdomen in response to changes in internal pressure. A sensitive electronic transducer thus can take and relay the pulsations in haemocoelic pressure merely from the tip of the abdomen.

In order to see whether the pulsations occur also in absolutely unrestrained insects, we have developed a special inductance method for their perception. The principle of this method (which will be described in detail elsewhere) can be found in Fig. 2. The beetle bears a small ferromagnetic object (f), which has been cemented to the distal abdominal segment (pygidium). The beetle is then placed into a plastic test tube (t) with two solenoids (s) forming a magnetic field using a low voltage 5 kHz AC current. The movement of the ferromagnetic object in the magnetic field causes small changes in the inductance and this is amplified and recorded.

The upper trace in Fig. 2 was only used to show a sample of low sensitivity recording, actually representing an actograph. It shows the periods of large "disturbance" movements, which are frequently present after disturbance. Their occurrence is usually diminished to one cycle per day after complete tranquilization. The lower trace in Fig. 2 shows the record obtained at the time when the insect remained motionless. It shows rhythmic pulsations of abdominal contractions whose amplitudes, frequencies and special patterns have been identical with the extracardiac haemocoelic pulsations revealed by the direct hydraulic method. Many specimens remained motionless for prolonged periods so that it was possible to record the pulsations for a long time. It appeared that these unrestrained diapausing adults were in a dynamic state always exhibiting a series of pulsations characteristic for the function of the coelopulse system. Disadvantages of this method are a large dependence of the signal on the location of the insect and its orientation to the magnetic field.

40

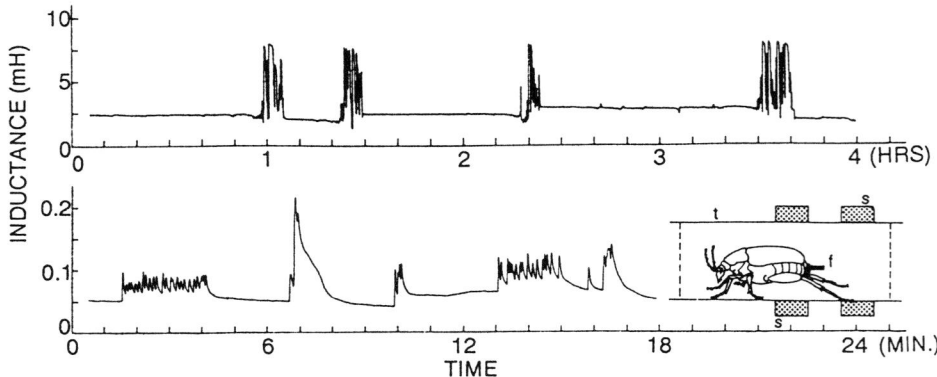

Figure 2. Actographic-like record of the movements of diapausing female of *B. affinis* in the magnetic field (above, 15-19 hrs after installation). The lower portion was taken at elevated sensitivity when the animal was motionless. It shows abdominal contractions associated with haemocoelic pulsations.

3.3. THE INDIRECT CONTACT METHOD

This appears to be the most suitable technique for prolonged monitoring of the coelopulse system in *B. affinis*. The technique becomes obvious from the scheme in Fig. 3 above. The insect to be measured was cemented by its dorsal side to a supporting rod (s). A small hook made from tungsten wire was glued to the distal end of the abdomen. The hook served for anchoring a filament leading to the membrane of the isotonic transducer (IT) (see Sláma, 1984a). A ball of polystyrene foam was placed between the legs (p) and a small arena (a) was used to keep the ball within reach of the insect. The whole experimental setup was placed in a dark compartment protecting the preparation from the movements of air, mechanical shocks or sudden changes of temperature.

The upper part of Fig. 3 shows a small fraction of a record which continued for several days. We can see relatively dense and regularly repeated series of abdominal contractions, corresponding exactly with the pulsations in haemocoelic pressure. This pattern has been repeated consistently for several days and it can be considered representative for diapausing males and females at room temperature. Fig. 3 B shows a small part of this pattern in more detail. We can distinguish here two different frequencies of the pulsations. There are pulsations of higher frequency and amplitude (10 or more pulses per min. at 25°C) occurring in 2 - 3 min. intervals. Between these pulsations we have often found more or less numerous pulses of smaller and usually irregular frequency which were completely absent in some cases.

Measurements with a large number of the diapausing adults revealed that the pulsations of *B. affinis* have all the common features with the pulsations found in other species and developmental stages (c.f. Sláma, 1989). The pulsations undergo large and almost immediate changes with respect to the changes in ambient temperature. With decreasing temperature the frequency of the pulsations correspondingly diminishes, but their amplitude may become greater. Fig. 3 C shows an example of the pulsations recorded at + 10°C, for comparison with the record in Fig. 3 B.

It is beyond the scope of this contribution to mention all results and experimental designs we have used. We may perhaps conclude that the functions of the coelopulse system reveal very accurately the extant physiological state of the animal, presence or absence of diapause, metabolic status, developmental situation, critical water reserves, etc. Its knowledge brings new aspects and introduces new possibilities for further inves-

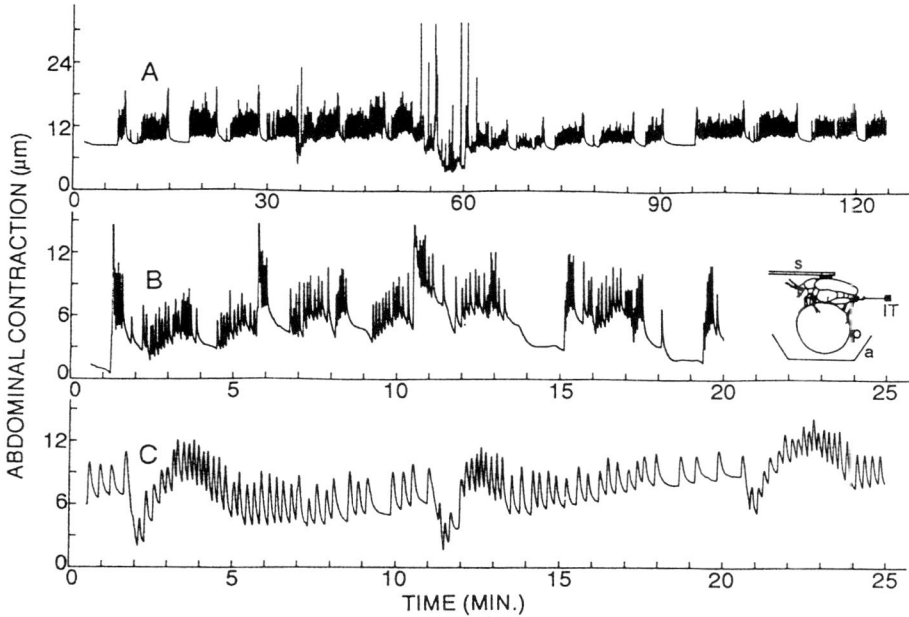

Figure 3. Recordings of extracardiac pulsations of diapausing female *B. affinis* by an indirect contact method. The pulsations have been received and relayed from the last abdominal segment by an isotonic transducer (IT). The A and B records were realized at 25°C, the record C at 10°C.

tigations, including pharmacodynamic action of many natural and synthetic chemicals whose mechanisms of action are virtually unknown.

3.4. THE AUTONOMIC, BRAIN-INDEPENDENT FUNCTIONS

Fig. 4 shows the normal pattern of the extracardiac pulsations several hrs before (upper record) and several hrs after decapitation (lower record). It is quite obvious that the pattern of the pulsations has not been substantially influenced by removal of the brain plus all cephalic parts of the nervous system. In addition, the pulsations also persisted in the body fragment without the head and prothorax. This is consistent with the earlier conclusions of Sláma et al. (1979) that the actual center of the autonomic nervous system in Coleoptera may be located in the mesothoracic region. The work with the decapitated body fragments appears very practical and suitable for the most accurate measurements. The pulsations are very stable over periods of 30 days or more.

3.5. ULTRA-MICRORESPIROMETRIC SCANNING

According to current physiological theories, insects respire by the simple principle of gaseous diffusion. Thus, oxygen is supposed to diffuse into the body through the opened spiracles while CO_2 and N_2 should go the opposite way. Recent studies show, however, that this may not be so simple. Some insects keep the spiracles hermetically closed most of the time. They are able to actively regulate the passage of air through some selected spiracles. These programmed inspirations and expirations of air, as well as the decision which spiracle will open at what time is also under control of the coelopulse system (Sláma, 1988). In this work we have employed the scanning ultra-microspirograph for

42

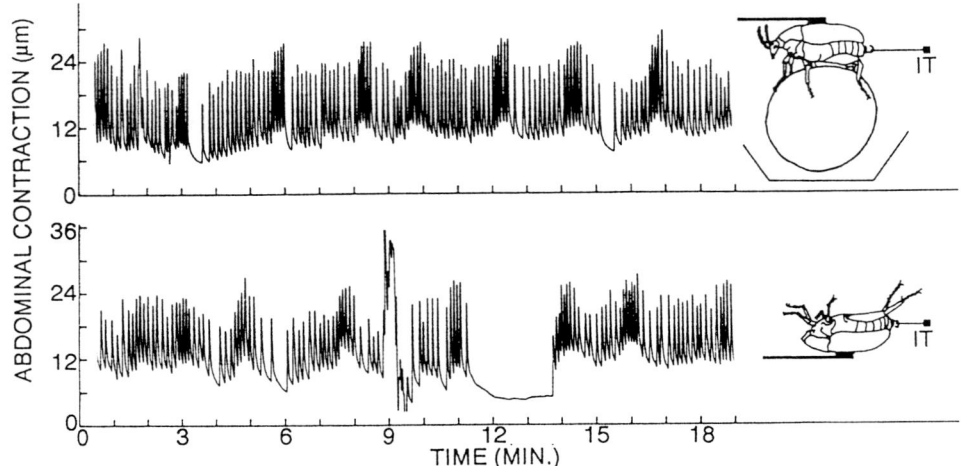

Figure 4. Pulsations in a normal diapausing females of *B. affinis* (above) and a few hrs following decapitation (below), 20°C.

monitoring the respiratory scenarios of diapausing *B. affinis*. The unusual results are exemplified by two fragments of the authentic records in Fig. 5.

The record in Fig. 5 A shows a relatively constant rate of O_2 consumption of approximately 36 nl/min., which is regularly interrupted in 2 or 3 min. intervals by a sudden release of gas. In most cases we have recorded, at this temperature, the appearance of 30 to 50 nl of gas. Each of these microcycles lasted about 10 sec. In the presence of CO_2 absorbent, such as in the case of Fig. 5A, the gas produced was completely absorbed

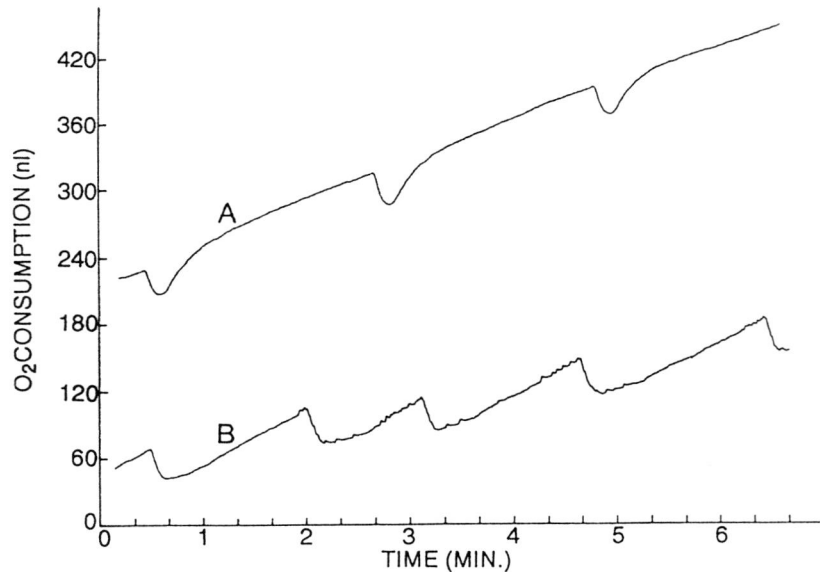

Figure 5. Ultrarespirographic recording of respiration in diapausing female of *B. affinis* in the presence of CO_2 absorbent (A) and without the CO_2 absorbent (B). The recording was done in vessels of 3 ml capacity at medium sensitivity at a constant 25°C.

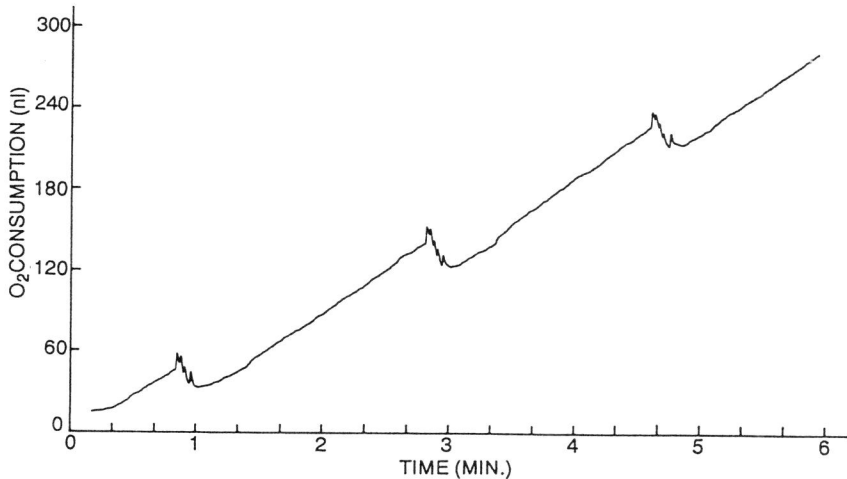

Figure 6. Ultra-microrespirographic record showing the respiration of diapausing female of *B. affinis* at increased sensitivity (without CO_2 absorbent). The electronic sensors of the respirograph recorded indirectly the extracardiac pulsations associated with the outbursts of CO_2.

within 20 - 30 sec., suggesting that it was 100 % pure CO_2. By contrast, in the absence of the absorbent, such as in the case of Fig. 5 B, the gas produced remained in the vessel and the respirograph recorded the O_2 values minus the CO_2. The practical implication of the described findings is that the metabolic CO_2 remains dissolved in haemolymph or tissues. It is actively produced in concentrated form by as yet unknown chemical process and it is rapidly released outside.

Extensive respirographic scannings with many individuals and during long periods of time revealed the presence of the described CO_2 outbursts in all diapausing males and females measured. The average range of O_2 consumption was from 450 to 1100 μl O_2/g/hr and the average respiratory quotient was 0.72. During the above discussed peaks of increased mobility, the O_2 consumption was temporarily raised 2 to 3 fold and the outbursts of CO_2 were also less distinctive at this period. The decapitated body fragments fully retained the ability to produce the concentrated CO_2 in regular intervals. Some measurements that were taken with high sensitivity revealed the presence of pulsations during the cycles of CO_2 production (Fig. 6). This provides direct evidence for the conclusion that the extracardiac pulsations of higher frequency may be directly involved in expulsion of the sublimated CO_2 out of the body. All the described data together make it clear that the curious respiratory fate of CO_2 in these insects is regulated by the coelopulse system.

REFERENCES

Bashar, A., Fabres, G., Hossaert, M., Valero, M., and Labeyrie, V. (1987) *Bruchus affinis* and the flowers of *Lathyrus latifolius*: an example of the complexity of relations between plants and phytophagous insects, Insects-Plants, Dr. W. Junk Publishers, Dordrecht, pp. 189-194.

Coquillaud, M. -S. and Sláma, K. (1989) *Bruchus-Lathyrus* relationships: effects of exogenous hormones, Proc. Symp. Insects-Plants, Budapest (in press).

Huignard, J., Germain, J. F., and Monge, J. P. (1987) Influence of the inflorescence and pods of *Vigna unguiculata* Walp (Phaseolinae) on the termination of the reproductive diapause of *Bruchidius atrolineatus* (Pic) Coleoptera Bruchidae, Insects-Plants, Dr. W. Junk Publishers, Dordrecht, pp. 183-188.

Sláma, K. (1984a) Recording of haemolymph pressure pulsations from the insect body surface, J. Comp. Physiol. B 154, 635-643.

Sláma, K. (1984b) Microrespirometry in small tissues and organs, in T. J. Bradley and T. A. Miller (eds.), Measurement of Ion Transport and Metabolic Rate in Insects, Springer, New York, pp. 101-129.

Sláma, K. (1988) A new look at insect respiration, Biol. Bull. 175, 289-300.

Sláma, K. (1989) Role of the autonomic nervous system (Coelopulse) in insect reproduction, Regulation of Insect Reproduction IV, Academia Praha, pp. 23-38.

Sláma, K., Baudry-Partiaoglou, N., and Provansal-Baudez, A. (1979) Control of extracardiac haemolymph pressure pulses in *Tenebrio molitor* L., J. Insect Physiol. 25, 825-831.

ROLE OF PHYSICAL AND CHEMICAL STIMULI OF LEGUME HOST SEEDS IN COMPARATIVE OVIPOSITIONAL BEHAVIOUR OF *CALLOSOBRUCHUS MACULATUS* (FAB.) AND *C. CHINENSIS* (LINN.) (COLEOPTERA: BRUCHIDAE)

V. G. GOKHALE[1], HIROSHI HONDA, AND IZURU YAMAMOTO
Laboratory of Pesticide and Bio-organic Chemistry, Faculty of Agriculture, Tokyo University of Agriculture, Tokyo 156 Japan

[1] Present address: *Department of Entomology, Agricultural Research Station, Rajasthan Agricultural University, Kota-324001, India*

ABSTRACT. In India, pulses like *Cajanus cajan* (pigeon pea), *Vigna radiata* (mung bean), *Cicer arietinum* (chickpea) and *Vigna mungo* (Urd bean) are split into two halves known as "Daal" and have been stored in this manner since time immemorial. Split pulses with or without seed coats thus stored afford excellent protection against bruchid attack. This fact, coupled with the reported observations that *C. chinensis* exhibits a preference for curvature and observations that aqueous fractions of the seed coats of *Phaseolus vulgaris* and *Vigna unguiculata* contain ovipositional stimulant activity for *C. maculatus*, formed the basis for the present investigation. We assessed the role of physical (curvature) and chemical stimuli (aqueous and/or methanol fractions of cowpea seed coat) in a quantitative and on a comparative basis for ovipositional preference of *C. chinensis* and *C. maculatus*. For standardization of the bioassay, several natural and artificial oviposition substrates were tested and glass beads with 6 mm diameter were found most suitable. The concentration of the stimulant material was expressed in terms of bean equivalent. The oviposition responses of gravid *C. maculatus* revealed that the curvature of oviposition substrate does not have any role in preferential oviposition and that the gravid females are solely guided by a chemical stimulus perceived from the oviposition substrate. Whereas for *C. chinensis* a physical stimulus (curvature) is a prerequisite for normal oviposition. However, once the requirement for the physical stimulus is met, then the chemical stimulus alone exerts its influence on oviposition. The chemical stimulus alone does not trigger normal ovipositional response.

1. Introduction

Ishii (1952) suggested physical factors (size and curvature of the host seeds) influenced host seed selection for oviposition by *Callosobruchus chinensis*. Subsequently, Avidov et al. (1964) evaluated the effect of curvature and surface area on the ovipositional responses of *C. chinensis* and concluded that curvature alone is responsible for the ovipositional preference. In *C. maculatus*, Gokhale (1971) observed that curvature of the oviposition substrate alone does not contribute towards the preferential oviposition since the gravid females oviposited very few eggs on steel balls as compared to host seeds. Thus characteristics of the seed other than curvature were considered more important in determining the ovipositional preference by *C. maculatus*. Gokhale and Srivastava (1973) demonstrated the presence of ovipositional stimulant activity in the

K. Fujii et al. (eds.), Bruchids and Legumes: Economics, Ecology and Coevolution, 45–51.

aqueous extract of *Phaseolus vulgaris* (French bean) seed coat for *C. maculatus* and postulated the chemical ovipositional stimulant for the insect as fairly common, if not universal. Gokhale and Sharma (unpublished) further observed ovipositional stimulant activity in the aqueous extract of cowpea seed coat for *C. maculatus*. In view of the complexities involved in the ovipositional behaviours of *C. maculatus* and *C. chinensis*, it was considered desirable to evaluate it on a comparative and quantitative basis in relation to physical and chemical stimuli.

2. Materials and Methods

2.1. MAINTENANCE OF INSECT CULTURES

Cultures of *Callosobruchus maculatus* (Fab.) and *C. chinensis* (L.) were maintained on azuki bean, *Vigna angularis* (Dainagon from Hokkaido) at 27°C and R.H. 70 ± 5%.

2.2. EXTRACTION OF STIMULANT MATERIALS

Seed coats of French bean, azuki bean and cowpea were manually separated after soaking them overnight in distilled water. The seed coat of each type of bean was pulverized into a fine flour. The extraction procedures followed for the determination of oviposition stimulant activity in the seed coat of French bean and azuki bean were as follows: the seed coat flour was extracted with ether by Soxhlet extractor. The residue was extracted with cold water to give a water soluble fraction. The cowpea seed coat flour was extracted with pentane, ether by Soxhlet extractor, cold methanol and then cold water.

The concentration of the stimulant was expressed in terms of "bean equivalent" (B. E.). One B. E. is defined as the amount of the active principle present in the seed coat of one bean.

2.3. STANDARDIZATION OF BIOASSAY

2.3.1. *Determination of Optimum Oviposition Substrate.* The relative efficacy of different oviposition substrates like glass beads (6 mm in diameter), chickpea seeds with seed coat and seeds of azuki bean and cowpea without seed coat was tested separately. Fifty ether washed glass beads/seeds of chickpea of uniform size with seed coat/cowpea and azuki beans without seed coats were dipped in aqueous extract of either French bean or cowpea seed coats at the required concentration of the stimulant and were then dried over a draft of a fan. Fifty ether washed glass beads/seeds were marked by sketch pen ink to distinguish them from the stimulant coated beads/seeds.

Ten or twenty *C. chinensis* or *C. maculatus* (gravid females, 24 hr. old) were provided with a choice of laying eggs for a fixed duration, among 10 stimulant-coated glass beads/seeds and 10 ether washed glass beads/seeds placed randomly on a filter paper (15 cm in diameter) and covered with a petri dish. The total number of eggs laid on either categories of glass beads/seeds were then counted.

2.3.2. *Determination of Dose-response Curve.* In order to determine the stimulant dose and oviposition response curve for gravid *C. chinensis* and *C. maculatus*, five concentrations of the oviposition stimulant from the aqueous extract of French bean seed coat viz., 4.00, 0.4, 0.04, 0.004, and 0.0004 B. E. were prepared. For each concentration of the stimulant, bioassay experiments were conducted following the same proce-

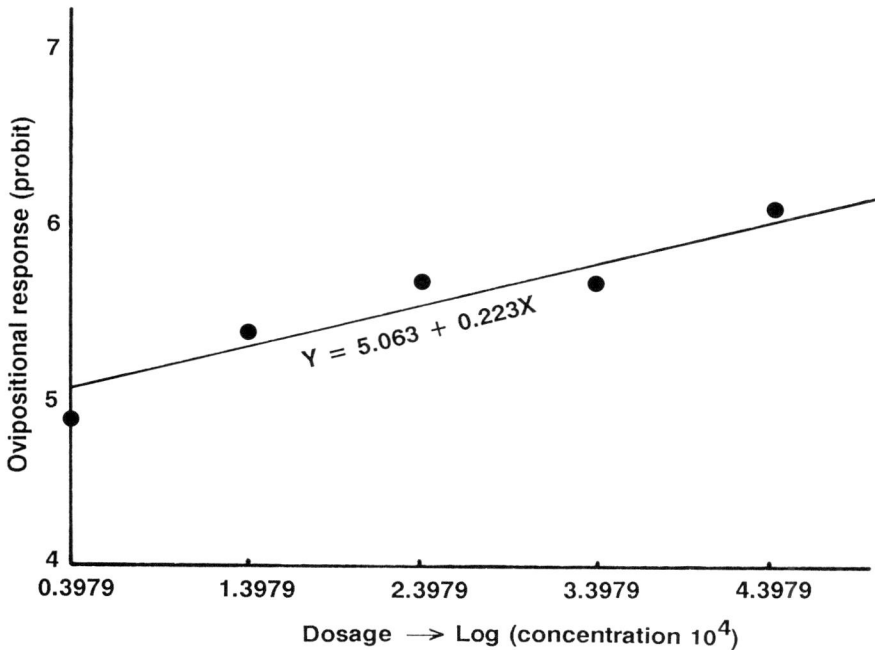

Figure 1.   Log concentration and ovipositional response regression line.

dure as described above.

The oviposition response (r) was calculated by using the following formula:

r = T / (T + C), where T denotes the number of eggs deposited on treated glass beads; C denotes the number of eggs deposited on untreated glass beads. The values of concentration were computed in dosage log concentration 10^4 and oviposition probit response and graphically depicted in Fig. 1.

2.4.   ASSESSMENT OF PHYSICAL AND CHEMICAL STIMULI FROM THE OVIPOSITION SUBSTRATE IN RELATION TO OVIPOSITIONAL RESPONSES OF GRAVID *C. CHINENSIS* AND *C. MACULATUS*

The bioassay experiments for the gravid females (10 females/replication) of *C. chinensis* and *C. maculatus* were conducted at a constant temperature of 27°C and R.H. 70 ± 5%.   The duration of bioassay was 24 h.   Glass beads of 6 or 12 mm diameter were coated with the stimulant from methanol extract of cowpea seed coats.   The concentration of the stimulant was 1 B. E.   A set of 10 glass beads with 6 mm and another with 12 mm diameter were offered together to the gravid females in the following combinations:

(a) Ten glass beads of 6 mm diameter plus 10 of 12 mm diameter.

(b) Ten stimulant coated glass beads of 12 mm diameter plus 10 of 6 mm diameter.

(c) Ten glass beads of 12 mm diameter plus 10 stimulant coated glass beads of 6 mm diameter.

(d) Ten stimulant coated glass beads plus 10 glass beads without stimulant both of 12 mm diameter.

Table 1. Comparative efficacy of the aqueous extract of French bean seed coat as an ovipositional stimulant for gravid *C. maculatus* and *C. chinensis*

Insect	Treatment	Number of eggs deposited on 10 glass beads Replications						Duration of bioassay hours	Number of females/ replication
		1	2	3	4	5	Total		
C. maculatus *	Treated	30	53	31	85	21	220	2	20
	Untreated	2	7	29	25	0	63		
	Total	32	60	60	110	21	283		
*C. chinensis***	Treated	34	53	60	40	43	230	1	10
	Untreated	17	37	22	17	28	121		
	Total	51	90	82	57	71	351		

* Concentration of the stimulant - 2 B.E.
** Concentration of the stimulant - 4 B.E.

3. Results and Discussion

Comparative oviposition responses of gravid *C. maculatus* and *C. chinensis* on glass beads (6 mm) coated with stimulant from the aqueous extract of French bean seed coat received significantly more numbers of eggs than untreated beans (Table 1) confirming the earlier finding of Gokhale and Srivastava (1973), while for *C. chinensis* stimulant activity is reported here for the first time.

Chickpea seeds are inert so far as oviposition by *C. maculatus* is concerned (Gokhale and Srivastava, 1973). Natural as well as artificial oviposition substrates were assayed for their suitability to *C. maculatus*. Stimulant-coated chickpea seeds were found to be as effective as stimulant-coated glass beads (Tables 2 and 3).

It was, however, difficult to obtain chickpea seeds with uniform substrate characteristics like size, curvature, smoothness etc.; these were thus not considered suitable for subsequent bioassays. Likewise, stimulant from the aqueous extract of cowpea seed coat when applied to azuki bean and cowpea seeds without seed coat did not evoke appreciable oviposition response from gravid *C. maculatus* (Table 3), probably due to interaction and interference of some unknown substances present in these seeds. Stimulant-

Table 2. Oviposition response of *C. maculatus* on chickpea seeds coated with aqueous extract of French bean seed coat

Treatment*	Number of eggs deposited on 10 chickpea seeds ** Replications					Total
	1	2	3	4	5	
Treated	55	85	102	78	74	309
Untreated	0	4	0	1	4	9
Total	55	89	102	79	78	318

* Concentration of the stimulant - 4 B.E.
** Ten females per replication in a 24 hr bioassay

Table 3. Ovipositional response by gravid *C. maculatus* on different oviposition substrates treated with ovipositional stimulant from the aqueous extract of cowpea seed coat

| Oviposition substrate | Treatment | Number of eggs deposited Replications | | | | | | Mean number of eggs/ replication |
		1	2	3	4	5	Total	
Azuki bean seeds without seed coat[1]	Treated	34	44	39	37	43	197	66.2
	Untreated	20	27	25	33	29	134	
Cowpea seeds without seed coat[2]	Treated	15	33	25	25	22	120	44.0
	Untreated	9	26	20	27	18	100	
Glass beads[3]	Treated	63	58	84	70	56	331	68.4
	Untreated	0	0	1	0	10	11	

* Concentration of stimulant - 15 B.E.
1) & 2) Ten females/replication; duration of bioassay -2 hr. 3) Ten females/replication; duration of bioassay - 1 hr

coated glass beads, however, evoked highly significant oviposition responses.

Thus, after considering the feasibility of various kinds of oviposition substrates, it was found that glass beads of 6 mm diameter could be used as the most ideal oviposition substrate because (i) physical characteristics/physical stimuli like curvature, surface area, smoothness etc. remain fairly uniform and (ii) physical and chemical stimuli could ideally be partitioned. Credland and Wright (1988) also concluded that glass beads provided a suitable physical arrangement, are an inert and reproducible substrate for *C. maculatus* which can be manipulated for experimental purposes.

The ovipositional stimulant dose-response curve for gravid *C. maculatus* could not be obtained (Table 4). However, a comparison of the data (Tables 4 and 5) for both the species of *Callosobruchus* revealed that in case of *C. maculatus*, the threshold of concentration of the stimulant was much higher (4 B. E.) than in *C. chinensis* (0.004 B. E.). Thus the former needed a stronger stimulus than the latter for triggering the oviposition

Table 4. Oviposition response of gravid *C. maculatus* on the glass beads treated with the aqueous extract of French bean seed coat in a certain range of concentrations *

| Concentration in B.E. | Number of eggs deposited in five replications ** | | Mean number of eggs/ replication | Standard error |
	Treated	Untreated		
4.00	214	125	67.8	± 36.9
0.40	32	24	11.2	± 19.6
0.04	6	0	-	-
0.004	2	9	-	-
0.0004	0	2	-	-

* Duration of bioassay - 4 hr; 20 females/replication
** A replication consisted of 10 treated and 10 untreated glass beads (6 mm diameter)

Table 5.   Effect of different concentrations of the ovipositional stimulant from the aqueous extract of French bean seed coat on oviposition response of gravid *C. chinensis* *

Concentration in B.E.	Number of eggs deposited in five replications **		Oviposition response (r)	Mean number of eggs/ replication	Standard error
	Treated	Untreated			
4.00	259	48	0.844	61.4	± 12.0
0.40	229	75	0.753	60.8	± 14.4
0.04	180	60	0.750	48.0	± 11.3
0.004	108	60	0.642	33.6	± 27.0
0.0004	111	142	0.439	50.6	± 36.0

 * Ten females per replication in one hr bioassay on glass beads
** A replication consisted of 10 treated and 10 untreated glass beads (6 mm diameter)

response.   Below the threshold concentration, the oviposition became scarce or stopped altogether.

The above results tentatively suggest that the role of the chemical stimulus might be more decisive for *C. maculatus* than the physical stimulus of the oviposition substrate.

In order to test the aforementioned possibility, an additional comparative bioassay experiment for *C. maculatus* and *C. chinenses* was undertaken, the results of which will be discussed later.

During the bioassay experiment for determining the dose-response curve (Table 5 and Fig. 1) for gravid *C. chinensis*, it was often observed that the gravid females tended to deposit eggs optimally on a suitable oviposition substrate (e. g. glass beads) having an optimum curvature. Our bioassay experiments have amply demonstrated that among the physical and chemical stimuli of the oviposition substrate, the latter dominates over the former when both kinds of stimuli are operating simultaneously. In the bioassay system of the two kinds of stimuli, one stimulus (i. e. curvature) always remains constant (that is intensity of the stimulus remains almost uniform), while the intensity of another stimulus (i. e. chemical stimulus) can be altered by increasing or decreasing the concentration of the stimulant material. When the chemical stimulus is withdrawn (as happens in suboptimum concentrations - Table 5), the pattern of oviposition by the gravid females changes and oviposition becomes random by deposition of eggs on either categories of glass beads. In other words, chemical stimulus is switched over to physical stimulus.

The role of physical and chemical stimuli of the oviposition substrate (glass beads) was evaluated by providing physical stimuli alone and by providing physical and chemical stimuli together on a comparative basis to the two species of *Callosobruchus*.

In the case of *C. maculatus* the results of bioassay experiments revealed that (i) the gravid females did not oviposit on the glass beads of 6 or 12 mm diameter and (ii) glass beads with 12 mm diameter and coated with the stimulant (1 B. E. methanol extract of cowpea seed coats) are significantly preferred over the glass beads of 6 mm diameter without a stimulant for oviposition. These results amply demonstrate (Table 6) that curvature does not have any role in preferential oviposition by the gravid females and that they are solely guided by the chemical stimulus from the oviposition substrate.

In case of *C. chinensis* the results revealed (Table 6) that (i) the gravid females tend to prefer glass beads of 6 mm diameter for oviposition over glass beads of 12 mm diameter and the curvature of the oviposition substrate has a major role in preferential oviposition; (ii) glass beads of 12 mm diameter and coated with the stimulant are not significantly preferred over the glass beads of 6 mm diameter without a stimulant, fur-

Table 6. Role of physical and chemical stimuli on the comparative ovipositional responses of gravid *C. maculatus* and *C. chinensis* **

Insect	Number of eggs diposited[1] Glass bead diameter[2]								
	C_{12}	C_6	Mean*	T_{12} C_6	Mean	T_6 C_{12}	Mean	T_{12} C_{12}	Mean
C. maculatus	0	13	2.6	437 11	87.4 ± 42.1	582 11	118.6 ± 37.3	- -	-
C. chinensis	8	724	146.4 ± 47.5	37 894	186.2 ± 44.0	1178 0	235.6 ± 57.9	21 14	7.0 ± 9.5

* Mean number of eggs per replication.
** Duration of bioassay, 24 hr; 10 females/replication.
1) Number of eggs in 5 replications
2) Number accompanying letter C and T indicate the diameter of the glass beads in mm. C - Glass beads without stimulant. T - Glass beads coated with stimulant.

ther indicating that physical stimulus dominates over the chemical stimulus and (iii) when equal numbers of stimulant coated and uncoated glass beads of 12 mm diameter were offered to the gravid females, both categories failed to receive eggs. However, when 6 mm diameter glass beads were used, the stimulant-coated beads were significantly preferred for oviposition over the uncoated (control) glass beads implying that physical stimulus (curvature) is a prerequisite for normal oviposition. However, once the requirement for physical stimulus is met, then chemical stimuli alone elicited oviposition. Chemical stimuli alone do not trigger normal oviposition.

Acknowledgments. V. G. Gokhale is greatly indebted to the Japan Society for Promotion of Science for providing a post-doctoral scholarship.

REFERENCES

Avidov, A., Berlinger, M. J., and Applebaum, S. W. (1965) Physiological aspects of host specificity in the Bruchidae: III. Effect of curvature and surface area on oviposition of *Callosobruchus chinensis* L., Anim. Behav. 13, 178-180.

Credland, P. F. and Wright, A. W. (1988) The effect of artificial substrates and host extracts on oviposition by *Callosobruchus maculatus* (F.) (Coleoptera: Bruchidae), J. stored Prod. Res. 24(3), 157-164.

Gokhale, V. G. (1971) Ovipositional and nutritional studies on *Callosobruchus maculatus* (F.) (Coleoptera: Bruchidae), Ph. D. thesis.

Gokhale, V. G. and Srivastava, B. K. (1973) French bean seed coats as an ovipositional attractant for the pulse beetle, *Callosobruchus maculatus* (Fabricius), Experientia 29, 630-631.

Ihsii, S. (1952) Studies on the host preference of the cowpea weevil (*Callosobruchus chinensis* L.), Bull. Nat. Inst. Agr. Sci. 1(C), 185-256.

CHEMICAL ECOLOGY OF BRUCHIDS

IZURU YAMAMOTO
Laboratory of Pesticide and Bio-organic Chemistry, Faculty of Agriculture, Tokyo University of Agriculture, Tokyo 156, Japan

ABSTRACT. Two important bruchids were studied: the azuki bean weevil, *Callosobruchus chinensis*, and the cow pea weevil, *C. maculatus*. Various ecochemicals are involved in their ecology. *C. chinensis* prefers azuki beans and green grams in the field for oviposition. Whether or not such host preference is due to chemical stimuli in plants is not known. *C. chinensis* deposits eggs not only on beans but also on surfaces of certain curvature, while *C. maculatus* oviposits by responding to chemical stimuli present in the seed coat. There are two prominent phenomena that are associated with the oviposition behavior of the two bean weevils. Under low population density, the females oviposit evenly among beans, being guided by a marking pheromone. Under conditions of high density, the number of eggs on each bean becomes high and egg distribution becomes random. Nevertheless, only several eggs per bean hatch and grow normally inside the bean, leaving the rest to die by the ovicidal action of higher doses of the marking pheromone. These weevils thus have developed a strategy to reduce competition among larvae and to maximally utilize the host beans by using the same substance at different levels. *C. chinensis* can grow in many different beans under experimental conditions, but not in kidney beans. The same is true for *C. maculatus* due to the presence of a substance which is toxic to the weevils but not to man. Immediately after emerging from beans, the adults of *C. chinensis* mate. A sex attractant released from the females attracts the males. When they are in close proximity, the second sex pheromone which does not attract, functions to induce erection and insertion of the male genital organ and ejaculation. This copulation-release pheromone was thus named erectin. Erectin is released from both male and female, but more from the latter. Some of the above ecochemicals have been isolated, characterized and identified. This information provides potential use of such ecochemicals for pest insect control.

1. Introduction

Our studies on two important bruchids are reviewed: *C. chinensis* and *C. maculatus*. Various ecochemicals are involved in their ecology.

2. Host Plant Finding

C. chinensis prefers azuki beans and green grams in the field for oviposition. Whether or not such host preference is due to chemical stimuli in plants has not been elucidated.

3. Sex Attraction

A male *C. chinensis* either standing or walking, when in the proximity of a female, raises

K. Fujii et al. (eds.), Bruchids and Legumes: Economics, Ecology and Coevolution, 53–62.

his antennae and runs toward her in a zigzag pattern. A flying male hovers over a female, then lands and runs toward her in the same zigzag pattern. As he contacts the female, he lowers his antennae and bends his head. The female will sometimes take several forward steps, but the male follows her with his antennae lowered. When she stops, the male bends his head, extends the tip of his abdomen toward the female and extrudes his genital organ. The female raises her abdomen slightly and accepts his organ, thereby copulating. The behavior which takes place from either the stationary, walking or flying position to the zig zag running is termed "sex attraction", and that from lowering the antennae to the extrusion of the genital organ as "copulation release" (CR). The behavior suggests the involvement of a sex attractant. The volatile material from the virgin females was trapped by Porapak Q, extracted with pentane and fractionated on column chromatography and gas chromatography (PEG-HT, OV-17). A concentrated fraction showing sex attractancy was isolated and was estimated to be of a magnitude of 37 pg/female, hampering further investigation (Yamamoto, 1984).

4. Copulation Release

The presence of a copulation release pheromone in *C. chinensis* is indicated from biological observation and chemical evidences (Tanaka et al., 1981b). When the attractive fraction is placed at the tip of a glass rod, the male is attracted to it and walks around the sample in frustration, but does not extrude his genital organ. However, when offered the dead body of a female or even the tip of her abdomen, the male is attracted, extrudes his genital organ and copulates. Once extracted with ether, the female body shows neither attractancy nor CR activity. Thus, the presence of a chemical factor other than sex attractant is indicated. The extract from the virgin female body could have been expected to contain CR factor. However, though it provoked some attractancy, the male showed little CR activity. The extract was very sticky and would have a masking effect. On the other hand, the ether extract of the filter paper, with which the male were kept, showed both strong sex attractancy and CR activity. Separating CR activity from sex attractancy was achieved by warming the extract at 90°C under an air stream. The residue had no attractancy but retained a strong CR activity. Table 1 indicates the presence of both sex attractant and CR pheromone.

When the extract showing only CR activity was fractionated into an acidic fraction and a neutral fraction, each showed no activity, but the original activity was restored by recombination. Therefore, the CR activity is due to the synergistic action between two

Table 1. Attractancy and copulation release activity of the female *Callosobruchus chinensis* and the extracts

Sample	Attractancy	CR activity
1. Virgin female	+ +	+ +
2. The dead body	+ +	+ +
3. The body after extraction	-	-
4. The extract of 2	±	±
5. Extract from filter paper	+ +	+ +
6. The extract 5 after heating	-	+ +
7. The volatile from 1	+ +	-

Dose: one female equivalent.

Table 2. Copulation release activity of the fractions from the extract of the female *Callosobruchus chinensis*

	CR activity
1. Whole extract	+ +
2. Neutral fraction	-
3. Acidic fraction	-
4. 2 + 3	+ +
5. Hexane eluate of 2	-
6. 5 + 3	+ +
7. 5% Ether/hexane eluate of 2	-
8. 7 + 3	-
9. 15% Ether/hexane eluate of 2	-
10. 9 + 3	-
11. 2% Methanol/ether eluate of 2	-
12. 11 + 3	-

Dose: one female equivalent.

fractions. To isolate the factors involved the neutral fraction was first fractionated on column chromatography, and each fraction was combined with the acidic fraction for bioassay. Only the eluate with hexane showed CR activity (Table 2).

The hexane eluate was a complex mixture of hydrocarbons. Each hydrocarbon appeared to more or less contribute to the synergistic activity, but not specifically. For fractionation of the acidic fraction, its methylated product was submitted to gas chromatography and each fraction was hydrolyzed, then combined with the neutral fraction for bioassay. The CR activity centered at one peak, which was eventually identified as a novel carbon 10 dicarboxylic acid and named "Callosobruchusic acid" (C-acid). The mixture of the hydrocarbons and callosobruchusic acid has no attractancy, but can induce all the other mating responses, thus it was named "Erectin" (Table 3).

Erectin is released from both male and female, more from the female, but the CR activity is shown only in the male. This explains the attempted copulation often observed between the males of *C. chinensis*.

The complex hydrocarbon mixture can be substituted with a single hydrocarbon, e.g. octadecane. C-acid was synthesized and not (Z)- but (E)-3,7-dimethyl-2-octene-1,8-dioic acid was biologically active (Tanaka et al., 1982; Yamamoto, 1984). Also, each enantiomer of C-acid was synthesized, but both of them were biologically active, thus the absolute configuration remains to be solved (Mori et al., 1983). When presented with an aluminum foil tube (30-70 μm i.d.) bearing synthetic erectin, the male was able to effect

Table 3. Chemistry of erectin

A mixture of:
hydrocarbons * (ca. 12 μg/female) and callosobruchusic acid ** (ca. 15 ng/female)

* 3-Methylpentacosane; 11-methylheptacosane; 3-methylheptacosane; 11-methylnonacosane; 13-methylnonacosane; 11,15-dimethylnonacosane; 9,13-dimethylhentriacontane; 11,15-dimethyltritriacontane
** (E)-3,7-Dimethyl-2-octene-1,8-dioic acid

erection, insertion and ejaculation, suggesting its use for insect control.

5. Oviposition Preference

C. chinensis deposits eggs not only on beans but also on surfaces of certain curvature, while *C. maculatus* oviposits by responding to chemical stimuli present in the seed coat. This aspect is given by Gokhale et al. (1989) in the present symposium. D-Catechin was identified as one of the oviposition stimulants by Kuwahara et al. (1989).

6. Oviposition Distribution

Female *C. chinensis* have a habit of depositing eggs among host beans as evenly as possible by discriminating the oviposited beans from the fresh beans or the more oviposited beans from less oviposited beans (Utida, 1943). A mated female initially oviposits one egg on one bean and continues oviposition on most of the available beans, and then begins the second round of oviposition in the same manner (Nakamura, 1968). Such a behavior is also found in *C. maculatus* (Yoshida, 1960; Umeya, 1966). The mechanism is partly understood by the release of BCS (biological conditioning substance) from the weevils during creeping and ovipositing. The female prefers non- or less-BCS conditioned beans for further oviposition (Yoshida, 1961). BCS was eventually isolated and identified by us (Oshima et al., 1973; Sakai et al., 1986). For collecting the BCS, *C. chinensis* that oviposits on any surface of certain curvatures was utilized. The females were allowed to oviposit on glass beads of the size of the bean (3-5 mm i.d.) for 4 days. BCS was obtained at the ether washing. Comparison was made between *C. chinensis* and *C. maculatus* and between males and females (Table 4).

Two or three times more BCS came from females than males. The chemistry indicated that BCS was a mixture of lipids consisting of fatty acids, hydrocarbons and triglycerides including mono- and di-glycerides as minor components (Table 5, 6, 7, and 8).

Whether or not this extract functions as an oviposition marking pheromone was examined as follows (Sakai et al., 1986). Beans with different numbers of eggs and different treatments were prepared: oviposited; oviposited and then ether-washed; oviposited and then egg-removed; BCS treated. In a petri dish 10 beans of two different groups were alternately placed on filter paper, and then 10 gravid females were released for 1 hr in darkness. The total number of eggs deposited on one group (Ns) and the control group (Nc) was compared. For comparison of data obtained with *C. chinensis* and *C. maculatus*, the latter are shown in parentheses (Table 9).

In general, although *C. maculatus* showed less ability, the females of both weevils discriminate fresh beans from conditioned beans by eggs and BCS. Both eggs and BCS

Table 4. BCS from two weevils

| | BCS μg/insect | |
	C. chinensis	*C. maculatus*
Male	11.42	9.87
Female	36.92	21.26
Mixed sexes	29.70	28.80

Average of over fifty thousands adults during the life span.

Table 5. Constituents of BCS

| | Components (%) | | | |
| | C. chinensis | | C. maculatus | |
	Female	Male	Female	Male
Hydrocarbons	42.8	82.4	29.2	68.4
Triglycerides	32.2	5.3	44.1	3.1
Fatty acids	11.1	4.4	12.7	9.2
Diglycerides	4.0	1.8	5.6	13.3
Monoglycerides	1.8	1.8		
Others	8.1	4.3	8.5	6.1

Table 6. Composition of triglyceride fraction of BCS

Triglycerides *: C48, C50, C52, C54, C56
Fatty acids components: palmitic, stearic, oleic, linoleic, and linolenic acids

* Total number of carbon atoms in fatty acid moiety.

were conditioning factors, but under natural conditions both function synergistically. BCS from males had some activity, but BCS from females was several times more active, 2-3 times more in amount and played major roles. Statistical analysis of oviposition preference by C. chinensis was also made in terms of egg and BCS (Yamamoto, 1984; Honda et al., 1989). The time course study of egg deposition and distribution among host azuki beans indicated that both the eggs and BCS affected preferential oviposition by the weevils on the less conditioned eggs until the stage of 3-4 eggs and 20 µg of BCS per bean, then the subsequent oviposition was conducted randomly. Each of the lipid components showed more or less marking activity.

Table 7. Composition of fatty acid fraction of BCS

Palmitic, stearic, oleic, linoleic, and linolenic acids,
including capric, lauric, and myristic acids as minor components

For both C. chinensis and C. maculatus.

Table 8. Composition of hydrocarbon fraction of BCS

3-methylpentacosane (*); n-heptacosane (* #); 3-methylheptacosane (* #); 5-methylheptacosane (#);
9-methylheptacosane (#); 11-methylheptacosane (*); 3,7-dimethylheptacosane (#);
9,10-dimethylheptacosane (#); 9.13-dimethylheptacosane (#); 10,11-dimethylheptacosane (*);
11,15-dimethylheptacosane (*); n-octacosane (* #); 3-methyloctacosane (#); 12-methyloctacosane (*);
n-nonacosane (* #); 3-methylnonacosane (* #); 11-methylnonacosane (* #);
13-methylnonacosane (* #); 7,11-dimethylnonacosane (#); 9,13-dimethylnonacosane (#);
11,15-dimethylnonacosane (*); 12,13-dimethylnonacosane (#); n-hentriacontane (#);
13-methylhentriacontane (*#); 11,15-dimethylhentriacontane (*); 9,13-dimethylhentriacontane (* #);
13-methyltritriacontane (#); 11,15-dimethyltritriacontane (* #)

* C. chinensis, # C. maculatus.

Table 9. Oviposition preference between differently treated beans

| | | Ns/Nc | | |
| | | Initial number of egg/bean | | |
Choice (* control)	Implication	1	2	4
a) Oviposited bean vs. fresh bean *	egg + BCS vs. none	0.40 (0.60)	0.26 (0.39)	0 (0.19)
b) Oviposited bean vs. ether-washed oviposited bean *	egg + BCS vs. egg	0.41 (0.87)	0.11 (0.84)	0.01 (0.32)
c) Egg-removed bean vs. fresh bean *	BCS vs. none	0.70 (0.75)	0.48 (0.70)	0.64 (0.59)
d) Ether-washed oviposited bean vs. fresh bean *	egg vs. none	0.84 (0.84)	0.42 (0.67)	0.24 (0.39)

| | | Dose (μg/bean) | |
		75	20
e) BCS treated bean vs. fresh bean *	BCS vs. none	0.04 (0.04)	0.15 (0.41)

a) The females recognized even one egg. Discrimination became more distinct when more eggs had been initially deposited and the females oviposited on the initially fresh beans until the number of both side became almost equal.
b) The females preferred the control for oviposition. It seems that BCS associated with even one egg was enough for discrimination.
c) Oviposition was affected by BCS and the females preferred less marked beans.
d) The egg was also a factor for oviposition.
e) The females discriminated even 20 μg per bean.

7. Population Regulation

For both *C. chinensis* and *C. maculatus*, under high density conditions, the number of eggs on each bean becomes high and the egg distribution becomes random. Nevertheless, only several eggs per bean hatch and grow normally inside the bean, leaving the rest to die. With increasing numbers of pairs of the weevils the number of eggs oviposited increased, but the hatched eggs did not increase beyond a certain population density. The ratio of emerged adults to the hatched eggs stayed almost constant, indicating that there was no effect of population density after penetrating into the beans (Sakai, 1980).

Beans were preconditioned with different numbers of oviposition, then eggs were removed leaving only different levels of BCS. When provided to the adult pairs, BCS had no effect on further oviposition but affected the mortality of eggs (Table 10).

The activity of BCS for oviposition preference and ovicidal activity is shown in Table 11. BCS had essentially no effect on oviposition, but showed ovicidal activity beyond 75

Table 10. Effect of preconditioning of beans on oviposition and egg mortality

	Pre-conditioning *	Number of eggs deposited **	Egg mortality
C. chinensis	1	243	3.3
	5	246	7.6
	10	192	13.5
	50	232	21.1
	100	208	22.6
C. maculatus	1	175	15.4
	5	147	26.2
	10	213	36.6
	50	208	32.2
	100	238	74.7

* After deposition of the indicated number of eggs, all the eggs were removed from the beans.
** Five pairs were allowed to oviposit among 50 egg-removed beans.

Table 11. The activity of BCS

			Dose (μg/bean)			
		Control	150	75	38	19
		Oviposition preference (Ns/Nc)				
C. chinensis	Female BCS		0.05	0.39	0.45	0.49
	Male BCS		0.10	0.70	0.81	0.85
C. maculatus	Female BCS		0	0.09	0.21	0.26
	Male BCS		0.13	0.32	0.54	0.66
		Number of eggs deposited				
C. chinensis		1324				
	Female BCS		1140	1251	1341	1350
	Male BCS		888	1181	1212	1199
	Mixed BCS		1364	1415	1309	1080
C. maculatus		1012				
	Female BCS		1015	1050	1080	834
	Male BCS		793	820	993	1162
	Mixed BCS		951	1062	1162	951
		Egg hatchability (%)				
C. chinensis	Female BCS	93.1	0.1	1.6	28.1	80.4
	Male BCS		0	0.4	11.5	34.3
	Mixed BCS		7.3	39.6	86.0	90.1
C. maculatus	Female BCS	89.7	0	4.8	77.9	93.1
	Male BCS		0	8.4	35.6	85.9
	Mixed BCS		1.1	30.3	69.0	88.4

μg/bean. As shown there were differences between male and female BCS. Male BCS was more ovicidal but less in amount. However, under natural conditions both male and female BCS work together. A pair of *C. maculatus* is estimated to secrete about 58 μg of BCS during their life span. When 200 pairs were released among 200 beans (1 pair/bean), the egg hatchability was 40.9%. On the other hand the egg hatchability affected by 58 μg of BCS is estimated to be about 60%, indicating the correlation between BCS and hatchability of eggs under natural oviposition processes.

These weevils thus have developed a strategy to reduce competition among larvae and to maximally utilize the host beans by using the same substance at different levels.

The finding that the BCS lipids inhibit egg hatching at higher doses led us to investigate the effect of edible oils. As shown in Table 12, certain oils were effective, providing a simple technology for protecting stored beans. The physical state of the lipids seem to play an important role. When inactive triolein (liquid) and tristearin (solid) were combined, the resultant semiliquid gave a strong ovicidal activity (Sakai, 1980). Nature provides a semiliquid by blending lipid components.

8. Growth regulation

The azuki bean weevil, *C. chinensis* can grow in sixteen different beans under experimental conditions, but not in kidney beans, *Phaseolus vulgaris* L. The same happens for *C. maculatus* due to the presence of a substance which is toxic to the weevils but not to man (Ishii, 1952). The larvae die at the first instar in the bean. Janzen et al. (1976) claimed that the growth inhibition of *C. maculatus* occurring in black beans (*Phaseolus vulgaris*) was due to the presence of phytohemagglutinin (lectin), subsequently confirmed by Gatehouse et al. (1984), but not to trypsin inhibitor. Gatehouse et al. (1979) concluded that the resistance of a variety of cowpea, *Vigna unguiculata* L. was due to an

Table 12. Ovicidal effect of various edible oils

Oils	Hatched egg % C. maculatus	C. chinensis	Physical state
Control	84	95	
Lard	0	0	Semiliquid
Coconut oil	7	34	Semiliquid
Palm oil	39	8	Semiliquid
Margarine oil	3	26	Semiliquid
Butter oil	8	10	Semiliquid
Salad oil	5	26	Liquid
Rape seed oil	0	8	Liquid
Olive oil	8	53	Liquid
Sesame oil	10	10	Liquid
Rice oil	4	12	Liquid
Cotton seed oil	36	25	Liquid
Peanut oil	15	8	Liquid
Soybean oil	46	58	Liquid
Whale oil	21	53	Liquid

Dose: 150 μg/bean. Twenty adult pairs were allowed to oviposit on 300 azuki beans.

elevated level of trypsin inhibitor. Tanaka et al. (1981a) found that the crude lectin obtained from kidney bean did show growth inhibitory activity to *C. chinensis* and there was a correlation of the growth inhibitory activity with unit activity and contents of the crude lectin in kidney beans (resistant) and azuki beans (susceptible), but the lectin purified by an affinity chromatography had no activity. The α-amylase inhibitor from *Phaseolus vulgaris* seeds was known to inhibit the α-amylase of certain cereal insects *in vitro* (Powers and Culbertson, 1982), while Gatehouse et al. (1986) correlated the inhibition of α-amylase from *C. maculatus* by α-amylase inhibitor obtained from wheat with the inhibition of its larval development. Ishimoto and Kitamura (1988) also found that the lectin preparations from kidney beans did not affect *C. chinensis* larvae and isolated the α-amylase inhibitor which was highly toxic to the larvae. They claimed that the inhibitor was a glycoprotein having the molecular weight of about 100,000, which showed one major and several minor bands with molecular weight lower than 17,000 on SDS-PAGE gel (Ishimoto and Kitamura, 1989). Our group also has studied the problem independently for years and isolated from kidney beans (*Phaseolus vulgaris*) a growth inhibitor (a glycoprotein; molecular weight, about 48,000; isoelectric point, 4.46). It showed α-amylase inhibitory activity, but was not trypsin inhibitor (Samitanond et al., 1989).

Although there is an apparent discrepancy in the findings between the various laboratories, this may either represent varietal differences, or all these compounds may be part of a multimechanism for resistance.

Self-defense mechanisms in kidney beans would provide a chance for breeding weevil-resistant varieties of other beans.

Acknowledgments. The author is indebted to his coworkers listed in REFERENCES for making it possible for this review.

REFERENCES

Gatehouse, A. M. R., Gatehouse, J. A., Dobie, P., Kilminster, A. M., and Boulter, D. (1979) Biochemical basis of insect resistance in *Vigna unguiculata*, J. Sci. Food Agric. 30, 948-958.

Gatehouse, A. M. R., Dewey, F. M., Dove, I., Fenton, K. A., and Pusztai, A. (1984) Effect of seed lectin from *Phaseolus vulgaris* on the development of larvae of *Callosobruchus maculatus*, J. Sci. Food Agric. 35, 373-380.

Gatehouse, A. M. R., Fenton, K. A., Jepson, I., and Pavey, D. J. (1986) The effects of α-amylase inhibitors on insect storage pests: inhibition of α-amylase *in vitro* and effects on development *in vivo*, J. Sci. Food Agric. 37, 727-734.

Gokhale, V. G., Honda, H., and Yamamoto, I. (1989) Role of physical and chemical stimuli of legume host seeds in comparative ovipositional behaviour of *Callosobruchus maculatus* (Fab.) and *C. chinensis* (Linn), This proceedings.

Honda, H., Yamamoto, I., and Sakai, A. (1989) Analysis of the oviposition preference from the egg distribution in *Callosobruchus chinensis* L., J. Agr. Sci. Tokyo Univ. Agr. 33, 239-245.

Ishii, S. (1952) Studies on the host preference of the cowpea weevil (*Callosobruchus chinensis* L.), Bull. Nat. Inst. Agr. Sci. 1 (C), 185-256.

Ishimoto, M. and Kitamura, K. (1988) Identification of the growth inhibitor on azuki bean weevil in kidney bean (*Phaseolus vulgaris* L.), Jpn. J. Breed 38, 367-370.

Ishimoto, M. and Kitamura, K. (1989) Growth inhibitory effects of an α-amylase inhibitor from the kidney bean, *Phaseolus vulgaris* (L.) on three species of bruchids (Coleoptera: Bruchidae), Appl. Ent. Zool. 24, 281-286.

Janzen, D. H., Juster, H. B., and Liener, I. E. (1976) Insecticidal action of the phytohemagglutinin in black beans on a bruchid beetle, Science 192, 795-796.

Kuwahara, Y., Ueno, T., Fujii, K., and Suzuki, T. (1989) Oviposition stimulant of azuki bean weevil from the host azuki bean, Abst. of this Symposium, p. 102.

Mori, K., Ito, T., Tanaka, K., Honda, H., and Yamamoto, I. (1983) Synthesis and biological activity of optically active forms of (E)-3,7-dimethyl-2-octene-1,8-dioic acid (callosobruchusic acid); a component of the copulation release pheromone (erectin) of the azuki bean weevil, Tetrahedron 39, 2303-2306.

Nakamura, H. (1968) A comparative study on the ovipositional behavior of two species of *Callosobruchus* (Coleoptera: Bruchidae), Jpn. J. Ecol. 18, 192-197.

Oshima, K., Honda, H., and Yamamoto, I. (1973) Isolation of an oviposition marker from azuki bean weevil, *Callosobruchus chinensis* (L.), Agr. Biol. Chem. 37, 2679-2680.

Powers, J. R. and Culbertson, J. D. (1982) *In vitro* effect of bean amylase inhibitor on insect amylases, J. Food Protection 45, 655-657.

Sakai, A. (1980) Population regulation in the bean weevil and its chemical basis, Ph. D. Thesis.

Sakai, A., Honda, H., Oshima, K., and Yamamoto, I. (1986) Oviposition marking pheromone of two bean weevils, *Callosobruchus chinensis* and *Callosobruchus maculatus*, J. Pesticide Sci. 11, 163-168.

Samitanond, B., Tanaka, K., Honda, H., and Yamamoto, I. (1989) An ecochemical in kidney bean which is growth inhibitory to the azuki bean weevil, J. Pesticide Sci. (in press).

Tanaka, K., Honda, H., and Yamamoto, I. (1981a) unpublished.

Tanaka, K., Ohsawa, K., Honda, H., and Yamamoto, I. (1981b) Copulation release pheromone, erectin, from the azuki bean weevil (*Callosobruchus chinensis* L.), J. Pesticide Sci. 6, 75-82.

Tanaka, K., Ohsawa, K., Honda, H., and Yamamoto, I. (1982) Synthesis of erectin, a copulation release pheromone of the azuki bean weevil, *Callosobruchus chinensis* L., J. Pesticide Sci. 7, 535-537.

Umeya, K. (1966) Studies on the comparative ecology of bean weevils. I. On the egg distribution and the oviposition behaviors of three species of bean weevils infesting azuki bean, Res. Bull. Plant Prot. Serv. Japan 3, 1-11.

Utida, S. (1943) Studies on experimental population of the azuki bean weevil, *Callosobruchus chinensis* (L.). VIII. Statistical analysis of the frequency distribution of the emerging weevils on beans, Mem. Coll. Agr. Kyoto Imp. Univ. 54, 1-22.

Yamamoto, I. (1984) Eco-chemical approaches to control of pest insects of medical and agricultural importance, Res. Proj. in Review, Nissan Sci. Foundation 7, 209-224.

Yoshida, T. (1960) Effect of interspecific competition between two species of bean weevil upon fecundity and fertility, Mem. Fac. Lib. Arts Educ. (Natl. Sci.), Miyazaki Univ. 10, 17-31.

Yoshida, T. (1961) Oviposition behaviors of two species of bean weevils and interspecific competition between them, Mem. Fac. Lib. Arts Educ. (Natl. Sci.), Miyazaki Univ. 11, 41-65.

SOME PROMISING PHYSICAL, BOTANICAL AND CHEMICAL METHODS FOR THE PROTECTION OF GRAIN LEGUMES AGAINST BRUCHIDS IN STORAGE UNDER BANGLADESH CONDITIONS

MD. MAHBUBAR RAHMAN
Division of Entomology
Bangladesh Agricultural Research Institute
Joydebpur, Gazipur, Bangladesh

ABSTRACT. In a study conducted in 1988, dusts of deltamethrin @ 3 ppm (T_1), sevin @ 50 ppm (T_2) and fenitrothion @ 30 ppm (T_3), neem oil @ 10 ml/kg grain (T_4) and linseed oil @ 10 ml/kg grain (T_5), applied to cowpea seeds were fully effective up to 4, 3, 2, 3 and 2 months respectively against adults of *Callosobruchus chinensis* while lantana, tobacco and clay-coating were ineffective. In another evaluation of 9 treatments including the above T_1, T_2, T_3, T_4, and T_5, nogos 100 WSC applied @ 3 µl per 1 kg capacity container with 5 days' exposure (T_6), heat-treatments by boiled water for 10 minutes (T_7), sun-drying for 22 hours (T_8) and cold-treatment at -12.5°C for 48 hours (T_9) caused complete mortality to eggs, 1st-instar larvae, internal larvae and pupae of *C. chinensis* and thus ensured complete prevention of infestation from each of the 50 seeds containing any one of these 4 stages to the remaining fresh seeds in each 15 g lot of greengram. T_4 and T_5 applied against internal larvae also reduced adult emergence significantly. T_1, T_2, and T_3 were lethal only to the 1st-instar larvae. They prevented further infestation significantly, also killing emerged adults. Finally, in another trial, each 500 g cowpea seeds of which each 5% contained separately each of the above 4 stages of *C. maculatus* and also containing 5 pairs of adults, was treated separately with each of the above 9 treatments (T_1 to T_9) and was stored in sealed polythene bags. T_6, T_7, and T_9 ensured complete protection of grains while T_1, T_2, T_3, T_4, and T_5 offered 89.88%, 87.87%, 83.62%, 67.91%, 95.19%, and 90.69% grain protection respectively as against 92.39% grain infestation in the control experiment during 5 months of storage. The germination ability of seeds was not affected by any treatment.

1. Introduction

The significant annual losses of pulses due to bruchids (*Callosobruchus* spp.) as reported from many countries including Bangladesh (BARI, 1984) have led scientists to make concentrated efforts for developing control measures against them. Several reports are available on the efficacies of chemicals (Nisa and Ahmad, 1970; Tyler and Binns, 1977; Govindrajan et al., 1978; Suchwant et al., 1979; Yadav et al., 1983; Lekha et al., 1984; Rahman and Yadav, 1985; Yadav, 1987) and plant oils (Pereira, 1983; Yadav, 1985; Das, 1986; Khalique et al., 1988). Very few reports are available on the physical methods, particularly, for disinfestation of pulse grains from these pests (Khan and Borle, 1985; Stoyanova, 1983). A critical review of most of these studies reveal the following short-comings:

1. The studies were aimed mostly at protecting the fresh grains against infestation by adult beetles. This sort of protection could be obtained just through exclusion of adult entry using sealed non-porous/microporous containers i.e. beetle-proof containers. But

K. Fujii et al. (eds.), Bruchids and Legumes: Economics, Ecology and Coevolution, 63–73.
© 1990 *Kluwer Academic Publishers. Printed in the Netherlands.*

in reality the infestation dose occurs mostly from the hidden source of infestations consisting of different developmental stages already present in some portions of the stored grains rather than by the adults. The actual problem lies in protecting the grains from this type of hidden infestations.

2. Most of the studies did not reveal the actual persistence of toxicity since the treated grains were exposed to the adult insects only once just after treatment.

3. In most of the studies, the toxicity/efficacy was judged from reduction of adult emergence which did not confirm the toxicity to or effect on a particular stage.

4. Methods of disinfestation mostly involved the use of conventional fumigants which are not suitable for use by small and marginal farmers.

Thus the present studies were designed as a modest approach on the basis of some unpublished reports of the author to resolve the above mentioned short-comings.

2. Materials and Methods

Three experiments were conducted in the laboratory of the Bangladesh Agricultural Research Institute, Joydebpur, Gazipur, Bangladesh during 1988-89 to test the efficacy of some physical, botanical, and chemical methods against five stages of *Callosobruchus chinensis* Linn. and *C. maculatus* Fab. on cowpea (*Vigna unguiculata* Walp) and greengram (*V. radiata* Wilczek). The chemical methods included seed treatments with deltamethrin dusts @ 3.0 ppm (T_1), sevin dust @ 50.0 ppm a.i. (T_2), fenitrothion dust @ 30.0 ppm (T_3) and fumigation by nogos @ 3.0 μl/kg capacity container with 5 days exposure (T_6). Deltamethrin 0.3% dust was prepared from decis 2.5 EC using local clay (200 μm) as diluent. All other insecticides were used in their available formulations viz., fenitrothion 3.0% dust, sevin 10% dust and nogos 100 WSC. The botanical methods included seed treatments with oils of neem (*Azadirachta indica*) (T_4) and linseed (*Linum usitatissimum*) (T_5) each @ 10.0 ml/kg grains, leaf dust of tobacco (*Nicotiana tabacum*) @ 2.0 g/kg (T_{10}) and crude extracts of lantana (*Lantana camara*) leaves @ 10.0 ml/kg grain (T_{11}). Tobacco leaf dusts and lantana leaf extracts were prepared by manual crushing of the well-dried and green leaves respectively from the respective plants. The physical methods included heat-treatments by boiled water for 10 minutes (T_7), sun-drying for 22 hours (T_8), cold-treatments for 48 hours (T_9) and clay-coating (T_{12}).

For fumigation, nogos was applied on filter paper using a microapplicator and the filter paper was then hung in the air-tight glass boyum. For seed treatments, seeds were taken in plastic tube/container. The respective seed treating material was added to it. The tube/container was then shaken thoroughly closing the open end with butter paper. For heat-treatment, seeds were taken in sealed polythene bags. Then the bags were dipped in boiled water (100°C) for 10 minutes. For cold-treatment, seeds were placed in sealed polythene bags and were kept for 48 hours in an ice-chamber of the refrigerator regulated at -12.5°C. For sun-drying, seeds were kept spread out on black polythene during 9-30 AM to 3-00 PM (Temperature, 33°-43°C) for four consecutive days. For clay-coating, clay (200 μm) and water were added @ 50 g and 50 ml respectively to 1.0 kg seeds with thorough shaking.

For the 1st experiment, 200 g of well-dried and infestation free cowpea seeds were subjected to T_1, T_2, T_3, T_4, T_5, T_{10}, T_{11}, T_{12} (as mentioned above), or shaking only (as control). Each was replicated 3 times. From each, 10 g seeds were placed in 75 mm petridishes. The rest of the seeds of each treatment were stored in a sealed polythene bag in the laboratory under ambient temperature and R.H. (24°-32°C and 65-90% respectively). Five pairs of one-day old adults of *C. chinensis* were released into each petridish. Data on adult mortality were recorded after 3 days of release. Then the insects were removed. Subsequently, data on oviposition, egg mortality, 1st instar larval mortality and adult emergence were recorded. The same procedure was repeated at each one month interval for 4 months.

For the 2nd experiment, 50 seeds of greengram of which each contained one egg of *C. chinensis*, were placed in a plastic tube. Then required amounts of fresh and infestation free seeds were added to make it up to 15 g. Each 15 g lot was then subjected to T_1, T_2, T_3, T_4, T_5, T_6, T_7, T_8, T_9 (as mentioned above), or shaking only (as control). Each was replicated 3 times. The same procedure was repeated for 1st-instar larvae, internal larvae (4th instar larvae) and pupae. The treated grains were kept treatment-wise in individual plastic tubes closing the open end with cotton plugs and stored under room conditions (Temperature, 26°-32°C and R.H., 65-90%). Data on mortality of eggs and 1st-instar larvae were recorded using a dissecting microscope after 4 days of treatment. Data on adult emergence from each stage as 1st generation and also in F1, were recorded. Mortality of internal larvae and pupae were judged from the reduction of adult emergence in 1st generation from the respective stage. Data on the number of grains with bruchid hole (s) (infested grains) and the total number of grains for each treatment were recorded after emergence of F1 adults to calculate the rate of infestation.

For the 3rd experiment, 200 g cowpea seeds infested with eggs, 1st-instar larvae, 4th instar larvae and pupae of *C. maculatus* in equal proportions were placed in a plastic container. Another 300 g fresh and infestation free seeds were added to it. Five pairs of one-day old adults of *C. maculatus* were also added. Each of these 500 g lots was subjected to T_1, T_2, T_3, T_4, T_5, T_6, T_7, T_8, T_9 (as mentioned above), or shaking only (as control). Each was replicated 3 times. Each lot was then placed in a sealed polythene bag. The polythene bags were stored in the laboratory under ambient temperature and R.H. (20°-30°C and 50-85% respectively). Data on the number and weight of infested grains (grains having even a single bruchid hole), and healthy grains (grains without any holes) were recorded after 5 months of storage to calculate percent infestation and protection of grains. Healthy seeds were tested for germination before treatment and also at the end of the experiment.

The treatment mortality of each stage was corrected for control mortality (Abbott, 1951). All the data were analyzed statistically after appropriate transformations.

3. Results

As revealed from Table 1, deltamethrin @ 3.0 ppm and sevin @ 50.0 ppm were toxic causing 100% mortality of adults of *C. chinensis* even after the 5th and 4th months respectively. But fenitrothion @ 30.0 ppm and neem oil @ 10.0 ml/kg grain were only moderately toxic while tobacco, lantana and clay-coating were almost non-toxic to adults. Deltamethrin, unlike fenitrothion and neem oil caused complete reduction in oviposition up to 4th month. Sevin did so only initially. However, it retained its efficacy to cause high reduction in oviposition up to the 4th month. Fenitrothion or neem oil and linseed oil, caused a high reduction in oviposition up to the 4th and 3rd month respectively. Tobacco or lantana and clay-coating caused about 48 and 53 percent reduction in oviposition respectively, however the adult emergence was sufficient to destroy the grains in the following month. Neem oil and linseed oil were moderately ovicidal even after the 4th and 3rd month respectively while fenitrothion was so only initially. Similarly, they were toxic enough to cause complete mortality of 1st-instar larvae even after the 3rd month while fenitrothion was so only initially. Neem oil retained its very high toxicity to 1st-instar larvae up to the 4th month while fenitrothion was moderately toxic to 1st-instar larvae up to the 3rd month. Thus deltamethrin and sevin were effective for 4 and

Table 1. Toxicity of different materials after ageing for 1 hr, 2-, 3-, and 4 months on cowpea seeds to the adults and subsequent stages of *C.chinensis*.

Date of seed treatment/Date of exposure to insects	Treatments	Adult mortality (%)	Reduction in oviposition (%)	Egg mortality (%)	Mortality of 1st instar larvae (%)	Reduction of adult emergence (%)
20.6.88/ 20.6.88	Daltamethrin @ 3.0 ppm (T_1)	100.00A (90.00)	100.00A (90.00)	-	-	100.00A (90.00)
	Sevin @ 50.0 ppm (T_2)	100.00A (90.00)	100.00A (90.00)	-	-	100.00A (90.00)
	Fenitrothion @ 30.0 ppm (T_3)	100.00A (90.00)	97.80B (81.56)	44.21A (43.54)	100.00A (90.00)	100.00A (90.00)
	Neem oil @ 10 ml/kg (T_4)	100.00A (90.00)	97.12B (80.28)	48.32A (44.04)	100.00A (90.00)	100.00A (90.00)
	Linseed oil @ 10 ml/kg (T_5)	66.67B (54.96)	70.01C (56.81)	38.69A (38.43)	100.00A (90.00)	100.00A (90.00)
	Tobacco leaf dust @ 2.0 g/kg (T_{10})	0.0C (0.0)	48.52E (44.18)	4.12B (11.70)	1.32C (5.04)	53.71C (46.82)
	Lantana leaf extract @ 10.0 ml/kg (T_{11})	3.70C (6.49)	46.25E (43.59)	13.47B (21.50)	8.54B (16.97)	62.49B (52.26)
	Clay-coating (T_{12})	3.70C (6.49)	53.94D (47.26)	4.95B (12.79)	1.67C (7.49)	63.47B (52.86)
20.6.88/ 20.8.88	T_1	100.00A (90.00)	100.00A (90.00)	-	-	100.00 (90.00)
	T_2	100.00A (90.00)	98.88A (83.59)	100.00A (90.00)	-	100.00 (90.00)
	T_3	88.89AB (74.14)	96.93AB (79.96)	11.12D (19.43)	53.88B (47.25)	100.00 (90.00)
	T_4	81.48B (68.88)	91.43BC (70.84)	42.27B (40.55)	100.00A (90.00)	100.00 (90.00)
	T_5	66.67B (54.96)	82.57C (65.48)	32.96C (35.04)	100.00A (90.00)	100.00 (90.00)
20.6.88/ 20.9.88	T_1	100.00A (90.00)	100.00A (90.00)	-	-	100.00 (90.00)
	T_2	100.00A (90.00)	97.73B (81.41)	20.79B (27.03)	63.37B (52.81)	100.00 (90.00)
	T_3	85.19A (71.77)	66.52C (54.69)	9.39C (17.79)	13.45D (21.46)	71.93B (58.17)
	T_4	44.45B (41.77)	59.38D (35.95)	31.55A (34.16)	95.75A (78.70)	100.00A (90.00)
	T_5	7.41C (12.97)	9.33E (17.58)	7.17C (15.56)	37.23C (37.60)	48.19C (43.97)

(continued)

Table 1. (continued)

20.6.88/ 20.10.88	T$_1$	100.00A (90.00)	92.29A (73.93)	12.92A (21.04)	29.75A (28.17)	96.69A (79.55)
	T$_2$	77.78B (62.40)	45.51B (42.43)	9.84AB (18.23)	20.76A (27.10)	68.41B (55.82)
	T$_3$	37.04BC (37.19)	34.72BC (36.03)	5.98BC (14.05)	15.81AB (23.43)	53.53C (47.03)
	T$_4$	11.11CD (15.86)	29.09C (32.65)	11.96A (20.40)	16.66AB (24.06)	52.20C (46.26)
	T$_5$	0.0D (0.00)	7.81D (16.14)	3.03C (9.86)	11.46B (19.46)	18.75D (25.61)

Figures in parentheses are transformed (Arcsine) means of 3 replications.
Means followed by same letter(s) do not differ significantly by DMRT at 1% level.

3 months respectively in restricting further progeny multiplication just through killing adults released. On the other hand, neem oil and fenitrothion or linseed oil were effective for 3 and 2 months respectively in preventing 100% adult emergence through the successive lethal actions on subsequent stages.

As shown in Table 2, nogos, heat-treatment by boiled water and cold-treatment caused complete mortality of all stages and consequently resulted in full reduction of adult emergence. Thus they ensured complete prevention of infestation from each 50 seeds containing any one of the 4 stages to the additional fresh seeds in the lot. But sun-drying did so only in the case of eggs and 1st-instar larvae. However, it also caused high mortality of internal stages as judged from the reduction of adult emergence from the treatment of respective stages. Linseed oil and neem oil were similar in their toxicity to eggs and 1st-instar larvae as also shown in Table 1. But neem oil, unlike linseed oil resulted in high reduction of adult emergence from internal larvae while both failed to prevent adult emergence from pupae. Toxicity of deltamethrin, sevin and fenitrothion to any stage was low as compared to the other treatments. However, as in the first experiment, here also they were moderately toxic to 1st-instar larvae. Thus there was a good reduction in adult emergence from eggs and/or 1st-instar larvae and almost no reduction in adult emergence from internal larvae or pupae. But they prevented further progeny multiplication to F1 from emerged adults by killing them completely. Accordingly, they ensured significantly high reduction in infestation from eggs and 1st instar larvae. Neem oil ensured high reduction in infestation from all stages except pupae, while linseed oil did so from eggs and 1st-instar larvae. But they prevented further infestations to fresh seeds from emerged adults.

Results presented in Table 3, reveal that nogos, heat-treatment by boiled water and cold-treatment ensured full protection of cowpea seeds against *C. maculatus*. All other treatments except sun-dyring, also offered very high levels of protection of seeds during the period. Sun-drying, although offering significant protection of seeds (67.91%), was poor in its efficacy as compared to other treatments.

The original germination ability of cowpea seeds was not adversely affected by any treatment except heat-treatment, where a slightly higher reduction (4.39%) of germination occurred.

Table 2. Efficacy of some chemical, botanical and physical treatments on the mortality of eggs, 1st instar larvae, internal larvae and pupae and reduction of adult emergence of *C. chinensis* infesting greengram.

Stage of the insect treated	Treatments	Mortality of eggs (%)	Mortality of 1st instar larvae (%)	Reduction of adult emergence in 1st generation (%)	Reduction of adult emergence in F_1 (%)	Reduction in infestation (%)
Eggs (one-day old)	Deltamethrin (T_1)	4.35D (12.06)	32.95C (35.04)	56.56E (48.78)	100.00A (90.00)	81.33D (64.42)
	Sevin (T_2)	2.17D (8.46)	20.81E (27.15)	61.98DE (51.94)	98.00BC (82.05)	81.67D (64.42)
	Fenitrothion (T_3)	2.88D (9.68)	26.21D (30.77)	64.35D (53.38)	97.33C (80.64)	82.00D (64.42)
	Neem oil (T_4)	25.19C (30.12)	78.83A (62.69)	93.00B (74.85)	98.33BC (82.67)	95.33B (77.64)
	Linseed oil ((T_5)	39.76B (39.04)	68.05B (55.59)	86.08C (68.14)	98.67B (83.46)	92.67C (74.32)
	Fumigation by Nogos (T_6)	100.00A (90.00)	-	100.00A (90.00)	100.00A (90.00)	100.00A (90.00)
	Heat-treatment(T_7)	100.00A (90.00)	-	100.00A (90.00)	100.00A (90.00)	100.00A (90.00)
	Sun-drying (T_8)	100.00A (90.00)	-	100.00A (90.00)	100.00A (90.00)	100.00A (90.00)
	Cold-treatment(T_9)	100.00A (90.00)	-	100.00A (90.00)	100.00A (90.00)	100.00A (90.00)
First instar larvae (Stage just after hatcing and before enterning into the grain)	T_1	-	30.13E (33.22)	36.26D (36.99)	98.67B (71.46)	74.33D (59.60)
	T_2	-	47.36D (43.49)	60.15C (50.91)	98.00B (70.05)	84.00C (66.51)
	T_3	-	38.76D (38.49)	43.01D (40.96)	98.00B (82.05)	76.33D (60.92)
	T_4	-	95.36D (77.83)	96.56B (78.78)	100.00A (90.00)	98.67B (83.46)
	T_5	-	88.39C (70.17)	95.53B (77.99)	100.00A (90.00)	98.33B (82.67)
	T_6	-	100.00A (90.00)	100.00A (90.00)	100.00A (90.00)	100.00A (90.00)
	T_7	-	100.00A (90.00)	100.00A (90.00)	100.00A (90.00)	100.00A (90.00)
	T_8	-	100.00A (90.00)	100.00A (90.00)	100.00A (90.00)	100.00A (90.00)
	T_9	-	100.00A (90.00)	100.00A (90.00)	100.00A (90.00)	100.00A (90.00)

(continued)

Table 2. (continued)

Internal larvae (4th instar larvae)	T_1	-	-	13.57C (21.63)	98.67ABC (83.46)	59.67D (50.57)
	T_2	-	-	14.24C (22.09)	98.00BC (82.05)	60.00D (50.98)
	T_3	-	-	11.45C (15.79)	98.00DC (82.05)	58.00D (49.99)
	T_4	-	-	94.97A (77.07)	99.33AB (86.17)	98.00B (82.05)
	T_5	-	-	14.95C (22.20)	96.00C (78.72)	56.33D (48.66)
	T_6	-	-	100.00A (90.00)	100.00A (90.00)	100.00A (90.00)
	T_7	-	-	100.00A (90.00)	100.00A (90.00)	100.00A (90.00)
	T_8	-	-	95.02B (77.17)	90.67D (72.56)	91.67C (73.25)
	T_9	-	-	100.00A (90.00)	100.00A (90.00)	100.00A (90.00)
Pupae	T_1	-	-	29.88CD (33.05)	98.00B (82.05)	65.33D (53.94)
	T_2	-	-	27.61D (31.54)	98.00B (82.05)	64.33D (53.34)
	T_3	-	-	22.66D (28.15)	98.00B (82.05)	62.00D (51.95)
	T_4	-	-	39.95C (39.19)	98.67B (83.46)	70.67C (57.22)
	T_5	-	-	18.58D (25.69)	98.00B (82.05)	60.00D (50.77)
	T_6	-	-	100.00A (90.00)	100.00A (90.00)	100.00A (90.00)
	T_7	-	-	100.00A (90.00)	100.00A (90.00)	100.00A (90.00)
	T_8	-	-	91.46B (73.08)	92.33C (73.98)	88.33B (70.03)
	T_9	-	-	100.00A (90.00)	100.00A (90.00)	100.00A (90.00)

Figures in parentheses are transformed (Arcsine) means of 3 replications.
Means followed by same letter(s) do not differ significantly by DMRT at 1% level.

4. Discussion

The persistent toxicity of deltamethrin to the adults of pulse beetles was in agreement with the findings of Rahman and Yadav (1985). The quick knockdown effect of deltamethrin as reported in other cases (Chen et al., 1985) resulted in complete mortality of adults before they laid any eggs. The reduction of adult emergence from eggs (Rahman and Yadav, 1987) was actually due to the mortality of 1st-instar larvae and not for egg mortality as confirmed from the present study. Similar to the previous studies,

Table 3. Effect of chemical, botanical and physical treatments on infestation by and protection against *C. maculatus* and on germination of cowpea seeds during 5 months of storage (1.11.88 to 1.4.89) in Bangladesh.

Treatments	Infested seeds (%) (By count)	Protected (healthy) seeds (%) (By weight)	Reduction of germination from original rate (%)
Deltamethrin @ 3 ppm (T$_1$)	11.47 D (19.68)	89.88CD (71.53)	2.10 B
Sevin @ 50 ppm (T$_2$)	13.23 D (21.29)	87.87 D (69.63)	2.15 B
Fenitrothion @ 30 ppm (T$_3$)	19.32 C (26.08)	83.62 E (65.89)	2.35 B
Neem oil @ 10 ml/kg (T$_4$)	5.18 E (13.14)	95.19 B (77.35)	2.33 B
Linseed oil @ 10 ml/kg (T$_5$)	10.85 D (19.23)	90.69C (72.27)	2.19 B
Fumigation by Nogos @ 3 μl/kg (T$_6$)	0.0 F (0.0)	100.00 A (90.00)	2.27 B
Heat-treatment (T$_7$)	0.0 F (0.0)	100.00 A (90.00)	2.56 B
Sun-drying (T$_8$)	36.71 B (37.30)	67.91 F (55.50)	4.39 A
Cold-treatment (T$_9$)	0.0 F (0.0)	100.00 A (90.00)	2.57 B
Control (only shaking)	92.39 A (74.05)	9.37 G (17.82)	-

Figures in parentheses are transformed (Arcsine) means of 3 replications.
Means followed by same letter(s) do not differ significantly by DMRT at 1% level.

here also deltamethrin was ineffective against the internal stages.

Direct comparison of toxicity of sevin (carbaryl) and fenitrothion could not be made in the absence of available literature. However, sevin and fenitrothion were reported to show lower toxicity than deltamethrin (Yadav et al., 1983). The higher toxicity of sevin compared to fenitrothion to adult beetles was in conformity with the findings of Lekha et al. (1984). The initial toxicity of fenitrothion as observed here was comparable with that obtained by Tyler and Binns (1977). The failure of sevin to prevent progeny emergence from internal stages and its ability to cause complete mortality of 1st-instar larvae were expected (Yadav, 1980). Fenitrothion expectedly caused complete mortality of 1st-instar larvae of *C. chinensis* as it also did in the case of free living larvae of *Trogoderma granarium* (Chahal and Ramzan, 1982).

The initial toxicity of neem oil to cause 100% adult mortality was in conformity with the findings of other workers (Das, 1986; Khalique et al., 1988). But the complete mortality of eggs as previously claimed (Ali et al., 1985) contradicted the present findings. In the present study, neem oil caused 25% egg mortality and 95% mortality of 1st-instar larvae (the stage just after hatching of egg and before entering the grain). The possible explanation to this deviation was that the 1st-instar larval mortality might have been mistaken as the egg mortality. The internal larval (grubs) mortality observed here was similar to the previous findings (Ali et al., 1983). The toxicity of linseed oil could not be compared in the absence of available literature. The persistent toxicity of neem oil as observed here negated the findings of Das (1986) and others. The protection offered beyond 3 months as reported by them was not actually due to neem oil but due to the containers itself. The inefficacy of lantana to render sufficient inhibition to oviposition as expected (Pandey et al., 1986) might be due to the improper method of extraction and application. The failure of clay-coating in arresting the development of immature stages of *C. chinensis* contradicted the findings of Khan and Borle (1985). They reported that the activated clay arrested the development of bruchids. The possible reason for such failure was that the clay used in the present study was not activated. No reference was available to compare the performance of tobacco.

The efficacy of nogos (dichlorvos) in disinfesting greengram through killing of eggs-1st instar larvae was comparable but the findings against internal stages did not conform with previous results (Yadav, 1980). However, the killing of internal stages was possible by the vapors emitted from the impregnated nogos in filter paper as it did so in wheat due to its penetration ability into grains (Champ et al., 1969). However, its similar efficacy for large bulk still remains doubtful because of its reported inability to penetrate into the deep site (Monro, 1969).

The heat-treatment by boiled water (100°C) noticeably raised the grain temperature well above 60°C. In previous studies inflow of hot air (70°C) raised the temperature in wheat to 60°C (Claflin et al., 1984). Thus the efficacy of heat-treatment in killing all the stages of *C. chinensis* was comparable with the reports of other workers. Temperatures above 40°C were lethal to all the stages of *C. maculatus* (Mookherjee et al., 1964). Hayes et al. (1984) reported a complete kill of eggs and larvae of *Dacus dorsalis* in pawpaw fruits due to a hot water dip of 20 minutes at 48.7°C. The $LT_{100}S$ for eggs, larvae, pupae and adults of *Oryzaephilus surinamensis* were 20, 10, 30, and 7 minutes respectively (Al-Azawi et al., 1985). Sun-drying under a temperature range of 32°-43°C was similarly sufficient to kill the eggs-1st instar larvae but not all of the internal stages. Thus a fraction of the internal stages survived and gave progeny to cause further infestation. The efficiency of cold-treatment (-12.5°C) was in conformity with the reported mortality of *Acanthoscelides obtectus* at -9°C (Stoyanova, 1984) and of many other stored product pests at freezing temperatures (Smith, 1984).

The toxicities/efficacies of different treatments as observed in experiments 1 and 2 were well reflected in the 3rd experiment. The fore going discussions clearly suggested that the efficacies of neem oil, fenitrothion, sevin and deltamethrin in protecting cowpea seeds against *C. maculatus* during 5 months of storage should not be considered as the definite indication of their persistence over that period. This would be simply a mistake. Because the hidden infestations and emerged adults were killed within 2-months of treatments and there was no further entry of adults from outside to cause new infestations, the protection offered beyond 2 months was actually due to the prevention of adult entry by the containers. This was particularly so for the disinfestants (heat-treatment, cold-treatment and fumigation by nogos). However, the persistence of deltamethrin and sevin as observed here confirmed their reported efficacy (Nisa and Ahmad, 1970; Rahman and Yadav, 1985) to protect grains for 5 months when stored in containers accessible to the entry of adult bruchids.

Acknowledgements. The author is grateful to Dr. Mohammed Abdul Karim, Chief Scientific Officer and Head, Division of Entomology, Bangladesh Agricultural Research Institute (BARI), Joydebpur, Gazipur for showing keen interest in this work. The author is also grateful to the authority of BARI for providing necessary facilities for this study.

REFERENCES

Abbott, W. S. (1925) A method of computing the effectiveness of an insecticide, J. Econ. Entomol. 18 (1), 265-267.

Al-Azawi, A. F., El-Haidari, H. S., Aziz, F. M., and Al-Saul, H. M. (1985) Effect of reduced atmospheric pressure with different temperature on *Oryzaephillus surinamensis* (L.) (Coleoptera: Cucujidae), a pest of stored dates in Iraq, Date Palm Journal 4 (1), 77-90.

Ali, S. I., Singh, O. P., and Misra, U. S. (1983) Effectiveness of plant oils against pulse beetle, *Callosobruchus chinensis* Linn., Indian J. Entomol. 54, 6-9.

BARI (1984) Pulse beetles incidence in stored pulses and control measures, Annual Report, Legume Postharvest Technology (Bangladesh) Project, Bangladesh Agricultural Research Institute, Joydebpur, Gazipur, Bangladesh, pp. 6-13.

Chahal, B. S. and Ramzan, M. (1982) Relative efficacy of synthetic pyrethroids and some organophosphate insecticides against the larvae of khapra beetle (*Trogoderma granarium* Everts), J. Res. Punjab Agric Univ. 19(2), 123-126.

Champ, B. R., Steele, R. W., Genn, B. G., and Elms, K. D. (1969) A comparison of Malathion, Diazinon, Fenitrothion and Diclorvos for control of *Sitophilus oryza* (L.) and *Rhyzopertha dominica* (F.) in wheat, J. stored Prod. Res. 5, 21-48.

Chen, J. S., Lee, C. J., Yao, M. G., and Sun, C. N. (1985) Effect of pyrethroids on knockdown and lack of coordination responses of susceptible and resistant diamondback moth (Lepidoptera: Plutellidae), J. Econ. Entomol. 78, 1198-1202.

Claflin, J. K., Evans, D. E., Fane, A. G., and Hill, R. J. (1984) Thermal disinfestation of wheat in a sprouted bed, in Proceedings of the Third International Working Conference on Stored Product Entomology, October 23-28, Kansas State University, Manhattan, Kansas, USA., pp. 531-537.

Das, G. P. (1986) Pesticidal efficacy of some indigenous plant oils against the pulse beetles, *Callosobruchus chinensis* Linn. (Coleoptera: Bruchidae), Bangladesh J. Zool. 14(1), 15-18.

Govindrajan, R., Vadivelu, S., and Balasubramaniam, M. (1978) Efficacy of fenvalerate-a-candidate pyrethroid in the control of the pulse beetle, *Callosobruchus chinensis* L., Bull. Grain Technol. 16(2), 128-131.

Hayes, C. F., Chingon, H. T. G., Nitta, F. A., and Wang, W. J. (1984) Temperature control as an alternative to ethylene dibromide fumigation for the control of fruit flies (Diptera: Tephritidae) in papaya, J. Econ. Entomol. 77(3), 683-686.

Khalique, F., Ahmed, K., Afzal M., Malik, B. A., and Malik, M. R. (1988) Protection of stored chickpea, *Cicer arietinum* L. from attack of *Callosobruchus chinensis* L. (Coleoptera: Bruchidae), Tropical Pest Management 34(3), 333-334.

Khan, M. I. and Borle, M. N. (1985) Efficacy of some safe grain protectants against pulse beetle, *Callosobruchus chinensis* L. infesting stored Bengal gram (*Cicer arietinum* L), Punjabrao Krishi Vidyapeeth Research Journal 9(1), 53-55.

Lekha, C., Pandey, V. K., Srivastava, A. K., and Singh A. K. (1984) Relative toxicity of some insecticides to the adult of *Callosobruchus chinensis* Linn., Bull. Grain Technol., 20 (1), 60-61.

Monro, H. A. U. (1969) Manual of fumigation for insect control, Food and Agriculture Organization of the United Nations, Via delle Terme di Caracalla, Italy.

Mookherjee, P. B. and Chawla, M. L. (1964) Effect of temperature and relative humidity on the development of *Callosobruchus maculatus* Feb., a serious pest of stored pulses, Indian J. Entomol. 26, 345-351.

Nisa, M. and Ahmad, H. (1970) Laboratory evaluation of organic insecticides against pulse weevil in stored chickpea, Int. Pest Control (London) 12, 17-19.

Pandey, N. D., Mathur, K. K., Pandey, S., and Tripathi, R. A. (1986) Effect of some plant extracts against pulse beetles, *Callosobruchus chinensis* Linnaeus,' Indian J. Entomol. 48(1), 85-90.

Pereira, J. (1983) The effectiveness of six vegetable oils as protectant of cowpeas and bambara groundnuts against infestation by *Callosobruchus maculatus* (F.) (Coleoptera: Bruchidae), J. stored Prod. Res, 9, 57-62.

Rahman, M. M. and Yadav, T. D. (1985) Efficacy of Deltamethrin, Cypermethrin, Permethrin and Fenvalerate dusts on three seeds with different moisture contents stored in different containers up to 180 days against *Callosobruchus maculatus* (Fab.) and *C. chinensis* (Lnn.), Seeds and Farms 11(6), 49-50.

Rahman, M. M. and Yadav, T. D. (1987) Efficacy of Deltamethrin, Cypermethrin, Permethrin and Fenvalerate dusts against developmental stages of *Callosobruchus maculatus* Fab. and *C. chinensis* Linn., Indian J. Entomol. 49(3), 387-391.

Smith, L. B. (1984) Control of stored grain insects with low temperatures, in proceedings of the Thirty-first Annual Meeting, Canadian Pest Management Society, Winnipeg, Manitoba, 20-22 August, Canada, pp. 44-49.

Suchwant, S., Odak, S. C., and Singh, Z. (1979) Persistent toxicity of different insecticides on grain to the pulse beetle, *Callosobruchus chinensis* (L.), Pesticides 13(2), 29-31.

Stoyanova. S. (1984) Disinfestation of seeds by the use of low temperatures, Rasteniev "dni Nauki 21(39), 91-96.

Tyler, P. S. and Binns, T. J. (1977) The toxicity of seven organophosphorous insecticides and Lindane to eighteen species of stored product beetles, J. stored Prod. Res. 13, 39-43.

Yadav, T. D. (1980) Efficacy of insecticidal dusts against developmental stages of *Callosobruchus maculatus* (Fab.) and *C. chinensis* (Linn.,), Indian J. Entomol. 42(4), 798-802.

Yadav, T. D. (1985) Antiovipositional and Ovicidal toxicity of neem (*Azadirachta indica* A. Juss) oil against three species of *Callosobruchus*, Neem Newsletter 2(1), 5-6.

Yadav, T. D. (1987) Toxicity of deltamethrin, Cypermethrin, Permethrin against Thirteen stored product insects, Indian J. Entomol. 49(1), 21-26.

Yadav, T. D., Singh, S., Khanna, S. C., and Pawar, C. S. (1983) Toxicity of Dusts of organophosphorus insecticides against stored product beetles, Indian J. Entomol. 45(3), 247-252.

CARBON DIOXIDE AS A CONTROL AGENT FOR *CALLOSOBRUCHUS MACULATUS* (FAB.) IN STORED ADZUKI BEAN

W. K. PENG
Department of Plant Pathology and Entomology,
National Taiwan University,
Taipei, Taiwan 10764 China

ABSTRACT. Seeds of legumes, an important protein source for human beings, are frequently damaged by bruchids during storage. Adzuki bean was the favorite host for *Callosobruchus maculatus* (Fab.). The insects were reared in growth chambers at 30°C, 75% R.H., and in a photoperiod of 12 hr light and 12 hr darkness. Following the deposition of the weevils on 1st, 3rd, 6th, 9th, 12th, 16th, and 18th days respectively, beans containing eggs, larvae, or pupae were separately sealed in bags, made of nylon and surlyn, and filled with carbon dioxide. They were examined on the 33rd day. The results showed that in the controls, on average, 384.7 weevils were obtained per 200 beans; no weevils emerged from beans in bags treated with carbon dioxide. Exposing infested adzuki beans containing eggs and larvae in 30% CO_2 for 33 days resulted in no adult emergence.

1. Introduction

Legume seeds are an important protein and food oil source for human beings. Mung-beans or soybeans are also used to produce sprouts, an important vegetable in certain parts of the world. A survey in Taiwan carried out by Lin et al. (1975) showed that *Callosobruchus maculatus* (Fab.) was one of the insect pests present in stored beans. The author (unpublished laboratory study) conducted an experiment to test the preference of *C. maculatus* with regard to a variety of beans, and found that mungbeans and adzuki beans were the favorite hosts for this insect.

Most adzuki beans are harvested in Taiwan during January and February. They are usually packed in PE woven bags and stored in warehouses for continuing use. The beans so stored commonly become infested, resulting in heavy damage if no prevention or control measures are applied.

The increased public concern for possible contamination of food-stuffs with pesticide residues has led to greater emphasis on developing non-chemical methods for protecting stored products against insect damage. An intentional alternation of atmospheric gas concentration in store is attractive because insect deterioration can be controlled without use of pesticides. This method has been studied rather extensively in grain stores (Shejbal, 1980; Ripp 1984). It will not affect the stored commodities but will depress the development of, or have lethal effects on, the other living organisms. Such an atmosphere not only can help store the commodities for a longer period but is also a method of slow-acting fumigation that does not leave any toxic residues.

The normal atmospheric gas consists of 78% nitrogen, 21% oxygen, 1% rare gases such as argon, and 0.03% carbon dioxide. There are two ways to alternate this atmos-phere. One is to introduce into the store N_2 or CO_2, derived from a source outside the

K. Fujii et al. (eds.), Bruchids and Legumes: Economics, Ecology and Coevolution, 75–79.
© 1990 *Kluwer Academic Publishers. Printed in the Netherlands.*

storage premises. The other way is to store a viable commodity in an airtight facility until the respiration of the commodity and any infesting insects and associated organisms decreases the O_2 concentration and raises the CO_2 concentration (Howe, 1943; Howe and Oxley, 1944; Singh et al., 1977). Bailey (1955) found that in airtight storage the O_2 concentration would have to be reduced to about 2% to effect complete insect mortality.

AliNiazee (1971) reported that adults of *Tribolium castaneum* and *T. confusum* exposed to various mixtures of N_2 or He and O_2 were killed when the O_2 concentration dropped to 1.7% or below, whereas most adults exposed to $CO_2:O_2$ mixtures were killed due to the deteriorative effects of CO_2 itself. Therefore, the use of CO_2 is a preferred atmosphere for pest control.

The present experiments were undertaken to determine the lethal effects of CO_2 mixed with air on different developmental stages of *C. maculatus*. Tests also were conducted to determine survival of this insect in adzuki beans sealed in bags with relatively low permeability to CO_2 and O_2. These data provided basic information on the feasibility of using CO_2 gas in the control of insects on stored adzuki beans.

2.   Materials and Methods

The test insect was *C. maculatus*, which was obtained in adzuki beans bought from supermarkets. The insects reared on adzuki beans were kept in growth chambers throughout the experiment, at 30°C, 75% R.H., in a photoperiod of 12 hr light and 12 hr darkness.

Adzuki beans were bought from a wholesale store, separately packed in PE bags, 5 kg of each, and were kept in a -18°C refrigerator for at least 2 weeks before use, in order to eliminate possible contamination by undesired insects. The rest were stored for continuous use.

A Queue cell culture incubator was used to test the lethal effect of CO_2 on the tested insects. The temperature in this experiment was set at 30°C. The atmospheric combinations inside the room was set at the ratio of CO_2 to air = 10:90, 15:85, 20:80, 25:75, and 30:70. Carbon dioxide gas was introduced from a cylinder obtained from commercial sources.

Uninfested beans (400 g) were exposed to egg-laying adults (400) in a Mason jar for 6 hr to obtain a supply of test insects of known age. After removal of the adults, the infested seeds were kept in the growth chamber for the required period of time.

Since the immature stages develop within seeds, it was impossible to check if any changes occurred during exposure to various atmospheres. Therefore, the relationship between the age (days following oviposition) and developmental stage was determined by cutting and examining the infested seeds daily until adults emerged, and by the measurements of larval head capsules. Thus, for this study, ages of insects in the seeds were taken as follows: after 1 day and 3 days oviposition period-- eggs; 6 days after the start of oviposition-- 1st instar; 9 days-- 2nd instar; 12 days-- 3rd instar; 16 days-- 4th instar; 18 days-- pupa.

Samples of infested seeds (200) containing weevil progeny at the ages of 1, 3, 6, 9, 12, 16, and 18 days after egg-laying were separately confined in 180-ml cup and exposed to the atmosphere for each combination in an incubator for 33 days. Each test was replicated 5 times. Control experiments were conducted in a growth chamber.

Each of the same conditioned samples were wrapped with muslin and placed in transparent film bags (12.5 x 21.5 cm, 0.085 mm thick), made of nylon and surlyn. The bags were then filled with 100% CO_2 and heat-sealed. They were kept in a growth chamber at 30°C for 33 days storage. The control consisted of insects with adzuki beans which were placed in 180-ml cup.

The number of adults which emerged from the samples was recorded. Survival rate of adults from the treated immature stages was compared with that of the controls.

3. Results and Discussion

The results obtained in the case of exposure of infested seeds containing eggs, larvae, and pupae of *C. maculatus* to various concentrations of CO_2 are presented in Table 1. The data indicate that the emergence percentage of the weevils decreased as their immature stages exposed to the atmosphere containing CO_2 are progressively increased. The atmosphere containing CO_2 below 15% did not significantly alter the survival of eggs and larvae. Larvae appeared to be more resistant to lower concentrations of CO_2. The percentage of emergence from the treated and the untreated pupae was significantly different at the 5% level. When eggs, larvae, and pupae were treated at 20% CO_2 or higher, their emergence was significantly different from the untreated. Exposing eggs to 25% CO_2, or larvae to 30%, resulted in no adult emergence.

Oosthuizen and Schmidt (1942) found that eggs and adults of *C. chinensis* L. were very susceptible to CO_2, while mature larvae and pupae were decidedly resistant. AliN-iazee and Lindgren (1970) reported that 100% mortality of *T. confusum* and *T. castaneum* eggs was obtained by exposure to an air atmosphere containing 25% or more CO_2. The differential susceptibility of the developmental stages of insects to CO_2 atmosphere

Table 1. Percentage of *Callosobruchus maculatus* (Fab.) adults emerged from developing stages exposed to various combinations of atmosphere for 33 days at 30°C

Treatment	Age[1] or stage[2] of test insects						
(CO$_2$: air)	1 E	3 E	6 L1	9 L2	12 L3	16 L4	18 P
10:90	101	98	73		82	85	80
Control	100	100	100		100	100	100
	ns	ns	ns		ns	ns	s
15:85	91	98	102		96	96	82
Control	100	100	100		100	100	100
	ns	ns	ns		ns	ns	s
20:80	41	87	86		67	76	68
Control	100	100	100		100	100	100
	s	s	s		s	s	s
25:75	0	0	3	6	5	10	6
Control	100	100	100	100	100	100	100
	s	s	s	s	s	s	s
30:70	0	0	0	0	0	0	0.01
Control	100	100	100	100	100	100	100
	s	s	s	s	s	s	s

(1) days following deposition.
(2) E: egg, L1-L4: 1st to 4th instal larva, P: pupa.
(3) s and ns indicate that above two means are significantly and not significantly different, respectively, at the 5% level.

78

may be due to the variation in sensitivity of different developing organ systems at different stages.

AliNiazee (1971) showed that exposing larvae and pupae of *Tribolium* to 100% CO_2 gas resulted in a number of morphological abnormalities in both the pupae and emerged adults. Some adults emerging from treated pupae were wingless, extremely small, non- or half-pigmented and sometimes with half pupal characteristics. Most of the larvae exposed to CO_2 showed a significant degree of retarded development. Brooks (1957) reported that German cockroach nymphs exposed to high concentrations of CO_2 for 3 minutes weekly until maturity were retarded in growth by 14% - 53% compared to controls, as determined by the time required to moult to adult. Nymphs kept under carbon dioxide for 2.5 hr remained paralyzed and died in several days. She concluded that the retardation and lethal factors seemed to be carbon-dioxide specific rather than a general immobilization, nervous, or oxygen-deficit effect inasmuch as nitrogen was relatively innocuous. In the insect body, anaerobic metabolism usually resulted in buildup of lactic acid (Wiggleworth, 1974; Kennington and Cannell, 1967). The oxidation of such a substance is a more wasteful process than it is during normal respiration, and may be deferred until oxygen is restored.

After 33 days of storage, the number of emerged adults from the infested seeds in the bags containing eggs, larvae, or pupae were counted and are listed in Table 2. The data show that no adults were found in bags filled with CO_2, whereas in the controls, on average, 384.7 beetles emerged.

On the 2nd day following CO_2 treatment, the internal CO_2 was found to have been adsorbed by the seeds. The sealed bags developed low internal pressures during storage, resulting in bags that appeared to have been vacuumized. Mitsuda et al. (1973) showed that grains can absorb a significant amount of CO_2 gas. The mechanisms of the absorption were examined. They found that solubility of CO_2 gas into moisture, and the presence of lipids in the grains had a minor effect on this absorption; however, diffusion of CO_2 into the grain is important in this phenomenon.

Cline and Highland (1987) found that some insects survived for 12 weeks when sealed in 0.7 atmosphere airtight pouches containing dried food. Thornton and Sullivan (1964) found that 0.03-0.05 atmospheres killed insects in as short a time as 64 minutes. Calderon et al. (1966) tested six species of common stored product insects at 16-20 mm Hg. After 120 hr, 100% mortality of both larvae and adults was obtained except in the case of *C. maculatus* larvae; however, with longer treatment higher mortality might have been obtained. Therefore, a combination of CO_2 with low pressure may be an effective form for control of stored product insects in seeds.

Table 2. Number of *Callosobruchus maculatus* (Fab.) adults emerged from adzuki beans sealed in bag filled with 100% CO_2 for 33 days.

Treatment	Age[1] or stage[2] of test insects						
	1 E	3 E	6 L1	9 L2	12 L3	16 L4	18 P
CO_2	0	0	0	0	0	0	0
Control	379	366	350	338	379	422	459

(1) days following deposition.
(2) E: egg, L1-L4: 1st to 4th instal larva, P:pupa.

REFERENCES

AliNiazee, M. T. (1971) The effect of carbon dioxide gas alone or in combinations on the mortality of *Tribolium castaneum* and *T. confusum*, J. stored Prod. Res. 7, 243-252.

AliNiazee, M. T. and Lindgren, D. L. (1970) Egg hatch of *Tribolium confusum* and *T. castaneum* (Coleoptera: Tenebrionidae) in different carbon dioxide and nitrogen atmospheres, Ann. Entomol. Soc. Am. 63, 1010-1012.

Bailey, S. W. (1955) Airtight of storage of grain: its effect on insect - *Calendra garanaria*, Aust. J. Agric. Res. 6, 33-51.

Brooks, M. A. (1957) Growth retarding effect of carbon dioxide anesthesia on German cockroach, J. Insect Physiol. 1, 76-84.

Calderon, M., Navarro, S., and Donahaye, E. (1966) The effect of low pressures on the mortality of six stored-product insect species, J. stored Prod. Res. 2, 135-140.

Cline, L. D. and Highland, H. A. (1987) Survival of four species of stored product insects confined with food in vacuumized and unvacuumized film pouches, J. Econ. Entomol. 80, 73-76.

Howe, R. W. (1943) An investigation of the changes in a bin of stored wheat infested by insects, Bull. Entomol. Res. 34, 145-158.

Howe, R. W. and Oxley, T. A. (1944) The use of carbon dioxide production as a measure of infestation by insects, Bull. Entomol. Res. 35, 11-12.

Kennington, G. S. and Cannell, S. (1967) Biochemical correlates of respiratory and developmental changes in anoxic *Tribolium confusum* pupae, Physiol. Zool. 40, 403-408.

Lin, T., Tsai, W. S., Pung, T. H., Lin, W. H., Hung, T. F., Yen, F. C., and Chen, Y. M. (1975) A survey on losses of stored grains to insect pests and their fumigation, Pl. Prot. Bull. Taiwan 17, 142-149.

Mitsuda, H., Kawai, F., Kuga, M., and Yamamoto, A. (1973) Mechanisms of carbon dioxide gas adsorption by grains and its application to skin-packaging, J. Nutr. Sci. Vitaminol. 19, 71-83.

Oosthuisen, M. J. and Schmidt, U. W. (1942) The toxicity of carbon dioxide to the cowpea weevil, J. Entomol. Soc. S. Afr. 5, 99-110.

Ripp, B. E. (1984) Controlled Atmosphere and Fumigation in Grain Storages, Elsvier Science Publisher, Amsterdam.

Shejbal, J. (1980) Controlled Atmosphere Storage of Grains, Elsevier Scientific Publishing Company, Amsterdam.

Singh, N. B., Sinha, R. N., and Wallace, H. A. (1977) Change in O_2, CO_2 and microflora of stored wheat induced by weevils, Environ. Entomol. 6, 111-117.

Thornton, B. C. and Sullivan, W. N. (1964) Effects on a high vacuum on insect mortality, J. Econ. Entomol. 57, 852-854.

Wiggleworth, V. B. (1974) The Principles of Insect Physiology, Chapman and Hall, London.

EFFECT OF VEGETABLE OILS ON THE OVIPOSITIONAL BEHAVIOUR OF *CALLOSOBRUCHUS MACULATUS* (FABRICIUS)

NEETA BHADURI, D. P. GUPTA[1], AND SHRI RAM[2]
Department of Zoology, Dr.Hari Singh Gour University
SAGAR (M.P.) 470 003 India

[1] *Corresponding author*
[2] *Senior Scientist (Ent.), Krishi Bhavan, Indian*
Council of Agricultural Research, New Delhi-110 001. India.

ABSTRACT.   Effects of nine vegetable oils viz., *Pongamia glabra, Madhuca indica. Ricinus communis, Brassica* sp., *Azadirachta indica, Cymbopogon citratus, Cocos nucifera, Cymbopogon nardus,* and *Arachis hypogea* at three different concentrations (0.2, 0.4, and 0.8% w/w) were studied on the ovipositional behaviours of *Callosobruchus maculatus* (Fabricius) at 27 ± 1°C and 70% R.H. Oviposition was completely inhibited by the seed treatment of *Pongamia* oil in each dose just after the treatment but its efficacy sharply deteriorated at later stages of sampling (30, 60 and 90 days after treatment). *Cymbopogon nardus* and *C. citratus* oils were effective even up to 90 days of treatment. Dose to dose variations were significant at each stage of sampling.

1.   Introduction

Pulse beetle, *Callosobruchus maculatus* (Fabricius) is one of the most destructive pests of cowpea during storage in the Bundelkhand region of M.P., India. The indiscriminate use of pesticides to protect cowpea seed from the pulse beetle may cause serious health hazards. Considerable work has been carried out on the effects of various indigenous plant products against pulse beetle (Jotwani and Sircar, 1967 on neem seed, Yadav, 1971 on oil of *Acorus calamus*, Jilani and Mohammad, 1975 on neem, Singh et al., 1978 on ground nut oil and Varma and Pandey, 1978 on edible oils). In view of the above facts, an investigation was undertaken to find out the effect of some edible and nonedible vegetable oils on the ovipositional behaviour of *C. maculatus* (F.) at 27 ± 1°C and 70% R.H.

2.   Materials and Methods

One kg of cowpea seed (variety, 42-1) was mixed with various oils viz., *Pongamia glabra* Vent. (Karanjee), *Madhuca indica* Gmel. (Mahua), *Ricinus communis* Linn. (Castor), *Brassica* sp. (mustard), *Azadirachta indica* A. Juss. (neem), *Cymbopogon citratus* stapf. (lemon grass), *Cocos Nucifera* L. (coconut) *Cymbopogon nardus* Linn. (Citronella), and *Arachis hypogea* L. (groundnut) in glass jars at three different concentrations - 0.2, 0.4 and 0.8 parts/100 parts of cowpea seed (w/w). A mechanical shaker was used for mixing 20 gm of treated seed at each concentration and these were placed in glass tubes

81

K. Fujii et al. (eds.), Bruchids and Legumes: Economics, Ecology and Coevolution, 81–84.

measuring 4" X 1". In each tube a freshly emerged pair of pulse beetles (*Callosobruchus maculatus* Fabr.) were released immediately after treatment and after intervals of 30, 60, and 90 days following treatment of cowpea seed with various oils at three concentrations. The glass tubes were placed in desiccators with a 70% R.H. maintained by KOH solution and these desiccators were placed in incubators maintained at 27 ± 1°C. The treatments were replicated six times.

The fecundity of one female beetle was observed immediately, 30, 60, and 90 days after treatment. The number of eggs laid on the seeds by each female beetle in each tube was counted daily until death of the beetle.

3. Results and Discussion

3.1. JUST AFTER TREATMENT.

The data presented in Table 1 indicated that all the oils significantly inhibited the oviposition at each concentration. Karanjee oil (*Pongamia glabra*) was superior to the others and completely checked the egg laying. Even the low dose of karanjee oil was significantly superior to high doses of other oils, except *Cymbopogon citratus* and *C. nardus*, when seeds were used just after the treatment.

3.2. 30 DAYS AFTER TREATMENT.

All the oils in each concentration were found to be extremely effective in reducing the egg laying on treated seeds as compared to untreated. The lowest levels of egg lay were observed in seeds treated with *Cymbopogon nardus, C. citratus, Brassica* sp. and *Ricinus communis* oils. The egg laying was least effected by coating with *Arachis hypogea* oil. Coating with *Pongaima glabra* oil was also less effective at 0.2 and 0.4% concentrations. However, it proved to be effective at 0.8% concentration.

3.3. 60 DAYS AFTER TREATMENT.

When treated seed was subjected to oviposition after 60 days following treatment, *Cymbopogon nardus* and *C. citratus* were found to be most effective. Minimum oviposition was observed on cowpea seed coated with a high dosage of *C. nardus*.

3.4. 90 DAYS AFTER TREATMENT.

C. citratus oil was found to give most protection followed by *C. nardus* when treated seed was subjected to pulse beetles 90 days after treatment. A high dose of both these oils was more effective and significantly superior to the rest of the oils.

All the oils significantly inhibited the egg laying of pulse beetle up to the last stage of sampling. The data further indicated that with the passage of time the efficacy of *Pongamia* oil deteriorated drastically and remained less effective after 30 days of treatment. *Cymbopogon nardus* was quite effective up to sixty days, whereas *C. Citratus* lost its effectiveness gradually but persisted up to the last stage of sampling. *Arachis hypogea* and *Cocos nucifera* oils were least effective.

The highest dose of oils remained highly effective and persisted up to the last stage of sampling. The efficacy of moderate dose levels was observed up to sixty days and deteriorated thereafter, whereas the low dose lost its efficacy after 30 days of treatment.

Table 1. Effect of vegetable oils and their dosages on the oviposition of one pair of pulse beetle at 27 ± 1°C.

S. No. Vegetable oils	Just after treatment Dosages(%)				30 days after treatment Dosages (%)				60 days after treatment Dosages (%)				90 days after treatment Dosages (%)			
	0.2	0.4	0.8	Mean	0.2	0.4	0.8	Mean	0.2	0.4	0.8	Mean	0.2	0.4	0.8	Mean
1. *Pongamia glabra*	0.61*	0.61	0.61	0.61	8.23	7.91	5.23	7.12	8.69	8.43	7.93	8.35	9.50	9.22	8.39	9.04
2. *Madhuca indica*	6.14	4.14	3.48	4.59	7.93	6.97	6.12	7.01	8.65	8.38	7.28	8.11	9.53	9.37	8.46	9.11
3. *Ricinus communis*	5.07	4.79	3.26	4.37	7.66	6.44	5.81	6.64	8.90	7.79	7.56	8.08	9.50	9.28	8.57	9.12
4. *Brassica sp.*	5.84	4.58	3.87	4.76	7.78	6.48	5.66	6.64	8.32	8.00	7.55	7.96	9.80	8.98	8.39	9.05
5. *Azadirachta indica*	4.98	4.16	3.02	4.05	8.18	6.91	6.27	7.12	9.32	8.37	6.84	8.18	9.59	9.47	8.44	9.17
6. *Cymbopogon citratus*	4.31	3.71	0.61	2.88	7.42	6.70	5.68	6.60	8.84	8.00	6.95	7.93	9.09	8.52	7.13	8.24
7. *Cocos nucifera*	6.33	4.71	4.43	5.16	8.59	7.34	6.85	7.59	9.14	8.54	7.48	8.39	9.81	9.40	9.15	9.45
8. *Cymbopogon nardus*	3.70	3.33	0.61	2.55	7.37	6.71	5.50	6.53	7.98	7.10	6.65	7.25	8.98	8.61	7.77	8.45
9. *Arachis hypogea*	6.98	6.38	5.64	6.33	8.30	8.01	7.83	8.05	8.79	8.51	8.25	8.51	9.77	9.37	9.05	9.40
Control (No treatment)	-	-	-	10.37	-	-	-	10.55	-	-	-	10.44	-	-	-	10.75
mean	4.88	4.04	2.84	-	7.94	7.05	6.11	-	8.74	8.12	7.39	-	9.51	9.14	8.37	-
	oils	Dos-ages	oils x Dosages Interaction		oils	Dos-ages	oils x Dosages Interaction		oils	dos-ages	oils x Dosages Interaction		oils	dos-ages	oils x Dosages Interaction	
S. Em. (±)	0.113	0.065	0.196		0.113	0.066	0.197		0.092	0.053	0.159		0.085	0.049	0.148	
C.D. at 5%	0.314	0.181	0.543		0.315	0.182	0.546		0.254	0.147	0.440		0.236	0.136	0.409	

* Transformed values of mean number of eggs $(x + 0.37)$

84

The activity of the oils might be due to their repellency, chemical toxicity or physical properties rendering changes in surface tension (which could lead to protoplasm coagulation) and in affecting oxygen tension within eggs. The thin oil layer is believed to block the oxygen supply to the embryo and their mode of action is partially attributed to interference with normal respiration, resulting in suffocation. Similar ovipositional behaviour of bruchids in oil coated seeds has also been reported by Gunther and Jeppson (1960), Singh et al. (1978) and Varma and Pandey (1978).

REFERENCES

Gunther, P. A. and Jeppson, L. R. (1960) Mod. Insecticides and World Food Production, Wiley, N.Y,

Jilani, G. and Mohammad, M. M. (1975) Studies on neem plant as a repellent against stored grain insect-pest, Pak. J. Sci. Ind. Res. 16(6), 251-254.

Jotwani, M. G. and Sircar, P. (1967) Neem seed as a protectant against bruchid, *Callosobruchus maculatus* (Fabricius) infesting some leguminous seed, Indian J. Entomol. 24(1), 21-24.

Singh, S. R., Luse, R. A., Leuschner, K., and Nanju, D. (1978) Groundnut oil treatment for the control of *Callosobruchus maculatus* (F.) during cowpea storage, J. stored Prod. Res. 14 (2/3), 77-80.

Verma, B. K. and Pandey, O. P. (1978) Treatment of stored green gram seed with edible oils for protection from *Callosobruchus maculatus* (Fabr.), Indian J. Agric. Sci. 48(2), 23-30.

Yadav, R. L.(1971) Use of essential oil of *Acorus calamus* L. as an insecticide against pulse beetle, *Bruchus chinensis* L., Rev. Appl. Entomol. 62(2), 198.

EFFECT OF FIVE BOTANICALS AS PROTECTANTS OF GREENGRAM AGAINST THE PULSE BEETLE *CALLOSOBRUCHUS MACULATUS*

ROHAN H. S. RAJAPAKSE
Department of Agronomy, University of Ruhuna,
Mapalana Kamburupitiya, Sri Lanka

ABSTRACT. Powdered leaves from four plants and citrus peel were assayed for their ovicidal and protectant properties against the pulse beetle *Callosobruchus maculatus* infestation of greengram. The plants tested were leaves of *Piper nigrum, Annona reticulata, Dillenia retusa, Ocimum sanctum* and peel of *Citrus crematifolia*. Percent oviposition on seeds were significantly reduced when peel of citrus was applied at doses of 0.20 grams/50 seeds. High adult mortality was observed with leaves of *P. nigrum* at the doses of 0.10 grams/50 seeds. Prolonged protection of greengram seeds were mainly due to reduced oviposition, low hatching, and high adult mortality.

1. Introduction

Callosobruchus maculatus (F.) (Bruchidae:Coleoptera), the pulse beetle, is widely distributed throughout the tropics where it is a major pest of cowpea *Vigna unguiculata* (L.) Walp. and mungbean *V. radiata* (Dobie, 1981). Adult females of *C. maculatus* oviposit on pods and hence some loss occurs in the field. Loss that is caused to stored dried legume seeds is of economic importance (Credland, 1986). Therefore, farmers growing mungbean for their own consumption frequently store greengram in their pods thereby limiting losses. However, the pods are usually removed before entering the commercial market where the seeds are then susceptible to attack by *C. maculatus*. *Callosobruchus maculatus* adults lay most of their eggs on the more mature pods (Alzouma and Huignard, 1981) and progressive maturation of pre-existing pods remain accessible as an egg laying substrate for females (Huignard et al, 1985). *Callosobruchus maculatus* females can lay eggs on pods at various stages of maturation and their larvae are capable of developing on dry grains and on developing grains (Huignard et al, 1985). It has been reported that the bruchid weevils can destroy up to 87% of untreated cowpeas in a period of 9 months (Southgate, 1958).

C. maculatus is a serious pest of cowpea and mungbean in Sri Lanka, along with *C. chinensis* (Dharmasena and Subasinghe, 1986). The use of methyl pirimiphos (Actellic) either as a 0.25% solution sprayed on to the storage bags containing seeds meant for consumption or as a 2% dust where seeds are used as planting material, is the current method of control recommended by the Sri Lanka Department of Agriculture (Anon, 1986). However, indiscrimate use of pesticides by the farmers and traders to keep the pest population under control has given rise to a situation where there is an urgent need to develop economically and ecologically safer and more sound pest control techniques which could be used both by farmers and traders.

The effectiveness of vegetable oils in controlling bruchid infestations is well docu-

K. Fujii et al. (eds.), Bruchids and Legumes: Economics, Ecology and Coevolution, 85–90.

mented. Singh et al (1978) reported ovicidal action of peanut oil against *C. maculatus*. Messina and Renwick (1983) reported the effectiveness of three vegetable oils, mineral oil and polyethylene glycol in the protection of cowpea from *C. maculatus*. However, information on the use of oils from other botanical sources is limiting. Therefore the objective of the present study is to evaluate the efficacy of five botanicals available in peasant holdings in Sri Lanka as surface protectants against the cowpea bruchid, *C. maculatus*.

2.   Materials and Methods

This study was carried out at the Entomology laboratories of the Department of Agronomy, University of Ruhuna, Mapalana, Kamburupitiya, Sri Lanka during 1986-87. The plant materials, leaves of *Piper nigrum* (L.), *Annona reticulata* (L.), *Dillenia retusa* (L.), *Ocimum sanctum* (L.), and peel of *Citrus crematifolia* (Lush.) were obtained from plantations surrounding the University farm. They were dried in a hot air drier at 50-55°C and powdered to 60 mesh in a Raymonds Hammer mill.

The mungbean variety MI3 obtained from Regional Research Station at Mahaaillup-pallama was cleaned and disinfested by keeping the seed material inside a refrigerator at 0°C for 14 days. The disinfested samples were brought to an equilibrium moisture of 12% by conditioning at 70% R.H. prior to use for experimental purposes.

Callosobruchus maculatus adults were obtained from laboratory cultures maintained on mungbean at 26 ± 2°C and 60-80% R.H. The experiments were also conducted at the above temperatures and relative humidity.

Seed samples of mungbean were treated with varying quantities of each powder to give doses of 0.10, 0.20, and 0.40 g/50 seeds in order to evaluate the ovicidal activity of the different powders. The coated seeds of each treatment were divided into 3 replicates and kept in 170 ml bottles. Each control also had 3 replicates and there were separate controls for each powder. Five pairs of *C. maculatus* adults (0-24 hrs old) were placed in each replicate and covered with muslin cloth. Observations on the adult mortality were recorded periodically for seven days after which the adults were discarded. After 14 days, total number of hatched or unhatched eggs were counted. Eggs that turned white and opaque were treated as hatched, and emerging adults were counted every day. For each treatment the number of eggs counted is represented as a percentage of the control.

2.1. BIOASSAY WITH POWDERED *P. NIGRUM* FRUITS

Different weights of (1,2,4, and 10g) of *P. nigrum* powder obtained by preparation of freshly picked ripe fruits were added to 100 seeds of mungbean var. MI3 in 50 ml conical flasks. These gave concentrations of 5.3, 10.5, 21.0, and 42.0% respectively calculated from the formula

$$\text{Concentration} = \frac{\text{Weight of powder x 100}}{\text{Weight of powder + Weight of cowpea}}$$

The control treatment did not have *P. nigrum* powder. Newly emerged *C. maculatus* (5 pairs) were added to each flask. These were plugged with cotton wool and kept at ambient temperature and humidity. Each treatment was replicated three times. Dead insects were replaced daily for the first 5 days. Egg counts were made at 10 days after treatment (DAT) and all the adults, whether dead or alive, were removed. At 30 DAT, the emerged adults were counted. At the end of 60 DAT, the eggs and adults were counted and the seeds were dissected.

3. Results

3.1. EFFECT OF 5 DIFFERENT POWDERS ON ADULT MORTALITY AND OVIPOSITION

The treatments with *P. nigrum* leaves at a concentration of 0.10 g/50 seeds caused significant mortality of *C. maculatus* in comparison to treatment with either fine sand or the control (Table 1). The leaves of *D. retusa* also gave significant mortality of *C. maculatus* adults at a concentration of 0.10 and 0.20/50 seeds.

The leaves of *A. reticulata* and *O. sanctum* gave 40% mortality at 0.10 concentration, but the percentage mortality was low for the other two concentrations tested.

The peel of *C. crematofolia* significantly reduced percentage egg hatch at 0.20 g/50 seeds (Table 2). However varying the concentration had no significant effect on egg hatch in the case of *A. reticulata* and *P. nigrum*. Powder of *A. reticulata* at all three concentrations significantly reduced the expected adult emergence. Microscopic examination of the unhatched eggs revealed that the majority of eggs were killed at the very early stages of their embryonic development.

Table 1. Effect of five different botanicals on adult survival of *Callosobruchus maculatus*

Name of the botanical and its taxonomic family	Part of the plant used	Concentration g/50 seeds	% Adult mortality at 7 days[1]		
			Treated with botanical	Treated with fine sand	Control
Piper nigrum (L.) Piperaceae	leaf	0.10	53.3a	30	20
		0.20	46.6a		
		0.40	30.0c		
Citrus crematofolia (Lush.) Rutaceae	peel of fruit	0.10	30.0c	30	10
		0.20	33.3c		
		0.40	33.3c		
Annona reticulata (L.) Annonaceae	leaf	0.10	40.0b	30	40
		0.20	26.6c		
		0.40	26.6c		
Dillenia retusa (L.) Dilleniaceae	leaf	0.10	46.6a	20	30
		0.20	46.6a		
		0.40	33.3c		
Ocimum sanctum (L.) Labiatae	leaf	0.10	40.0b	20	20
		0.20	26.0c		
		0.40	30.0c		

[1] Means followed by same letter in each column are not significantly different in DMRT.

Table 2. Effect of five different botanicals on percentage egg hatch and emergence of *Callosobruchus maculatus*

Name of the botanical and its taxonomic family	Part of the plant used	Dosage in g/50 seeds[2]			
		control[1]	0.10	0.20	0.40
Piper nigrum	leaf	94.04a (100.00)	17.50c (41.0)	21.67c (30.75)	28.50c (31.60)
Citrus crematofolia	peel	98.18a (98.78)	34.68d (91.00)	3.69d (43.3)	5.75d (45.17)
Annona reticulata	leaf	96.94a (100.00)	13.50c (11.0)	17.61c (13.0)	45.67c (14.0)
Dillenia retusa	leaf	94.15a (92.75)	25.61c (40.0)	28.7c (45.0)	6.17b (68.0)
Ocimum sanctum	leaf	96.79a (91.80)	90.0a (92.0)	67.1b (78.0)	80.7a (80.0)

[1] Figures in parentheses represent adult emergence calculated as percentage of hatched eggs.
[2] Means followed by same letter in horizontal rows do not differ significantly (P ≥ 0.05).

3.2. BIOASSAY WITH POWDERED *P. NIGRUM*

Powder of *P. nigrum* significantly reduced oviposition and adult emergence in *C. maculatus* (Table 3). At 10 DAT, no eggs were laid on seeds treated at 42%, while only an average of 43 eggs were laid on the same amounts of seeds treated at 5.3%. However at 60 DAT different observations were noted; each seed in the control had a mean of 18.6 eggs. No adult emergence was observed in the seeds treated at 42% concentration while up to 54 and 230 adults were obtained in the controls at 10 and 60 DAT respec-

Table 3. Oviposition and adult emergence of *Callosobruchus maculatus* on cowpea treated with powder of *Piper nigrum*

Concentration of *P. nigrum* W/W %	Eggs[1]		Adult emergence[1]	
	10 DAT	60 DAT	10 DAT	60 DAT
5.3	4.3b	4.9b	5.1b	5.3b
10.5	3.1b	3.7b	4.9b	4.7b
21.0	3.9b	3.9b	2.1b	2.6b
42.0	0.0c	0.0c	0.0c	0.0c
control	72.3a	18.63a	54.1a	230.2a

[1] Means followed by same letter are not significantly different at 5% level by Duncans Multiple Range Test.

tively. When the seeds were dissected at the end of 2 months the seeds had a moldy infection inside the holes.

4. Discussion

Treatment with *P. nigrum* leaves caused significant adult mortality of *C. maculatus*. This could be attributed to contact toxicity with the bruchid adults. Ivbijaro and Agbaje (1986) also reported that powder of *P. guineense* protected cowpea seeds from *C. maculatus*.

The peel of *C. crematofolia* powder significantly reduced percentage egg hatch at 0.20g/50 seeds. Taylor (1975) was the first to report that orange peel powder significantly depressed oviposition and progeny emergence of *C. maculatus*.

The results of the bioassay study showed that powder of *P. nigrum* fruits protected greengram seeds from *C. maculatus*. Applied at a level of 5.3%, the powder significantly reduced oviposition, and at 42% completely suppressed it. A similar observation was made by Olaifa and Erhun (1988) with *P. guineense* fruits and oils. Olaifa and Erhun (1988) reported that both oils and powder of *P. guineense* can be recommended for *C. maculatus* control. Woode et al (1984) reported that *P. guineense* contains naturally occurring piperne type alkaloids such as wisanine and okolasin. Therefore processing *P. guineense* in powder form by the simple method of drying and grinding could be practiced more easily at the household level. There is no doubt that *P. nigrum* will be readily accepted as a pesticide to preserve greengram from seed beetle attack. Unlike other botanicals tested in this study, *P. nigrum* is traditionally used as an important spice by Sri Lankans, and is found in most of the households as a supporting crop. Therefore the use of local medicinal plants and spice plants such as *P. nigrum* offers a cheap control method for those categories of people who cannot afford the present high cost of synthetic pesticides.

Acknowledgements. The author gratefully acknowledges the technical assistance of F. K. Lal.

REFERENCES

Alzouma, I. and Huignard, J. (1981) Dounes preliminaires sur la biologie et comportement de ponte dans la nature de Bruchidius atrolineatus dans une zone sud sahelienne an Niger, Acta. oecol. 2, 391-400.

Anonymous, (1986) Major crops pests and their control with pesticides, Publication Unit, Education and Training Div., Dept. Agr., Peradeniya, Sri Lanka.

Credland, P. F. (1986) Effect of the host availability on reproductive performance in *Callosobruchus maculatus*, J. stored Prod. Res. 22, 49-54.

Dharmasena, C. M. D. and Subasinghe, S. M. C. (1986) Resistance of Mungbean (*Vigna radiata*) to *Callosobruchus* spp., Trop. Agr. 142, 1-6.

Dobie, P. (1981) Storage, in Pest Control in Tropical Grain Legumes, Centre for Overseas Pest Research, London, pp. 37-45.

Huignard, J., Leroi, B., Alzouma, I., and Germain, J. F. (1985) Oviposition and development of *Bruchidius atrolineatus* (Pic) and *Callosobruchus maculatus* in *Vigna unguiculata* cultures in Niger, Insect Sci. Applic. 6, 691-699.

Ivbijaro, M. F. and Agbaje, M. (1986) Insecticidal activities of *Piper guineense* and *Capsicum* species on the cowpea bruchid *Callosobruchus maculatus*, Insect Sci. Applic. 7, 521-527.

Messina, F. J. and Renwick, J. A. A. (1983) Effectiveness of oils inprotecting stored cowpea from the cowpea weevil, J. Econ. Entomol. 76, 634-637.

Olaifa, J. I. and Erhun, W. O. (1988) Laboratory evaluation of *Piper guineense* for the protection of cowpea against *Callosobruchus maculatus*, Insect Sci. Applic. 9, 55-59.

Singh, S. R., Luse, R. A., Leuschner, K., and Nangju, D. (1978) Groundnut oil treatment for the control of *Callosobruchus maculatus* during cowpea storage, J. stored. Prod. Res. 14, 77-80.

Southgate, B. J. (1958) Systemic importance on species of *Callosobruchus* of economic importance, Bull. Entomol. Res. 49, 591-599.

Taylor, T. A. (1975) Effects of orange and grape fruit peels on *Callosobruchus maculatus* infestation of cowpea, Ghana J. Agric. Sci. 8, 169-172.

Woode, K. A., Philips, F. L., Addae-Mensha, I., Bart, J. C. J., and Chaudhuri, S. (1984) X-ray crystal structure of naturally occurring trans-2-cis-isomer of wisanine, a piperine type alkaloid from *Piper guineense*, J. Nat. Prod. 47, 1024-1027.

INSECTICIDAL ACTIONS OF SEVERAL PLANTS TO *CALLOSOBRUCHUS CHINENSIS* L.

B. MORALLO-REJESUS, H.A. MAINI, K. OHSAWA,
AND I. YAMAMOTO
Dept. of Entomology, College of Agriculture
Univerisity of the Philippines at Los BaÇos
College, Laguna, Philippines and
NODAI Research Institute, Tokyo, Japan

ABSTRACT. Nine plants were evaluated for their insecticidal actions on *Callosobruchus chinensis* L: *Ageratum conyzoides, Blumea balsimifera, Chrysanthemum indicum, Coleus amboinicus, Vitex negundo, Azadirachta indica, Cocos nucifera, Capsicum frutescens* and *Piper nigrum*. By filter paper impregnation method, the oils of the first seven plants exhibited contact toxicity with mortality ranging from 66 to 100% at 100 mg/ml, 48 hrs after exposure. At the same concentration, the oils were more toxic when mixed with the seeds, giving 100% mortality at 24 hours. *C. amboinicus* oil was the most toxic causing 93% mortality at 10 mg/ml within 15 minutes exposure, while the rest of the oils gave mortalities ranging from 43 to 100% 24 hours after exposure. Seed treatment at 5 mg inhibited egg laying by 71 to 100%. Mungbean seed treated with ground *P. nigrum* and *C. frutescens* at 600 ppm was very toxic (91% mortality) after 48 hours exposure and was residually toxic 6 months after treatment to the weevils. *P. nigrum* inhibited the development of F1 progenies. Exposure of the bean weevil to mungbean treated with oils/extracts/powders from the test plants exhibited one or a combination of the following actions: toxicity, repellancy, antioviposition and growth inhibition. GC-mass infrared analysis of the volatile oil from sambong and its fractions showed the presence of sesquiterpenoidal compounds. Purification and identification of the active compounds are in progress.

B. balsimifera, C. indicum, C. amboinicus, C. frutescens and *A. conyzoides* are being reported for the first time as being insecticidal against *C. chinensis*.

1. Introduction

The use of plant materials in the control of storage pests is an ancient measure in many parts of the world. The mixing of the plant parts (leaf, bark, seeds) and vegetable oils were traditionally practiced especially in Asia and Africa. This practice was abandoned with the advent of modern or synthetic insecticides. However, due to the problems encountered with the use of the petroleum-based insecticides, interest in the use of biocides from plants has been revived. This is because insecticides derived from plants like pyrethrum and rotenone have been found to be non-persistent in the environment, of low mammalian toxicity, and relatively safe to other non-target organisms.

This paper discusses the evaluation of the insecticidal actions of seven plant oils and two ground peppers against cowpea weevil, *Callosobruchus chinensis* L., a very destructive insect pest of mungbean [*Vigna radiata* (L.) Wilczek] in the Philippines.

K. Fujii et al. (eds.), Bruchids and Legumes: Economics, Ecology and Coevolution, 91–100.
© 1990 *Kluwer Academic Publishers. Printed in the Netherlands.*

2. Materials and Methods

2.1. TEST PLANTS

The leaves of sambong (*Blumea balsimifera*), oregano (*Coleus amboinicus* Dour), lagundi (*Vitex negundo* L.), manzanilla (*Chrysanthemum indicum* L.) and flowers of bulak-manok (*Ageratum conyzoides* L.) were collected from the botanical garden maintained at the Department of Entomology, UPLB and air-dried for a period of 4-9 weeks. The dried materials were powdered to 40 mesh using a Wiley mill prior to extraction of the oil.

The neem oil was provided by Dr. R. Saxena of the International Rice Research Institute and the coconut oil (Golden Fry) was bought from the market. The black pepper seeds and red pepper fruits obtained from the market were oven dried at 45°C for one week then ground to about a size of 841 μm (10 mesh) using a Wiley Mill.

2.2. REARING OF TEST INSECTS

One hundred pairs of weevils were introduced into each plastic or glass container containing 200 gms of mungbean. The containers were fitted with fine mesh screen covers and stored in a room at 27°C with moisture level of 60-70%. The insects were allowed to oviposit for five days and later discarded. Adult emergence was observed at 20 days after the introduction, and daily, thereafter. The adults were sieved and transferred daily to another container with a few mungbeans before using them for bioassay. One to 2-day old adults were used for all the tests.

2.3. VOLATILE OIL EXTRACTION

One hundred grams of the dried, powdered material were placed in a 2L round-bottom flask. One liter of distilled water was added so that it is approximately 2 cm above the level of the plant material and soaked overnight. The clevenger apparatus was connected and fitted to the flask, and the mixture was gently heated using a heating mantle for 4 hours until the volatile oil has been completely separated from the plant or until such time that the volatile oil no longer collects into the graduated tube of the clevenger apparatus. The collected volatile oil was dehydrated with anhydrous sodium sulfate, stored in the refrigerator and protected from light before use.

2.4. FRACTIONATION AND CHARACTERIZATION OF THE POSSIBLE ACTIVE COMPONENTS

2.4.1. *Sambong.* The fractionation of the oils and bioassay of the fractions were carried out at NODAI Research Institute, Tokyo University of Agriculture. Five hundred and eighty-five (585) mg sample of the volatile oil was column chromatographed using silica gel (Wakogel C-200) as adsorbent and n-hexane-chloroform mixtures as eluant. The column was allowed to run at a flow rate of 5 ml/min and eluates of 100 ml each were collected, after which the column was washed with acetone. Eluates were chromatographed by thin layer chromatography (TLC) using prepared silica gel TLC plates which were first eluted with acetone overnight prior to use. N-hexane and n-hexane-chloroform mixtures were used as the solvent system. Spots were detected by 95% sulfuric acid. Eluates having the same chromatograms were combined and bioassayed. Preparative TLC using prepared silica gel plates washed with acetone prior to use was made on the fraction exhibiting insecticidal action and the same detecting reagent was used to visualize the spots.

2.4.2. *Oregano.* A sample (646 mg) of the volatile oil was column chromatographed using silical gel (Wakogel C-200) as adsorbent and n-hexane, n-hexane-ether mixtures and acetone as eluants. The column was allowed to run at 5 ml/min and eluates of 100 ml each were collected. The eluates were concentrated and analyzed by TLC as described above. The resulting eluates were bioassayed.

2.5. PREPARATION OF SOLUTION FOR BIOASSAY

A stock solution was prepared from the volatile oil by diluting it with acetone. Solutions of lower concentrations (100, 50, 10, 5 and 1 mg/ml) were prepared by diluting the stock solution with acetone prior to testing.

2.6. BIOASSAY OF THE OIL

2.6.1. *Contact Toxicity.* The contact toxicity of the oils were evaluated by filter paper impregnation method (FPIM) and by direct mixing with the grain, known as admixture or protectant treatment.

Whatman #1 (9.0 cm) filter papers were placed on petri dishes. The filter papers were impregnated with the test solution and were allowed to dry at room temperature (RT). Five pairs of one day old adults of *C. chinensis* were introduced into the treated filter paper. Each treatment was replicated three times. Mortality was observed at 24 and 48 hours after treatment.

Mungbean seeds (50 g) inside a small jar were mixed with the oil at 100, 50, 5 mg concentrations. Uniform wetting of the seeds' surface was achieved by shaking the jar vigorously for three to five minutes and letting it stand for an hour. This procedure was repeated to all treatments, including a control, three times. Five pairs of adult weevils were exposed to the treated seeds. Mortalities were noted 24 and 48 hours after treatment.

2.6.2. *Effect on Oviposition.* Seeds were treated with the different oils at 100, 50, and 5 mg concentrations as in the contact toxicity experiment. Five pairs of the 2 day-old adult weevils were exposed to the treated beans for 1-2 days. The number of eggs oviposited on the seeds were counted.

2.6.3. *Effect on Development.* Another batch of mungbean seeds were treated with coconut oil at 1, 5, 10 and 20 ml/kg seed. Wetting was insured by shaking the jar vigorously as described earlier. For each concentration, 600 g of seeds were treated and 200 g of these seeds were placed in each jar (half-gallon) replicated three times.

Twenty pairs of adult weevils were introduced to each jar. Five days after exposure, the adult weevils were removed and the eggs deposited on the surface were counted. After counting, the seeds were returned to their respective jars and kept at room temperature for adult emergence of F_1 progenies.

2.6.4. *Repellancy.* Two concentrations (10, 1 mg) were evaluated for repellancy against one to two-day old weevils using ten cm long glass tubing with both ends open. One end of the tubing was plugged with treated cotton (1 x 1 cm). Five males and five females were introduced individually into glass tubings, with the other end being plugged with untreated cotton after each introduction. Each sex was replicated five times. The insects repelled were counted at 10 minute intervals for 2 hours.

2.7. BIOASSAY OF THE GROUND MATERIALS

2.7.1. *Admixture Toxicity.* Three concentrations (300 ppm, 600 ppm, 1200 ppm) of both the ground black pepper (GBP) and red pepper (GRP) were evaluated against the cowpea bean weevil and the bean weevil (*Acanthoscelides obtectus*). The ground materials were directly mixed with 100 grams of mungbean in a ball jar. In addition to the control treatment a standard treatment (10 ppm malathion mixed with the grains) was included. All treatments were replicated three times.

Ten pairs of weevils were introduced into the jar containing mungbean. Mortality was noted 48 hours after the introduction by sieving the live and dead insects. The latter were discarded while the live insects were returned to the jar. Exactly one week after the second observation was made, however, the live insects were not returned to the bottle but the jars with the mungbeans were kept to determine the F1 progenies. Emergence was noted 28 days after the removal of the parents, and continued for four weeks. From this set-up the effect of the treatments on the development of the insect was determined.

2.7.2. *Residual Toxicity.* The residual toxicity of GBP and GRP were observed by exposing ten pairs of weevils to treated mungbeans at different storage intervals after treatment. The ground materials at 300, 600 and 1200 ppm were mixed with the mungbeans inside a jar and were stored at a room temperature of 27°C. Insects were introduced into three jars each at 2, 4 and 6 months after treatment. A control and a standard (treated with 10 ppm malathion) were included. All treatments were replicated three times.

3. Results

3.1. CONTACT TOXICITY

3.1.1. *Oils.* By filter paper impregnation assay, all the oils at 100 mg/ml gave more than 60% mortality at 24 and 48 hours after exposure except the coconut oil (Table 1). Of these oils, sambong, oregano, lagundi, and neem were the most toxic with mortality ranging from 80-90%. At 24 hours, sambong and oregano already exhibited 80% mortality, while lagundi and neem attained 90 and 80% mortality, respectively, after 48 hours. Coconut oil was the least toxic among the oils.

All oils at 10 mg/ml caused mortality ranging from 52 to 64%, 48 hours after treatment.

The exposure of the weevils to mungbean treated with oils at 100 mg caused mortalities ranging from 73 to 100% (Table 4). Neem and coconut oil were the least toxic showing mortalities of 37 and 23% at 50 mg/ml while the rest of the oils exhibited 100% mortalities 24 hours after exposure to the treated beans.

3.1.2. *Oil Fractions.* The two most toxic oils, sambong and oregano, were fractionated. Five fractions SA, SB, SC, SD and SE weighing 713, 5.0, 2.5, 85.4 and 88.8 mg, respectively, were recovered by GLC from sambong. Among the fractions only SB, SD and SE exhibited positive reactions to 95% H_2SO_4 and SD was repellant to the weevils. Further fractionation of SD resulted in fractions SD-1, SD-2, SD-3 and SD-4 weighing 22.9, 15.3, 24.3 and 11.0 mg, respectively. Only SD was repellant. GC-mass spectral

Table 1. Toxicity[1] (percent mortality) of plant oils against *C. chinensis*[2].

Treatments	Conc. (mg/ml)	Corrected 24 h	Mortality 48 h
Oregano	100	80	100
	10	64	64
	1	48	56
Sambong	100	80	100
	10	30	52
	1	0	18
Lagundi	100	60	90
	10	40	54
	1	0	23
Manzanilla	100	76	76
	10	30	56
	1	6	32
Bulak-manok	100	66	66
	10	38	62
	1	32	50
Neem	100	67	80
	10	30	58
	1	17	43
Coconut	100	40	73
	10	10	55
	1	17	43
Control			

[1] By filter paper dry film method Conc-concentration.
[2] Ten one-day old adult weevils per replicate, replicated three times.

analysis of the fractions revealed the presence of sesquiterpenoidal compounds.

Five fractions (42 mg DA, 420 mg DB, 23 mg OC, 7 mg OD, and 15 mg DE) were recovered when oregano oil was fractionated by TLC and GLC. Except for OD, all fractions were screened for contact toxicity by the filter paper impregnation method. Paralysis and convulsion were observed on insects exposed to the paper treated with 10 mg of OB and 100% mortality was noted after 24 hours exposure. Further fractionation and bioassay studies will be done on this toxic fraction OB.

3.1.3. *Ground Peppers.* The black pepper was more toxic to both *C. chinensis* and *A. obtectus* than the red pepper (Table 2). However, the former was more sensitive to the ground materials than the latter. GBP at 300 ppm was two times more toxic than 10 ppm malathion.

Table 2. Toxicity of ground black (GBP) and red pepper (GRP) mixed with mungbean seeds after 48 hours and its effect on the rate of development and F1 progenies.

Dosage (ppm)		Percent Mortality		Number of Adult Emergence[1]		Developmental Period[1] (Days)	
		C. chinensis	A. obtectus	C. chinensis	A. obtectus	C. chinensis	A. obtectus
GBP	300	78 ab	75 b	342 bc	17 cd	26	26
	600	92 a	78 b	288 c	12 cd	25	24
	1200	93 a	98 a	152 d	8 d	25	26
GRP	300	42 c	7 cd	444 ab	24 bc	25	26
	600	55 bc	8 cd	437 ab	14 cd	24	26
	1200	62 bc	13 c	182 d	12 cd	25	26
Malathion (10 ppm)		31 c	8 cd	382 abc	34 ab	26	24
Control		7 d	3 d	472 a	46 a	23	23

[1] Means without letter or followed by the same letter(s) on the same column are not significantly different at 5% level of probability (DMRT).

3.2. RESIDUAL TOXICITY

Mungbean treated with ground black and red pepper were significantly toxic to the adult weevils even after 6 months of storage (Table 3). There was even more residual toxicity than with malathion after 2 to 6 months storage.

Table 3. Residual toxicity (% mortality at 2-5 days exposure)[1] to C. chinensis[2] of treated mungbean stored for 2, 4 and 6 months.

Dosage (ppm)		Storage Time					
		2 Months		4 Months		6 Months	
		2	5	2	5	2	5
GBP	300	45 cd[2]	100 a	48 d	100 a	68 abc	100 a
	600	65 bc	100 a	73 ab	100 a	78 abc	100 a
	1200	95 a	100 a	77 a	100 a	80 a	100 a
GRP	300	42 d	100 a	62 c	100 a	33 bcd	100 a
	600	48 cd	100 a	65 bc	100 a	33 cd	100 a
	1200	53 bcd	100 a	68 abc	100 a	72 abc	100 a
Malathion(10 ppm)		73 b	100 a	17 e	76 b	55 d	33 b
Control		3 e	17 b	10 f	30 c	3 d	17 c

[1] Means without letter or followed by the same letter(s) on the same column are not significantly different at 5% level of probability (DMRT)
[2] Ten pairs of one-day old weevils per replicate, replicated 3 times.

Table 4. The mortality and oviposition of *C. chinensis*[1] exposed to mungbeans treated with plant oils.

Treatments	Conc. (mg)	Percent Mortality		No. of Egg Sites	
		24 hrs	48 hrs	24 hrs	48 hrs
Oregano	100	100	100	0	0
	50	100	100	0	0
	5	97	97	5	11
Sambong	100	100	100	0	0
	50	100	100	0	0
	5	43	73	10	19
Lagundi	100	100	100	0	0
	50	100	100	0	0
	5	70	80	1	2
Manzanilla	100	100	100	0	0
	50	100	100	0	0
	5	100	100	0	0
Bulak-manok	100	100	100	0	0
	50	100	100	0	0
	5	97	100	0	0
Neem	100	80	80	1	1
	50	37	37	4	0
	5	13	23	12	0
Coconut	100	73	73	4	4
	50	23	23	2	2
	5	0	10	1	2
Control		10	5	55	65

[1] Replicated 3 times using five pairs of 2 day old adults per replicate.

3.3. EFFECT ON OVIPOSITION AND DEVELOPMENT

3.3.1. *Oils.* All the oils tested prevented the adult weevil from laying eggs on the treated mungbeans (Table 4). At 5 mg, egg laying was reduced by 91 to 100%. Further tests with coconut oils showed similar results (Table 5). The number of eggs laid decreased with higher concentration. At 5 ppm, a reduction of 89% was already observed. No F1 adults emerged from beans treated with 5 to 20 ppm oil. At 1 ppm only 12.6% of the eggs laid reached the adult stage.

3.3.2. *Ground Peppers.* The two different peppers were not only toxic but affected the development of the larva as shown by the reduction in adult emergence of F1 progenies (Table 2). At high concentration (1200 ppm) both peppers drastically reduced the adult emergence but at lower concentrations of 300 and 600 ppm, the black pepper

98

Table 5. Effect of treating mungbean with coconut oil on the oviposition of *C. chinensis* and F_1 adult emergence.

Treatment (ppm)	Mean Number[1] Eggs Laid	Adult Emergence
20	2	0
10	5	0
5	31	0
1	84	11
Control	297	185

[1] From three replicates with ten pairs of one to 2-day old weevils per replicate.

was more potent than the red pepper. Neither of the materials affected the rate of development.

3.3.3. *Repellancy Effect.* Seventy to 80 percent of the weevils were repelled by the treated mungbean (Table 6). The oils were considerably more repellant at 10 mg/ml than at one mg/ml.

Table 6. Repellancy of plant oils against *C. chinensis*[1].

Treatments	Conc. (mg)	Percent Repelled 10 min.	1 h	2 h
Oregano	10	90	90	80
	1	80	70	80
Sambong	10	100	80	80
	1	100	70	60
Lagundi	10	40	60	70
	1	30	30	10
Manzanilla	10	60	40	30
	1	20	20	0
Bulak-manok	10	40	50	30
	1	20	20	0
Coconut	10	10	40	70
	1	30	10	20
Control		0	0	0

[1] Replicated 3 times with five pairs of adult weevils per replicate.

4. Discussion

Four of the nine plants evaluated in the present study, namely: *V. negundo*, *A. indica*, *C. nucifera* and *P. nigrum* are among the 48 plant species that have been found to be toxic to *C. chinensis* as compiled by Grainge and Ahmed (1987). Golob and Webley (1980) also reported these plants to be traditionally used as protectants of stored products in Africa and Asia.

Our findings support the reports that vegetable oils protect stored legumes from *Callosobruchus* infestation, prevents egg laying and inhibits the multiplication of this weevil. Sangappa (1977) reported that neem oil applied at 0.75 to 1 (% by weight) protected redgram seeds for 161 days from weevil infestation while Pandey et al. (1981) reported that coconut oil and cotton seed oil prevent egg laying and act as ovicides by interfering with the late stage of embryogenesis.

The present study showed that few eggs were laid on any of the oil-treated mungbeans. Of the eggs laid in mungbean treated with coconut oil at 1 ppm, only 12.6% reached the adult stage. The oils prevented the hatching of the eggs which Tikku et al. (1981) attributed to the interference of water balance which is shown by the undifferentiated and shrunken eggs. The residual toxicity of the black pepper also confirms the findings of Javier and Morallo-Rejesus (1981) on corn against *Sitophilus zeamais*.

B. balsimifera, *C. amboinicus*, *C. indicum* and *A. conyzoides* and ground fruits of *C. frutescens* are being reported for the first time to be toxic to the cowpea weevil, inhibit egg laying and prevent development.

5. Conclusions

The seven plant oils and two ground peppers evaluated were found to exhibit any one or a combination of the following insecticidal actions: toxicity, repellancy, growth inhibition and antioviposition. Of these plants, five (*B. balsimifera*, *C. incidum*, *C. amboinicus*, *C. frutescens* and *A. conyzoides*) are being reported for the first time to be insecticidal against *C. chinensis* and have been added to the 48 species previously reported by Grainge and Ahmed (1988). The potential of using these materials in the warehouse or storage facilities needs to be further evaluated. The isolation and identification of the active principle(s) is necessary for the determination of mammalian toxicity of these materials, the effects on the quality of the commodities to be treated with it, and for future synthesis.

REFERENCES

Grainge, M. and Ahmed, S. (1987) Handbook of Plants with Pest Control Properties, John Wiley & Sons Publ., N.Y.

Golob, B. and Wohley, D. J. (1980) The use of plants and minerals as traditional protectants of stored products, Bull. Stored Prod. Inf.

Javier, P. and Morallo-Rejesus, B. (1986) Insecticidal activity of black pepper (*Piper nigrum* L.) extracts, Philipp. Entomol. 6, 517-525.

Pandey, G. P., Doharey, R. B., Varma, U. K. (1981) Efficacy of some vegetable oils for protecting greengram against the attack of *Callosobruchus maculatus* (F.), Indian J. Agric. Sci. 51(12), 910-912.

Sangappa, S. K. (1977) Effectiveness of oils as surface protectants against the bruchid, *C. chinensis* (L.) infestation on redgram, Mysore J. Agric. Sci. 11, 391-97.

Tikku, K., Koul, I., Saxena, B. P. (1981) Possible mode of action of vegetative oils to protect *Phaseo-*

lus aureus Roxb. from bruchid attack, Science and Culture 47(3), 103-105.

INSECTICIDAL CONTROL OF COWPEA WEEVIL, *CALLOSOBRUCHUS MACULATUS* F., A PEST OF MUNGBEAN

P. VISARATHANONTH, M. KHUMLEKASING,
AND C. SUKPRAKARN
Stored Product Insect Research Group,
Entomology and Zoology Division,
Department of Agriculture,
Bangkok 10900, Thailand

ABSTRACT. The efficacy of pirimiphos-methyl, chlorpyrifos-methyl, fenitrothion and methacrifos on cowpea weevil, *Callosobruchus maculatus* F. were evaluated during February 1986 to April 1987 at Chainat Field Crops Research Center by admixing to mungbean grain (11.26% M.C.) at the rate of 5 and 10 ppm. The percentage of infested grain, number of insects present and insecticide residues were recorded every 4 weeks for 48 weeks. The results showed that pirimiphos-methyl at 10 ppm was the most effective in controlling the insect up to 36 weeks while chlorpyrifos-methyl, fenitrothion and methacrifos at both rates and 5 ppm pirimiphos-methyl were effective from 20 to 28 weeks.

The residues of insecticide treated mungbean were analysed and indicated that pirimiphos-methyl at 5 and 10 ppm were below the Maximum Residues Limit (MRL) within 36 and 48 weeks respectively. Chlorpyrifos-methyl and fenitrothion at 5 and 10 ppm were 12 and 20 weeks. However, methacrifos at both rates decreased rapidly, the residues were found to be below MRL within 1 day after treatment. The details of this experiment will be discussed.

1. Introduction

Cowpea weevil (*Callosobruchus maculatus* F.) is the most destructive pest of stored mungbean in Thailand. This weevil can also attack mungbean in the field before harvest. It has been reported to infest a wide variety of legumes in this country such as mungbean (*Vigna radiata* (L.) Wilezek), black seeded race (*Vigna unguiculata* (L.) Walp.), pigeon pea (*Cajanus cajan* (L.) Millsp.), and winged bean (*Psophocarpus tetragonolobus* (L.) DC.). The life cycle of the cowpea weevil is 25-27 days with several generations in a year.

Prevention of insects damaging stored mungbean is dependent on the intended use of the seed, quantity and duration of storage. For small amounts of seed and short time storage, the application of vegetable oils such as palm oil, bran oil, olive oil, and coconut oil admixing seed at the rates of 5 and 10 ml per 1 kg are effective for controlling the weevil for 6 months. Plant products such as the stem of the milkbush (*Euphobia tirucalli* L.) and pepper (*Piper nigrum* L.) at the respective rates of 40 and 20 g per 1 kg of seeds for 6 months protection are also used. Rice husk, saw dust, rice husk ash, and wood ash are used as a top layer to the seed at the rate 20 g per 1 kg. However, these methods are not suitable for a large amount of seeds and for a long period of storage. Therefore, some specific insecticides were evaluated in their ability to protect mungbean for long time storage. Insecticides which have low mammalian toxicity and are effective

101

against the insect for long periods of time are desirable for this purpose. This paper describes the effectiveness of four insecticides in protecting stored mungbean from cowpea weevil and evaluates residues remaining on the grain after treatment.

2. Materials and Methods

This experiment was conducted at Chainat Field Crops Research Center (194 km north of Bangkok) during February 1986 to April 1987 with an average temperature 28.35°C and 66.99% relative humidity.

The insecticides tested in the present investigation were pirimiphos-methyl, chlorpyrifos-methyl, fenitrothion and methacrifos. They were tested at the rates of 5 and 10 ppm and each was separately admixed to 20 kg of mungbean grains. Distilled water was applied to the grain as a control standard for comparison. The treated grains were placed in cloth bags; each treatment having three replications.

Two hundred gram samples were taken from each bag every 4 weeks for 48 weeks in order to examine the insect population, and insect damage was evaluated by 1000 grain counts. For residue analysis the samples were ground in a Wiley Mill and 25 g extracted with 100 ml ethyl acetate and blended for 3 - 5 minutes. The liquid was decanted through a funnel and the filtrate concentrated in a rotating cooperator at a bath temperature of 45°C. A gas-liquid chromatograph (Varian-Vista: 6,000) equipped with flame photometric detector (FPD) was used for analysis. The chromatographic conditions were gas column id 4 mm containing 3% OV-101 on 100-120 mesh WHR. Temperatures used were: injection port 220°C, column 200°C, detector 300°C.

3. Results

The average percentage of mungbean damage and number of *Callosobruchus maculatus* F. infestation in every 4 weeks are given in Tables 1 and 2. The 10 ppm pirimiphos-

Table 1 Average percentage of mungbean damage caused by *Callosobruchus maculatus* F.

Insecticides	Dosage (ppm)	% mungbean damage at indicated week								
		16	20	24	28	32	36	40	44	48
Pirimiphos-methyl	5	0	0.37ab	2.27ab	10.87ab	22.17ab	47.23ab	61.53a	71.57	80.33
	10	0	0.07c	0.27b	0.10c	0.40c	3.70c	3.75b	21.63	39.83
Chlorpyrifos-methyl	5	0	0.70a	14.00a	27.07a	33.30a	36.43ab	54.87a	70.79	84.33
	10	0	0.23ab	2.23ab	14.00ab	47.00a	52.13ab	65.67a	79.80	84.27
Fenitrothion	5	0	0.33ab	4.73ab	28.97a	30.80a	69.87a	68.60a	78.66	81.60
	10	0	0.33ab	0.50b	9.73ab	9.70bc	27.67bc	73.57a	71.03	84.60
Methacrifos	5	0	0.07c	1.30ab	8.45ab	19.73ab	43.77ab	51.07a	60.43	62.67
	10	0	0.23ab	1.50ab	13.77ab	33.77ab	40.93ab	41.70a	48.87	54.73
Untreated	0	0.27	0.73a	9.47a	20.73ab	26.87ab	54.00ab	66.63a	75.87	89.07

Table 2. Average number of *Callosobruchus maculatus* F. infested mungbean

Insecticides	Dosage (ppm)	number of insects at indicated week								
		16	20	24	28	32	36	40	44	48
Pirimiphos-	5	0	10	45	365 ab	603 ab	1325 a	1160 ab	960	1272
methyl	10	1	0	1	5 c	9 c	57 bc	436 c	582	845
Chlorpyrifos-	5	1	9	80	548 ab	721 ab	1385 a	1042 abc	1136	1009
methyl	10	1	5	53	476 ab	1354 a	1004 ab	911 abc	947	1108
Fenitrothion	5	1	4	71	906 a	1419 a	2979 a	2187 a	2422	2067
	10	0	1	6	72 bc	479 abc	420 ab	1181 ab	1045	1279
Methacrifos	5	0	3	24	63 bc	259 abc	407 ab	575 bc	641	921
	10	0	2	20	72 bc	242 abc	528 ab	339 c	437	766
Untreated	0	4	15	109	526 ab	809 a	1319 a	1501 ab	1254	1873

methyl was the most effective for controlling cowpea weevil for 36 weeks with some damage and a small number of insects present. Methacrifos at both rates and 10 ppm of fenitrothion were effective against cowpea weevil for 28 weeks. Both pirimiphos-methyl and fenitrothion 5 ppm and chlorpyrifos-methyl 10 ppm could afford protection from the insect for 24 weeks. Chlorpyrifos-methyl 5 ppm was effective for 20 weeks.

The chemical analysis of pirimiphos-methyl, chlorpyrifos-methyl, fenitrothion, and methacrifos residues on the mungbean samples are presented in Table 3. The results obtained after admixing the four insecticides on the mungbean at both rates are as follows (5 and 10 ppm); pirimiphos-methyl was persistent on grain and the remaining residues after 48 weeks were 0.04 ppm at the lower level of application (5 ppm) and 0.25 ppm at the high level (10 ppm). Chlorpyrifos-methyl and fenitrothion left residues

Table 3 Residue of insecticides on stored mungbean during 48 weeks (ppm)

Insecticide	Dosage (ppm)	Residue of insecticide (ppm) at indicated week												
		0	4	8	12	16	20	24	28	32	36	40	44	48
Pirimiphos-	5	4.86	4.50	2.75	2.46	1.27	1.26	1.28	0.96	0.78	0.47	0.39	0.28	0.04
methyl	10	9.60	9.71	8.11	6.59	3.63	3.31	3.44	3.15	2.90	1.34	1.26	0.53	0.25
Chlorpyrifos-	5	1.76	1.02	0.15	0.07	0.04	0.03	0.03	0.03	0	0	0	0	0
methyl	10	3.06	1.64	0.30	0.10	0.05	0.04	0.04	0.02	0	0	0	0	0
Fenitrothion	5	2.48	0.54	0.24	0.07	0.03	0.03	0.02	0.01	0	0	0	0	0
	10	5.79	1.84	0.29	0.23	0.21	0.07	0.05	0.04	0	0	0	0	0
Methacrifos	5	0.23	0.05	0.01	0	0	0	0	0	0	0	0	0	0
	10	0.87	0.12	0.02	0.01	0	0	0	0	0	0	0	0	0

of 0.02 - 0.04 ppm at the 28th week after treatment. Methacrifos was the least persistent and at 5 ppm after 8 weeks only 0.01 ppm was recorded. The residue of 10 ppm of methacrifos was 0.01 ppm at 12 weeks.

The level of protectant residues must be lower than the maximum residue limits (MRL) recommended by the Codex Alimentarius Commission of the United Nations (Anon, 1978). The residues of pirimiphos-methyl at the rate of 10 ppm was lower than MRL within 48 weeks. Pirimiphos-methyl 5 ppm and fenitrothion 10 ppm were 36 and 20 weeks respectively. Chlorpyrifos-methyl at both rates and fenitrothion 10 ppm were below the limit at 12 weeks. The residues for methacrifos at both rates were below MRL within 1 day after treatment.

4. Discussion

This study indicated that only pirimiphos-methyl at 10 ppm could protect the weevil for 36 weeks and its residue was higher than the MRL. Appropriate grain protectants should be affective for a long time in controlling insect pests but with lower residue levels. Fenitrothion and methacrifos were suitable for short time storage. Higher concentrations of these two insecticides should be adjusted for the length of storage and storage conditions. In the Sahelian countries, pirimiphos-methyl treated stored cowpea at the rate of 12.5 ppm has been used for controlling *Callosobruchus maculatus* for 6 months (Pierrard, 1986).

Storage conditions such as relative humidity, temperature, moisture content and size of grain have been reported as factors influencing the biological efficacy of residue insecticides (Pradhan, 1949; Strong and Sbur, 1960; Kadoum and La Hue, 1979; Water, 1959; Weaving, 1975). Pirimiphos-methyl and deltametrin were used to treat wheat against *Sitophilus oryzae* stored at three different temperatures. The lower rate of population increase for pirimiphos-methyl and deltamethrin were respectively achieved at 32°C and 21°C (Longstaff and Desmarchelier, 1983). Fenitrothion was the most adversely affected by moisture content and chlorpyrifos-methyl was the least among the 5 organophosphorus compounds when applied to control *Tribolium castaneum* (Samson et al., 1988).

REFERENCES

Anon. (1978) Guide to Codex maximum residue limits, F.A.O. of U.N. W.H.O., Rome, 209 p.

Kadoum, A. M. and La Hue, D. W. (1979) Degradation of malathion on wheat and corn of various moisture contents, J. Econ. Entomol. 72, 228-229.

Longstaff, B. C. and Desmarchelier, J. M. (1983) The effect of the temperature-toxicity relationships of certain pesticides upon the population growth of *Sitophilus oryzae* (L.) (Coleoptera: Curcul ionidae), J. stored Prod. Res. 19, 15-29.

Pierrard, G. (1986) Control of the cowpea weevil *Callosobruchus maculatus*, at the farmer level in Senegal, Tropical Pest Management 32, 197-200.

Pradhan, S. (1949) Studies on the toxicity of insecticide films III - Effect of relative humidity on the toxicity of films, Bull. Entomol. Res. 40, 431-444.

Samson, P. R., Parker, R. J., and Jones, A. L. (1988) Comparative effect of grain moisture on the biological activity of protectants on stored corn, J. Econ. Entomol. 81, 949-954.

Strong, R. G. and Sbur, D. E. (1960) Influence of grain moisture and storage temperature on the effectiveness of malathion as a grain protectant, J. Econ. Entomol. 53, 341-349.

Watters, F. L. (1959) Effect of grain moisture content on residual toxicity and repellency of malathion, J. Econ. Entomol. 52, 131-134.

Weaving, A. J. A. (1975) Grain protectants for use under tribal storage conditions in Rhodesia-l, Comparative toxicities of some insecticides on maize and sorghum, J. stored Prod. Res. 11, 65-70.

PEA WEEVIL (BRUCHUS PISORUM L.) AND CROP LOSS - IMPLICATIONS FOR MANAGEMENT

A.M. SMITH
Department of Agriculture and Rural Affairs
Plant Research Institute
Swan Street, Burnley 3121
Victoria, Australia

ABSTRACT. Development of pea weevil and weight loss of infested seed of field peas were examined in relation to damage thresholds for spraying adults and timing of harvest. Dissection cf medium-sized seed revealed the mean weight loss was 1.2, 9.0 and 26.2% of dry weight equivalent seed for 3rd and 4th instar larvae, and pupae respectively. Between 1986 and 1988, infestation levels and age structure of populations were studied in the field. At the earliest possible harvest date (EPHD), 3rd and 4th instar larvae were predominant. The % weight loss was usually below 4%, which is the loss equivalent to the 'break-even' cost for spraying adults with insecticide. However, this level was often exceeded when harvest was 1-3 weeks after EPHD. The % splits in harvested peas increased with infestation level and larval growth. Weight loss could have been minimised, and grain quality maximised, if the crops were harvested at the EPHD. The risk of not adopting a border spray strategy for adults is that unforeseen events may delay harvest and fumigation.

1. Introduction

Pea weevil (*B. pisorum*) is the primary pest of field peas (*Pisum sativum* L.) in scuthern Australia and it is well established in the major pea growing districts (Smith et al., 1987). Attempts to control pea weevil usually involve two insecticide treatments: one applied to flowering crops to kill adults before they lay eggs on pods, and a second for fumigation of infested peas in storage. The aim is to minimise the weight loss, and reduction in quality and germination, of seed which occurs when larvae feed (Skaife, 1918; Brindley and Hinman, 1937). Field peas are primarily exported as dry stock feed and shipments are rejected if they contain live pea weevil. Therefore, fumigation of infested seed is commonplace but the value of routine application of insecticides in flowering crops is being questioned. How much damage is caused by feeding larvae? At what level of damage is spraying the adults warranted? Can efficient use of cultural control practices, such as early harvest, reduce the need for sprays?

Brindley and Hinman (1937) found that germination of infested seed was 98% just after harvest but decreased rapidly afterwards. When seeds from which adults had emerged were planted, only 6% germinated (Skaife, 1918). This problem could be overcome by harvesting the seed for future crops from the centre of the paddock where pea weevil infestations are usually very low (Smith et al., 1987). Problems associated with weight and quality losses are more difficult to solve.

Estimates of maximum weight loss per seed have varied from 25-30% (Reichart, 1964) to 43% (Yao Kang et al., 1966). Weight loss was correlated with insect develop-

105

K. Fujii et al. (eds.), Bruchids and Legumes: Economics, Ecology and Coevolution, 105–114.
© 1990 *Kluwer Academic Publishers. Printed in the Netherlands.*

ment and the greater part of the damage was accomplished by fourth instar larvae (Brindley and Hinman, 1937). However, the exact relationship between weight loss of pea seed and insect development was not characterised fully in these studies. A large proportion of weight loss can occur in the stored seed after harvest (Brindley and Hinman, 1937; Smith et al., 1982). The effects of the harvesting process on yield and quality of pea weevil infested seed have not been examined, except for a study of seed threshing and cleaning methods by Smith et al. (1982).

In this study, the following relationships are investigated: (a) pea weevil development and weight loss of pea seed; (b) effect of the harvesting process on yield loss and seed quality; and (c) influence of harvest and fumigation times on estimates of yield loss in different field populations. Economic considerations are discussed in relation to the costs of insecticidal control of adult populations.

2. Materials and Methods

2.1. MONITORING LARVAL GROWTH AND % SEED INFESTATION

To monitor the age composition and infestation levels of pea weevil, plots of field peas (cv Dun) were sampled at Dooen (36°19' S, 145°41' E) in north-western Victoria from late-November until late-January in 1987-88 and 1988-89. Three plots (sowing dates: 10 May, 2 June, and 12 July) in 1987 and one plot in 1988 (sowing date: 29 June) were sampled every 4-8 d by randomly collecting about 20 vines along the edge of each plot. Pea pods were stripped from vines and stored at -5°C for later dissection. Seeds were examined for larval entrance marks - small, usually discoloured, scars where newly hatched larvae had entered the young seed. Larvae were dissected out and classified to instar according to maximum width of the head capsule (Brindley, 1933). If a larva or pupa was dry and shrivelled, it was considered to have died before being frozen.

2.2. RELATIONSHIP BETWEEN WEIGHT LOSS AND STAGE OF INSECT

Pea weevil infested seed was collected from the edge of the 1987 June-sown plot on the 25 November 1987 by randomly sampling about 200 vines. Pea seeds were separated from the pods, sorted into pea weevil infested and non-infested seed, and separated further into 3 size classes using a 7.5 mm square hole and a 6.35 mm round hole sieve. For convenience, the 3 size classes of <6.35 mm, between 6.35 and 7.5 mm and >7.5 mm were called 'small', 'medium' and 'large' pea seed respectively. Only those infested seeds with one entrance mark were used in the following analyses.

Seed was held in the laboratory at 25 ± 1.0°C and subsamples of about 40 seeds from each size class of infested seed were dissected after 10, 18, 23 and 45 d until only the adult stage was detected. If a seed had a "window" - the area of thin integument through which the adult emerges - and it contained a 4th instar larva, the pea weevil was deemed to be a 'prepupa'. Since the prepupae do not feed and weight loss is maximised, they were not distinguished from the pupae. Weights of the infested seed and pea weevil were taken separately.

Subsamples of approximately 100 non-infested seeds from each size class were also weighed on each day of the dissections. The last subsamples were placed at 50°C and weighed daily until the weight had stabilised.

2.3. WEIGHT AND QUALITY LOSSES IN A HARVEST TRIAL

A harvesting trial was conducted at Dooen in the 1988 plot. The experimental design was a split block with one block of 4 plots (each 4 m x 25 m) at either end of the paddock. After preliminary sampling, the blocks were placed on the edge to ensure a reasonably high and even infestation into the paddock. Using a small-plot harvester fitted with a 1.85 m wide pea front, two randomly selected plots in each block were harvested on 3 December 1988. This was the earliest possible date for harvest (EPHD) when all pods were dry and brown and the seed was dry. The other plots were harvested on the 3 January 1989. A 0.5 x 0.5 m quadrat was used to randomly sample (10 samples/plot) the pea seed left in the stubble after harvest; seed was sorted into whole and split seed and counted. Harvested seed for each plot was weighed and a 1.0 kg subsample sieved (round hole with 6.35 mm diameter). Whole and split seed fractions were separated and weighed. About 0.5 kg of larger whole seed was separated into infested and non-infested seed, and infested seed dissected to determine the stage and survival of pea weevil.

3. Results

3.1. MONITORING BRUCHID POPULATIONS AND % SEED INFESTATION

Pea weevil populations were dominated by 3rd instar larvae in the first samples in all plots (Fig. 1). At EPHDs, in early to mid-December, the populations largely comprised 3rd and 4th instar larvae in most plots. The age composition in the July-sown plot in

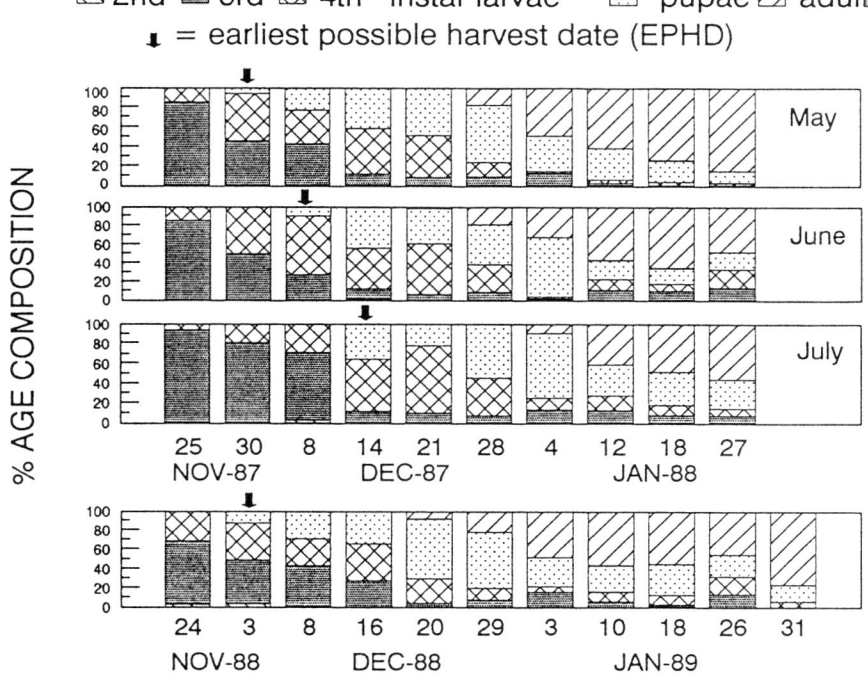

Figure 1. Age composition of *B. pisorum* populations in field peas, Victoria 1987-89.

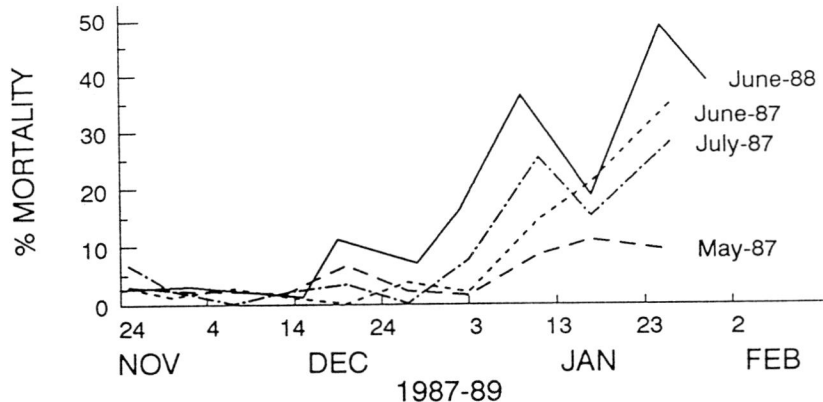

Figure 2. Mortality of *B. pisorum* populations in field peas, Victoria 1987-89.

1987 included 37% pupae. In 1987, there were 9 wks between sowing the earliest and latest plots but only about 2 to 3 between the EPHDs for all plots.

Pea weevil appeared to develop through a steady progression of stages except towards the end of the sampling period when the age composition remained relatively constant. This was due to the inclusion of dead insects in the calculation of % age composition; this was necessary for later estimates of % total weight loss. Numbers from multiple infested seed were only included if all insects were dead. Mortality was low up until late-December to early-January in all plots but then it rose sharply with the highest value recorded (49%) in the 1988 plot in late-January (Fig. 2). These high levels of mortality produced the relatively fixed age distributions later in the sampling period (Fig. 1).

The 1988 plot had the highest level of infested seed (84.3%) and the highest proportion of infested seed with multiple infestations (43.1%). Later sowings in 1987 resulted in lower levels of both these variables: 52.1, 36.9 and 22.6% of seed infested and 9.0, 3.4 and 1.1% of infested seed with multiple infestations for plots sown in May, June and July respectively.

3.2. RELATIONSHIP BETWEEN WEIGHT LOSS AND BRUCHID DEVELOPMENT

The dominant stage in the first sample (at 10 d) was 3rd instar larva and no live 1st or 2nd instar larvae were found. Mean wet weight of the insect's body increased until the pupal stage and then dropped at the adult stage (Table 1). For each stage, mean body weight did not differ significantly between size classes of seed, except for 3rd instars in the large seed (P<0.01). Insect body weight contributed substantially to the overall seed weight, ranging from 1.4% for 3rd instar larvae in large seeds up to 17.9% for pupae in small seeds (Fig. 3a).

In the first subsamples of non-infested seed, mean weights (± s.e.) were significantly different (P<0.01) for small, medium and large seeds: 153.8 mg (± 2.5 mg), 205.4 mg (± 4.2 mg) and 246.0 mg (± 3.5 mg) respectively. Mean seed weight dropped by 4.8 mg or 2.3% of the initial weight during the course of the experiment and the final water content was 8.8% by weight.

Within each seed class, the mean weight of infested seed, minus insect body weight, decreased as the insects grew from the 3rd larval instar to pupal stage (Table 1). However, there was no significant difference between weights of infested seed at the

Table 1. Mean ± S.E. weight of *B. pisorum* larvae, pupae and adults, in 3 size classes of infested pea seeds.

	Size class of seed	Stage of insect development			
		3rd instar larva	4th instar larva	pupa	adult
Insect	small	3.1 ± 0.2	14.3 ± 1.6	20.9 ± 0.5	17.1 ± 0.4
(wet)	medium	3.1 ± 0.2	12.5 ± 1.3	21.1 ± 1.3	16.9 ± 1.0
	large	4.1 ± 0.3	11.5 ± 1.3	21.5 ± 0.5	17.7 ± 0.4
Seed	small	146.5 ± 2.8	130.3 ± 3.6	116.9 ± 3.4	103.6 ± 4.1
(less wt.	medium	206.0 ± 4.3	199.9 ± 3.9	172.9 ± 4.0	171.6 ± 2.8
of insect)	large	240.2 ± 3.5	231.4 ± 5.3	216.0 ± 4.7	208.5 ± 5.0

pupal and adult stages. The greatest difference in weight loss occurred between 4th and prepupal/pupal stages which suggests that most damage was caused by late 4th instars.

The % weight loss of seed was calculated using the mean of the weight of infested seed minus the weight of the insect, and the mean weight of uninfested seed for each size class. Although, the weight of a live insect can contribute substantially to total weight of seed, it was considered to be unimportant to the ultimate loss of seed weight. It was assumed that water loss in a non-infested seed was not significantly different from an infested seed, and all seed weights were standardised to a water content of 8.8%. The % weight loss increased from 3rd instar larval to pupal stages in all classes of seed size (Fig. 3b). At comparable stages, % weight loss for small seed was higher than for other seed sizes.

3.3. WEIGHT AND QUALITY LOSSES IN A HARVEST TRIAL

The mean weights of seed harvested on the two harvest dates were not significantly dif-

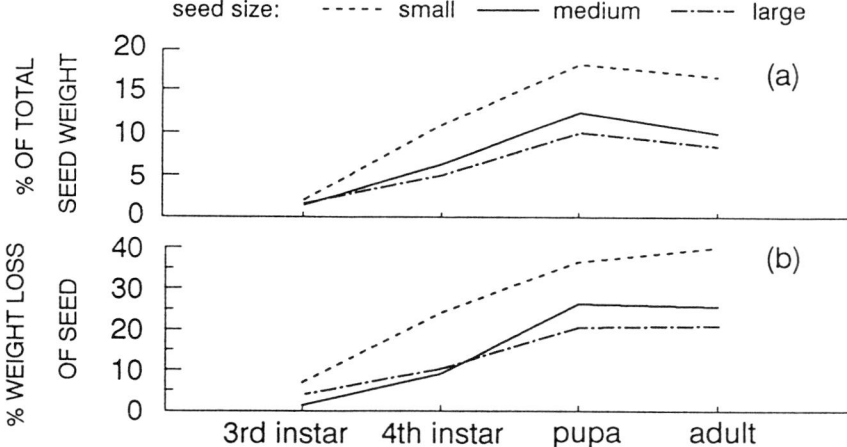

Figure 3. (a) Weight of insect, expressed as % of total seed weight, and (b) % weight loss of pea seed for various stages of *B. pisorum* development and 3 seed sizes.

Table 2. Statistics for harvest trial of pea seed infested with *B. pisorum* at Dooen 1988-89.

Harvest date	Mean ± S.E. weight of seed/plot (kg)	% whole seed infested	% split seed by weight	% mortality	% age composition (live larvae)	
					3rd	4th
7.XII.88	20.7 ± 2.2	51.6	6.4	56.3	85.7	14.3
3.I.89	19.4 ± 0.7	51.5	25.2	75.8	26.1	73.9

ferent (Table 2). Most of the harvested seed (82.1%) was > 6.35 mm (i.e. in the medium to large size seed classes). The percentage of infested seed (51.6%), based on the whole seed fraction, was less than for the hand-picked seed (84.3%).

Of the live pea weevil, 3rd instar larvae dominated the first harvest while 4th instar larvae dominated the second (Table 2). Relatively few pupae or adults (12.6%) were found inside seed on the second harvest and all of them were dead. In comparison, pupae and adults comprised 77.2% of the population in hand-picked seed. Mortality was high (Table 2), especially in the second harvest, and was about 3 to 5 times higher than in seed collected by hand on the same dates (cf Fig. 2).

On both harvest dates, the amount of seed on the ground after harvest was negligible, less than 0.1% by weight of the total seed harvested.

3.4. ESTIMATING WEIGHT LOSS IN FIELD POPULATIONS

To examine the influence of changing agronomic practices, such as harvest and fumigation times, the relationship between % weight loss and insect development (derived from dissections in the laboratory) was used to estimate yield loss in field populations. Using % infested seed, % age composition (Fig. 1) and mean % weight loss caused by each stage (Fig. 3b), the % total weight loss can be estimated for each plot and sampling occasion. Only values of mean % weight loss for medium size seed class were used because most seeds appeared to fall into this category. For simplicity, all infected seeds were assumed to be single infestations.

The % weight loss with time is presented for each plot in 1987 and 1988 in Figure 4. The values ranged from 2.9% to 5.8% at the EPHD and then rose to an upper plateau, the level of which depended upon the % infested seed.

4. Discussion

4.1. GENERAL

The % weight loss of field pea seed caused by *B. pisorum* depended upon the age of the larva and size of the seed (Fig. 3b). If third instars caused approximately 2% weight loss, then weight loss was presumably negligible for earlier instars. Maximum weight loss for a medium sized seed was approximately 26%. The different values which have been reported [e.g. 43% by Yao Kang et al. (1966)] are probably due to variations in seed size, such as occur with poor growing conditions or with different varieties.

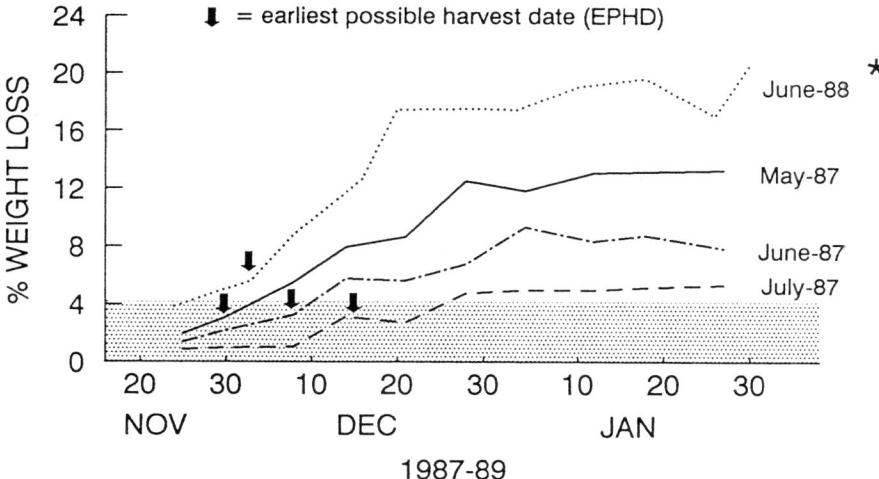

Figure 4. Estimated weight loss of seed caused by *B. pisorum* infesting plots of peas sown at different times*, Victoria 1987-89.

Field populations of pea weevil were usually dominated by 3rd and 4th instar larvae at the EPHD (Fig. 1) and substantial weight loss had already begun in most plots (Fig. 4). By contrast, Pajni (1981) stated that freshly harvested seed usually contained 2nd and 3rd instar larvae; potentially, weight loss would have been less. Studies of weight loss by Brindley and Hinman (1937) also suggest that larval populations can be younger at harvest because the weights of infested and non-infested seeds were similar. If one assumes that the peas were harvested at the same stage of maturity in all of these studies, then the relationship between rate of insect growth and plant development appears to vary with agronomic and environmental conditions. For example, larval populations at the EPHD may be younger in earlier maturing varieties. This study suggests the timing of sowing, and therefore flowering and pod formation, is not an important factor in the relationship.

After the EPHD's, pea weevil continued to develop and estimates of % weight loss increased until pupation was complete and maximum losses were reached for each plot (Fig. 4). Maximum weight loss per plot ranged from 5.3 to 20.5%. Brindley and Hinman (1937) recorded final losses between 14.4 and 16.4% in post-harvest seed. The effect of multiple infestation of seed on weight loss was not considered in this study. In a multiple infested seed, only one larva survives while the others tend to die at the 3rd instar stage (Skaife, 1918). Therefore, calculations of weight loss were underestimates, especially in the 1988 plots which had a high percentage of multiple infested seed.

Ultimate weight losses were somewhat curtailed by larval mortality which was often considerable (Fig. 2). Mortality was initially low but rose sharply from early- to late-January. No parasitoids or pathogens were observed. It is proposed that the chief cause of larval mortality was the high lethal temperature which can be experienced in summer. Skaife (1918) exposed infested seeds to a maximum ambient temperature of 54°C and no larvae died. Maximum temperatures in the canopy of a pea crop at Dooen rarely exceeded about 48°C in summer but temperatures inside a pea pod were recorded up to 12°C higher than ambient (Smith, unpublished data). Larvae may also be killed by rapid hardening of the seed resulting from hot dry weather (Schopp et al., 1955). Another cause of mortality was operative in the harvest trials: mortality was 3 to 5 times higher in machine harvested seed compared with seed harvested by hand. Pea weevil may have been killed by severe shaking which occurs during threshing.

Lower infestation levels in the later-sown plots in 1987 was consistent with findings by Brindley (1933) and Newman and Elliot (1938). The latter authors found the % infested seed reduced progressively from 92.5% to 9% with successive monthly sowings. On the other hand, Gerding et al. (1987) reported no effect of sowing date on infestation level after trials over 5 years. The practice of sowing late to minimise the size of infestations would not necessarily preclude the application of an insecticide. Even if the cost of the spray was avoided, the benefits would be lost because late-sown crops generally have lower yields.

Infested seeds were efficiently recovered during the harvesting (threshing) process; negligible amounts were lost in the stubble. However, % split seed in harvested seed was quite high. This was particularly evident in the second harvest probably because larvae were older, more damage had been caused to seeds and seeds were more susceptible to splitting during threshing. Smith et al. (1982) examined the effects of different methods of seed threshing and cleaning methods on the efficiency of recovery of infested seed. They found the chaff that remained after cleaning was approximately 9% of the total weight of threshed material. Although, the composition of the chaff was not specified, it presumably contained a large proportion of split seed. Thus, in a harvest where only 17% of the seed was infested, the split seed component was probably quite high.

High % splitting of seed may partly explain the lower % infested seed and fewer pupae and adults in machine-harvested seed relative to hand-harvested seed: estimates of age composition were based on dissections of the whole seed fraction only. Another possible reason for lower % infested seed was the differences in methods of sampling the seed. Infestation levels are generally higher on the edge of crops and tend to drop off further into the crop (Smith et al., 1986). Therefore, machine harvesting to 25 m into the crop may have collected seed over a gradient from high to lower densities while hand-picking concentrated on the higher densities on the edge.

4.2. IMPLICATIONS FOR MANAGEMENT

In Victoria, growers can either deliver their peas immediately after harvest to large storage depots or store them on the farm. At the depots, growers are paid for the weight and quality of seed at the time of delivery and any losses due to pea weevil afterwards are borne by the buyer. Until the stores are fumigated, larvae will continue to feed and reduce the seed weight. One factor that may not be considered by buyers is the weight loss that occurs when insect inside the seed has been killed and dries out. This study revealed that, depending on the stage, the live pea weevil can contribute up to 17.9% of the wet weight of infested seed. Therefore, buyers may unknowingly incur costs for weight loss after harvest if they do not check infestation levels in seed and discourage the delivery of heavily infested loads.

The timing of harvest and development of adults is an important consideration for reducing the size of future infestations. Adults emerge from the seed, find shelter locally and hibernate, and move into crops in the following spring (Brindley, 1933). Adults were beginning to develop inside the seed in late December at Dooen (Fig. 1). If the peas had been harvested at the EPHD, most of the infested seed would have been removed from the paddock before the adults had emerged. Immediate fumigation of infested seed would have negated the stores as sources of hibernating adults. Infested seed can also be found in the stubble after harvest. Any insects which are inside the seed in the stubble would be exposed to high lethal temperatures on the soil surface (Smith, unpublished data).

The harvest trial demonstrated that infested seed was prone to splitting during threshing. Seed, especially from the later harvest, would not have been acceptable by any of the current standards for seed quality. The recommended standard for the lowest grade, 'Australian Feed Grade', states that the peas should not contain more than 15% by weight of sprouted, grub-eaten and otherwise damaged grain and foreign matter. Since the heaviest infestations are on the edge of the crop, a grower may selectively harvest and isolate the seed from the edge, perhaps for on-farm use as feed, and maintain a high quality for the rest of the crop.

There were 3 major factors which determined the extent of quality and weight losses due to infestation by *B. pisorum*: % infested seed, age composition of the population and the timing of harvest and fumigation. The % infested seed will largely depend on the size of adult populations in pea crops in spring. A grower must decide if the adult density will ultimately result in a weight loss of seed which costs more than the price of an insecticide application (i.e. exceeds the 'break-even' level). The cost of weight loss varies with yield per hectare, market value of the peas and cost of the insecticide application. In Victoria, these are currently about 1.5 t/ha, $A 200/t and $A 12/ha respectively. Assuming that an insecticide spray kills all the adults and no seed is infested, then the break-even for spraying is 4% weight loss. In 1987-89 at Dooen, the estimated % weight loss was less than 4% at the EPHD in 3 out of the 4 plots (Fig. 4). Therefore, insecticide applications to control adults would not have been economically justified in most plots, if harvest had occurred at EPHD and the peas had been fumigated immediately.

When the infestation levels, yields or prices are lower than above, a grower may decide not to spray but harvest at the EPHD and fumigate immediately. The problem with not spraying is that unforeseen events may delay harvest. For instance, if harvest had been delayed in the plots under study, the break-even level would have been exceeded in only 1 to 2 weeks after the EPHD. The 'ideal' situation of early harvest and immediate fumigation may be difficult, if not impossible, to achieve in everyday farm practice. Some reasons why a grower would have trouble achieving the ideal situation include: (a) delays in the harvest due to inclement weather, waiting for a harvesting contractor or because other crops, such as the cereals, are harvested first; (b) delays in fumigation due to completing harvest in other crops, the time required to transport to stores and prepare stores for fumigation.

Growers may decide not to spray because the density of adults is considered to be too low; the resultant infestation level could not exceed the break-even cost, irrespective of the timing of harvesting and fumigation. Most growers will currently spray as a matter of course because it is difficult to predict infestation levels from adult densities. Research needs to define the reliability and efficiency of sweep nets - the sampling method most commonly used - and to provide knowledge about the relationship between adult densities, oviposition and infestation levels.

Acknowledgments. I wish to thank Ms A. Leith and Ms S. A. Soerono, for assistance in sampling and processing samples. Experimental plots were kindly supplied by Dr J. Mahoney and staff at the Cereal Experimental Centre, Dooen. I am also indebted to Dr P. Stahle, Mr D. Williams and Mrs J. Horne for providing helpful criticisms of the manuscript. This research was funded by grants from the Grain Legumes Research Council and the Victorian Government's Economic Strategy.

REFERENCES

Brindley, T. A. (1933) Some notes on the biology of *Bruchus pisorum* L. (Coleoptera, Bruchidae) at Moscow, Idaho, J. Econ. Entomol. 26, 1058-1062.

Brindley, T. A. and Hinman, F. G. (1937) Effect of growth of pea weevil on weight and germination of seed peas, J. Econ. Entomol. 30, 664-670.

Gerding, M. P., Juan, T. U. and y Mario, P. C. (1987) Incidence of *Bruchus pisorum* L. (Coleoptera: Bruchidae) on pea, according to seeding date and density, Agricultura Tecnica (Chile) 47, 160-162.

Newman, L. J. and Elliot, H. G. (1938) The pea weevil, *Bruchus pisorum* (Linn), J. Dept. Agr. Western Austr. 15, 156-158.

Pajni, H. R. (1981) Trophic relations and ecological status of the adults of *Bruchus pisorum* L. and allied species of Bruchidae (Coleoptera), in V. Labeyrie (ed.), The Ecology of Bruchids Attacking Legumes (Pulses), Proceedings of the International Symposium held at Tours (France), April 16-19, 1980, Dr W. Junk Publishers, The Hague.

Reichart, G. (1964) Comprehensive study on the pea weevil (*B. pisorum*) and its control, Kiserl. Közl. 57, 149-168.

Schopp, R., Brindley, T. A., and Hinman, F. G. (1955) Factors affecting pinhole injury to dry peas by the pea weevil, J. Econ. Entomol. 48, 693-695.

Skaife, S. H. (1918) Pea and bean weevils, Union of South Africa, Dept. Agr. Bull. 12, 1-32.

Smith, A. H., O'Keefe, L. E., and Muehlbauer, F. J. (1982) Methods of screening dry peas for resistance to the pea weevil (Coleoptera: Bruchidae), J. Econ. Entomol. 75, 530-534.

Smith, A. M., Comery, J., and Chaffey, B. (1986) Review of DARA Recommendations for the Control of Pea Weevil in Victorian Field Peas, Research Report Series, No. 4, Department of Agriculture and Rural Affairs, Melbourne.

Yao Kang, H., Sueh-sien, T., Liang-bing, and Dung, W.-S. (1966) Biology and control of the pea weevil in Lotien, Hupeh, Acta Entomol. Sinica 15, 288-293.

PLANT QUARANTINE AND FUMIGATION OF IMPORTED GRAIN LEGUMES IN JAPAN

MASANORI YONEDA, MASASHI KANEDA,
AND HIROSHI AKIYAMA
Yokohama Plant Protection Station,
Naka-ku, Yokohama, 231, Japan

ABSTRACT. A total of 150,000-200,000 tons of beans (excluding soybeans and peanuts) are imported annually into Japan from many countries. These grain legumes are inspected by the plant quarantine inspectors prior to importation to check for pest infestation. In 1987, 22% of imported grain
legumes were fumigated with methyl bromide at the port of entry to eradicate insect pests; about 27%
of the fumigated commodities were infested with bruchid beetles such as *Callosobruchus chinensis,*
and *C. maculatus.* Insect pests which do not occur, or have limited distribution, in Japan are designated as "major pests of quarantine significance" and special procedures are taken when these pests are
found on the legume commodities. Fumigation facilities, i.e. commercial warehouses, barges, silos, etc.
that are used for quarantine treatment should be approved by the Plant Protection Station, and fumigation operations should be supervised by approved, qualified persons from the fumigation company.
A list of intercepted bruchid beetles, construction specifications and performance standards of fumigation facilities, fumigation schedules, and a description of the plant quarantine system in Japan, are
included in this paper.

1. Plant Quarantine System in Japan and Import Plant Quarantine

The plant quarantine service in Japan was initiated in 1914 at the time that the Plant
Quarantine Law was promulgated. At first, only import and export quarantine services
were carried out. But now domestic quarantine measures are also applied under the
current Plant Protection Law.

Plant Protection Stations are located in all major sea- and airports in Japan. The
headquarters of stations are located in Yokohama, Nagoya, Kobe, Moji and Naha and
they have branches and subbranches where plant inspectors are stationed. At present,
the total number of   stations is 90 with about 800 plant inspectors engaged in plant
quarantine inspection.

Imported plants and plant products are inspected by the plant inspectors to check for
pest infestation at the port of entry. Cereals are usually imported by ship in bulk or
bagged. The cereals imported in bulk are inspected on ship and those bagged are inspected on barges prior to unloading. Those cereals imported by sea containers are
inspected at the container yard. Grain legumes are usually imported bagged and inspected on barges.

When insect pests are found during import inspection, the commodities must be
fumigated at the port of entry in facilities approved by the Plant Protection Station.
Commodities that are contaminated with plant diseases, such as ergot or sclerotium, are

115

K. Fujii et al. (eds.), Bruchids and Legumes: Economics, Ecology and Coevolution, 115–120.
© 1990 *Kluwer Academic Publishers. Printed in the Netherlands.*

given a heat treatment or are destroyed. Under this plant quarantine system, the entry of foreign pests into Japan has been largely prevented.

2. Importation of Grain Legumes

A total of 150,000-200,000 tons of grain legumes, including green gram, adzuki beans, common peas, etc. are imported annually into Japan from many countries. Figure 1 shows the quantity of major grain legumes imported in 1984-1988. Some commodities, such as soybeans and peanuts are excluded from Fig. 1 because bruchid beetles are rarely found on these legumes during import inspection.

The quantity of major legumes imported in 1988 was as follows: green gram 62,636 tons, adzuki beans 32,491 tons, common peas 22,422 tons, kidney beans 18,898 tons, bamboo beans (rice beans) 18,472 tons and broad beans 10,284 tons. In addition to these legumes, lima beans, cowpeas, etc. were also imported into Japan. Of these legumes, green gram, adzuki beans, bamboo beans and broad beans were largely imported from China and South East Asian countries, and common peas and kidney beans from Canada, the U.S.A., New Zealand and China (Fig. 2).

3. Bruchid Beetles Intercepted by Import Inspection

More than 10 species of bruchid beetles have been found in imported legumes and

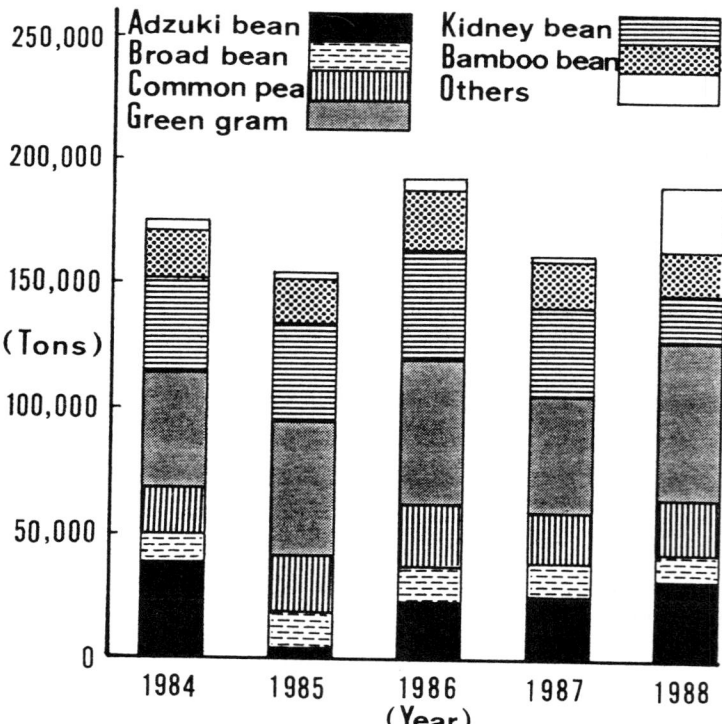

Figure 1. Quantities of major grain legumes imported into Japan in 1984-1988.

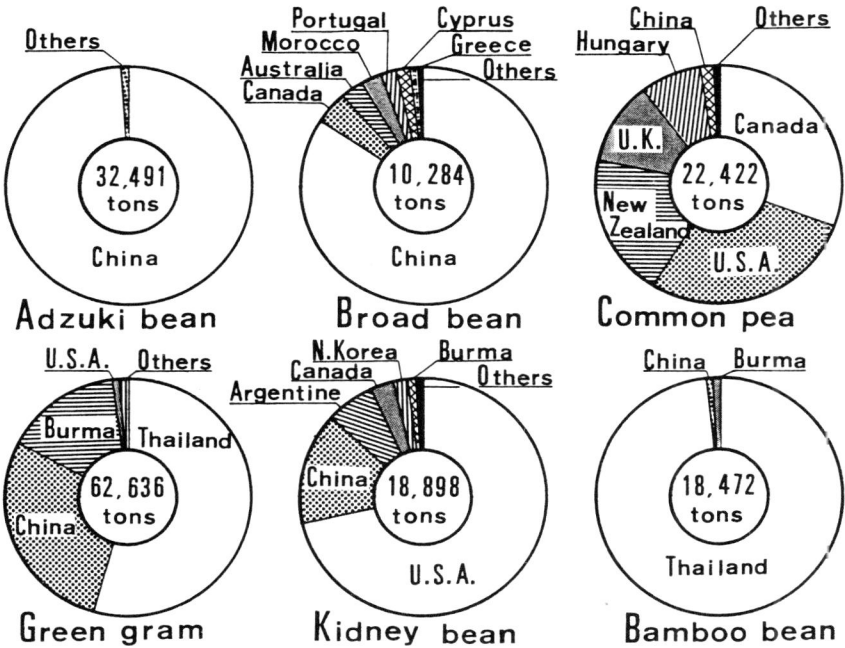

Figure 2. Quantities of major grain legumes imported into Japan and exporting countries (1988).

some of these are important pests of stored legumes all over the world. Table 1 shows the number of times that bruchid beetles were intercepted during import inspection during the period 1983-1987. *Callosobruchus chinensis*, *C. maculatus* and *Zabrotes subfasciatus* were the species most frequently found, however, the number of intercep-

Table 1. The number of interceptions of bruchid beetles at Japanese ports in the years 1983-1987.

Bruchid beetles	Year					Total
	1983	1984	1985	1986	1987	
Acanthoscelides obtectus	13	24	11	4	11	63
Bruchidius hibisci		21				21
Bruchidius japonicus		68	1	1		70
Bruchus pisorum	4	16	18	8	1	47
Bruchus rufimanus	5	21	17	28	6	77
Callosobruchus analis	23	3	48	7	3	84
Callosobruchus chinensis	273	310	141	134	47	905
Callosobruchus maculatus	211	121	152	79	90	653
Callosobruchus rhodesianus	1	2	2	3	13	21
Caryedon serratus	10	15	36	17		78
Zabrotes subfasciatus	173	78	113	62	55	481

tions of these species has decreased recently, possibly because of treatment of the legumes with pesticides at the exporting ports.

The percentage of legumes fumigated relative to the total amount imported in 1987 was: green gram 60.8%, bamboo beans 59.5%, cowpeas 52.1%, adzuki beans 44.0% (Table 2). Insect pests that were found included other stored product pests in addition to bruchid beetles. The percentage of legumes infested with bruchid beetles compared to total importation was 32.7% on cowpeas, 19.4% on lima beans, 14.7% on green gram and 10.6% on bamboo beans. Some bruchid beetles, which are not known to occur in Japan, are frequently intercepted during import inspection.

The important pests which do not occur or are of limited distribution in Japan are designated as "major pests of quarantine significance". For the bruchid beetles, these include *Acanthoscelides obtectus*, *C. analis*, *C. maculatus*, *C. rhodesianus*, *Caryedon serratus* and *Zabrotes subfasciatus*. Of these, *A. obtectus* and *C. maculatus* are known to have been introduced into Japan; *A. obtectus* was reported in Okinawa in 1951 (Sato, 1951) but no information on damage by this pest has been obtained recently and the current status of it is not known. *C. maculatus* is distributed over the south west of Japan in warehouses and mill factories but it does not occur in the field (Nagayasu and Matsushita, 1981; Watanabe, 1986).

4. Quarantine Treatment of Imported Cereals

When imported cereals are infested with insect pests, they should be fumigated at

Table 2. Quantities of major grain legumes imported into Japan (1987). Proportion treated by plant quarantine fumigation, proportion infested with bruchid beetles and species intercepted on each legume commodity.

Commodity	Imported (Tons)	Fumi-gated (%)	bruchids infesta-tion(%)	Intercepted Bruchid beetles*
Adzuki bean	25,354	44.0	3.2	*Acanthoscelides obtectus* *Callosobruchus chinensis* *C. maculatus*
Broad bean	13,957	21.7	9.4	*Bruchus rufimanus* *C. chinensis*
Common pea	45,643	0.7	0.2	*B. pisorum, C. maculatus*
Green gram	48,248	60.8	14.7	*C. analis, C. chinensis* *C. maculatus, C. rhodesianus* *Zabrotes subfasciatus*
Kidney bean	14,541	15.3	8.8	*A. obtectus, C. maculatus* *Z. subfasciatus*
Bamboo bean	17,831	59.5	10.6	*C. maculatus, Z. subfasciatus*
Lima bean	19,696	27.9	19.4	*C. maculatus, C. rhodesianus* *Z. subfasciatus*
cowpea	3,073	52.1	32.7	*C. analis, C. chinensis* *C. maculatus*

* Interceptions of some bruchid beetles on unavailable legumes for their hosts were due to contamination with the beetles from other commodities in cargo ship.

warehouses, silos, etc. approved by the Plant Protection Station. Facilities for quarantine fumigation are divided into three classes, "A", "B" and "C", according to their gastightness. To test for gastightness, methyl bromide is introduced into an empty facility at a dose rate of 10 g/m^3 and the concentration is measured after a period of 48 hours. The facilities are classified "A" when the concentration of methyl bromide is more than 7 g/m^3 48 hours after dosing; "B" when it is 5.5-7 g/m^3 and "C" when it is 4-5.5 g/m^3. Most facilities used for quarantine fumigation are classified "A".

Recently refrigerated warehouses have been used for storage of legumes to maintain quality and to prevent reinfestation by insect pests. The fumigants used for plant quarantine treatment are methyl bromide, hydrogen cyanide and aluminum phosphide. Of these fumigants, methyl bromide is widely used for grains, fresh vegetables and fruits and also for fumigation of grain legumes. Hydrogen cyanide is used for fresh vegetables and fruits, and aluminum phosphide is used mainly for feed pellets such as alfalfa pellets and beet pulp pellets.

Legumes imported in bags are transported from the ship to warehouses by barges and the inspection is conducted on the barges prior to unloading. The legumes infested with designated bruchid beetles are fumigated on the barges prior to unloading to prevent their escape on the shore, while legumes infested with other insect pests are fumigated in the warehouses.

Table 3. Dosage of methyl bromide and exposure period for quarantine fumigation of grain legumes in the warehouse at a load rate of 0.29 ton/m^3 or less.

Commodity	Exposure period (hours)	Temperature (°C)	Dosage (g/m^3) Class of facility		
			A	B	C
Green gram Common pea	24	20 and above	15	16	18
		10 - 19	20	22	24
		below 10	25	28	30
	48	20 and above	9	10	12
		10 - 19	12	14	16
		below 10	14	17	20
	72	20 and above	8	9	11
		10 - 19	11	13	15
		below 10	14	17	19
Adzuki bean Broad bean Cowpea	24	20 and above	19	21	23
		10 - 19	26	28	31
		below 10	32	35	38
	48	20 and above	13	15	18
		10 - 19	17	21	24
		below 10	21	26	30
	72	20 and above	12	14	16
		10 - 19	16	20	22
		below 10	21	26	29

The current dosages for the plant quarantine fumigation with methyl bromide in the facilities are determined on the basis of type of commodity, length of exposure period, temperature, class of fumigation facilities, etc. Commodities are divided into four groups according to their degree of methyl bromide absorption. Absorption of methyl bromide by green gram, common peas, etc. is relatively low and that of adzuki beans, broad beans, etc. is relatively high. The dosages of methyl bromide and exposure period for fumigation of legumes are shown in Table 3.

The schedule for barge fumigation of the legumes with methyl bromide in warm and cool seasons is as follows:

Month	Dosage (g/m^3)	Exposure period (hours)
May-October	24	24
November-April	32	24

The quarantine fumigation of imported cereals is performed by commercial fumigation companies with employees that have been trained in a special training course. The course includes the following subjects: principles of fumigation, detection and analysis, toxicity and hazards, safety equipment and use.

Acknowledgements. We gratefully thank Dr. Edwin J. Bond of London Research Centre Agriculture Canada and Dr. Angharad M.R. Gatehouse, Department of Biological Sciences, University of Durham, for their invaluable advice and critical reading of the manuscript. We also thank Dr. Naoshi Watanabe, Kobe Plant Protection Station, and Dr. Fumihiko Ichinohe, Yokohama Plant Protection Station, for their suggestions and critical comments.

REFERENCES

Nagayasu, M and Matsushita, K. (1981) Faunal survey of bean weevils in processing mills, Res. Bull. Pl. Prot. Japan 17, 88-91.

Sato, K. (1951) Notes on quarantine pests in Ryukyu Island, Shokubutsu boeki 5, 269-270 (in Japanese).

Watanabe, N. (1986) Quarantine significance against expansion of insect pests caused by agricultural diffusion of their host plants; with special reference to the seed beetles, Res. Bull. Pl. Prot. Japan 22, 1-9.

NEW WORLD BRUCHIDAE PAST, PRESENT, FUTURE

JOHN M. KINGSOLVER
Systematic Entomology Laboratory
PSI, ARS, U.S. Department of Agriculture
Room 1, Building 004
Beltsville Agriculture Research Center-West
Beltsville, MD 20705 U.S.A.

ABSTRACT. The New World beetle family Bruchidae is diverse and extensive but largely endemic to the Western Hemisphere. More than 1000 species names have been proposed of which an estimated 750 are considered to be valid. These are assigned to 40 genera, of which 11 are monotypic. The fauna of the West Indies, Canada, the United States, Mexico and Central America, and the northern countries of South America is fairly well known, but the southern part of South America is largely unexplored. Five research scientists and three graduate students are currently working on the classification of this important family. Three of the scientists will soon retire. Since 1929, about 250 papers relating to taxonomy and biology of New World bruchids have been published of which 29 are generic monographs. A substantial foundation now exists on which to further explore the New World fauna.

1. Introduction

The New World extends from Alaska and Greenland in the north to Tierra del Fuego at the southern tip of South America. The largest number of bruchids are found in the Subtropical and Tropical zones but the United States has a fairly extensive temperate fauna extending into Canada. One species is found as far north as Alaska. Temperate areas in the southern hemisphere are likewise diverse, especially in the xeric regions, and at least one species is found on Tierra del Fuego less than 500 miles from Antarctica. The faunas of the Galapagos Islands and Hawaiian Islands are recently acquired from mainland areas and have no endemic species, whereas the West Indies, although sharing many species with the mainland, exhibit some endemism.

The family Bruchidae in the New World is diverse and extensive, but nearly all of the genera and species are endemic to the Western hemisphere. By the end of 1988, more than 1000 species names had been proposed for this hemisphere, of which about 750 are now considered valid. These are assigned to 40 genera. During the last 60 years, 462 of the 750 species have been described or redescribed for the two continental masses and the West Indies. Of the 40 genera, four include species immigrant from the Old World. Eleven of the remaining genera are monotypic, and an additional six include species that have emigrated to the Old World, or have been spread through man's activities.

Europeans dominated the early history of bruchidology (late 1700's to early 1800's) because most of the collections were made by Europeans, deposited in European museums, and studied by European taxonomists. The type specimens of the majority of the species described by early workers are deposited in the British Museum (Natural History), London; the Museum National d'Histoire Naturelle, Paris; the University

121

K. Fujii et al. (eds.), Bruchids and Legumes: Economics, Ecology and Coevolution, 121–129.
© 1990 *Kluwer Academic Publishers. Printed in the Netherlands.*

Zoological Museum, Copenhagen; and the Naturhistoriske Riksmuseet, Stockholm. Motschulsky's collection is housed in the Zoological Museum, University of Moscow. Not until LeConte, Say, and Horn in North America and Philippi and Philippi in Chile began publishing were types placed in New World museums.

Bruchid taxonomy in the New World is currently being studied in four principal centers, each with a taxonomist specializing in the family, and with an extensive collection at hand.

1. National Museum of Natural History, Washington, D.C., has the most representative collection of New World Bruchidae. Most of Bridwell's and Schaeffer's collections are included. J.M. Kingsolver, U.S. Department of Agriculture, is curator. About 90,000 specimens are included. Types of species described by Bridwell, Schaeffer, Johnson, and Kingsolver are deposited here.

2. C.D. Johnson Collection, Flagstaff, Arizona, a private collection, is rich in representatives from the United States, Mexico, and Central America, and northern South America, most of which Johnson personally has collected. He estimates his holdings at about 400,000 specimens. No primary types are deposited in his collection.

3. Fundacion Miguel Lillo, Tucuman, Argentina, specializing in South American species. A.L. Teran is curator of the Bruchidae. No estimate of numbers of specimens is available. Types of species described by Teran and some by Kingsolver are deposited here.

4. G.S. Pfaffenberger Collection, Portales, New Mexico, a private collection, is the center for immature forms of bruchids, principally of the New World fauna. Approximately 5000 specimens are included. No type specimens are deposited in this collection.

The Canadian National Collection, Ottawa, is the repository for the extensive L.J. Bottimer Collection of Bruchidae, mostly collected in the United States. It is conservatively estimated at 30,000 specimens. No one at the CNC is presently conducting research on bruchids. Bottimer's types are deposited there.

The Museum of Comparative Zoology, Cambridge, Massachusetts, is the depository for the collections and type specimens of Fall, LeConte, and Horn. No one is currently working on bruchids at this museum.

2. Chronology of Research Activities

Taxonomic research in the New World can be divided into four broad periods but with considerable overlap between periods. Principal workers are given for each phase:

1. Descriptive period: 1767-1874. Isolated descriptions only with rudiments of groups.

2. Faunal study period: 1873-1929. Descriptions but with detailed or broad diagnostic keys for a restricted faunal area.

3. Organizational period: (1929-1960). Development of higher classification.

4. Monographic period: (1962-present). Generic revisions with keys, descriptions of new and old species, host plant associations, and geographical distribution.

These phases reflect a trend from the generalist, working with many different groups of beetles, to the specialist concentrating his research on a family or a genus, or even parts of a genus.

2.1. DESCRIPTIVE PERIOD

New World bruchidology began in 1767 with Linnaeus' placing two species into his new genus *Bruchus*: *Bruchus bactris* from Colombia, now in *Pachymerus*, and *Bruchus gleditsiae* from southern United States, now in *Caryobruchus*. For several decades, the genus *Bruchus* was used as the principal repository for isolated descriptions of species by generalists- Blanchard, Boheman, Chevrolat, Erichson, Fabricius, Fahraeus, Faldermann, Gyllenhal, Jekel, Motschulsky, Philippi and Philippi, Schoenherr, and Thunberg. Although several genera had been described in the early 1800's and were available (*Amblycerus*, *Caryedes*, *Eubaptus*, *Megacerus*, *Pachymerus*, and *Rhipibruchus*), they were ignored, and nearly all species were described in *Bruchus*, *Caryoborus*, or *Spermophagus*. The Bruchidae was included in the first comprehensive treatment of the world Curculionidae by Schoenherr (1833-1846), with contributions by Boheman, Chevrolat, Gyllenhal, and Fahraeus.

2.2. FAUNAL STUDY PERIOD

Horn opened the second period of taxonomic history with his revision of United States Bruchidae (1873), adding thirty new species to the North American list. Although he recognized only three genera (*Bruchus*, *Caryoborus*, and *Spermophagus*), he set up groups of species that agree in a broad sense with the genera proposed by Bridwell (1946).

The Russian coleopterist Motschulsky, in his treatise on world Bruchidae (1874), added 45 specific names for the New World described in *Bruchus*, *Kytorhinus*, *Pachymerus*, and *Spermophagus*. Most of his species remained unrecognized until only recently because his collection was unavailable for study.

Sharp followed in 1885 with his pivotal monograph of the Bruchidae of Mexico and Central America in the Biologia-Centrali Americana series, recognizing only the three genera used by Horn in 1873. This fine work with one color plate is basic to most of the subsequent taxonomic studies for this faunal area. The distribution of many of his species were subsequently found to extend northward into the United States and southward into South America.

Horn proposed the genus *Zabrotes* with six new species in 1885. The notorious *Spermophagus subfasciatus* Boheman was subsequently transferred to this genus.

Schilsky in 1905 proposed two generic names, *Acanthoscelides* and *Bruchidius*, that would later become repositories for hundreds of bruchid species names. *Acanthoscelides* now includes the largest number of specific names in the New World, and *Bruchidius* assumes a similar role for the Old World.

To the list of American bruchids, Schaeffer (1904, 1907, 1909) added 13 new specific names to the genera *Bruchus* and *Spermophagus*.

Fall published the second synopsis on North American bruchids in 1910, adding 17 new species names but generally following the groups proposed in the first synopsis by Horn (1873).

Pic's (1913) catalog of the Bruchidae of the world, despite some shortcomings, became the standard reference for bruchid names for many years and is often cited in modern papers. Pic was the reigning authority on Bruchidae during the first quarter of the twentieth century. It is estimated that he described more than 20,000 species of Coleoptera in his 65 year career. He added 218 bruchid species names to the New World list, most of them so poorly described that type specimens must be seen to determine the status of the species.

Seven years later, Leng (1920) published a "Catalogue" of North American Coleoptera. In reality, it was no more than a checklist. He inexplicably transferred to *Mylabris* all of the American species names previously assigned to *Bruchus*, and used Mylabridae for the family. Some supplements to the Leng Catalogue continued to use Mylabridae while acknowledging that Bruchidae was preferred by taxonomists for the family. De-

124

spite its flaws, Leng's Catalogue was an indispensible reference for American coleopterists.

Bridwell published on the Bruchidae of the Hawaiian Islands in 1918-1920, but no new species were described from that area until 1929.

2.3. ORGANIZATIONAL PERIOD

In 1929, the third period of the evolution of New World bruchid taxonomy opened with Bridwell's durable world classification of the subfamily Pachymerinae (in the New World, the "palm bruchids"). He established a new subfamily, three new tribal names, and one new generic name with six new species, and included keys to all taxa. Although several additional genera have been described in the Pachymerinae in the Old World, Bridwell's basic classification is still valid.

Bridwell continued his ventures into higher classification of the family in 1932 with a comprehensive arrangement of world genera into five subfamilies (Amblycerinae, Bruchinae, Eubaptinae, Kytorhininae, and Pachymerinae), which brought together all of the previously described genera. Some of these had been in obscurity for nearly a century. For each genus he gave bibliographic references, cited or proposed a type species, and listed synonyms. Although some elements of his classification are probably artificial, it is used by modern workers because no one else has proposed a better scheme, although Borowiec (1987) combined two of Bridwell's tribes and proposed two others.

A crucial Bridwell paper published in 1946 included 12 new genera for the New World, a key to genera, the proposal of three new tribes, and discussions of the validity of several generic names. The three papers (1929, 1932, 1946) are seminal to bruchid classification in the New World. He published only a few small papers after 1946, then lost all of his notes and manuscripts when his home burned in 1955. His insight into the classification of the family was far ahead of his time.

The year 1946 also saw publication of Blackwelder's Checklist of Coleoptera of Mexico, Central America, the West Indies, and South America. This significant compilation included many new bruchid genus/species combinations probably originating upon advice from Bridwell. The majority of New World species were transferred to *Acanthoscelides*, but none of the genera Bridwell proposed in 1946 were included, although his paper was published three months prior to the checklist.

Bottimer was a protege of Bridwell and knew well the North American Bruchidae but his publishing career never really developed (12 papers in 30 years). His principal contribution was his meticulous field work for which he kept accurate records of host plants and localities. Following the deposition of his collection in the Canadian National Collections in 1965, he published six papers, one of which (1968) is an extensive summary of the status of bruchid classification for the world.

2.4. MONOGRAPHIC PERIOD

In the decade beginning with 1960, three New World bruchid taxonomists initiated their careers. In 1962, Teran began a series of publications on the Argentine bruchids; in 1963, C.D. Johnson published on the genus *Stator* for the United States; and J.M. Kingsolver characterized the genus *Neltumius* in 1964. The next decade brought in D.R. Whitehead, G.S. Pfaffenberger, and S.M. L'Argentier. Whitehead and Kingsolver concentrated on the Mexico and Central America fauna, especially since D.H. Janzen was supplying large quantities of Costa Rica specimens from his studies on seed production. Pfaffenberger began studies on immature forms (Pfaffenberger and Johnson, 1976), and L'Argentier co-authored several papers on Argentine bruchids.

From 1960 to the present time, these six taxonomists, either individually or in coop-eration, have published 140 papers directly impinging on classification of North Ameri-can Bruchidae, of which 29 are generic monographs, 27 are on biological aspects includ-ing host plant records, and 12 are on immature forms. Also, during this productive period, Prevett produced two papers, both in 1966, detailing two South American Pachymerinae genera, and Borowiec (1987) described two more genera from South America along with his summary of bruchid genera of the world. Silva has published three papers on Brazilian bruchids.

Kingsolver is compiling a handbook of the Bruchidae of the United States (including Hawaii) and Canada that will characterize approximately 145 species of bruchids and will provide taxonomic history, geographical distribution, lists of host plants and parasites, and references to sources of information on biology, morphology, and physiology where these are known.

Johnson and Kingsolver (1982) published a checklist of the Bruchidae of the North American continent and the West Indies summarizing the status of bruchid classification.

2.4.1. *West Indies.* The West Indian taxonomic history is much the same as that of the mainland with single species descriptions contributed by Gyllenhal, Jacquelin du Val, Suffrian, and others. The first comprehensive papers were by Kingsolver (1970) on the subfamily Amblycerinae (1970) and on the genus *Stator* (1972). Subsequent revisionary studies by Johnson and Kingsolver have included West Indian species. A thorough faunal study would help elucidate relationships of the islands with mainland faunas.

2.4.2. *Hawaiian Islands.* The bruchid fauna of this state comprises 16 species accidently or deliberately introduced into the islands. Bridwell (1918, 1919, 1920) was the first to concentrate on this fauna. Hinckley (1960) included a list of 12 species, but no thor-ough taxonomic study has been made. The Hawaiian fauna will be included in the Handbook of Bruchidae of the United States and Canada mentioned above.

2.4.3. *Galapagos Islands.* The fauna of this group of islands apparently has received its fauna from the mainland, mostly from Ecuador. Decelle (1976) summarized the three species then known, but Kingsolver (1983) recorded from these islands *Scutobruchus ceratioborus* (Philippi), a species widespread in South America.

2.4.4. *Collectors.* Special note should be taken of some entomologists who made exten-sive and intensive special collections of Bruchidae in the New World not only collect specimens but also to establish authentic host plant associations. Without these collec-tions, our knowledge of the fauna and its biological associations would indeed be mea-ger.

Gregorio Bondar lived in Bahia, South America. His studies were mostly on Curcul-ionidae but he published three important papers on Brazilian Bruchidae (1931a, 1931b, 1936) that comprise nearly all that is known of the biologies of the family in that coun-try, although he described only one bruchid species. His biological notes were meticu-lous and well documented, and his specimens were for the most part described or identi-fied by Pic. Pedrito Silva for years has reared palm bruchids in Brazil. His collection is located in Salvador, Bahia.

D.H. Janzen's specialized collections of bruchids in Costa Rica played a large part in increasing the faunal list for that country from 10 species according to the Blackwelder Checklist (1946) to the current number of 135. Most of the specimens he collected were reared from host seeds. Several of the monographic studies of North American Bruchi-dae were based on his material.

C.D. Johnson is both collector and taxonomist. He has published several papers detailing host associations of bruchids mostly reared from his own collections of host seeds in western United States, Mexico, Central America, and northern South America.

L.J. Bottimer was an avid collector who kept careful notes on localities and host associations. Most of his collections were made in Texas and Arizona.

3. Future of Bruchidology in the New World

During the past 60 years of New World bruchidology, a firm foundation for future taxonomic studies has been established, especially for Canada, the United States, Mexico, and Central America, the West Indies, and northern South America. Much is yet to be done, especially in South America. Several new genera will undoubtedly be needed to accomodate species that cannot be logically assigned to existing genera. Many species are yet undescribed, and relatively little is known of immature forms, or of South American host associations. The higher classification will no doubt have to be reviewed as more is learned about the fauna.

Who is to continue the work? Three of the principal workers, Kingsolver, Johnson, and Teran, will soon retire, and will be without institutional support if they wish to continue their research on the family. Whitehead changed his research emphasis to Curculionidae 12 years ago. L'Argentier can contribute only a small portion of her time to bruchid study because of other duties, and Pfaffenberger likewise is burdened with teaching duties and research on animal parasites so that his research time for immature bruchids is limited.

Unless one or more of the positions that are to be vacated are filled within a reasonable period, the impetus to continue bruchid research will be lost or muted. Problems of funding exist everywhere, and the filling of vacant taxonomic positions is far down the list of priorities. Collections in the principal centers will undoubtedly continue to be well cared for, but without specific curation to keep them current, they will become static and eventually obsolete unless they are revived by a specializing taxonomist.

Three graduate students are at present working on the systematics of New World Bruchidae. Jan Nilsson, a student of Johnson in Arizona, is planning to revise the subfamily Pachymerinae; C.S. Ribeiro in Curitiba, Brazil, is revising the genus *Amblycerus* of Brazil, and J. Barriga in Santiago, Chile, is writing a thesis on the agricultural bruchids of that country. No one else in the New World has shown any interest in pursuing taxonomic studies in the family, and prospects for research to continue at the present level of production are rather discouraging.

4. Interpretive Summary

Bruchidae are efficient in the destruction of seeds, and many are serious pests of agricultural seed stock and stored comestible seeds, some species are beneficial to man since they destroy seeds of weedy plants, whereas others apparently have little impact on man's activities. A brief history of studies in the classification of the beetle family Bruchidae (seed beetles) in the New World is presented in Table 1, including some predictions about future research in this important economic group. Research studies in each of the 40 New World genera during the past 60 years is summarized including data on general distribution, numbers of species predicted, and the principal investigators for each genus.

Table 1. Summary of Generic Research in the New World.

This table gives the generic names, the area in which the genus is found, the number of species now assigned to the genus, the projected number of species that will be described, and authors of the monographs, revisions or original descriptions treating each genus. Abbreviations in Column 2 are (W) for West Indies, (N) for North America, (M) for Mexico, (C) for Central America, and (S) for South America. An asterisk indicates that the genus occurs in the Hawaiian Islands. Author-date references are not necessarily listed in the bibliography.

Genus	Area	Now	Future	Authorities
1. Abutiloneus	N	1	10	Kingsolver 1965; Johnson 1978
2. *Acanthoscelides	WNMCS	340	550	Johnson 1970, 1983, 1989, numerous smaller papers; Kingsolver, small papers
3. *Algarobius	NMCS	6	6	Kingsolver 1972, 1986
4. Althaeus	N	3	3	Kingsolver 1989
5. Amblycerus	WNMCS	102	250	Kingsolver 1970, small papers
6. Bonaerius	S	1	3	Bridwell 1952
7. Bruchidius	NS	3	?	Bottimer 1931, 1968; Aldridge and Pope 1986
8. *Bruchus	WNS	3	3	Kingsolver 1964
9. Butiobruchus	S	1	2	Prevett 1968; Nilsson and Johnson, in progress
10. *Callosobruchus	WNMCS	5	?	Southgate 1958, Kingsolver 1964
11. Caryedes	NMCS	23	40	Kingsolver and Whitehead 1974; Kingsolver 1989
12. *Caryedon	MCS	1	1	Johnson 1966, 1986; Kingsolver 1970, Southgate and Pope; Decelle 1960
13. Caryoborus	CS	2	10	Bridwell 1929; Nilsson and Johnson, in progress
14. Caryobruchus	WNMCS	9	25	Bridwell 1929; Nilsson and Johnson, in progress
15. Cosmobruchus	C	1	3	Bridwell 1931
16. Ctenocolum	MCS	8	18	Kingsolver and Whitehead 1974
17. Dahlibruchus	C	1	3	Bridwell 1931
18. Eubaptus	S	3	3	Teran 1964, 1967; Kingsolver 1969
19. Gibbobruchus	NMCS	13	25	Whitehead and Kingsolver 1975
20. Kytorhinus	N	1	1	Johnson 1976
21. *Lithraeus	S	2	20	Bridwell 1952
22. *Megacerus	WNMCS	51	65	Teran and Kingsolver 1977
23. Megasennius	C	1	1	Whitehead and Kingsolver 1975
24. Meibomeus	WNMCS	15	30	Kingsolver and Whitehead
25. Merobruchus	WNMCS	25	30	Kingsolver 1980, 1988
26. *Mimosestes	WNMCS	15	25	Kingsolver and Johnson 1979
27. Neltumius	NM	3	4	Kingsolver 1964; Johnson 1978
28. Pachymerus	CS	6	25	Bridwell 1929; Prevett 1966; Nilsson and Johnson, in progress
29. Palpibruchus	S	1	1	Borowiec 1987
30. Pectinibruchus	S	1	1	Kingsolver 1967
31. Penthobruchus	S	2	2	Kingsolver 1973
32. Pseudopachymerina	S	1	2	Decelle 1966

128

Table 1 (Cont.)

33. Pygiopachymerus	CS	2	2	Kingsolver 1970
34. Rhipibruchus	S	7	10	Kingsolver 1973, 1982; L'Argentier and Kingsolver 1984
35. Scutobruchus	S	6	10	Kingsolver 1968, 1983
36. Sennius	WNMCS	35	75	Johnson and Kingsolver 1973
37. Spatulobruchus	S	1	1	Borowiec 1987
38.*Stator	WNMCS	31	40	Johnson 1963; Kingsolver 1971 (W) Johnson and Kingsolver 1976 (N); Johnson, Kingsolver and Teran 1989 (S)
39. Stylantheus	N	1	1	Johnson 1976
40.*Zabrotes	WNMCS	20	50	Horn 1885; Kingsolver 1970, 1989

REFERENCES

Blackwelder, R. E. (1946) Checklist of the coleopterous insects of Mexico, Central America, the West Indies, and South America, Part 4, Bull. U.S. Nat. Mus. 185, 551-763.

Bondar, G. (1931a) Notas biologicas sobre alguns Bruchideos brasileiros do genero *Pseudopachymerus*, Rev. Entomol. 1, 417-422.

Bondar, G. (1931b) Notas biologicos sobre bruchideos brasileiros do genero *Spermophagus*, O Campo 11, 86-88.

Bondar, G. (1936) Notas biologicas sobre bruchideos observados no Brasil, Archiv. Inst. Biol. Veget. 3, 7-44.

Borowiec, L. (1987) The genera of seed beetles (Coleoptera, Bruchidae), Pol. Pismo Entomol. 57, 1-207.

Bottimer, L. J. (1968) Notes on Bruchidae of America north of Mexico with a list of world genera, Can. Entomol. 100, 1009-1049.

Bridwell, J. C. (1918) Notes on the Bruchidae and their parasites in the Hawaiian Islands, Proc. Hawaii. Entomol. Soc. 3, 465-509.

Bridwell, J. C. (1919) Some additional notes on the Bruchidae and their parasites in the Hawaiian Islands, Proc. Hawaii. Entomol. Soc. 4, 15-20.

Bridwell, J. C. (1920a) Insects injurious to the algaroba feed industry, Hawaii. Planter's Rec. 22, 337-343.

Bridwell, J. C. (1920b) Notes on the Bruchidae (Coleoptera) and their parasites in the Hawaiian Islands, 3rd paper, Proc. Hawaii. Entomol. Soc. 4, 403-409.

Bridwell, J. C. (1929) A preliminary generic arrangement of the palm bruchids and allies with descriptions of new species, Proc. Entomol. Soc. Wash. 31, 141-160.

Bridwell, J. C. (1932) The subfamilies of the Bruchidae, Proc. Entomol. Soc. Wash. 34, 100-106.

Bridwell, J. C. (1946) The genera of beetles of the family Bruchidae in America north of Mexico, Jour. Wash. Acad. Sci. 36, 52-57.

Decelle, J. (1976) Bruchidae (Coleoptera Chrysomeloidea) recoltes aux Iles Galapagos par N. et J. Lelup, Miss. Zool. Belge Iles Galapagos et en Ecuador 3, 326-334.

Fall, H. C. (1910) Miscellaneous notes and descriptions of North American Coleoptera, Trans. Amer. Entomol. Soc. 36, 160-189.

Hinckley, A. D. (1960) The klu beetle, *Mimosestes sallaei* (Sharp), in Hawaii (Coleoptera:Bruchidae), Proc. Hawaii. Entomol. Soc. 17, 260-269.

Horn, G. H. (1873) Revision of the Bruchidae of the United States, Trans. Amer. Entomol. Soc. 4, 311-342.

Horn, G. H. (1885) Contributions to the coleopterology of the United States, Trans. Amer. Entomol. Soc. 12, 128-162.

Johnson, C. D. (1963) A taxonomic revision of the genus *Stator* (Coleoptera:Bruchidae), Ann. Entomol. Soc. Amer. 56, 860-865.

Johnson, C. D. and Kingsolver, J. M. (1982) Checklist of the Bruchidae (Coleoptera) of Canada, United States, Mexico, Central America, and the West Indies, Coleop. Bull. 35, 409-422.

Kingsolver, J. M. (1964) The genus *Neltumius* (Coleoptera:Bruchidae), Coleop. Bull. 18, 105-111.

Kingsolver, J. M. (1970) A synopsis of the subfamily Amblycerinae Bridwell in the West Indies, with descriptions of new species. (Coleoptera: Bruchidae), Trans. Amer. Entomol. Soc. 96, 469-497.

Kingsolver, J. M. (1972) Synopsis of the genus *Stator* Bridwell in the West Indies, with descriptions of new species, Proc. Entomol. Soc. Wash. 74, 219-229.

Kingsolver, J. M. (1983) A review of the genus *Scutobruchus* Kingsolver (Coleoptera:Bruchidae), with descriptions of four new species, and new synonymy, Proc. Entomol. Soc. Wash. 85, 513-527.

Leng, C. W. (1920) Catalogue of the Coleoptera of America, north of Mexico, John D. Sherman, Jr., Mt. Vernon, NY.

Linnaeus, C. (1767) Systema naturae per regna tria naturae. Edition 12, Volume 1, pars 2, Holmiae.

Motschulsky, V. (1874) Enumeration des nouvelles especes de coleopteres rapportes de ses voyages, Bull. Soc. Imp. Natur. Moscou 46, 203-252.

Pfaffenberger, G. S. and Johnson, C. D. (1976) Biosystematics of the first-stage larvae of some North American Bruchidae (Coleoptera), U.S. Dept. Agric. Tech. Bull. 1525, Washington, D.C.

Pic, M. (1913) Bruchidae. Coleopterorum Catalogus, 55, 1-74, W. Junk, Berlin.

Prevett, P. F. (1966a) A new genus and species of Pachymerinae (Coleoptera: Bruchidae) from South America, Proc. Roy. Entomol. Soc. London (B) 35, 81-83.

Prevett, P. F. (1966b) The identity of the palm kernel borer in Nigeria, with systematic notes on the genus *Pachymerus* Thunberg (Coleoptera, Bruchidae), Bull. Entomol. Res. 57, 181-192.

Schaeffer, C. F. A. (1904) New genera and species of Coleoptera, Jour. New York Entomol. Soc. 12, 197-236.

Schaeffer, C.F.A. (1907) New Bruchidae with notes on known species and list of species known to occur at Brownsville, Texas, and in the Huachuca Mountains, Arizona, Science Bulletin, Museum of the Brooklyn Institute of Arts and Sciences 1, 291-306.

Schaeffer, C. F. A. (1909) New Coleoptera chiefly from Arizona, Science Bulletin, Museum of the Brooklyn Institute of Arts and Sciences 1, 375-386.

Schilsky, J. (1905) Synonymische bemerken zur Gattung *Bruchus* L., Deutsche Entomologische Zeitschrift 1904, 55-456.

Schoenherr, C. J. (1833) Genera et species curculionidum, cum synonymia hujus familiae. Volume 1(1). Lipsiae, Paris.

Schoenherr, C. J. (1839) Genera et species curculionidum, cum synonymia hujus familiae. Volume V(1). Lipsiae, Paris.

Sharp, D. (1885) Bruchidae, Biologia Centrali-Americana, Coleoptera 5, 437-504.

Teran, A. (1962) Observaciones sobre Bruchidae (Coleoptera) del noroeste Argentino, Acta Zool. Lilloana 18, 211-242.

A SYNOPSIS OF THE BRUCHID FAUNA OF JAPAN

KATSURA MORIMOTO[1]
Entomological Laboratory, Faculty of Agriculture,
Kyushu University, Fukuoka 812, Japan

Abstract. A total of 27 species of bruchids known from Japan is revised, of which 14 species are recognized as indigenous or established including an undescribed one. Five species have not been properly verified as to their establishment in the field, 6 species are newly synonymized, and 3 species are adventitious or their localities are mislabelled and thus excluded from our fauna. Keys for the identification of genera and species of *Spermophagus* and *Bruchidius* are given.

1. Introduction

Japanese species of the Bruchidae were monographed by Chûjô in 1937 as a fascicle of Fauna Nipponica. But it has not been quoted by any entomologists outside Japan owing to its limited distribution by the outbreak of World War II and lack of citation in the Zoological Record. Since then little progress has been made in the taxonomy of our bruchids except for the illustrations in two coloured books on Japanese Coleoptera.

In this paper I revise the Japanese species based on detailed morphological characters and by examining the type materials.

The terminology of parts of the male genitalia and the posterior legs follows that proposed by Kingsolver (1970) and by Johnson and Kingsolver (1973), respectively. The body length is measured from anterior margin of the head to apex of the pygidium.

I thank the following persons and institutions for the generous loan of specimens for this study: Prof. S. Takagi and Mr. M. Ohara, Entomological Institute, Hokkaido University; Dr. K. Baba, Niigata Pref.; Dr. S. Uéno, National Science Museum; Mr. Y. Hirano, Odawara City; Mr. T. Nohira, Forest Research Institute of Gifu Pref.; Prof. H. Sasaji, Fukui University; Mr. I. Matoba, Wakayama Pref.; Prof. M. Miyatake, Entomological Laboratory, Ehime University; Mr. M. Sakai, Laboratory of Medical Zoology, Ehime University; Mr. N. Gyotoku, Fukuoka Pref.; and Mr. S. Imasaka, Nagasaki Pref. I also thank Dr. N. Berti, Museum National d'Histoire Naturelle, Paris, and Dr. C. N. C. Lyal, British Museum (Natural History), London, for the loan of type specimens described by Pic and Sharp from Japan.

I am deeply indebted to Dr. J. M. Kingsolver for his kindness in reading the manuscript.

---

[1] Contribution from the Entomological Laboratory, Faculty of Agriculture, Kyushu University, Fukuoka (Ser. 4, No. 1).

K. Fujii et al. (eds.), Bruchids and Legumes: Economics, Ecology and Coevolution, 131–140.
© 1990 *Kluwer Academic Publishers. Printed in the Netherlands.*

2. Systematic and Synonymic Notes

1. *Bruchus rufimanus* Boheman, 1833
Infestation of the broad bean by this species was first discovered in 1925 in Kumamoto Pref., and had rapidly spread over the broad bean growing areas in Honshu, Shikoku and Kyushu.

2. *Bruchus pisorum* (Linnaeus, 1758)
This species also invaded Japan from America about 1887 (or about 1900 by some authors), and became distributed all over the country some ten years later.

3. *Bruchus loti* Paykull, 1800
Bruchus maculatipes Pic, 1927: 153 (Japon:Kioto). - Chûjô, 1937b: 31,84(Sado I.; adult fig.). - Nakane, 1963: 319, pl.160, fig. 3 (Honshu; host-plant: *Lathyrus maritimus*). - Morimoto, 1984: 225. **New synonymy.**
I have compared the type of *Bruchus maculatipes* from Kioto, Japan, with specimens from England and confirmed their identity. This species is commonly found on *Lathyrus maritimus* along the seashore in northern Japan in late June to August.
Distribution: Europe, Siberia, Russian Far East, Sakhalin, Kuril Isls., Hokkaido, Honshu.

4. *Callosobruchus ademptus* (Sharp, 1886)
This is a common species and weevils emerge from beans of *Pueraria lobata* in October and early December. They feed on the pollen of the Umbelliferae in spring and summer.
Distribution: Honshu, Shikoku, Kyushu, Tsushima and Taiwan.

5. *Callosobruchus chinensis* (Linnaeus, 1758)
A common and most injurious species among Japanese species, it is distributed all over the country.

6. *Callosobruchus maculatus* (Fabricius, 1775)

7. *Callosobruchus analis* (Fabricius, 1781)

8. *Callosobruchus phaseoli* (Gyllenhal, 1853)
These three species have frequently been intercepted in imported beans, and often collected from beans in markets in such big cities as Tokyo and Osaka, but they have never been collected in the field up to the present except for a single specimen of *C. analis* which was taken from Iriomote I. in the Ryukyus according to Nakane (1989).

9. *Sulcobruchus sauteri* (Pic, 1927) (Fig. 1)
The generic definition was given and the genitalia were illustrated on a species of this genus by Borowiec (1987). But, the generic status needs revision, because the type-species of the genus, *sauteri*, has peculiar genitalia with sclerotized parts at the apices of both median and lateral lobes, and are very different from that of Borowiec's species.
Biological studies were done by Watanabe (1985) on its host-plant *Caesalpinia sepiaria* var. *japonica*.
Distribution: Japan (Honshu, Shikoku), Taiwan.

10. *Acanthoscelides obtectus* (Say, 1831)

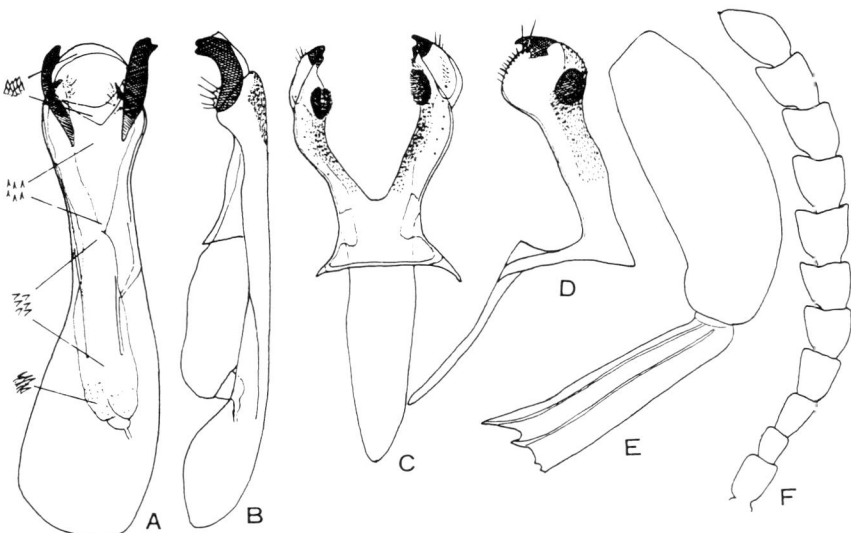

Figure 1. *Sulcobruchus sauteri.* A, B: median lobe, dorsal and lateral aspects. C, D: tegmen, dorsal and lateral aspects. E: hind leg. F: male antenna.

This famous pest has frequently been intercepted in Japan, and once caused serious damage in Okinawa (Hino, 1958). But, its establishment in Japan has not been verified.

11. *Acanthoscelides pusillimus* (Sharp, 1885)

Record of this species from Japan is adventitious, and only once found in cargo by Lewis (Sharp, 1886).

12. *Bruchidius kiotoensis* (Pic, 1913)

Bruchus kiotoensis Pic, 1913: 14 (Japon: Kioto).
Bruchidius kiotoensis: Chûjô, 1937a: 194. - Chûjô, 1937b: 58.
= *Acanthoscelides griseolus* (Fall, 1910). **New synonymy.**

I have examined the type of *B. kiotoensis*, which is a species of *Acanthoscelides* and corresponds well with the descriptions of *A. griseoulus* made by Johnson (1970, 1983).

13. *Bruchidius atriceps* (Pic, 1927)

Bruchus atriceps Pic, 1927: 12 (Japon: Kioto)
Bruchidius atriceps: Chûjô, 1937a: 193. - Chûjô, 1937b: 57.
= *Sennius lebasi* (Fahraeus, 1839). **New synonymy.**

The type agrees well with the description of *Sennius lebasi* (=*celatus*) in the revision by Johnson and Kingsolver (1973).

These two species noted above are distributed from Mexico to Central America, and their localities in Japan are apparently erroneous.

14. *Bruchidius lautus* (Sharp, 1886) (Fig. 2)

This species is characteristic in having long antennae in the male, and a slender aedeagus. The lateral lobes are fused to form a slender strap-like structure and are shallowly cleft at apex.

134

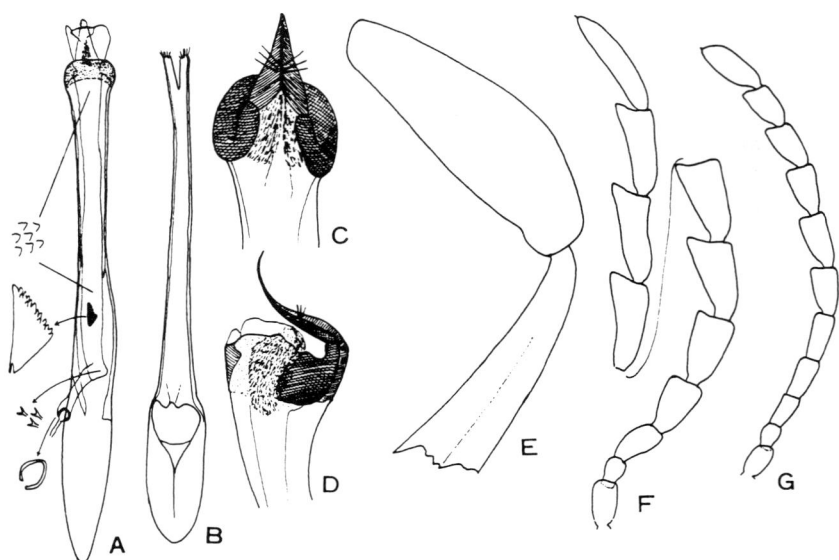

Figure 2. *Bruchidius lautus.* A: median lobe. B: tegmen. C, D: apex of median lobe, ventral and lateral aspects. E: hind leg. F: male antenna. G: female antenna.

The adult is well illustrated by Chûjô (1937), Lukjanovitsh and Ter-Minassian (1957), Nakane (1963), Tan et al. (1980), and Morimoto (1984). It has a wide distribution in the Russian Far East, Northern China, and Japan (Hokkaido, Honshu, Shikoku, Kyushu). The adults have been collected from *Vicia cracca.*

15. *Bruchidius japonicus* (Harold, 1878) (Fig. 3)
    Mylabris japonica Harold, 1878: 87 (Japan: Hagi).

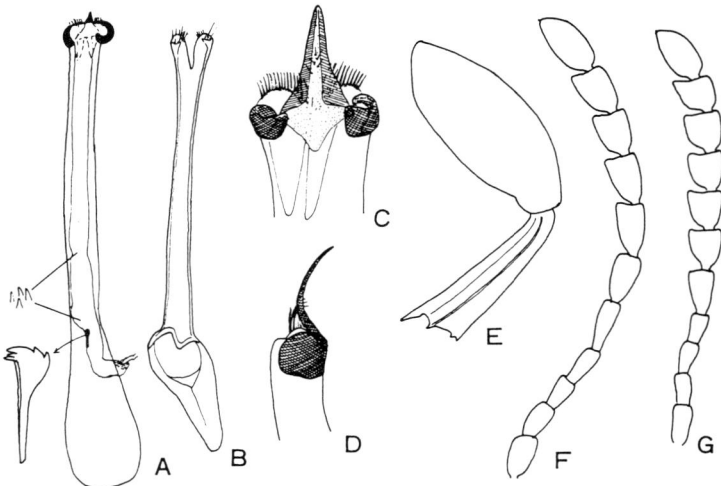

Figure 3. *Bruchidius japonicus.* A: median lobe. B: tegmen. C: apex of median lobe, ventral and lateral aspects. E: hind leg. F: male antenna. G: female antenna.

Bruchus japonicus: Sharp, 1886: 36 (Japan: Nagasaki, Junsai, Hagi; = *fulvipes*).

Bruchidius japonicus: Schilsky, 1905: Z, 84. - Chûjô, 1937b: 52 (Hokkaido, Honshu, Kyushu; Adult fig.). - Chûjô, 1950: 1253, fig. 3611. - Nakane, 1963: 319, pl.160, fig. 11. - Tan et al., 1980: 39, pl.II, fig. 20 (China). - Morimoto, 1984: 226, pl. 44, fig. 7.

Bruchus fulvipes Roelofs, 1879: lv (Japon). - Roelofs, 1880: 30.

Bruchidius fulvipes: Chûjô, 1937a: 194 (Honshu). - Chûjô, 1937b: 55 (Honshu; Adult fig.). - Tan et al., 1980: 38 (China). - Egorov et Ter-Minassian, 1984: 43 (Primorsky kray). - Morimoto, 1984: 226, pl. 44, fig. 9.

Bruchus comptus Sharp, 1886: 36 (Japan: Kobe, Hosokute).

Bruchidius comptus: Chûjô, 1937a: 193 (Honshu). - Chûjô, 1937b: 53 (Honshu; Adult fig.). - Tan et al., 1980: 37 (China). - Morimoto, 1984: 226, pl. 44, fig. 8. **New synonymy.**

Bruchidius japonicus, fulvipes and *comptus* are no more than individual variations of the same species. Chûjô(1937a,b) distinguished these three species by the presence or absence of the frontal carina as a key character, but the carina is more or less discernible in all specimens. The frons of the male is densely clothed with whitish broad hairs and the carina is usually concealed by them, whereas in the female, the frons is clothed with darker fine hairs and the carina is readily discernible. *B. comptus* was named for the smallest individual of the species.

This species emerges from seeds of *Lespedeza bicolor*, and has a wide distribution in the Russian Far East, Korea, northern China, Hokkaido, Honshu, Shikoku and Tsushima.

The genitalia are similar to those of *B. lautus*.

16. *Bruchidius terrenus* (Sharp, 1886) (Fig. 4)

Bruchus terrenus Sharp, 1886: 35 (Japan: Yokohama, Nagasaki, Ichiuchi, Yuyama).

Bruchidius terrenus: Chûjô, 1937a: 194. - Chûjô, 1937b: 61 (Honshu, Kyushu; Adult fig.; Host-plant: *Albizzia julibrissin*). - Nakane, 1963: 319, pl. 160, fig. 11. - Tan et al., 1980: 38, pl. II, fig. 19 (China). - Morimoto, 1984, Col. Jap., IV: 226, pl. 44. fig. 10.

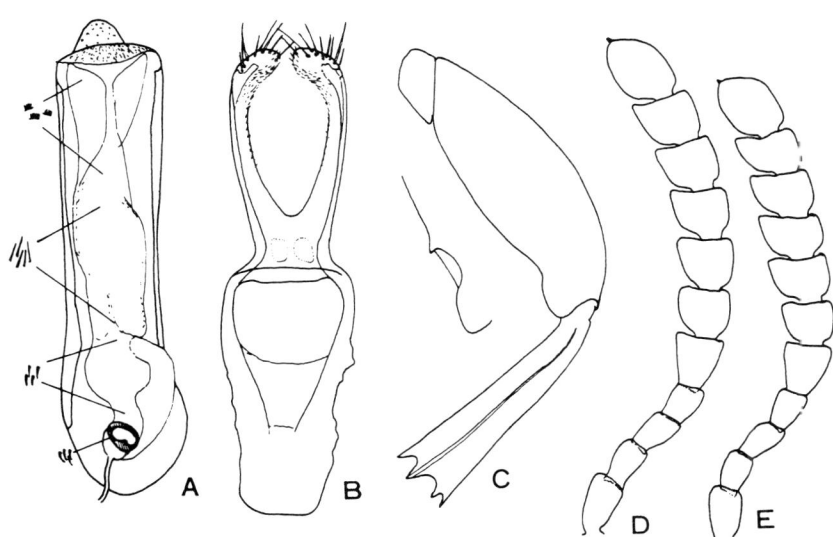

Figure 4. *Bruchidius terrenus.* A: median lobe. B: tegmen. C: hind leg. D: male antenna. E: female antenna.

Bruchidius notatus Chûjô, 1937a: 196 (Taiwan). - Chûjô, 1937b: 64 (Taiwan; Adult fig.). - Tan et al., 1980: 39 (China). **New synonymy.**

This species is widely distributed in China from Peking to Kwangton, Taiwan, Honshu, Shikoku, Kyushu, Tsushima and Amami-Oshima, and has been reared from seeds of *Albizzia julibrissin* and *Robinia pseudoacacia*.

17. *Bruchidius urbanus* (Sharp, 1886)

This species also attacks the same host-plants as *B. terrenus* and easily distinguished by its characteristic scaly marking on the elytra. Arora (1977) recorded *Albizzia procera* as a host-plant in India.

Distribution: Honshu, Shikoku, Kyushu, Tsushima, Korea, India.

18. *Bruchidius* sp.

This is the second species that has serrate margins of the pronotum as in *B. coreanus* Chûjô, 1937. Specimens were collected from Kyoto and Kumamoto, and biology is not known.

19. *Megabruchidius dorsalis* (Fahraeus, 1839)

The new genus *Megabruchidius* was named based on this species by Borowiec (1984; *M. bifovelatus* = *dorsalis*). This species has a wide distribution from India, China, Japan to Papua New Guinea. Host-plant is *Gleditsia japonica*.

20. *Spermophagus rufiventris* Boheman, 1833 (Fig. 5)
Spermophagus rufiventris Boheman, 1833: 107.
Spermophagus complectus Sharp, 1886: 37. (**New synonymy**)
Spermophagus japonicus Schilsky, 1906: 94.
Spermophagus kiotensis Pic, 1918: 8.
Spermophagus albonotatus Chûjô, 1937a: 198.

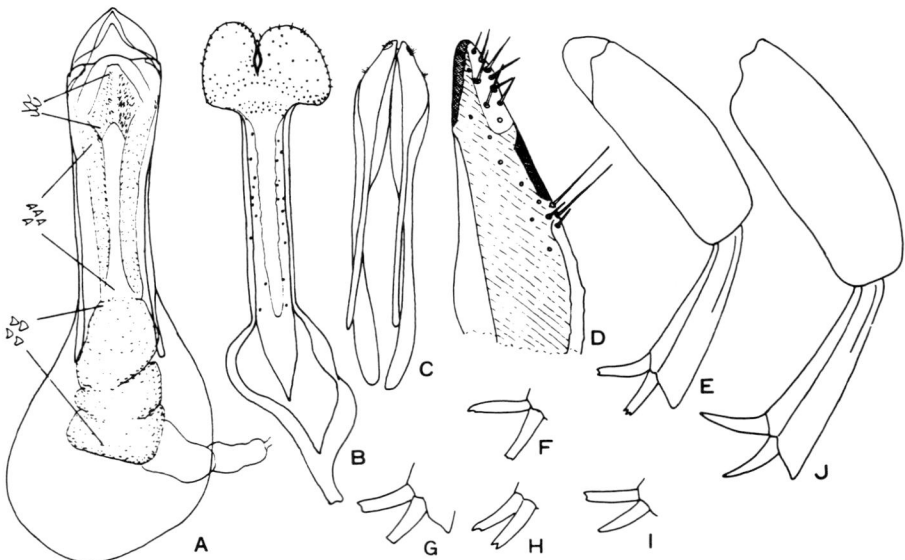

Figure 5. *Spermophagus rufiventris*. A: median lobe. B: tegmen. C: ovipositor. D: apex of left ovipositor, enlarged. E: hind leg (Niigata). F - I: tibial spurs of hind legs (Niigata). J: ditto (Taiwan).

? = *Spermophagus testaceicolor* Pic, 1917: 11.

(For other synonyms see Kingsolver and Borowiec, 1988)

This species occurs in the Belgian Congo, Caucasus, south-eastern Asia from India to Sumatra and Taiwan, China and Japan, and is extremely variable in size, colour, vestiture and structure of the antenna, but constant and characteristic in the structure of the genitalia according to the recent review by Kingsolver and Borowiec (1988); *S. japonicus* Schilsky, *kiotensis* Pic, *albonotatus* Chûjô and *undulatus* Chûjô were synonymized with this species by them. *S. complectus* is also a synonym of this species according to the examination of the type. *S. undulatus* from Taiwan is a valid species as noted in the key. Specimens from Japan, the Ryukyus and Taiwan are, however, apparently classified into two types by the structures of the tibial spurs of the hind legs; namely, the spurs are truncate or truncate and notched at the apices in the specimens from Hokkaido, Honshu and Kyushu (*complectus*-type), whereas they are sharply pointed in the specimens from Amami-oshima, Okinawa, Ishigaki and Iriomote Isls. and Taiwan (*albonotatus*-type).

S. testaceicolor Pic, 1917, described from Kioto, Japan, is probably a teneral form of *rufiventris* judging from the description, but the type is not present in the collection of the Museum in Paris (personal communication from Dr. Berti).

21. *Zabrotes subfasciatus* (Boheman, 1833)

This species has often been collected from some hosts in markets in Kyushu (Sakaguchi, 1952; Hamada, 1953; Fukamachi, 1954), and is also commonly intercepted from imported beans, but is not yet established in Japan.

22. *Kytorhinus senilis* Solsky, 1867 (Fig. 6)
　　Kytorhinus senilis Solsky, 1867: 310.
　　Pygobruchus scutellaris Sharp, 1886: 38 (Male; Kobe).

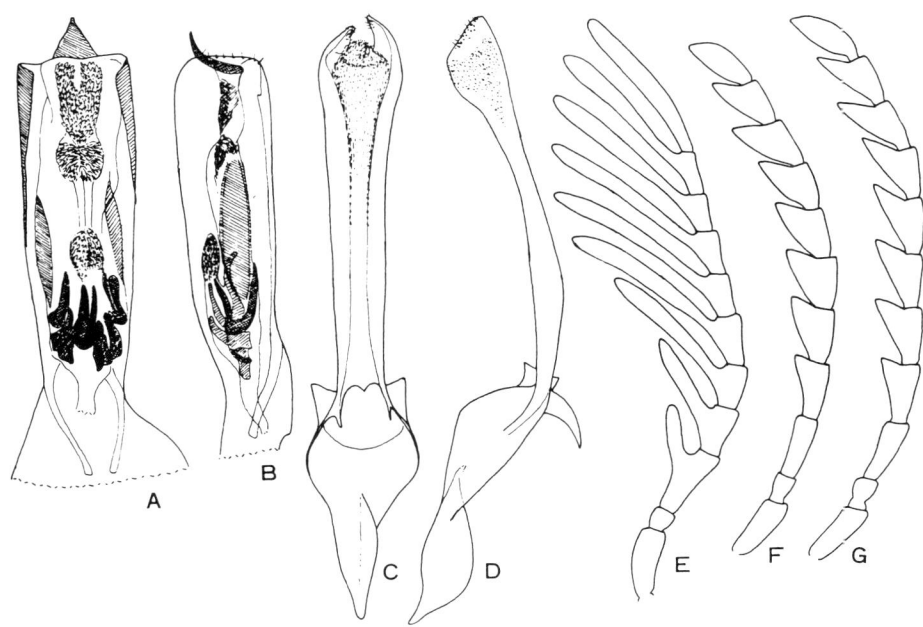

Figure 6. *Kytorhinus senilis*. A, B: median lobe, dorsal and lateral aspects. C, D: tegmen, dorsal and lateral aspects. E: male antenna. F, G: female antenna.

Kytorhinus sharpianus Bridwell, 1932: 106 (New name for *P. scutellaris* Sharp, nec Motschulsky, 1874). - Chûjô, 1937b: 69 (Honshu; Male fig.). - Chûjô, 1950: 1252, fig. 3609. - Morimoto, 1984: 225. **New synonymy.**

Japanese specimens agree well with the description of *senilis* made by Lukjanovitsh and Ter-Minassian (1957) and Egorov and Ter-Minassian (1983).

It occurs in eastern Siberia, the Russian Far East, northern China, and Japan (Honshu, Shikoku, Kyushu). Its host-plant is *Sophora flavescens* in Japan and China (Ue, 1934; Gyotoku, 1954; Tan et al., 1980) and biological studies were made by Shimada (1988).

3. Keys to Genera and Species of Japanese Bruchids

Genera
1(4) Hind tibia with two long movable spurs at apex.   Subfamily Amblycerinae
2(3) Elytron with tenth stria extending nearly to apex. Prosternal process acuminate, completely separating fore coxae. . . . . *Spermophagus*
3(2) Elytron with tenth stria abbreviate, not extending beyond middle. Prosternal process short, triangular, separating fore coxae only for half of their length. . . . . *Zabrotes*
4(1) Hind tibia without movable spurs, often with a mucro and spinules at apex.
5(6) Mesepisternum and mesepimeron nearly of the same size. Pygidium and one or two tergites exposed behind elytra. Hind tibia without apical spine. Antennae sexually dimorphic, strongly serrate or pectinate in male.
                                        Subfamily Kytorhininae
. . . . *Kytorhinus*
6(5) Mesepimeron very narrow. Only pygidium exposed behind elytra. Hind tibia with a mucro and spinules at apex.   Subfamily Bruchinae
7(8) Pronotum square or trapezoidal, emarginate on lateral margins near middle and with a denticle before the emargination. Middle tibia in male with apical spines or plates. . . . . *Bruchus*
8(7) Pronotum conical or campanulate, without lateral emargination and denticle.
9(10) Hind femur with a spine or denticle on both internal and external carinae. . . . . *Callosobruchus*
10(9) Hind femur dentate on internal carina only, or not dentate.
11(12) Hind femur sulcate on ventral margin, both external and internal margins carinate and unarmed. . . . . *Sulcobruchus*
12(11) Hind femur not sulcate on ventral margin, internal margin carinate and often dentate.
13(14) Pygidium in female with a pair of large foveae. First ventrite with a median velvety patch in male. Larger species more than 4.0 mm in length. . . *Megabruchidius*
14(13) Pygidium and first ventrite simple. Usually smaller than 4.0 mm in length.
15(16) Hind femur usually with one large and two small subapical denticles. . . . . *Acanthoscelides*
16(15) Hind femur usually with a small subapical denticle or unarmed. . . . . . *Bruchidius*

Spermophagus (including Taiwanese species)
1(4) Hind tibia with both lateral and lateroventral carinae in entire length. Smaller than 2.8 mm in length. Lateral lobes slender.
2(3) Vesture brownish with whitish two bands on elytra. Median lobe with a sclerite on each side at apex; lateral lobes slender, originating from a broad plate. Body

2.7 mm in length. (Taiwan)..................... *S. undulatus* Chûjô, 1937

3(2) Vesture blackish brown with whitish variable patches on elytra, usually with two bands. Median lobe pointed at apex; lateral lobes broad at base and tapered apically, originating from the part of ordinal structure. Body 2.2-2.5 mm in length. (Taiwan, China, India)........................ *S. niger* Motschulsky, 1866

(= *formosanus* Pic, 1917)

4(1) Hind tibia with a sharp lateroventral carina, and abbreviate lateral carina, the latter present on basal half. Body larger than 2.8 mm in length. Lateral lobes oval.

5(6) Body larger, 4.2-4.4 mm in length. Lateral lobes distant at base. (Taiwan, the other locality see Kingsolver and Borowiec, 1988)... *S. titivilitius* Boheman, 1833

6(5) Body 2.8-3.6 mm in length. Lateral lobes close to each other at base.......
.... *S. rufiventris* Boheman, 1833

Bruchidius

1(4) Pronotum with 3-5 denticles on each side behind the middle.

2(3) Body larger, 5.0-5.5 mm in length. Vesture predominantly greyish white or yellowish brown, with small reddish brown spots on 3, 5, 7 and 9 intervals arranged transversely before and behind the middle of elytra. (Korea)................
.... [*B. coreanus* Chûjô, 1937]

3(2) Body 2.8-3.9 mm in length. Vesture of elytra predominantly blackish, with three greyish bands and first two intervals greyish................. *Bruchidius* sp.

4(1) Pronotum not denticulate at sides.

5(8) Terminal segment of antenna distinctly longer than broad. Median lobe very slender, with a triangular and warped sclerite at apex and a pair of sclerites surrounding apex; lateral lobes very slender, shallowly cleft at apex.

6(7) Vesture concolorous grey. Antenna dark reddish brown except for reddish basal segments, much longer than body in male. Hind femur almost blackish. Internal sac with a triangular sclerite. Body 2.3-4.2 mm in length... *B. lautus* (Sharp, 1886)

7(6) Vesture of elytron greyish with a black mark each at base, middle and apex in general, often almost greyish with three brownish bands instead of marks. Antennae and legs reddish yellow, male antenna not longer than body. Internal sac with a slender sclerite. Body 1.4-3.0 mm in length....... *B. japonicus* (Harold, 1878)

8(5) Terminal segment of antenna almost as long as broad. Median lobe robust, simply terminate at apex; lateral lobes deeply cleft.

9(10) Abdomen reddish except for blackish basal margin. Elytron black, with two greyish bands and greyish sutural intervals. Internal sac with a pair of small sclerites at ostium. Body 3.0-4.1 mm in length. .... *B. urbanus* (Sharp, 1886)

10(9) Abdomen and thorax concolorous, usually black. Elytron variegated with greyish and brownish vesture, the latter forming irregular and indefinite bands before and behind the middle and apex. Internal sac without sclerites at ostium. Body 2.7-3.7 mm in length. .... *B. terrenus* (Sharp, 1886)

REFERENCES

Arora, G. L. (1977) Taxonomy of the Bruchidae (Coleoptera) of Northwest India, Part 1 (Adults), Oriental Insects, Suppl. 7, 1-132.

Borowiec, L. (1984) Two new genera and species of seed-beetles from the oriental region. (Coleoptera, Bruchidae, Bruchinae), Polsk. Pismo Entomol. 54, 115-129.

Borowiec, L. (1985) On the oriental *Spermophagus* Schoenherr (Coleoptera, Bruchidae, Amblycerinae), with description of four new species, Polsk. Pismo Entomol. 55, 781-790.

Borowiec, L. (1987) The genera of seed-beetles (Coleoptera, Bruchidae), Polsk. Pismo Entomol. 57, 2-207.

Borowiec, L. (1988) Fauna Polski, 11, Bruchidae (Insecta: Coleoptera), Polska Akademia Nauk, Instytut Zoologii, Warsawa (in Polish).

Bridwell, J. C. (1932) The subfamilies of Bruchidae (Coleoptera), Proc. Entomol. Soc. Washington 34, 100-106.

Chûjô, M. (1937a) Some additions and revisions of Bruchidae (Coleoptera) from the Japanese empire, Trans. Nat. Hist. Soc. Formosa, 27, 189-201.

Chûjô, M. (1937b) Fauna Nipponica, vol.10, fasc. 8, no. 9, Bruchidae, Sanseido, Tokyo (in Japanese).

Chûjô,M. (1950) Bruchidae, in Esaki et al. (eds.), Iconographia Insectorum Japonicorum (ed. II), 1252-1254.

Egorov, A. B. and Ter-Minassian, M. E. (1983) Bruchidae of eastern Siberia and Russian Far East, Vladivostok (in Russian).

Harold, E. v. (1978) Beiträge zur Käferfauna von Japan (Viertes Stück), Deut. Entomol. Zeit. 23, 321-365.

Hirano, I. (1959) The list of Japanese references on insects, 203, Bruchidae, Osaka Plant Protection 7, 317-348.

Hoffmann, A. (1965) Observations sur les *Kytorrhinus* et description d'une espèces inédite de la Mongolie centrale (Col. Bruchidae), Ann. Soc. Entomol. Fr. (N. S.) 1, 63-70.

Johnson, C. D. (1970) Biosystematics of the Arizona, California, and Oregon species of the seed beetle genus *Acanthoscelides* Schilsky (Coleoptera: Bruchidae), Univ. Calif. Publ. Entomol. 59.

Johnson, C. D. (1983) Ecosystematics of *Acanthoscelides* (Coleoptera: Bruchidae) of Southern Mexico and Central America, Misc. Publ., Entomol. Soc. Amer. 56.

Johnson, C. D. and Kingsolver, J. M. (1973) A revision of the genus *Sennius* of North and Central America (Coleoptera: Bruchidae), Tech. Bull. U.S.D.A. 1462.

Kingsolver, J. M. (1970) A study of male genitalia in Bruchidae (Coleoptera), Proc. Ent. Soc. Wash. 72, 370-386.

Kingsolver, J. M. and Borowiec, L. (1988) The genus *Spermophagus* in the New World (Coleoptera, Bruchidae), Elytron 2, 81-84.

Lukjanovitsh, F. K. and Ter-Minassian, M. E. (1957) Fauna SSSR, XXIV (1), Bruchidae (in Russian).

Morimoto, K. (1984) Bruchidae, in Hayashi et al. (eds.), Coleoptera of Japan in Color, IV, Hoikusha, Osaka, pp. 225-226, pl. 44 (in Japanese).

Nakane, T. (1963) Bruchidae, in Nakane et al. (eds.), Iconographia Insectorum Japonicorum Colore Naturali Edita. II, Hokuryukan, Tokyo, pp. 319-320, pl. 160 (in Japanese).

Nakane, T. (1989) Classification and discrimination of beetles: Bruchidae, Nature and Insects 24(1), 2-7 (in Japanese).

Pic, M. (1913) Espèces et variétés nouvelles appartenant a diverses familles, Mélanges exotico-entomologiques 6, 8-16.

Pic, M. (1927a) Nouveautés diverses, Mélanges exotico-entomologiques 48, 1-32.

Pic, M. (1927b) Coléoptères nouveaux de Chine et du Japon, Bull. Soc. Entomol. France 152-154.

Roelofs, W. (1879) Diagnoses de nouvelle espèces de Curculionides, Brenthides, Anthribides et Bruchides du Japon, C. R. Soc. Entomol. Belg. 23, liii-lv.

Roelofs, W. 1880) Additions à la faune du Japon, nouvelles espèces de Curculionides et familles voisines, observations sur les espèces deja publiées, Ann. Soc. Entomol. Belg. 24, 5-31.

Schilsky, J. (1905) Die Käfer Europa's, 41, A-MM + 1-100.

Schilsky, J. (1906) Die Käfer Europa's, 42, 1-100.

Sharp, D. (1886) On the Bruchidae of Japan, Ann. Mag. Nat. Hist. (5) 17, 34-38.

Shimada, M. (1988) Dry bean infestation and oviposition without feeding by a wild multivoltine bean weevil, *Kytorhinus sharpianus* (Bridwell) (Coleoptera: Bruchidae), Appl. Entomol. Zool. 23, 459-467.

Tan, J., Yu, Y., Li, H., Wang, S., and Jiang. S. (1980) Economic Insect Fauna of China, fasc. 18, Coleoptera: Chrysomeloidea (I), Beijin, China (in Chinese).

Watanabe, N. (1985) Oviposition habit of *Sulcobruchus sauteri* (Pic) and its significance in speculation on the pre-agricultural life of seed beetles attacking stored pulses (Coleoptera: Bruchidae), Kontyû 53, 391-397.

DIVERSITY IN LIFE CYCLE PATTERNS OF BRUCHIDS OCCURRING IN JAPAN (COLEOPTERA: BRUCHIDAE)

NAOSHI WATANABE
Kobe Plant Protection Station,
1-1 Hatobacho, Chuo-ku, Kobe 650 Japan

ABSTRACT. By original observations and a review of the literature, I found seven types of life cycles of bruchids that occur in Japan. I studied *Bruchus rufimanus, Callosobruchus ademptus, Megabruchidius dorsalis, Sulcobruchus sauteri, Kytorhinus senilis, C. chinensis* and *C. maculatus. B. rufimanus, C. ademptus,* and *M. dorsalis* are univoltine and infest only soft seeds in maturing pods. The remainder are multivoltine and infest both soft seeds in pods and hardened ripe seeds. The adult is the overwintering stage in *B. rufimanus,* the larva in *C. ademptus, M. dorsalis,* and *K. senilis,* and both adult and larva in *S. sauteri.* Imaginal feeding before reproduction is indispensable for *B. rufimanus, C. ademptus,* and *S. sauteri.* The adults of *K. senilis, C. chinensis* and *C. maculatus* can deposit viable eggs without feeding, but their life span is lengthened by taking food as adults. The adults of *M. dorsaris* does not drink or eat. The species *C. chinensis* and *C. maculatus* live both in the field and in storage, but the Japanese population of the latter live only in stored seeds.

1. Introduction

Past ecological studies on the bruchids in Japan have been concentrated on the species of *Callosobruchus* and *Bruchus* of economic importance. The former are called storage bruchids and have several generations a year (multivoltine). They infest legume seeds both in the field and in storage. The latter are called field bruchids and have but one generation per year (univoltine), feed on soft seeds in pods in the field, and overwinter as adults. The former contrast so much to the latter in their patterns of life cycle that they have been often cited as having typycal life histories of the Bruchidae and species of *Bruchus* are regarded as representative of wild bruchids. Center and Johnson (1974), however, suggested that there are several types of coevolution between bruchids and their hosts especially in terms of ovipositional habits of the beetles. Watanabe (1986) and Shimada (1988) reported that the wild bruchids *Sulcobruchus sauteri* (Pic) and *Kytorhinus senilis* (Solsky)(= *K. sharpianus* Bridwell) are both essentially different from species of *Bruchus* in their mode of life cycle.

Because of these differences, I will report and discuss here the diversity in the life cycles of the bruchids occurring in Japan through both original observations and a review of the literature. The field observations were made in Okayama and Kanagawa Prefectures and supplementary studies were conducted in a laboratory in 1985 and 1986. Okayama and Kanagawa Prefectures have similar climatic conditions throughout the year.

K. Fujii et al. (eds.), Bruchids and Legumes: Economics, Ecology and Coevolution, 141–147.
© 1990 *Kluwer Academic Publishers. Printed in the Netherlands.*

2. Life Cycle Patterns

I have confirmed seven representative species whose life cycles are different from each other. These are summarized in Table 1.

2.1. *BRUCHUS RUFIMANUS* (PIC)

The region covering West Asia and the Meditteranean Basin is widely known to be the center of *Bruchus* beetles and their legume hosts of the Tribe Vicieae. *B. rufimanus* and *B. pisorum* are the major introduced pests of horsebeans and peas, respectively, in Japan. The adults of *B. rufimanus* that overwinter in the field feed on the leaves and flowers of the host plants in the spring. The female beetles lay their eggs on the pods, and the larvae feed on the growing seeds in the pods. The adults usually emerge in September, but sometimes overwinter in stored seeds and emerge in the spring (Kamito et al., 1977).

2.2. *CALLOSOBRUCHUS ADEMPTUS* (SHARP)

This beetle is widely known to infest kudzu, *Pueraria lobata* (Wild.) Ohwi but its life cycle has been little studied. The larvae were observed to overwinter in the seed in pods which remained on the defoliated vines, and the adults to emerge in March and visit wild flowers. Under laboratory conditions of 24°C, the adults died within a week

Table 1. Seven representative bruchids showing different life cycle patterns in Japan

Feeding on soft (S) or hard (H) seed				Time of ☙:flowering, ⌀:pod ripening, 🪲:adult emergence							
Imaginal feeding for reproduction (+ : indispensable, — : unnecessary)											
No. of generations a year											
Wintering stage (A: adult, L: larva)											
Representative bruchid				APR	MAY	JUN	JUL	AUG	SEP	OCT	
Bruchus rufimanus [1]	A	1	+	S	☙		⌀			🪲	
Callosobruchus ademptus [2]	L	1	+	S		🪲	⌀		☙		⌀
Megabruchidius dorsalis [3]	L	1	—	S			☙	🪲		⌀	⌀
Sulcobruchus sauteri [4]	AL	2+?	+	SH	🪲	☙		🪲		⌀	⌀ 🪲
Kytorhinus senilis [5]	L	3	±	SH		☙	🪲	⌀	🪲	🪲	
Callosobruchus chinensis (in the field) [6]	L	1+?	+	S(H?)	🪲				☙	⌀	
(in storage) [7]*	L	4	—	H	🪲		🪲		🪲	🪲	
Callosobruchus maculatus (in the field, no record in Japan) (in storage) [8]*	L	4	—	H	🪲		🪲		🪲	🪲	

Host plants: [1] *Vicia faba*, [2] *Pueraria lobata*, [3] *Gleditsia japonica*, [4] *Caesalpinia sepiaria* var. *japonica*, [5] *Sophora flavescens*, [6] *Vigna* spp. and *Dumbaria villosa*, [7], [8] Many kinds of stored pulses.
* Wintering stage and time of adult emergence are changeable according to the storage conditions.

when they were starved, but they survived 2 to 3 months when supplied with dehydrated yeast, sugar, and water; or the pollen of *Cameria japonica* L. and water.

These data indicate that the adults can lenghten their life span by taking pollen, honey etc., until August when the kudzu vine bears pods.

2.3. *MEGABRUCHIDIUS DORSALIS* (FAHRAEUS)

This species (= *Bruchidius dorsalis*) has the largest body of all the bruchids in Japan and had been merely known to infest *Gleditsia japonica* Miquel. The beetle was observed to overwinter in the larval stage (as was *C. ademptus*) and the adults emerged in late June to early July.

The emerged adults were put in a container at 24°C. They took neither water nor solid food (dehydrated yeast and sugar) and died within 10 days. They did, however, lay many eggs on the inner surface of the container; some of which showed embryonic development. The adults emerge not before, but after the host plants bear young pods. This might be the reason why *M. dorsalis* lives for a short period and does not require imaginal food, which is unlike *C. ademptus*.

2.4. *SULCOBRUCHUS SAUTERI* (PIC)

Previous observations (Watanabe, 1986) and this study proved that this beetle lays its eggs on green pods as well as hardened seeds. The imaginal feeding is indispensable for oviposition. The females laid their eggs even on glass balls placed in a container when there were no host seeds in the container (Table 2).

2.5. *KYTORHINUS SENILIS* (SOLSKY)

This beetle (= *K. sharpianus* Bridwell) infests *Sophora flavescens* Aiton known as a Chinese herb. Shimada (1988) found that the beetles oviposit on both the green pods and hardened seeds as does *S. sauteri*. The first generation larvae developed within the fresh growing seeds and emerged as adults in mid August. These adults laid eggs and the second generation larvae fed on matured, dry beans on the standing plants. Adults emerged at the end of September. These adults laid eggs directly on dry beans

Table 2. Ovipositional tests for *Sulcobruchus sauteri*. The adults were held in a container supplied with food and objects for oviposition

No. adult pairs	Objects for oviposition		No. eggs deposited				
	Kind	Number	Total	Per object			
				Max	Min	Mean	m[1]
4 {	*Caesalpinia sepiaria*	50	50	6	0	1	2.08
	Vigna angularis	50	0	-	-	-	-
10	*Vigna angularis*	50	162	17	0	3.24	6.36
10 {	Large glass balls[2]	20	61	11	0	3.05	4.89
	Small glass balls[3]	20	1	1	0	0.02	-

[1] Iwao's (1968) 'mean crowding' for eggs per seed. [2] 8.0 cm in diameter. [3] 5.5 cm in diameter.

inside the split pods and thus beans were infested by the third generation larvae.

The most significant difference between *S. sauteri* and *K. senilis* is that the latter does not always require imaginal food for their reproduction unlike the former.

2.6. *CALLOSOBRUCHUS CHINENSIS* (L.)

This is the most important bruchid in Japan, because it attacks stored pulses as well as pods on standing crops. Even though harvest of infested seeds from farm gardens is the primary cause of storage infestation, the field life of this beetle had rarely been studied till Shinoda and Yoshida (1985) initiated their intensive studies. Shinoda and Yoshida (1989) confirmed the univoltine field life of this species. The larvae overwinter in seeds of a wild legume, *Dunbaria villosa* (Thunb.) Makino or *Vigna angularis* var. *nipponensis* (Ohwi) Ohwi and Ohashi. The adults emerged in early summer and survived by feeding in the field till the legume bore pods in autumn.

I found that the adults deposited their eggs on the pods of an early ripening variety of *V. angularis* or *V. unguiculata* and the resulting offspring deposited their eggs on the seeds which remained in the shattered pods or on the ground. This indicates that the beetle is not always univoltine in the field.

2.7. *CALLOSOBRUCHUS MACULATUS* (F.) 'FLIGHTLESS STRAIN'

This species is likely to have originated in Tropical western Africa (Watanabe and Sugimoto, 1988), where the beetle was found flying and infesting wild and cultivated cowpeas (Alzouma, 1981). Like *C. chinensis*, it feeds on the legume seeds both in pods and in storage in its original home, but it has been found exclusively in storage in Japan (Nagayasu and Matsushita, 1980). I confirmed that the two popuations collected from Kanagawa Prefecture and Kobe City, respectively, produced only the 'flightless form' (Utida, 1954, 1972) even when the rearing conditions were made suitable for producing the 'flight form'. This retrogression in flying ability (Utida, 1970) might be due to the longterm life in stored seeds during its travel from its original home to other lands.

3. Discussion

3.1. THREE TYPES OF INSECTS FEEDING ON SEEDS

Bruchids and some other insects develop in basically two types of seeds: 'soft' (growing) and 'hard' (matured). Thus these insects can be classified into three groups based on the condition of their host seeds: 'S-type' feeding only on soft seeds, 'SH-type' feeding on both soft and hard seeds, and 'H-type' feeding only on hard seeds.

The S-type is represented by the snout beetles (*Curculio* spp.) infesting growing acorns; SH-type by the grain moth (*Sitotroga cerealella* (Oliv.)) which attacks both immature and harvested wheat seeds; and the H-type by the granary weevil (*Sitophilus granarius* (L.)) which lacks the ability to fly and is found exclusively in storage at present.

3.2. VOLTINISM AND FOOD HABIT

In Table 1, the upper three univoltine species are the S-type and the lower three multivoltine species are the SH-type. The bottom species, *C. maculatus* is the SH-type in its original home (Alzouma, 1981), but the introduced Japanese population seems to have lost its ability to fly and to lack the ability to live in the field. Although this strain was observed to oviposit on green pods in the laboratory, its life cycle is actually the same as an insect of the H-type.

The S-types inevitably have a univoltine life cycle when it is monophagous. A multivoltine life is possible in S-types when the insect is polyphagous but such species are unkown in the bruchids.

3.3. PRE-AGRICULTURAL EVOLUTION IN LIFE CYCLES

Crowson (1981) maintains that the Bruchidae belong to the Chrysomeloidea and probably appeared in the Upper Jurassic Period when the cotyledonous plants probably also appeared. Probably bruchids evolved from common ancestor(s), the Chrysomeloidea. Thus it appears bruchid larvae adapted to mine into and to feed on the immature kernels of newly evolved cotyledonous plants.

An early bruchid was presumably the S-type and adapted to synchronize their larval period with the podding stage of their host plants so that this mode of life cycle inevitably became univoltine and monophagous. In a physiological sense, this does not always mean that the larva can develop in only one host species but that the female chooses one species of seed or pod for its oviposition. Thus, a larva that emerges feeds on that seed or pod. This habit can be defined here with the term of 'monoovipositional'.

Among the S-type, there supposedly appeared the SH-type. They were multivoltine and their lives as larvae were in dry seeds as well as soft seeds. These SH-type species might have evolved in different taxonomic groups of the bruchids at about the same time (parallel evolution). For example, both the genera *Kytorhinus* and *Callosobruchus* have SH-types but only distantly related to each other phylogenetically. Such a species might have evolved in an area where adult bruchids had difficulty in lengthening their life spans, such as an area with a dry season.

3.4. POST-AGRICULTURAL EVOLUTION IN LIFE CYCLES

The most important environmental changes for the bruchids in the period of post-agriculture were 1) the extension of geographic distribution of cultivated legumes; 2) the storage of harvested seeds; and 3) the development of trade of harvested pulses.

Some SH-type bruchids which were attacking these edible pulses and had the same mode of life as *K. sharpianus*, became the the 'storage species'. They could extend their host range to some allied legume species introduced from other regions. For example *C. chinensis* in Asia infested the cowpea, *Vigna unguiculata* from Africa as a new field host (Watanabe and Sugimoto, 1988). This may not be attributable to a change in ovipositional behaviour, but merely to ovipositing on a substrate with a substance common to species belonging to the genus *Vigna*.

In storage, these bruchids were free from the yearly life cycle of the host plants, which enabled the bruchids to find new hosts, even if they were distantly related to the original host(s). A most surprising example is that *C. chinensis*, whose proper host is supposed to be the Phaseoleae legume(s), can infest the lotus seed, *Nelumbo nucifera* in storage (Kingsolver, 1979; Furusawa, 1987). This might be, of course, triggered by the storage of the different kind of seeds in the same place and by the retrogression of host selectivity in the ovipositional behavior rather than the change in the larval food preference (Watanabe, 1986). The fact that even a wild bruchid *S. sauteri* oviposited on glass balls in the laboratory indicates that the SH-type bruchids might have been preadapted, to some extent, to oviposit on seeds other than their original hosts. Thus the extension

146

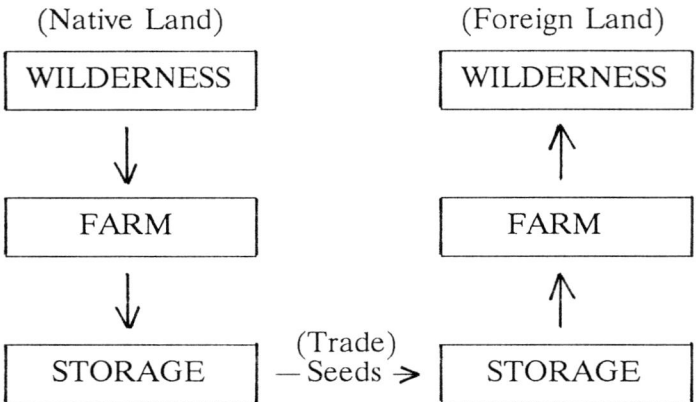

Figure 1. Schematic illustration for the usual extension of life space in 'storage' type bruchids.

of the kinds of possible hosts and behavior ('polyovipositional' habit) in the 'storage' type bruchids might be acquired by the agricultural activity of humans (Fig. 1) rather than any positive change in the habits of bruchids. The H-type strains such as the flightless *C. maculatus* are developing in storage in new lands where the the field hosts are hardly or not available (Watanabe and Sugimoto, 1988).

The H-type strain of bruchids is a strain undergoing rapid evolution which adapts them exclusively to the manmade conditions. Such a strain, however, can be easily controlled by man because the indoor conditions are easily treated by man unlike the outdoor ones. Therefore, it is possible that such evolution could endanger the continued existence of a species.

3.5. CONCLUSION

Presumably the early bruchids were univoltine and monophagous which adapted them to have a different mode of life, especially the adult stage. Then some multivoltine species appeared before the initiation of agriculture and among such species some bruchids have become polyphagous due to post-agricultural dispersal into new hosts.

Acknowledgements. For his aid in observing wild bruchids, I am deeply appreciative to Mr. T. Okamoto of Nariwa Highschool, and to Dr. S.Utida, Prof. Emer., Kyoto Univ. for his advice on this study. I am also much grateful to Dr. C. D. Johnson, Northern Arizona Univ. and to Dr. J. M. Kingsolver, USDA National Museum of Natural History, for their reading and correcting the manuscript.

REFERENCES

Alzouma, I. (1981) Observation of the ecology of *Bruchidius atrolineatus* Pic and *Callosobruchus maculatus* F. (Coleoptera: Bruchidae), in the Ecology of Bruchids Attacking Legumes (pulses), Series Entomologica Vol. 19, Junk, The Hague, pp. 205-221.

Center, T. D. and Johnson, C. D. (1974) Coevolution of some seed beetles (Coleoptera: Bruchidae) and their hosts, Ecology 55, 1096-1103.

Crowson, R. A. (1981) The Biology of the Coleoptera, Academic Press, London, New York, Toronto, Sydney, San Fransisco.

Furusawa, K. (1987) The lotus seed, *Nelumbo nucifera* Gaertn as a new host plant of the adzuki bean weevil, *Callosobruchus chinensis* L. (Coleoptera: Bruchidae), Appl. Entomol. Zool. 22, 388-389.

Iwao, S. (1968) A new regression method for analyzing the aggregation pattern of animal populations, Res. Popul. Ecol. 10, 1-20.

Kamito, A., Sakai, K., and Obi, A. (1977) Studies on broad bean weevils, *Bruchus rufimanus* Boheman with reference to its bionomics and control, Agricultural Chemicals Institute, Nihon Tokushu Noyaku Seizo K. K., Hinoshi, Tokyo (in Japanese with an English summary).

Kingsolver, J. M. (1979) A new host record of *Callosobruchus chinensis* (L.) (Coleoptera: Bruchidae), The Coleopterists Bulletin 33(4), 438.

Nagayasu, M. and Matsushita, K. (1980) Faunal survey of bean weevils in bean processing mills, Res. Bull. Pl. Prot. Japan 17, 87-91. (in Japanese with an English summary).

Shimada, M. (1988) Dry bean infestation and oviposition without feeding by a wild multivoltine bean weevil, *Kytorhinus sharpianus* (Bridwell) (Coleoptera: Bruchidae), Appl. Entomol. Zool. 23, 459-467.

Shinoda, K. and Yoshida T. (1985) Field biology of the azuki bean weevil, *Callosobruchus chinensis* (L.) (Coleoptera: Bruchidae). Seasonal prevalence and assessment of field infestation of akia-zyki, autumn variety of *Phaseolus angularis* W, Jpn. J. appl. Zool. 29, 713 (in Japanese with an English summary).

Shinoda, K. and Yoshida, T. (1990) (in this volume)

Utida, S. (1954) Phase dimorphism observed in the laboratory population of the cowpea weevil, *Callosobruchus quadrimaculatus*, Jpn. J. appl. Zool. 18, 161-168 (in Japanese with an English summary).

Utida, S. (1970) Secular change of persent emergence of the flight form in population of southern cowpea weevil, *Callosobruchus maculatus*, Jpn. J. appl. Ent Zool. 14, 71-78 (in Japanese with an English summary).

Utida, S. (1972) Density dependent polymorphism in the adult of *Callosobruchus maculatus* (Coleoptera: Bruchidae), J. stored Prod. Res. 8, 111-126.

Watanabe, N. (1986) Oviposition habit of *Sulcobruchus sauteri* and its significance in speculation on the preagricultural life of seed beetles attacking stored pulses (Coleoptera, Bruchidae), Kontyu, 53, 391-397.

Watanabe, N. and Sugimoto, S. (1988) Geographic variation in male antenna of the azuki bean weevil, *Callosobruchus chinensis* L. (Coleoptera: Bruchidae), Appl. Ent. Zool. 23, 282-290.

LIFE HISTORY OF THE AZUKI BEAN WEEVIL, *CALLOSOBRUCHUS CHINENSIS* L., (COLEOPTERA: BRUCHIDAE), IN THE FIELD

KAZUTAKA SHINODA[1] AND TOSHIHARU YOSHIDA
Laboratory of Applied Entomology, College of Agriculture,
Okayama University, Okayama 700, Japan

[1] *Present address: Laboratory of Pest Control, Toyo Sangyou Co., Ltd.,*
Sin'yashiki-cyo 3-19-20, Okayama 700, Japan

ABSTRACT. The life history of the azuki bean weevil, *Callosobruchus chinensis* L., was studied in the field. The weevil had two wild leguminous hosts: the annual *Vigna angularis* var. *nipponensis* (Ohwi) Ohwi & Ohashi and the perennial *Dunbaria villosa* (Thumb.) Makino. The adults were observed on the legumes from mid-August to mid-October and oviposited on pods from mid-September to mid-October. This is also true of the biology in the field of their usual host, the azuki bean, *V. angularis* (Willd.) Ohwi & Ohashi. After hatching, first instar larvae penetrated the developing seeds in the pods and grew to the 2nd instar or pupae by the time of pod dehiscence. The weevils overwintered in the seeds scattered on the ground and emerged from the seeds in late May. By feeding on pollen, water and fungi, the adults prolonged their longevity and oviposition period into late October. The azuki bean weevil had one generation a year in the field.

1. Introduction

In Japan, the azuki bean weevil, *Callosobruchus chinensis* L., is the most serious pest of the azuki bean, *Vigna angularis* (Willd.) Ohwi & Ohashi. The weevil was thought to be a multivoltine species which migrated between azuki bean fields and barns where beans were stored. In barns the weevils overwintered and bred for some generations (Chujyo, 1937; Kiritani, 1956; Umeya and Shimizu, 1968; Harada, 1969). This life history was hypothesized based on observations in the field and in storage. A migration from the barn to the bean field had never been observed. They are moved from the bean field to the barn through the storage of infested beans. It is difficult for the weevils to migrate from barns to the bean fields, because the beans are usually stored in a container with a cover or cap (unpublished data). Many azuki bean weevils, however, infested the bean fields (Shinoda and Yoshida, 1985, 1987). This suggests that the azuki bean weevil overwinters somewhere other than barns.

The weevil may have two types of life histories. The Weevils could overwinter in azuki beans scattered on the ground in a bean field and breed there. Then the adults lay their eggs on the pods of the new crop of azuki beans. The azuki bean weevil does not do this because few beans drop to the ground when the beans are harvested. Furthermore, even the few beans scattered on the ground are unavailable for the weevils, the majority being buried under ground by the cultivation of the bean field.

Another possible life history is that the weevils may have wild hosts, on which they

149

K. Fujii et al. (eds.), Bruchids and Legumes: Economics, Ecology and Coevolution, 149–159.
© 1990 *Kluwer Academic Publishers. Printed in the Netherlands.*

breed and overwinter. In this case, many of the seeds may be available at least as overwintering sites, because they are not buried under the ground by cultivation.

The wild life history of the azuki bean weevil has not been investigated. All Bruchids infesting commercial beans had wild hosts before they could infest beans in a bean field. This simple fact indicates that the azuki bean weevil may have some wild hosts and a wild life history.

To reveal the wild life history, we investigated the seasonal trends in a number of adult weevils, their biology on wild hosts, longevity, fecundity and overwintering in the field.

2. Seasonal Prevalence of the Number of Adults in the Field

The field biology of the azuki bean weevil had not been recorded until Shinoda and Yoshida (1985, 1987) presented the biology in a bean field, though it had been recognized that the first infestation of the beans by the weevils occurred in the bean field (Chujyo, 1937). Their studies show that the azuki bean weevils were observed at least from mid-August to late October. The studies, however, did not show whether the adult weevils could be observed during the months earlier than mid-August.

To reveal the entire seasonal abundance of the number of adult weevils in the field, we recorded the number of the weevils migrating into a bean field by the following procedure. Summer and autumn varieties of azuki beans were cultivated in each half of a bean field. Thus we extended the period when the pods of the azuki bean were available for the weevils (Doria and Raros, 1975). The number of the weevils observed were recorded by a strip census in the bean field.

The entire seasonal changes in the number of adults in the bean field are presented in Fig. 1. The first incidence of adults was observed in mid-July and thereafter the adult weevils were observed continuously until late October. During the period, two clear peaks of the number of adults were observed. The first peak was in the summer variety

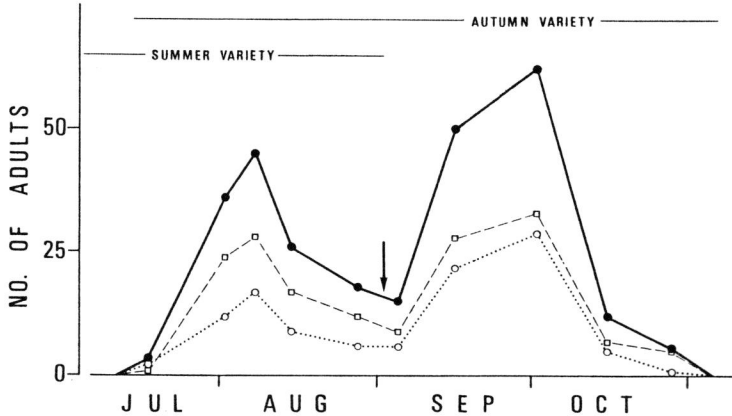

Figure 1. Seasonal changes in the number of adult weevils observed in the azuki bean field, when summer and autumn varieties the azuki bean were cultivated continuously in the same field: male (□), female (○) and total (●). The arrow shown in the figure indicates harvest of the summer variety of the azuki bean.

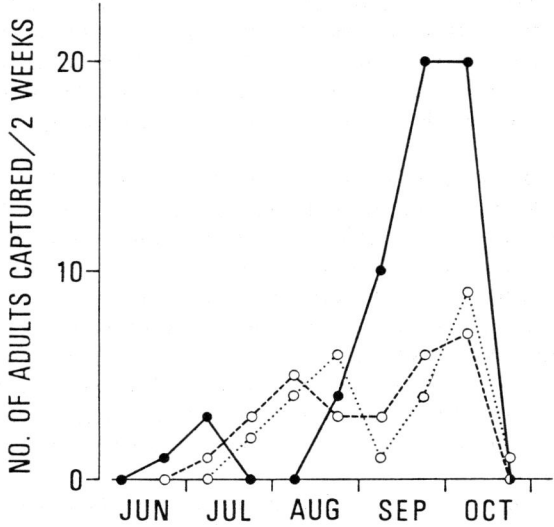

Figure 2.   Seasonal changes in the number of adults captured by the migration trap in the field; ● in 1987, ○ in 1988.

of the azuki beans in mid-August and the second peak in the autumn variety in late September.   Both peaks coincide with the season when the azuki beans were beginning to produce mature pods.

Migration of adults into the bean field suggests that they should be observed in other places.   To reveal the seasonal trend in the number of adults in the other places, we set the migration traps used by Gyotoku et al. (1987) and captured adult weevils migrating in the field.

Figure 2 shows the seasonal trend of a number of adults captured by the migration trap in the field.   The first adults were captured in late June 1987 and in early July 1988.   These were earlier than the time of first incidence observed by Shinoda and Yoshida (1985 and 1987).   After the first capture, the adults were captured continuously by the traps until late October.   Seasonal trends after mid-August 1987 was similar to that shown by Shinoda and Yoshida (1985 and 1987).   This was because the trap was placed near the bean field.   On the other hand, the traps used in 1988 were placed at a distance at least 100m away from the bean field.   Thus the seasonal trends from mid-August to late October in 1988 differed from those observed in the bean field.   These results indicate that the adults of the azuki bean weevils migrate continuously from mid-June to late October in the field.

3.   The Biology of the Azuki Bean Weevils and Their Wild Hosts

Some Bruchids infesting commercial legumes have wild hosts: *Callosobruchus maculatus* F. (Alzouma, 1981), *Zabrotes subfasciatus* (Boh.) (Pimbert and Pierre, 1983) and *Acanthoscelides obtectus* Say (Biemont, 1981).   No wild host of *C. chinensis* has been recorded and the commercial legumes: *V. unguiculata* (Back, 1922), *Cajanus cajan* (L.) Millsp (Raina, 1971), *V. radiata* (L.) Wilczek (Bato and Sanchez, 1972), *V. angularis* (Shinoda and Yoshida 1985) were reported as hosts.

To find out the wild hosts of the azuki bean weevil, we collected seeds of common wild legumes in Okayama prefecture during the period when the weevils were observed in the field. The collected seeds were introduced into a laboratory controlled at 25°C and were checked to see whether the weevils emerged from the seeds. Thereafter we tested the ability of weevils to develop in the seeds, by introducing weevils on seeds free from infestation.

Of the 11 species of wild legumes, adult weevils were observed on the plants of only two legumes in the field: *Vigna angularis* var. *nipponensis* (Ohwi) Ohwi & Ohashi and *Dunbaria villosa* (Thumb.) Makino (Table 1). When newly emerged adults from the azuki beans were allowed to oviposit on seeds of wild legumes, they oviposited on all of the seeds. But a fewer number of eggs were laid on *Aeschynomene indica* L., *Indigofera pseudo-tinctoria* Matsum., *Millettia japonica* A. Gray and *Albizia julibrissin* Durazz. Apparently, this was because these seeds were smaller or larger than preferred size of the azuki beans (Ishii, 1952). Almost all of the eggs laid on the seeds of the wild legumes hatched. The weevils, however, emerged only from the seeds of *V. angularis* var.*nipponensis*, *D. villosa*, *Amphicarpaea edgeworthii* Benth. var. *japonica* Oliver and *Pueraria lobata* (Willd.) Ohwi.

Population parameters of the weevils introduced on *V. angularis* var. *nipponensis* were nearly equal to those of the weevils introduced on *V. angularis* (Table 2)(control). On the other hand, the weevils introduced on *A. edgeworthii* and *P. lobata* had a much lower percentage of emergence, longer developmental period, shorter adult longevity and lower fecundity than those of the control weevils. Thus, the net reproductive rate of the weevils was lower than that of the control weevils, and the rate of the weevils introduced on *P. lobata* was lower than 1.00.

Table 1. List of wild legumes collected in a field; record of the presence of adults observed on growing legumes and availability of the legumes for the azuki bean weevil.

Species[a]	Date of collecting	Observation of adults	Emergence from seed	No. of eggs	Hatch-ability(%)	Percent of emergence
Vigna angularis (azuki bean)	1984, Oct 25	Yes	Yes	17.2	92.7	86.9 ± 10.1
Vigna angularis var. nipponensis	1985, Oct 20	Yes	Yes	11.0	91.8	86.1 ± 12.0
Dunbaria villosa	1985, Oct 20	Yes	Yes	17.0	97.6	85.7 ± 15.5
Amphicarpaea edgeworthii	1985, Oct 25	No	No	16.6	90.9	20.0 ± 9.3
Pueraria lobata	1983, Nov 8	No	No	16.1	91.0	16.1 ± 13.0
Vicia sativa	1983, May 26	No	No	16.1	92.5	0.0
Robinia pseudo-Acacia	1983, Aug 3	No	No	16.0	91.9	0.0
Glycine Soja	1983, Oct 7	No	No	13.9	97.8	0.0
Aeschynomene indica	1983, Oct 25	No	No	4.6	95.2	0.0
Indigofera pseudo-tinctoria	1983, Oct 25	No	No	4.1	93.9	0.0
Milletia japonica	1983, Oct 10	No	No	0.5	100.0	0.0
Albizzia julibrissin	1983, Sep 23	No	No	0.5	88.9	0.0

a) All species were collected in Okayama prefecture in Japan.

Table 2. Population parameters of adults emerged from each legumes.

Species	Mean developmental period (days)		Mean longevity (days)		Mean oviposition period (days)	Total number of eggs	Net repro -ductive rate (R_0)
	Male	Female	Male	Female			
V. angularis	32.8 ± 1.7	33.8 ± 2.0	11.5 ± 2.7	9.5 ± 2.3	6.6 ± 1.8	72.3 ± 10.1	30.64
V. angularis var. nipponensis	34.8 ± 2.8	34.6 ± 2.1	15.9 ± 1.2	10.6 ± 1.7	6.3 ± 1.4	70.5 ± 12.3	30.79
D. villosa	34.9 ± 2.1	35.8 ± 2.6	10.3 ± 2.9	7.9 ± 1.8	4.8 ± 1.1	45.1 ± 17.2	19.35
A. edgeworthii	68.0 ± 9.4	70.0 ± 7.8	7.8 ± 2.2	7.4 ± 2.1	5.0 ± 1.3	42.2 ± 16.8	3.82
P. lobata	64.0 ± 4.7	65.4 ± 5.2	3.7 ± 0.8	4.4 ± 1.7	2.4 ± 0.9	4.1 ± 5.7	0.33

In the field, many colonies of both *A. edgeworthii* and *P. lobata* are distributed among or near the fields of *V. angularis* var. *nipponensis* and *D. villosa*. None of the adults, however, were observed on the former colonies, though many adults were observed on the latter. This is good evidence that *A. edgeworthii* and *P. lobata* are not wild hosts of the azuki bean weevil but that *V. angularis* var. *nipponensis* and *D. villosa* are.

V. angularis var. *nipponensis* is an annual plant and *D. villosa* a perennial. Both legumes are distributed along a stream and produce mature pods at the same time as *V. angularis*: from mid-September to mid-October. Colonies of *D. villosa* were much more numerous than those of *V. angularis* var. *nipponensis*. We found 23 colonies of *D. villosa* in our study area which was 5 Km along a stream. Only two colonies of *V. angularis* var. *nipponensis* were in the same area. This suggests that *D. villosa* is a more important food resource than *V. angularis* var. *nipponensis*.

Seasonal changes in the number of adults observed on the colonies of *V. angularis* var. *nipponensis* and *D. villosa* (Fig. 3) were similar to the results of Shinoda and Yoshida (1985, 1987). Oviposition on the mature pods by the weevils was observed from mid-September to mid-October. Yearly changes in the percentages of seeds infested by the weevils differed among colonies (Fig.4). Many of the colonies were infested in the range of 1 to 5% every year.

The wild hosts, that fruited only in autumn, were found in this study (Table 1), though the weevils were observed during June and October in the field. This indicates that some intermediate hosts may exist in the field during the time when the first weevils were observed and when they migrated into the colonies of the wild hosts. Table 3 shows the frequency of the pod maturing seasons of 34 species of the wild legumes which are common in Okayama prefecture. Many of the legumes produce mature pods in autumn and the other legumes in spring and summer. Most of the legumes which produce mature pods in spring and summer are *Lespedeza* and *Aruncus*, whose seed size is so small that the weevils cannot develop in them. Even if some of the other legumes which were not investigated (Table 3) are available for the weevils, they are insufficient as a food resource because the legumes are uncommon species in limited areas. We therefore conclude that the weevils have no intermediate hosts in the field in Okayama prefecture.

4. Field Longevity and Fecundity

The azuki bean weevils have no intermediate host until the wild hosts mature their

154

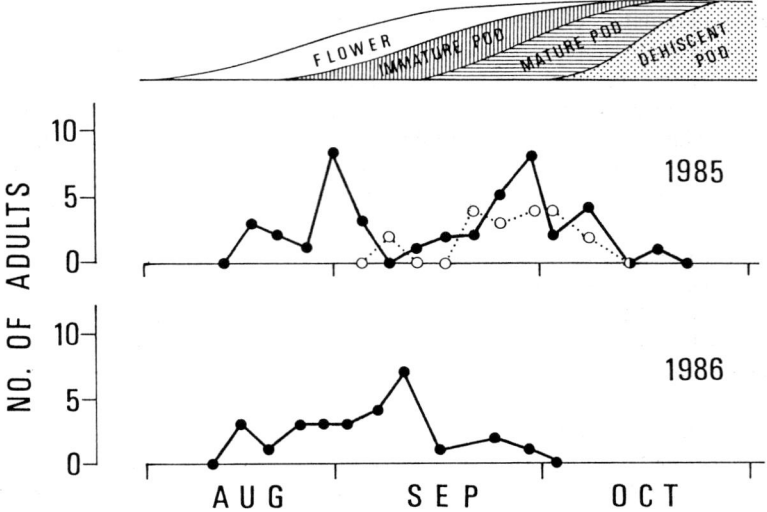

Figure 3. Seasonal prevalence in the number of adults observed on the colony of *V. angularis* var. *nipponensis* (○) and *D. villosa* (●). Observations were carried out in 1985 (upper) and 1986 (lower).

pods. The question, then, is how do they maintain their population without intermediate hosts during June to October?

Umeya and Shimizu (1968) and Shinoda and Yoshida (1984) suggested that the azuki bean weevils take food to lengthen their longevity and oviposition period in field. Shinoda and Yoshida (1987) confirmed this. On the basis of their study, the longevity and oviposition period of the weevils in the bean field were much longer than those

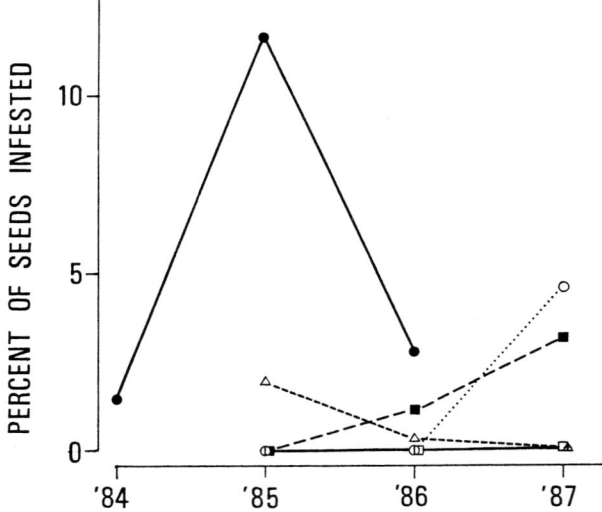

Figure 4. Yearly changes in percentages of seeds infested by azuki bean weevils in the field. *V. angularis* var. *nipponensis* (○), and *D. villosa* (the other marks).

Table 3. Frequency of pod maturing seasons of 34 wild legumes which are common in Okayama prefecture. Values shown in the table indicate the number of species of legumes that produce mature pods. If a legume produces pods during two or more months, it is counted in each month.

Month	No. Species	Main Genera of Leguminosae
May	3	Aruncus, Vicia
Jun.	5	Lathyrus, Aruncus, Vicia
Jul.	6	Lathyrus, Vicia, Trifolium
Aug.	11	Vicia, Trifolium, Sophora, Lespedeza
Sep.	29	Cassia, Crotalaria, Aeschynomene
		Lespedeza, Desmodium, Vigna, Indigofera
		Amphicarpaea, Pueraria, Dunbaria
Oct.	24	Lespedeza, Desmodium, Indigofera
		Vigna, Amphicarpaea, Pueraria, Dunbaria

shown by Utida (1971) under laboratory conditions (25°C). However, their study did not show the consequences of the weevils observed earlier than mid-August in the field.

To determine the maximum longevity and oviposition period under laboratory conditions (25°C, 70 ± 5 % RH), we supplied the newly emerged adults with foods which were found most commonly in the field. When pollen and water were supplied at the same time, their adult longevity and oviposition period was more than ten times as long as those of the control weevils without a food supply (Fig. 5). On the contrary, when the weevils were supplied with either pollen or water, they could not lengthen their longevity and oviposition period. Honey lengthened adult longevity but not oviposition period. These foods are commonly found in the field and therefore, are staple food resources for the adult weevils. Some observations on adult feeding in the field are summarized in Table 4. Many of the adults were observed on various kinds of flowers and leaves wet with dew. This indicates probably that the weevils usually feed on pollen and water. Fungi may also be available as food in the field (Shinoda and Yoshida, 1987).

Results of adult feeding on pollen and the other foods suggest that the weevils can survive over 5 months in the field. To confirm this, we released about 60,000 weevils marked with enamel in the field during May to July when the first incidence was observed, and tried to recapture the weevils in the bean field in late October. As a result, only one marked male, which was released in June, was recaptured in late October. The weevil migrated a distance of 2km in a straight line from the release point. This confirmed clearly that the azuki bean weevils could survive at least 5 months in the field without intermediate hosts in the field. Even if the weevils have intermediate hosts, we believe they would have only one generation per year.

5. Overwintering in the Field

The above results, however, indicate that a wild population of azuki bean weevils may exist in the field. Overwintering of the weevils should completely confirm the presence of a wild population. But no evidence of overwintering has been recorded, though Umeya et al. (1970) speculated that the weevils overwinter in the field on the basis of their results that the weevils could survive for a long time (about 60 days) at 5°C.

156

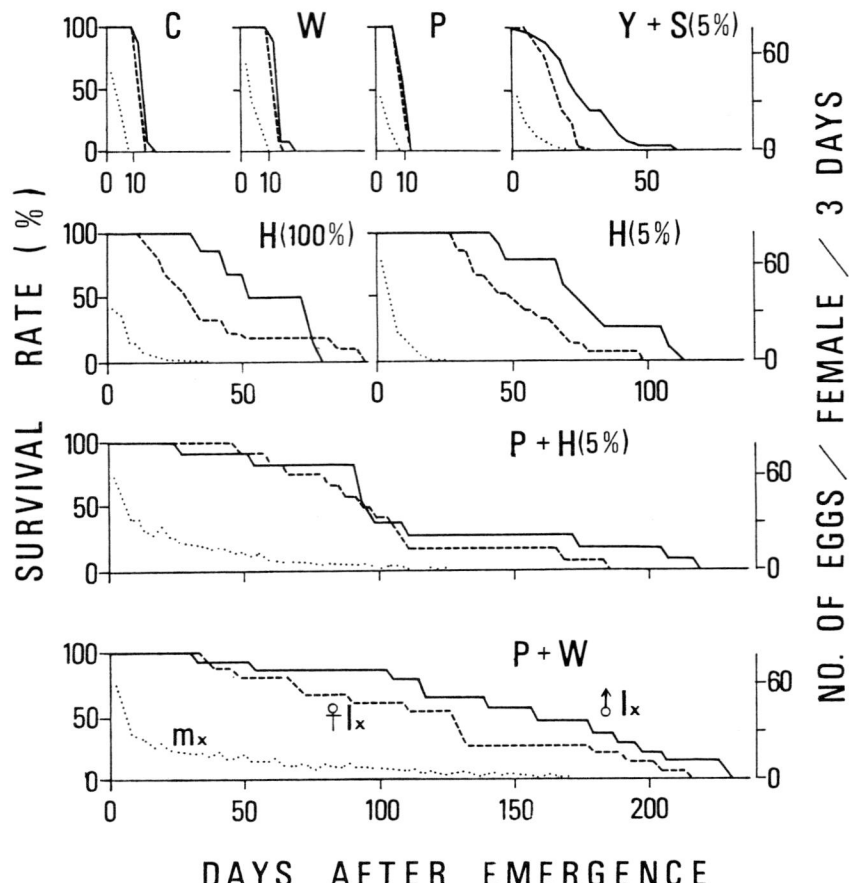

Figure 5. Effect of taking of foods on longevity and fecundity. C; control (without food), W; water, P; pollen, Y; yeast, S(5 %); Saccharose solution (5 %), H(100 %); honey (100 %), H(5 %); honey (5 %).

To test this, we carried out the following experiments in the bean field. Plants of the azuki bean which had infested and uninfested mature pods were left as they were in the field. The pods did not dehisce after their maturation. The weevils in the larval stage remained in the seeds in the pods during the winter season. The seeds were not harvested in February (which is the coldest season in Japan) but in late March. Because the weevils in seeds did survive until late March, we believe they are likely to overwinter in seeds in the field. After harvesting, the seeds taken to the laboratory at 25°C and the number of adults emerged from the seeds were counted.

The weevils could survive until late March (Table 5). The number of adults that emerged from the seeds and the percentage of seeds infested by the weevils were lower than the number observed at harvest time (Shinoda and Yoshida, 1985). Presumably, this difference was caused by low temperatures in winter. The low temperatures decreased the survival rate of the weevils in the seeds (unpublished data).

To demonstrate overwintering in the wild hosts, we covered a colony of *D. villosa* with victoria lawn and counted the number of adults that emerged from the seeds that dropped to the ground. Five adults were captured in the colony in late May. This

Table 4. Field observations of adult feeding

Time		Plants observed	Position Observed	Foods
1977	(a)	*Cajanus cajan*	Flower	-
1979	Aug.	*Arachis hypogaea*	Leaf	Pollen
	Sep.	*Chrysanthemum morifolium*	Flower	Pollen
	Sep.	*Erigeron annuus*	Flower	Pollen
	Jul.-Sep.	*Vigna angularis*(b)	Flower	Pollen
	Aug.-Oct.	*V. angularis*(c)	Flower	Pollen
	Aug.-Sep.	*V. angularis*	Leaf	Water
	Aug.-Oct.	*V. angularis*	Leaf	Rust(d)
	Aug.-Oct.	*V. angularis*	Leaf	Powdery mildew(e)
1980	Sep.	*Oryza sativa*	Flower(f)	Pollen
1981	(g)	*Chrysanthemum frutescens*	Flower	Pollen
1984	Jul.-Sep.	*Dunbaria villosa*	Flower	Pollen
	Jul.-Sep.	*V. angularis* var. *nipponensis*	Flower	Pollen

(a) and (g) were cited from Williams (1977) and Umeya (1981), respectively. (b) and (c) are summer and autumn variety of *V. angularis* (Willd.) Ohwi & Ohashi, respectively. (d): *Sphaerotheca fuliginea*, (e): *Uromyces azukicola*. (f): Adults were captured in the rice field during its flowering season.

confirms that the weevils can overwinter in the field and that there is a wild population of azuki bean weevils in the field.

6. Discussion

This study shows the entire life history of the azuki bean weevil in the field (Fig. 6) and that the weevil has two wild hosts: *V. angularis* var. *nipponensis* and *D. villosa* which mature their pods during mid-September to mid-October. Adult weevils, which migrate into the colony during mid-August to mid-October, increase their longevity and oviposition period by feeding on pollen, water and fungi. They lay their eggs on mature pods. After the eggs hatch, the larvae overwinter in the seeds on the ground and thereafter emerge in late May or June. The weevils then migrate into the colonies of their

Table 5. Number of the azuki bean weevils emerged from the azuki beans which were left in the pods without harvest in the bean field until late March.

	Replication				
	1	2	3	4	5
Weight of beans sampled	108	114	102	122	113
Number of the beans	650	686	613	734	680
Number of males emerged	10	3	6	7	4
Number of females emerged	7	7	12	7	8
Number of the adults/100g of beans	15.7	8.8	17.6	11.6	10.7
Percent of beans infested	2.6	1.5	2.9	1.9	1.8

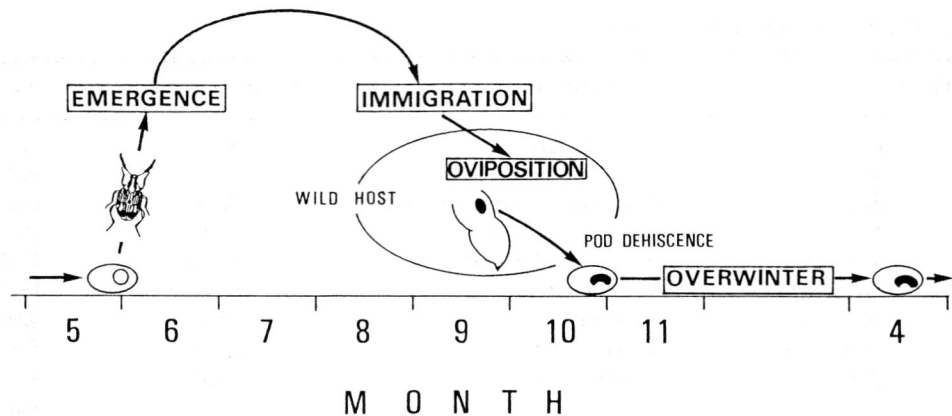

Figure 6. Life history of the azuki bean weevil in the field.

wild hosts again.

Our study confirms that the azuki bean weevil lives in the wild as was speculated by Watanabe and Sugimoto (1988). We did not observe any data which were presented in their speculations; i.e., that the weevils breed on the host seeds scattered on the ground and that the newly emerged adults from stored beans migrate into the bean field. We are convinced that the weevils emerge from the seeds on the ground and migrate into the bean field because the weevils released in the field in June were recaptured in the bean field in late October.

The life history revealed here indicates that the azuki bean weevil has characteristics of univoltine insects in the field. On the other hand, the same population is multivoltine in storage without variety in their genetics. Then, what is the cause of the different life history? Kiritani (1956) considered the multivoltinism characteristics to be habits that the weevils acquired as a result of domestication for storage conditions. However, Watanabe and Sugimoto (1988) and Shimada (1988) revealed that some wild Bruchids had acquired the multivoltine characteristics in the field, though the weevils had never bred under storage conditions. Since the azuki bean weevils do not need to live under conditions of storage as essential factors for their existence, we believe they probably acquired the multivoltine characteristics in the field long before they became a storage pest.

Almost all bean weevils have the habit of laying their eggs on the pods of their hosts in the field and a few bean weevils have the habit of ovipositing on dry beans in storage (Kiritani, 1956) or in the field (Shimada, 1988). This suggests that all bean weevils have the former oviposition habit at first, and thereafter, some of the weevils acquire the latter habit.

REFERENCES

Alzouma, I. (1981) Observations on the ecology of *Bruchidius atrolineatus* Pic. and *Callosobruchus maculatus* F. (Coleoptera: Bruchidae) in Niger, in V. Labeyrie (ed.), Series Entomologica Vol. 19, Dr. W. Junk Publishers, The Hague, pp. 205-213.

Back, E. A. (1922) Weevils in beans and peas, Fmrs' Bull. U.S.D.A. 1275, 1-35.

Bato, S. M. and Sanchez, F. F. (1972) The biology and chemical control of *Callosobruchus chinensis* (Linn.)(Coleoptera: Bruchidae), Phillip. Entomol. 2(3), 176-182.

Biemont, J. C. and Bonet, A. (1981) The bean weevil populations from the *Acanthoscelides obtectus* Say. group living on wild or subspontaneous *Phaseolus vulgaris* L. and *Phaseolus coccineus* L.

and *Phaseolus vulgaris* L. cultivated in the tepoztlan region state of Morelos-Mexico, in V. Labeyrie (ed.), Series Entomologica Vol. 19, Dr. W. Junk Publishers, The Hague, pp. 23-41.

Doria, R. C. and Raros, R. S. (1975) Varietal resistance of mungo to the bean weevil, *Callosobruchus chinensis* (Linn.) and some characteristics of field infestation, Philipp. Entomol 2(6), 399-408.

Chujyo, M. (1937) Family Bruchidae, in Fauna Nipponica, Vol. 10, Fas. 8, No. 9. Sanseidou, Tokyo (In Japanese).

Gyotoku, M., Ito, K. and Nakasuji, F. (1987) Monitoring of Lepidopterous insect migrations using the migration trap, Jpn. J. Appl. Entomol. Zool. 31, 350-358 (in Japanese with an English summary)

Harada, T. (1969) Stored-product pests, Kourinsyoin, Tokyo (in Japanese).

Ishii, S. (1952) Studies on the host preference of the cowpea weevil (*Callosobruchus chinensis* L.), Bull. Nat. Inst. Agr. Sci. C-1 185-256 (in Japanese with an English summary)

Kiritani, K. (1956) Bruchid biology and domestication. II, Sinkoncyu 9(6), 7-11 (in Japanese).

Pimbert, M. P. (1983) A model of host plant change of *Zabrotes subfasciatus* Boh. (Coleoptera: Bruchidae) in a traditional bean cropping system in Costa Rica, Biol. Agric. Hort. 3, 39-54

Raina, A. K. (1971) Observation on Bruchids as field pests of pulses, Indian. J. Entomol. 32(2), 194-197.

Shimada, M. (1988) Dry bean infestation and oviposition without feeding by a wild multivoltine bean weevil, *Kytorhinus sharpianus* (Bridwell) (Coleoptera: Bruchidae), Appl. Entomol. Zool. 23(4), 459-467.

Shinoda, K. and Yoshida, T. (1984) Relationship between adult feeding and emigration from beans of azuki bean weevil, *Callosobruchus chinensis* Linne (Coleoptera: Bruchidae), Appl. Entomol. Zool. 19(2), 202-211.

Shinoda, K. and Yoshida, T. (1985) Field biology of the azuki bean weevil, *Callosobruchus chinensis* (L.)(Coleoptera: Bruchidae). I. Seasonal prevalence and assessment of field infestation of aki azuki, autumn variety of *Phaseolus angularis* W., Jpn. J. Appl. Entomol. Zool. 29, 14-20 (in Japanese with an English summary)

Shinoda, K. and Yoshida, T. (1987) Effect of fungal feeding on longevity and fecundity of the azuki bean weevil, *Callosobruchus chinensis* (L.)(Coleoptera: Bruchidae), in the azuki bean field, Appl. Entomol. Zool. 22(4), 465-473.

Umeya, K. and Shimizu, K. (1968) Studies on the comparative ecology of bean weevils. III. Effect of feeding on the life span and oviposition of the adult of three species of bean weevils, Res. Bull. Plant Prot. service Japan 5, 39-49 (in Japanese with English summary).

Umeya, K., Kato, T. and Sekiguchi, Y. (1970) Studies on the comparative ecology of bean weevils. IV. Tolerance to low temperature (5°C) of the five species of bean weevils infesting azuki bean, Res. Bull. Plant Prot. Service Japan 8, 39-50 (in Japanese with English summary).

Umeya, K. (1981) Biology of bruchids--Life span of the adults, Insectarium 18(7), 16-23 (in Japanese).

Utida, S. (1971) Influence of temperature on the number of eggs, mortality and development of several species of Bruchids infesting stored beans, Jpn. J. appl. Entomol. Zool. 15(1), 23-30 (in Japanese with an English summary).

Watanabe, N. and Sugimoto, S. (1988) Geographic variation in male antenna of the azuki bean weevil, *Callosobruchus chinensis* L. (Coleoptera: Bruchidae), Appl. Entomol. Zool. 23, 282-290.

Williams, I. H. (1977) Behaviour of insects foraging on pigeon pea (*Cajanus cajan*) in India, Trop. Agric. 54, 353-363.

BIOCHEMICAL ADAPTATIONS BY THE BRUCHID BEETLE, *CARYEDES BRASILIENSIS*

GERALD A. ROSENTHAL
T. H. Morgan School of Biological Sciences, University of Kentucky, Lexington, Kentucky 40506 U.S.A.

ABSTRACT. This communication provides an evolutionary perspective on the biochemical adaptations that enables only the bruchid beetle, *Caryedes brasiliensis* Thunberg to utilize the seeds of *Dioclea megacarpa* [Leguminosae] as an ovipositional and food resource. This plant enjoys invulnerability to other seed predators because it has created a formidable chemical barrier to predation by storing up to 13 % of its seed dry matter as L-canavanine, L-2-amino-4-(guanidinooxy)butyric acid, a potentially insecticidal nonprotein amino acid.

1. Introduction

The deciduous lowland forest of northwestern Guanacaste Province of Costa Rica is home to a large assemblage of higher plants whose seeds are a food and ovipositional resource for bruchid beetles (Janzen and Liesner, 1980) (Table 1).

Among the seed predators is the bruchid beetle, *Caryedes brasiliensis* Thunberg. About a month before the pod matures in the late Fall, the female oviposits her egg clusters on the outside of the ovary wall of *Dioclea megacarpa* [Leguminosae]. The

Table 1. Some canavanine-containing seeds from Santa Rosa National Park (Northwestern Guanacaste Province, Costa Rica) and their associated bruchid seed predators

Plant[*]	Insect[#]
Sesbania emerus	*Acanthoscelides griseolus*
Indigofera suffruticosa	*Acanthoscelides kingsolveri*
Calopogonium mucunoides	*Acanthoscelides puellus*
Calopogonium caeruleum	*Acanthoscelides hectori*
Dioclea megacarpa	*Caryedes brasiliensis*
Centrosema pubescens	*Caryedes helvinus*
Centrosema pubescens	*Caryedes incensus*
Galactia striata	*Caryedes juno*
Calopogonium caeruleum	*Caryedes paradisensis*
Centrosema plumieri	*Caryedes quadridens*
Sesbania emerus	*Stator pruininus*

[*]Adapted from Bell et al., 1978.
[#]Adapted from Janzen, 1980.

K. Fujii et al. (eds.), Bruchids and Legumes: Economics, Ecology and Coevolution, 161–169.
© 1990 *Kluwer Academic Publishers. Printed in the Netherlands.*

newly hatched larvae bore the testa and then feed upon and develop within the cotyledonous tissues of the seeds. The larvae metamorphose and eventually the young adults emerge to complete the cycle.

What makes this bruchid-plant interaction so interesting to study is that *C. brasiliensis* is the sole seed predator of *D. megacarpa*. This legume enjoys freedom from other seed predators because it has established a formidable chemical barrier to herbivory: up to 13 % of its dry matter with the nonprotein amino acid, L-canavanine, a potentially insecticidal nonprotein amino acid (Rosenthal, 1977a).

2. Aberrant Protein Production

L-Canavanine, L-2-amino-4-(guanidinooxy)butyric, is an arginine antimetabolite that can elicit potent insecticidal properties in canavanine-sensitive organisms (Rosenthal, 1977b). This nonprotein amino acid is a substrate for arginyl-tRNA synthetase (Allende and Allende, 1964; Mitra and Mehler, 1967). This enzyme normally links arginine to arginyl-tRNA, which then carries this amino acid to the protein assembly site. Canavanine-sensitive insects commonly possess an arginyl-tRNA synthetase that fails to differentiate between these two amino acids (Rosenthal et al., 1987) and thereby erroneously charges and carries canavanine. Canavanine, which is much less basic than arginine, is incorporated into the growing polypeptide chain. When canavanine replaces arginine in proteins, the resulting decrease in residue basicity can alter the interaction between amino acids that is responsible for producing the three dimensional conformation unique to the protein (Rosenthal et al., 1989a). Such alteration in tertiary and/or quaternary structure may explain why canavanine-containing proteins can exhibit biochemically altered function (Rosenthal, 1986; Rosenthal et al., 1989b). Thus, insectan production of structurally aberrant, functionally impaired proteins is an important and perhaps the principal biochemical basis for the potent insecticidal properties of this arginine antagonist.

If this seminal hypothesis is valid, then a comparative examination of canavanine-sensitive and canavanine-resistant organisms must demonstrate that the former incorporate canavanine into protein while the latter avoid production of canavanyl proteins. This point has been addressed by examining several insects for their relative incorporation, under optimal condition for each amino acid, of L-[*guanidinooxy*-14C]canavanine and L-[*guanidino*-14C]arginine into larval protein. These insects include the canavanine-sensitive tobacco hornworm, *Manduca sexta* [Sphingidae]; the canavanine-resistant tobacco budworm, *Heliothis virescens* [Noctuidae] and two canavanine-utilizing insects: *Caryedes brasiliensis* and the weevil, *Sternechus tuberculatus* [Curculionidae]. The weevil feeds upon *Canavalia brasiliensis*, a legume whose seeds contain nearly 8% canavanine, on a dry weight basis (Bleiler et al., 1988).

Canavanine-sensitive, *M. sexta* readily incorporates canavanine into its proteins (Table 2). *Heliothis virescens*, which is naturally resistant to canavanine (Berge et al., 1986), has a higher arginine to canavanine ratio but this value is far less than that for *S. tuberculatus* and *C. brasiliensis*.

These findings provide evidence consistent with the view that canavanine consumption and/or resistance is associated with little canavanyl protein production. The ability of *Caryedes brasiliensis* to consume a large amount of seed canavanine results in part from a critically important biochemical adaptation. Namely, the production of an arginyl-tRNA synthetase able to discriminate between arginine and canavanine, and thereby minimize erroneous incorporation of canavanine into *de novo* - synthesized proteins. This finding is supported by a direct comparison of the ability of *C. brasiliensis* and *M.*

Table 2. Comparative incorporation of L-[*guanidino*-^{14}C]arginine and L-[*guanidinooxy*-^{14}C]canavanine into larval protein[*]

Insect	[14C]arginine to [14C]canavanine incorporation ratio
Canavanine-utilizing	
Caryedes brasiliensis	365:1
Sternechus tuberculatus	> 500:1
Canavanine-resistant	
Heliothis virescens	75:1
Canavanine-sensitive	
Manduca sexta	3.3:1

[*]Rosenthal et al. (1987).

sexta to distinguish between arginine and a series of arginine analogs as potential substrates for newly synthesized larval proteins (Table 3).

Evaluation of these arginine analogs for their assimilation into bruchid beetle proteins discloses the acutely discriminatory nature of this bruchid's protein-synthesizing system. The marked ability of this seed predator to distinguish molecules structurally akin to arginine confers general resistance to the incorporation of such nonprotein amino acids (Rosenthal and Janzen, 1983). Indeed, the evolution of a discriminatory arginyl-tRNA synthetase may have been of paramount importance in the adaptation of *C. brasiliensis* to canavanine-storing legumes.

3. Canavanine Utilization via Arginase

Dioclea megacarpa allocates much of its nitrogenous resources to the production and storage of canavanine. This legume sequesters more than 95% of the nitrogen allocated to free amino acid storage in the seed to canavanine (Rosenthal, 1977a).

Nitrogen utilization by *C. brasiliensis* commences with the enzyme arginase (EC

Table 3. Incorporation of arginine and certain of its structural analogs into larval proteins[*]

Substrate	[14]C incorporation (pCi/mg soluble protein)	
	Manduca sexta	*Caryedes brasiliensis*
arginine	20,278	12,393
canavanine	3,611	34
homoarginine	264	< 5
5-hydroxyhomoarginine	480	< 5
2-amino-4-guanidinobutyrate	262	< 5
2-amino-3-guanidinopropionate	112	< 5

[*]Rosenthal and Janzen (1983).

3.5.3.1), which hydrolyzes L-canavanine to L-canaline and urea.

$$H_2N\text{-}C(NH_2)\text{=}N\text{-}O\text{-}CH_2\text{-}CH_2\text{-}CH(NH_2)COOH \longrightarrow$$
$$\text{L-canavanine}$$

$$H_2N\text{-}O\text{-}CH_2\text{-}CH_2\text{-}CH(NH_2)COOH \quad + \quad H_2N\text{-}C(\text{=}O)\text{-}NH_2$$
$$\text{L-canaline} \qquad\qquad\qquad\qquad \text{urea}$$

This process is completed by hydrolysis of urea to carbon dioxide and ammonia, a reaction mediated by urease (EC 3.5.1.5).

$$H_2N\text{-}C(\text{=}O)\text{-}NH_2 \longrightarrow CO_2 + 2NH_3$$
$$\text{urea}$$

Arginase is widely distributed among insects. Plants that contain canavanine typically possess an arginase that uses canavanine as a substrate much more effectively than does the arginase of canavanine-free plants (Rosenthal, 1970; Downum et al., 1983). It is possible, therefore, that *C. brasiliensis* may have evolved an arginase with a marked affinity for and reactivity with canavanine. Analysis of the maximum reaction velocity (rate of product formation) for arginase-mediated utilization of canavanine in a diversified array of insects discloses no significant difference between the arginase of *C. brasiliensis* and the other insects tested (Table 4).

The majority of insects studied in this experiment have had no significant evolutionary or ecological exposure to canavanine; yet, they have an arginase that effectively degrades canavanine. Possession of this trait must be a fortuitous circumstance. The ability of these insects, including *C. brasiliensis*, to catabolize canavanine via arginase may not have been selected specifically as a part of their adaptation to canavanine-containing seeds. That is, *C. brasiliensis* did not have to evolve a distinct biochemical capacity to be able to catabolize canavanine via the action of arginase.

4. Canavanine Utilization via Urease

In order to use the nitrogen of urea, this metabolite must be hydrolyzed via urease to release the nitrogens as ammonia. This factor instigated a comparative examination of the urease of *C. brasiliensis* and a number of other insects. This study included evaluation of another Costa Rican bruchid beetle that feeds on canavanine-containing seeds. Like its congener *C. brasiliensis*, *Caryedes quadridens* Jekel develop within canavanine-containing seeds; namely, *Centrosema plumieri*. However, *Centrosema plumieri* has only minute amounts of canavanine (<1% dry mass). Thus, it is possible to compare the urease activity in *C. brasiliensis* and *S. tuberculatus*, insects that feed upon canavanine-rich seeds, with *C. quadridens*, which feeds upon a seed depauperate in canavanine. Even with a highly sensitive radiometric assay, most of the tested insects have little detectable urease (Fig. 1). This stands in sharp contrast to *C. brasiliensis* and *S. tuberculatus* whose larvae produce massive amounts of urease. On the other hand, *C. quadridens*, which does not feed on a canavanine-rich seed, possess little urease activity (Fig. 1).

Larvae that feed on canavanine-containing seeds benefit from elevated urease activity because this enables them to produce ammonia from urea. Ammonia, in turn, can be fixed into newly synthesized amino acids. The enzyme urease has rarely been reported in insects (Rosenthal et al., 1977). Our studies suggest that development of pronounced

Table 4. Maximum arginine- or canavanine-dependent arginase reaction velocity for various insects[*].

| Insect | Maximum velocity (μmol min^{-1} mg protein^{-1}) | | Ratio |
	Arginine	Canavanine	
Canavanine-feeding insects			
Coleoptera			
Caryedes brasiliensis (A)[#]	4.5	2.5	1.8:1
Caryedes brasiliensis (L)	4.8	1.8	2.7:1
Caryedes quadridens (L)	5.2	3.4	1.5:1
Sternechus tuberculatus (L)	2.7	1.8	1.5:1
Non-canavanine feeding insects			
Coleoptera			
Callosobruchus chinensis (A)	2.8	1.4	2.0:1
Callosobruchus chinensis (L)	6.3	2.6	2.4:1
Callosobruchus maculatus (A)	4.5	3.3	1.4:1
Callosobruchus maculatus (L)	4.2	1.8	2.3:1
Acanthoscelides obtectus (A)	2.7	1.8	1.5:1
Acanthoscelides obtectus (L)	2.5	1.3	1.9:1
Lepidoptera			
Manduca sexta (L)	1.5	0.3	5.0:1
Heliothis virescens (L)	2.6	1.1	2.4:1
Odonata			
Ischnura verticalis (L)	8.8	4.0	2.2:1
Orthoptera			
Periplaneta americana (A)	9.0	3.0	3.0:1
Hymenoptera			
Polistes sp. (L)	0.5	0.2	2.5:1
Hemiptera			
Oncopeltis fasciatus (A)	6.1	1.6	3.8:1

[*]Rosenthal et al. (1987).
[#]A = adult, L = larva.

urease activity is an important biochemical adaptation to dietary canavanine.

It is not presently known if the elevated urease activity noted in these canavanine-utilizing insects results from a high enzyme content or rather the presence of a urease molecule with a high turnover number (i.e., moles of substrate converted to product per minute per mole of enzyme).

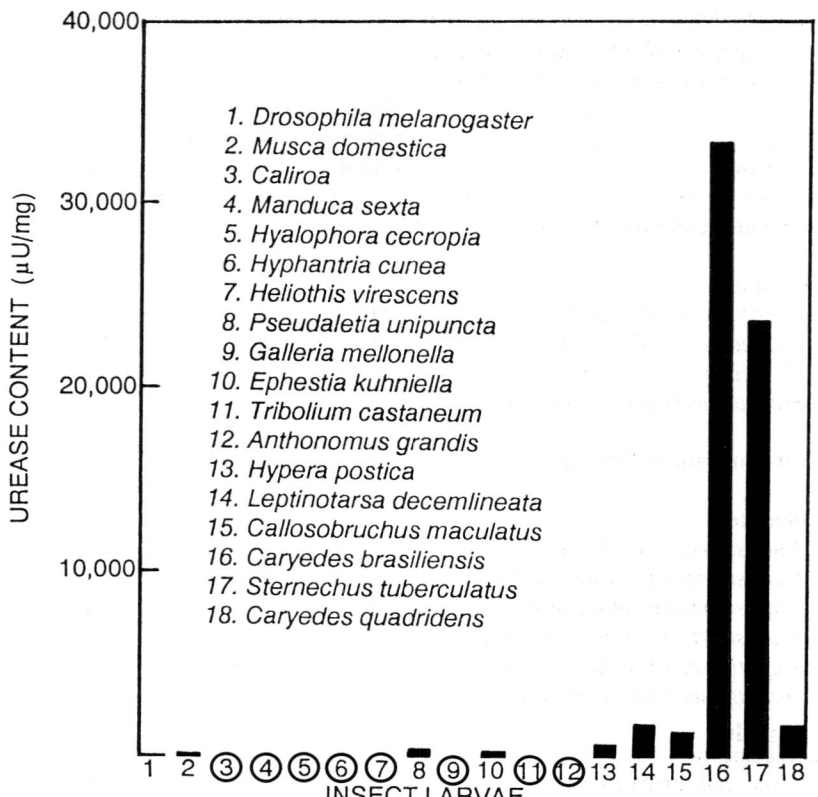

Figure 1. The urease content of various insects. Urease activity was measured as previously described (Rosenthal et al., 1977). One microunit of urease is that amount of enzyme that forms 1 pmole of $^{14}CO_2$ per minute at 37°C.

5. Coping with Ammonia

The above considerations combine to make *C. brasiliensis* an effective producer of ammonia. Ammonia is toxic to most insects, particularly terrestrial forms (Chefurka, 1965). Ammoniotelic and ureotelic, as compared to uricotelic nitrogen excretion, is rare in insects and generally limited to aquatic or meat-eating individuals (Bursell, 1967). Analysis of the contribution of ammonia, urea, and uric acid to the total nitrogen of the frass of *C. brasiliensis* disclosed that uric acid accounts for only 11 % of the fecal nitrogen (Rosenthal and Janzen, 1981). Excess nitrogen is eliminated by the bruchid beetle larvae by fecal excretion of ammonia and urea.

This raises the question of whether *C. brasiliensis* is capable of eliminating ammonia metabolically. Automated amino acid analysis of the hemolymph of the larva reveals the presence of appreciable proline and glutamine. This suggests that this seed eating beetle may have an active glutamine synthetase (EC 6.3.1.2), an enzyme mediating the reaction of L-glutamic acid and ammonia to form L-glutamine (Tate and Meister, 1969). Additional removal of ammonia can be achieved by reaction of 2-oxoglutaric acid with ammonia to form L-glutamic acid. This reaction is catalyzed by glutamic acid dehydrogenase (EC 1.4.1.2). Working in concert, these reactions would provide a means of trapping ammonia by forming glutamic acid and then further sequestering ammonia via glutamine formation.

These two enzymes were analyzed from 16 insects in addition to *C. brasiliensis* (Table 5). All of the tested insects contain glutamine synthetase activity, but *C. brasiliensis* larvae possess twice the enzyme activity of any of the tested insects, and over 40 times more than the American cockroach, *Periplaneta americana*; the saltmarsh caterpillar, *Estigmene acraea*; and the Colorado potato beetle, *Leptinotarsa decemlineata*.

Analysis of the glutamic acid dehydrogenase activity reveals the disproportionately high activity of this enzyme in the larvae of *C. brasiliensis*. Except for *Drosophila melanogaster*, the glutamic acid dehydrogenase activity of *C. brasiliensis* runs from 3.5 times more in *Callosobruchus maculatus*, a closely related beetle, to 34 times more in

Table 5. Glutamine synthetase and glutamic acid dehydrogenase of various insects[*]

| Organism | Enzyme activity | | Apparent K_m |
| | Glutamine synthetase | Glutamic acid dehydrogenase | |
	(nmol/min/mg protein)		(mM)
Diptera			
Drosophila melanogaster	25.4	108	600
Musca domestica	11.5	71	400
Orthoptera			
Periplaneta americana	0.8	13	190
Hymenoptera			
Vespula spp.	8.1	35	230
Polistes sp.	13.0	8	105
Lepidoptera			
Manduca sexta	10.9	49	430
Papilio polyxenes	4.3	28	290
Ephestia kuhniella	9.3	nd	-
Heliothis zea	10.6	nd	-
Heliothis virescens	14.9	nd	-
Hyphantria cunea	21.6	16	210
Galleria mellonella	17.2	nd	-
Estigmene acraea	1.0	21	190
Coleoptera			
Tribolium castaneum	18.6	33	245
Callosobruchus maculatus	21.2	78	240
Leptinotarsa decemlineata	1.2	48	160
Caryedes brasiliensis (adult)	12.7	62	345
Caryedes brasiliensis	49.5	27	70

[*]Rosenthal and Janzen, 1985
nd = lacks detectable activity.

168

Polistes spp., a wasp (Table 5).

Another salient parameter of enzyme function is the apparent Michaelis-Menten constant (K_m) of its substrate. This constant is inversely proportional to the enzyme-substrate affinity. Insectan glutamic acid dehydrogenase is characterized by a limited affinity (high K_m) for ammonia (Table 5). In several insects, the K_m value exceeds 300 mM and in *Drosophila melanogaster*, it reaches 600 mM. Except for *Polistes* spp., all of the examined insects, other than *C. brasiliensis* larvae, possess a K_m for ammonia near 200 mM. In *Polistes* spp., although the K_m value is low, enzyme activity is among the lowest of the values obtained. Only *C. brasiliensis* larvae exhibit both elevated enzyme activity and high substrate affinity. Relying upon these two enzymes, *C. brasiliensis* is capable of employing metabolites such as products of the tricarboxylic acid cycle to sequester ammonia in glutamate and glutamine.

The above findings suggest that development of the biochemical ability to avoid canavanyl protein production must occur or be acquired early in the process of a seed predator's adaptation to plants containing significant canavanine. The genetic capacity to degrade canavanine, via arginase, to use urease to obtain ammonia, to produce its amino acids from ammonia, and to efficiently process and eliminate ammonia could have evolved subsequently. These investigations raise the important point that some of the metabolic capacities required by *C. brasiliensis* may not have been acquired as part of its adaptation to canavanine. Rather, they may have existed before exposure to dietary canavanine and facilitated this predator's ability to breech the canavanine-supported defensive barrier of the seed.

Acknowledgments. The investigations described in this communication were supported by grants from the National Science Foundation (DCB-8901749), National Institutes of Health (DK-17322), NIH Biomedical Research Grant (2-S07-RR07114-20), and the Graduate School of the University of Kentucky.

REFERENCES

Allende, C. C. and Allende, J. E. (1964) Purification and substrate specificity of arginyl-ribonucleic acid synthetase from rat liver, J. Biol. Chem. 239, 1102-1106.

Bell, E. A., Lackey, J. A., and Polhill, R. M. (1978) Systematic significance of L-canavanine in the Papilionoideae (Faboideae), Biochem. System. Ecol. 6, 201-212.

Berge, M. A., Rosenthal, G. A., and Dahlman, D. L. (1986) Tobacco budworm, *Heliothis virescens* [Noctuidae] resistance to L-canavanine, a protective allelochemical, Pest. Biochem. Physiol. 25, 319-326.

Bleiler, J. A., Rosenthal, G. A., and Janzen, D. H. (1988) Biochemical ecology of canavanine-eating seed predators, Ecology 69, 427-433.

Chefurka, W. (1965) The Physiology of Insects, M. Rockstein (ed.), Academic Press, New York, pp. 669-768.

Downum, K. R., Rosenthal, G. A., and Cohen, W. S. (1983) L-Canavanine and L-arginine metabolism in the jack bean, *Canavalia ensiformis* (L.) DC. and soybean, *Glycine max*, Plant Physiol. 73, 965-968.

Janzen, D. H. (1980) Specificity of seed-attacking beetles in a Costa Rican deciduous forest, J. Ecol. 68, 929-952.

Janzen, D. H. and Liesner, R. (1980) Annotated check-list of plants of lowland Guanacaste Province, Costa Rica, exclusive of grasses and non-vascular cryptograms, Brenesia 18, 15-90.

Mitra, S. K. and Mehler, A. H. (1967) The arginyl transfer ribonucleic acid synthetase of *Escherichia coli*, J. Biol. Chem. 242, 5490-5494.

Rosenthal, G. A. (1970) Investigations of canavanine biochemistry in the jack bean plant, *Canavalia ensiformis* (L.) DC. I. Canavanine utilization in the developing plant, Plant Physiol. 46, 273-

276.

Rosenthal, G. A. (1977a) Nitrogen allocation for L-canavanine synthesis and its relationship to chemical defense of the seed, Biochem. System. Ecol. 5, 219-220.

Rosenthal, G. A. (1977b) The biological effects and mode of action of L-canavanine, a structural analogue of L-arginine, Q. Rev. Biol. 52, 155-178.

Rosenthal, G. A. (1986) Biochemical insights into the insecticidal properties of L-canavanine, a higher plant protective allelochemical, J. Chem. Ecol. 12, 1145-1156.

Rosenthal, G. A. and Janzen, D. H. (1981) Nitrogenous excretion by the terrestrial seed predator, *Caryedes brasiliensis*, Biochem. System. Ecol. 9, 219-220.

Rosenthal, G. A. and Janzen, D. H. (1983) Discrimination against nonprotein amino acid incorporation into protein by the seed predator, *Caryedes brasiliensis*, J. Chem. Ecol. 9, 1353-1361.

Rosenthal, G. A. and Janzen, D. H. (1985) Ammonia detoxification by the bruchid beetle, *Caryedes brasiliensis*, J. Chem. Ecol. 11, 539-544.

Rosenthal, G. A., Janzen, D. H., and Dahlman, D. L. (1977) Degradation and detoxification of canavanine by a specialized seed predator, Science 196, 658-660.

Rosenthal, G. A., Reichhart, J. M., and Hoffmann, J. A. (1989a) L-Canavanine incorporation into vitellogenin and macromolecular conformation, J. Biol. Chem. 264, 13693-13696.

Rosenthal, G. A., Lambert, J., and Hoffmann, D. (1989b) L-Canavanine incorporation into protein can impair macromolecular function, J. Biol. Chem. 264, 9768-9771.

Rosenthal, G. A., Berge, M. A., Bleiler, J. A., and Rudd, T. (1987) Avoidance of aberrant protein production and an organism's ability to utilize or tolerate L-canavanine, Experientia 43, 558-561.

Tate, S. S. and Meister, A. (1969) The Enzymes of Glutamine Metabolism, S. Pruisiner and E. R. Stadtman (eds.), Academic Press, New York, pp. 77-127.

COEVOLUTIONARY RELATIONS BETWEEN BRUCHIDS AND THEIR HOST PLANTS. THE INFLUENCE ON THE PHYSIOLOGY OF THE INSECTS

J. HUIGNARD, P. DUPONT, AND B. TRAN
Institut de Biocénotique Expérimentale des Agrosystèmes
Université de Tours, URA CNRS 1298, Av.Monge, Parc Grandmont
37200 Tours, France

ABSTRACT. Most Bruchidae (Coleoptera) are specialists developing on a limited number of species of Leguminosae. Host selection is by females ovipositing on pods which usually are available only during a short period of the year. Thus, there is a precise synchronization between the reproductive cycles of the plants and bruchids. Experiments in the field and laboratory show that *Bruchidius atrolineatus* (Pic) and *Bruchus rufimanus* (Boh.) are in reproductive diapause when the pods of their host plant are absent. The appearance of flowers induces diapause termination and make insects sexually active at the beginning of the fructification period. Chemicals produced by flowers probably stimulate the development of oogenesis. Chemical or tactile stimuli perceived in direct contact with the pods stimulate oviposition on the trophic substratum of the larvae. These interactions increase the reproductive fitness of the insects and explain the high levels of bruchid infestations, particularly in crops. Some structures observed in wild Leguminosae could represent defenses against bruchids. The texture of the pods could modify egg-laying behaviour ; the size, the structure and the chemical composition of the seeds could influence larval development. The importance of these host plant characteristics is analyzed in relation with the hypothesis of coevolution.

1. Introduction

Bruchid beetles are specialist insects. According to Johnson (1981), Johnson and Slobodchikoff (1979), Varaigne-Labeyrie and Labeyrie (1981), and Rasplus (1988), 80 to 90% of species of Bruchidae have only 1 to 3 host species. As in numerous specialists, neonate larvae alone cannot find their trophic substrate, thus the adults seek out the host plant pods or seeds on which the larvae develop (Labeyrie and Huignard, 1973).

As in other specialist phytophagous insects (Visser, 1986) some volatile compounds produced by the Leguminosae (leaves or pods) could be involved in finding the host plant. Electroantennographic studies (Pouzat, 1981) have shown that females of *Acanthoscelides obtectus* (Say) can perceive the odors of *Phaseolus vulgaris* (especially the odors of green pods). However the chemical nature of these volatile compounds influencing the behaviour of this bruchid are unknown. When *A. obtectus* females discover *P. vulgaris* pods, gustatory stimulations produced by the host plant and tactile informations are perceived by the sensory receptors of the maxillary and labial palps, and the ovipositors. This information stimulates oogenesis and induces egg-laying (Pouzat, 1981). A liposoluble compound and a water-soluble compound present on the surface of the pods or the seeds of *Phaseolus vulgaris* influence *A. obtectus* egg-laying. The lipo-soluble

171

K. Fujii et al. (eds.), Bruchids and Legumes: Economics, Ecology and Coevolution, 171–179.
© 1990 *Kluwer Academic Publishers. Printed in the Netherlands.*

compound only stimulates vitellogenesis (Monge, 1983).

The egg-laying behaviour of bruchid females probably plays a major role in the specificity of the relationships between each species of insect and their host plant (Johnson and Kistler, 1987). When the oviposition substrate is unavailable, various adaptations enable the insect to wait until conditions favorable for reproduction and larval development are again present. Reproductive quiescence (Huignard and Biémont, 1978) or reproductive diapause (Huignard et al., 1987) have been described in various species of Bruchidae. They lead to a more or less precise synchronization between the reproductive cycle of the insect and that of the host plant (Huignard et al., 1987).

When oviposition occurs on a pod, post-embryonic development will occur only if the larva is capable of eating, assimilating or detoxifying the abundant secondary compounds of the seeds. According to Janzen (1981) the ability of Bruchid larvae to detoxify these compounds is the major cause of host specificity. Each species of seeds has a different set of chemical defenses and the specialists of the seeds of one species cannot deal with the combination of chemicals in the seeds of other species. The ability to detoxify is often very highly developed. Extreme specialists such as *Stator generalis* (Johnson and Janzen, 1982) or *Caryedes brasiliensis* (Rosenthal et al., 1977) may have restricted themselves to the highly toxic seeds of one host plant by developing expensive physiological detoxification mechanisms.

These possibilities for adaptation, both at the reproductive and larval levels, explain the often very high attack levels observed on wild or cultivated Leguminosae (Pierre and Huignard, 1989; Huignard et al., 1985; Dupont, 1989). Janzen (1969) and Center and Johnson (1974) have discussed the possible evolutionary response to seed predation by seed beetles. Janzen emphasized the development of protective devices by the host plant which could reduce predation by bruchids.

Here we use several examples of the relationships between bruchids and their host plants and their evolutionary consequences. Specifically we examined :
(1) adaptive mechanisms enabling these phytophagous species to reproduce and develop on their host plant, and
(2) adaptations developed by the plant, limiting losses due to the action of phytophagous species.

2. Analysis of Relationships between Bruchids and Their Host Plants

2.1. RELATIONS BETWEEN A UNIVOLTINE SPECIES *BRUCHUS RUFIMANUS* (Boh.) AND ITS HOST PLANT *VICIA FAVA* (L.)

This European insect develops in some species of *Vicia*, especially *Vicia faba*. As in the case of most bruchids of temperate zones, *B. rufimanus* is in reproductive diapause during autumn and winter and hibernates in lichens or under the bark of trees (Genduso, 1978; Dupont, 1989). Adults colonize *V. faba* crops in the spring and we have analyzed the adult population throughout the period of flowering and fructification of *V. faba* (from May to July). Behavioural observations have shown that adults seek refuge at night and the early morning in resting sites (flowers, young leaves), where they are easily captured. The beetles are active during the day (flights, egg-laying) when the temperatures are favourable (20°C) (Dupont, 1989).

Studies carried out in three successive years have shown the importance of the flowering phase in insect-plant relationships (Fig. 1).

1) *B. rufimanus* adults consume large quantities of pollen and visit large number of flowers during the day. Throughout the entire period of flowering (phases 1-2), adults

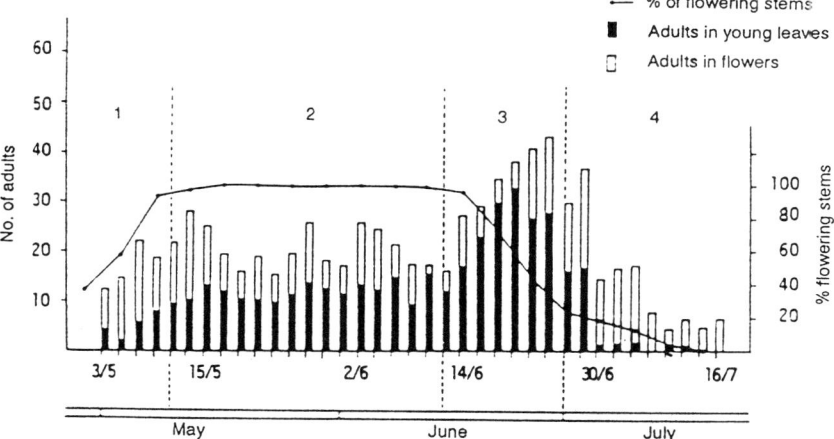

Figure 1. Variations in the number of *B. rufimanus* adults captured in the *V. faba* culture from 29 April 1987 to 16 July 1987 and in the percentage of flowering stems.

consume the pollen of only *V. faba*. At the end of the flowering phase (phase 3) adults congregate on the last available flowers (explaining the high rate of captures at this level), then seek out the pollen of other plant species still flowering.

The numerous visits to *V. faba* flowers in the period of maximal flowering can be favorable for the plant, since these bruchids pollinate this allogamous plant. Under natural conditions, *B. rufimanus* is only one of the pollinating species because a large number of others insects are pollinators (bees, bumblebees, etc.). Greenhouse experiments in the absence of all other pollinators, however, have shown that only those flowers visited by *B. rufimanus* form pods with seeds.

2) In the beginning of the colonization phase, males captured in crops are able to reproduce. Females, however, are still in reproductive diapause (table 1). Vitellogenesis starts later, during the phase of full plant flowering.

Laboratory experiments involved placing insects maintained in reproductive diapause for 4-5 months (at 10°C) with the pollen of various plant species encountered in the

Table 1. Maturation of the female reproductive organs captured in the *Vicia faba* culture during the four phases of the study. A : From 13 to 20 May 1987 ; B : From 20 May to 12 June 1987.

	Phase 1	Phase 2		Phase 3	Phase 4
		A	B		
No. of females	25	20	22	23	21
Diapausing females	23	0	2	0	0
Oocytes in vitellogenesis	2	20	13	0	0
Mature ovocytes	0	0	7	23	21

Table 2. Influence of feeding with pollen on the maturation of the reproductive organs of *B. rufimanus* in 16 h L 26°T/8 h D 23°C conditions for 20 days. All the insects were in reproductive diapause before the experiment. MG: mature gonads; MO: mature oocytes.

Feeding with pollen	Males		Females	
	No.	% M.G.	No.	% M.O.
Vicia faba	37	100	32	30
Vicia sepium	23	100	24	17
Lathyrus latifolius	14	100	20	18
Heracleum sphondylium	12	100	18	0
Stellaria holostea	12	100	23	0
Prunus sp.	14	100	27	0

agroecosystem (Rosaceae, Caryophyllacae, Umbelliferae, Leguminosae). It was found that (Table 2):

a) a pollen-rich diet, regardless of the plant species, triggered the development of male reproductive organs after a period of 8 to 10 days. The male accessory glands produced secretions and new spermatogenesis phases were observed in the testes.

b) only a diet consisting of the pollen of Leguminosae (*Vicia* sp. or *Lathyrus* sp.) could induce vitellogenesis in females. The other pollens caused an increase of fat body reserves of females, but did not induce the maturation of oocytes.

Consuming pollen of species of Leguminosae can terminate reproductive diapause in females of *Bruchus pisorum* (Pajni and Sood, 1975) or *Bruchus affinis* (Bashar et al. 1987). The chemical compounds in pollen influence the reproductive physiology of these bruchids but the physiological mechanisms induced are unknown.

3) Although the feeding behaviour of bruchids in flowers can be favourable for *V. faba* by insuring pollination, at the end of flowering it can cause a concentration of adults on the last stems still flowering. This concentration of adults can increase oviposition on the pods located near these stems, leading to a high loss of seeds.

In order to analyze the importance of this phenomenon, the numbers of adults and eggs laid on the pods were followed in two plots inside the crop. One in which early flowering occurred in a short time interval (a densely planted plot) and the other in later flowering was spread over a longer period (a sparsely planted plot). Colonization of the crop always began in the zone of early flowering (Fig. 2). At the end of flowering in this plot, the adults tended to assemble in the sparsely planted plot where flowers were still available. This change in the spatial distribution of adults led to increased oviposition on the pods of stems in the late flowering area.

4) Of the pods of *V. faba* sampled, 44.8% did not produce seeds. These flat, small (in comparison to those with seeds) pods remained on the stems, or fell off in most cases, at the end of the fruiting period. The pods without seeds received only 5.7% of the eggs laid. This oviposition behaviour thus limits the number of eggs on pods where they have no chance of developing, thus contributing to optimization of the reproductive effort.

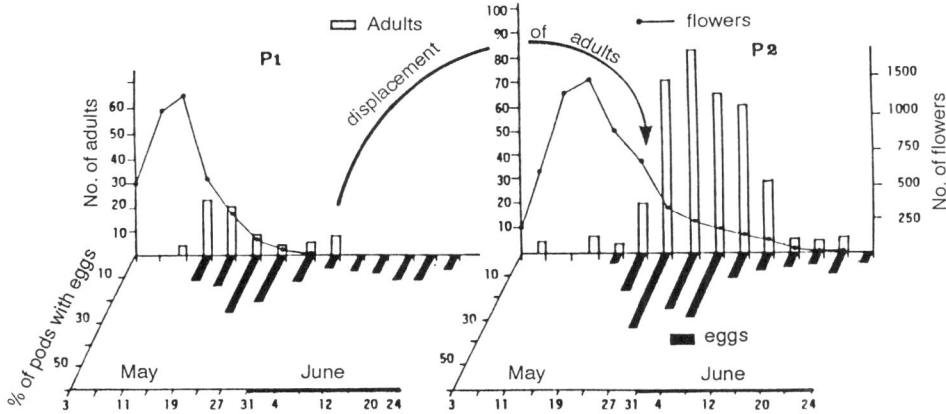

Figure 2. Variations of the number of *B. rufimanus* adults in two plots (P1 and P2) differing by the density of *Vicia faba* (P1 = 18 plants/m², P2 = 8 plants/m²) in relation with the number of available flowers and importance of egg-laying on pods.

These very close relationships between *B. rufimanus* and its host plant limit the adaptive possibilities of the species. Thus, it can reproduce only on green pods, a phenological stage which is available only for a short period. Under these conditions, reproductive diapause probably enables populations to survive during the long period when pods are not available. Stored energy is used for the survival of individuals and not for gametogenesis.

2.2. RELATIONS BETWEEN MULTIVOLTINE SPECIES OF BRUCHIDS AND THEIR HOST PLANTS

Multivoltine bruchids must constantly have suitable oviposition substrates available, as well as seeds for larval development (Southgate, 1981). This situation was observed in two cases:

a) When host plant pods are available in nature throughout the year. This was observed by Pierre and Huignard (1989) in the African Sahel zone on *Bauhinia rufescens* (Caesalpiniaceae). Four or five generations of *Caryedon serratus* develop during the year on this tree, which continually produces pods which remain for 8 to 10 months on the tree. No diapause (during post-embryonic or imaginal phase) was observed and it was primarily climatic factors which limited the number of insects in the course of different generations (fig. 3).

b) When the species have higher adaptive possibilities, of both adults and larvae, they are capable of reproducing and developing on the oviposition substrates and in different biocenotic conditions (pods at various stages of maturity, seeds, cultures, or seed stores). These species (*Callosobruchus* spp., *Acanthoscelides obtectus*) are the main pests of stored legume seeds and have a large geographical distribution as a result of human activity.

The permanent presence of food is not, however, in itself sufficient for multivoltinism. In a store containing *Vigna unguiculata* pods in Niger, *Bruchidius atrolineatus* (Pic) gave rise to only two generations during the dry season. The first was composed of sexually active adults. The second generation of adults were in reproductive diapause and rapidly left the stores which were often not bruchid-proof to reach unknown aestivation sites.

176

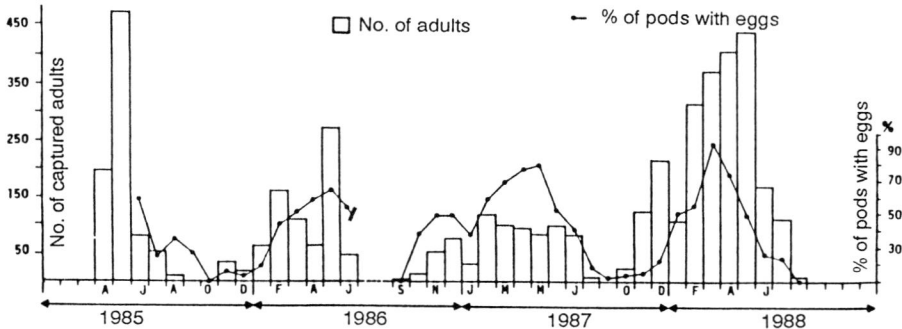

Figure. 3 Variations of the number of *Caryedon serratus* adults captured in a trap) on *Bauhinia rufescens* during three years (1985-1988) and variations of the percentages of pods receiving *C. serratus* eggs during the same time.

This reproductive diapause is induced during the post-embryonic development by variations of the temperature at the beginning of the dry season (Monge et al., 1989). As in univoltine species, this leads to the maintenance of *B. atrolineatus* populations in nature until a new fructification phase of its host plant *V. unguiculata*.

The biology of adults during the long dry season is unknown but they must survive during a long period of high temperatures (from March to August). Diapause termination is due to biocenotic factors. The consumption of *V. unguiculata* pollen is very limited and does not represent the major factor inducing maturation of the reproductive organs (Huignard et al., 1987). Experimental studies have shown that contact with *V. unguiculata* flowers or green pods induce termination of diapause after a short latency period (10 days). Sensory stimulations perceived by gustatory receptors of the palpi could be a factor inducing the reproduction of *B. atrolineatus*. Adults are sexually mature when the young pods of *V. unguiculata* appear at the end of the rainy season (fig. 4).

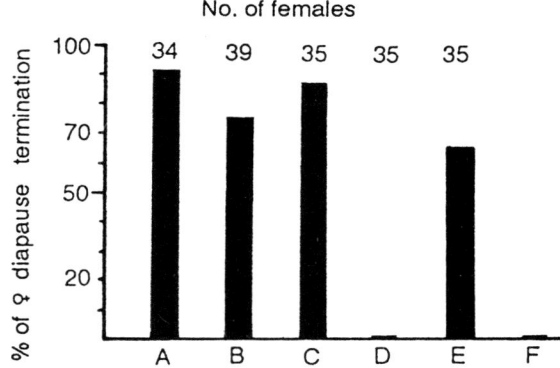

Figure. 4 Percentages of sexually active *B. atrolineatus* females; influence of: A: inflorescences of *V. unguiculata*; B: extrafloral nectaries; C: corolla; D: host plant leaves; E: green pods; F: mature pods. Results obtained in 12 h L 35°C/ 12 h D 25°C (after 20 days).

3. Adaptations Developed by Leguminosae against Bruchids

Considering the adaptations developed by bruchids in the presence of their host plants, seed destruction is often very high. The system of plant defense leading to limited seed loss and the retention of reproductive potential are highly varied and have been extensively studied (Janzen, 1969; Center and Johnson, 1974; Johnson and Slobodchikoff, 1979). Depending on the species, these plant defenses may affect the oviposition behaviour of the females, or their larval biology.

1) Morphological adaptations can prevent the egg-laying on the pod. Thus, species whose pods have a hairy pericarp are attacked little by Bruchidae of the genus *Bruchus*, which cannot deposit their eggs (Huignard, unpublished observation). When females oviposit only in cracks in the pods such as *Algarobius prosopis*, the production of a pod without surface cracks could limit egg-laying (Janzen, 1969).

2) Dehiscence (in *Leucaena* sp.), fragmentation (in *Mimosa* sp.) or explosion (*Canavalia* sp.) of pods, scattering the seeds to escape from larvae coming through the pod walls and from ovipositing females (Center and Johnson, 1974).

3) Mechanical or chemical barriers produced by the seed coat or pod pericarp may cause high mortality during larval penetration of pods or seeds (Janzen, 1981; Stamopoulos and Huignard, 1980). The existence of lignin in seed coats, which is very toxic to *A. obtectus* larvae (Stamopoulos, 1988) is an example of this high mortality.

4) The production of toxic compounds inside cotyledons may cause high mortality during larval development. As has been shown in a number of publications, seeds of Leguminosae are very rich in alkaloids, saponins, cyanogenetic glycosides, haemagglutinins, free amino acids (Smartt, 1971), which may be toxic to larvae (Janzen et al., 1977). Certain relationships between bruchids and seeds have been carefully analyzed and the toxicity of these compounds has been shown unequivocally (Gillon, 1986). In certain conditions they may confer effective protection, as reported by Southgate (1979) in Australia. Acacias are abundant but it seems there are apparently very few bruchids that feed in their seeds. Precise research has shown that Australian acacias have different chemical components in them than those from other parts of the world, thus suggesting that toxic chemicals account for the lack of bruchids in the seeds. Nevertheless, in the presence of secondary substances, bruchids can elaborate highly effective detoxication systems, as observed by Rosenthal et al. (1977) in *Caryedon brasiliensis*, which can use the plant chemicals as food. These adaptations are probably very important, particularly in crop protection. One can ask if the selection of chemical resistance in seeds in fact represents an effective system of protection against the bruchids.

4. CONCLUSION

The relationships between Bruchidae and Leguminosae are very complex and it may be asked if coevolution exists between these phytophagous species and their host plant. A number of studies have shown the importance of secondary compounds contained in the seeds for larval biology, but Gillon (1986) noted that the proof of an evolutionary process is not established on only these grounds. According to this author, nothing proves that the "raison d'être" of a substance such as canavanine is a protective system against insects. It can, in fact, be asked why Leguminosae such as *Dioclea megacarpa* form a substance which no longer protects it against *Caryedes brasiliensis*. Similarly, the hypothesis of protection against generalist insects is not well studied. In addition, it is highly unlikely, as noted by Jermy (1984) that selection pressures exerted by the Bruchidae are sufficient to influence the evolution of plant species such as perennial Leguminosae. These species produce a large number of seeds and even if attacks are numerous and decrease the reproductive potential of the plant species, there remain a sufficient number to assure there will be future generations.

The Bruchidae-Leguminosae model is a very good example of the analysis of physiological and evolutionary relationships between plants and insects. Caution is nevertheless required when interpreting these relationships in the context of coevolutionary hypotheses.

REFERENCES

Bashar, A. H., Fabres, G., Hossaert, M., Valero, H., and Labeyrie, V. (1987) *Bruchus affinis* and the flowers of *Lathyrus latifolius*: an example of the complexity of the relations between plants and phytophagous insects, in V. Labeyrie et al. (eds.), Insects Plants, Junk Publishers, Dordrecht.

Center, T. D. and Johnson, C. D. (1974) Coevolution of some seed beetles (Coleoptera Bruchidae) and their hosts, Ecology 55, 1096-1103.

Dupont, P. (1989) Contribution à l'étude des populations de Bruchus rufimanus(Boh.) dans le Centre de la France. Influence de la phénologie de la plante-hôte sur la répartition spatiotemporelle des adultes, Thèse de Doctorat, Univ. TOURS.

Genduso, P. (1978) Insectes nuisibles aux Légumineuses en Sicile et observations sur l'hivernage des bruches monovoltines, Boll. Int. Entomol. Agr. Oss. Fitopat. Palermo 10, 169-175.

Gillon, Y. (1986) Coevolution cumulative et coevolution substitutive, Acta Oecologia Oecol. Gener. 71, 27-36.

Huignard, J. and Biémont, J. C. (1978) Comparison of four populations of *Acanthoscelides obtectus* from different ecosystems. Assay of interpretation, Oecologia 35, 307-318.

Huignard, J., Germain, J. F. and Monge, J. P. (1987) Influence of the inflorescence and pods of *Vigna unguiculata* on the termination of the reproductive diapause of *Bruchidius atrolineatus*, in V. Labeyrie et al. (eds.), Insects Plants, Junk, Dordrecht, pp. 183-188.

Huignard, J., Leroi, B., Alzouma, I. and Germain, J. F. (1985) Oviposition and development of *Bruchidius atrolineatus* and *Callosobruchus maculatus* in *Vigna unguiculata* cultures in Niger, Insect Sci. Applic. 66, 691-699.

Janzen, D. H. (1969) Seed eaters versus seed size, number, toxicity and dispersal, Evolution 23, 1-27.

Janzen, D. H. (1981) The defenses of legumes against herbivores, in R. M. Polhill and P. H. Raven (eds.), Advances in Legume Systematics, pp. 951-977.

Janzen, D. H., Juster, H. B. and Bell, E. A. (1977) Toxicity of secondary compounds to the seed eating larvae of a bruchid beetle *Callosobruchus maculatus*, Phytochemistry 16, 223-227.

Jermy, T. (1984) Evolution of insect/host plant relationships, Amer. Nat. 124, 609-630.

Johnson, C. D. (1981) Relations of *Acanthoscelides* with their plant hosts, in V. labeyrie (ed.), The Ecology of Bruchids Attacking Legumes, Junk, The Hague, pp. 73-81.

Johnson, C. D. and Slobodchikoff, C. N. (1979) Coevolution of *Cassia* (Leguminosae) and its seed beetle predators, Environ. Entomol. 8, 1059-1064.

Johnson, C. D. and Janzen, D. H. (1982) Why are the seeds of Central American guanacaste tree not attacked by bruchids except in Panama, Environ. Entomol. 11, 373-377.

Johnson, C. D. and Kistler, R. A. (1987) Nutritional ecology of bruchid beetles, in F. Slansky and J.G. Rodriguez (eds.), Nutritional Ecology of Insects, Mites and Spiders, J. Wiley and Sons, New York, pp. 259-277.

Labeyrie, V. and Huignard, J. (1973) Relations trophiques et comportement reproducteur des insectes, Ann. Soc. Roy. Zool. Belg. 1031, 43-51.

Monge, J. P. (1983) Comportement de ponte de la bruche *Acanthoscelides obtectus* sur le substrat artificiel imprégné d'extrait de la plante hôte *Phaseolus vulgaris*, Biol. Behav. 3, 205-213.

Monge, J. P., Lenga, A., and Huignard, J. (1989) Induction of reproductive diapause in *Bruchidius atrolineatus* during the dry season in a sahelian zone, Entomol. Exp. Appl. (in press).

Pajni, H. R. and Sood, S. (1975) Effect of pea pollen feeding in maturation and copulation in the beetle *Bruchus pisorum*, Ind. J. Exp. Biol. 13, 202-203.

Pierre, D. and Huignard, J. (1989) Biological cycle of *Caryedon serratus* in presence of one of its host plant *Bauhinia rufescens*, Acta Oecologica (in press).

Pouzat, J. (1981) The role of sense organs in the relations between Bruchids and their host plants, in V. Labeyrie (ed.), The Ecology of Bruchids Attacking Legumes, Junk, The Hague, pp. 61-72.

Rasplus, J. Y. (1988) La communauté parasitaire des Coléoptères séminivores des Légumineuses dans une mosaique forêt-savane en Afrique de l'Ouest, Thèse, Université de Paris 11. 435 pp.

Rosenthal, G. A., Janzen, D. H., and Dahlman, D. L. (1977) Degradation and detoxification of canavanine by a specialist seed predator, Science 196, 658-660.

Smartt, J. (1977) Tropical Pulses, Longman Eds. London.

Southgate, B. J. (1979) Biology of the Bruchidae, Ann. Rev. Entomol. 24, 449-473.

Southgate, B. J. (1981) Uni and multivoltine cycles. Their significance, in V. Labeyrie (ed.), The Ecology of Bruchids Attacking Legumes, Junk, The Hague, pp. 17-22.

Stamopoulos, D. C. and Huignard, J. (1980) L'influence des diverses parties de la graine de haricot (*Phaseolus vulgaris*) sur le développement des larves d'*Acanthoscelides obtectus* Coléoptère Bruchidae, Entomol. exp. appl. 28, 38-46.

Stamopoulos, D. C. (1988) Toxic effect of lignin extracted from the tegument of *Phaseolus vulgaris* seeds on the larvae of *Acanthoscelides obtectus*, J. Appl. Entomol. 105, 317-320.

Varaigne-Labeyrie, C. and Labeyrie, V. (1981) First data on Bruchidae which attack the pods of legumes in Upper Volta of which eight species are man consumed, in V. Labeyrie (ed.), The Ecology on Bruchids attacking Legumes, Junk, The Hague, pp. 83-96.

Visser, J. H. (1986) Host odor perception in phytophagous insects, Ann. Rev. Entomol. 31, 121-144.

COEVOLUTION OF BRUCHIDAE AND THEIR HOSTS: EVIDENCE, CONJECTURE, AND CONCLUSIONS

CLARENCE DAN JOHNSON
Department of Biological Sciences
Northern Arizona University
Flagstaff, Arizona 86011-5640, U.S.A.

ABSTRACT. A review of the literature revealed that coevolution in the Bruchidae is very correlative and thus only presumed to exist. When the contemporary precepts of coevolution were scrutinized and compared with the coevolutionary research on the Bruchidae, it was found that only components of each study of all of the coevolutionary research conducted to date on the Bruchidae was experimental, the balance of each study was correlative. It is suggested that future research on bruchid-plant coevolution use cladistics and morphometrics in studies of systematics, these can then be used in conjunction with rigorous experimental and manipulative studies of chemistry, biochemistry, and detoxification in both plants and herbivores, edaphic factors, primary metabolic requirements, genetics, ecological factors, and variation of these over space and time to conclusively define the process of coevolution.

1. Introduction

Janzen (1969) was the first biologist to discuss coevolution in the Bruchidae and one of the first to publish on the topic after Ehrlich and Raven (1964) produced their classic paper on coevolution. Janzen's paper was not only a great stimulus to those who study bruchids and their interactions with their hosts, but also to other biologists interested in ecology and evolution. Janzens's work stimulated other studies on coevolution in bruchids (e.g. Center and Johnson, 1974; Smith, 1975; Bradford and Smith, 1977; Janzen, 1977; Janzen et al., 1977; Rosenthal and Bell, 1979; Johnson and Slobodchikoff, 1979; Johnson, 1981a,b; Rosenthal and Janzen, 1983, 1985; Nelson and Johnson, 1983a,b; Johnson, 1987). During the last decade the interesting and thought-provoking studies by Ehrlich and Raven, Janzen, and others have led to research on a variety of organisms that has resulted in the publication of more than 1,000 papers that have had coevolution in the title or abstract (Thompson, 1989). Coevolution indeed is a popular, sometimes controversial, and seemingly well-studied phenomenon.

Because the term coevolution had been used in a variety of ways by different authors (e.g., that coevolution can occur between organisms and abiotic factors in their environment), Janzen (1980) defined coevolution as "an evolutionary change in a trait of the individuals in one population in response to a trait of the individuals of a second population, followed by an evolutionary response by the second population to the change in the first". Thompson (1989) shortened the definition to "coevolution is reciprocal evolutionary change in interacting species".

K. Fujii et al. (eds.), Bruchids and Legumes: Economics, Ecology and Coevolution, 181–188.
© 1990 *Kluwer Academic Publishers. Printed in the Netherlands.*

2. Evidence and Conjecture on Coevolution in Bruchids

The model for studies on coevolution was, of course, the paper by Janzen (1969). His interpretation of the results of his research were that in the tropical lowlands of Costa Rica, large seeds were toxic and presumably protected from bruchids and that smaller seeds were less toxic and fed upon by bruchids. Consequently, seed fitness seems to be determined by seed size and toxicity. Thus, individual large seeds have a better chance of survival than many small seeds. If a plant invests energy in large, toxic seeds, after germination a seedling has a much better chance of surviving. On the other hand, small seeds have less stored energy and toxins, are dispersed in great quantities, thus escaping from predators, but after germination have less chance to develop into viable seedlings. Janzen termed escape from predation by small seeds, predator satiation. Center and Johnson (1974) listed and interpreted examples that presumably showed that bruchids had overcome the defense of predator satiation with countermeasures such as evolving the ability to feed in smaller seeds by becoming smaller in body size or feeding in more than one seed during their ontogenetic development. Nelson and Johnson (1983a) reported that in some species of *Astragalus* there was stabilizing selection for an optimal seed size because bruchids in species of *Astragalus* prefer large seeds over smaller seeds but larger seeds had a better chance of germinating and developing into a mature plant. Therefore, because of the opposing processes of predation rates and germination success, the intermediate seed-size classes were most likely to contribute more to the next generation.

Smith (1975) and Bradford and Smith (1977) studied the interactions between palm bruchids, palm seeds, and squirrels. They determined that there was probably selection pressure by squirrels and bruchids to cause the palm fruits to evolve different numbers of seeds (seed packaging) in their fruits depending upon whether squirrels or bruchids were the principal predators and that seed predation is a significant factor in maintaining multiseeded fruits in *Scheelea* palm populations.

Evolution of seed packaging may or may not be driven by bruchids and squirrels in palms. Mitchell (1977) proposed that seed packaging was influenced by bruchids in *Cercidium floridum*. Recently, however, Mitchell's hypothesis has been refuted by Siemens and Johnson (in press). We found that in species of *Cercidium*, seed defenses against bruchids (1) are toxins in the seed coats or (2) rapid maturation and dispersal of the seeds, not the number of seeds in the pods. Many other factors may influence numbers of seeds in pods including the number of ovules that are pollinated (see also Huignard et al. this volume).

Many legume seeds contain toxins. Toxins act in allelopathic interactions but probably the greatest benefit to seeds is that toxins protect seeds from predation by most, and in some cases all, organisms. Seed toxins are one of the most robust selective agents driving bruchids to specificity to their hosts. This has led to some bruchids evolving the ability to feed in seeds that are very toxic. Janzen (1977), Johnson (1970), Center and Johnson (1974), Janzen et al. (1977), and Rosenthal and Bell (1979), among others, have all presented examples of bruchids that feed in very toxic seeds. Trelease and Trelease (1937), Johnson (1970) and Nelson and Johnson (1983b) presented evidence that several species of *Acanthoscelides* are able to feed in seeds of *Astragalus* that contain selenium. Rosenthal and Bell (1979) stated that some plants can contain as much as 15,000 ppm of selenium and that Se-methyl-L-seleno-cysteine and selenocystathionine are the principal causal agents of the element's toxicity. The first compound occurs in at least one species of *Astragalus*. Nelson and Johnson (1983b) found that some species of bruchids are much more successful in feeding in selenium-containing *Astragalus* seeds than others. However, success in feeding in these seeds was probably due to the level of selenium compounds in seeds. One species was able to tolerate very high levels of selenium while others could tolerate only much lower concentrations, but concentrated enough to be toxic to mammals. Center and Johnson (1974) found examples of bruchids that either are able to tolerate toxins or to avoid them.

The most intensive studies of bruchids and toxins have been conducted by G. Rosenthal and coauthors. *Caryedes brasiliensis* feeds only in *Dioclea megacarpa* whose seeds are rich in the toxic amino acid canavanine. *C. brasiliensis* can detoxify canavanine to produce dietary nitrogen by converting canavanine to canaline and urea and then degrading urea to ammonia (Rosenthal and Bell, 1979; Rosenthal and Janzen, 1983, 1985). Bleiler et al. (1988), however, found that several species of larval beetles have become specialists on canavanine-containing seeds. Also, in laboratory feeding experiments, they found that a number of other insects could degrade canavanine to canaline. Thus they believe that "certain biochemical capacities required by canavanine-feeding insects may have existed prior to their exposure to dietary canavanine". Studies similar to those of Rosenthal et al., may reveal many pathways where bruchids use other toxic chemicals for their metabolism. For example, Rosenthal and Bell (1979) cite the toxic amino acid mimosine as being present in both *Mimosa pudica* and *Leucaena leucocephala*. The seeds of the latter may contain as much as 9% dry weight of mimosine. Both species have relatively small seeds and are fed upon by several species of bruchids (Johnson and Kingsolver, 1976; Johnson, 1983).

Much of the research on the feeding capabilities of bruchids has been in nonhost plants. This has enabled researchers to determine the ability of the insects to penetrate thick or hard or toxic seed coats or to develop in presumably toxic seed contents. Janzen (1977) ascertained that the bruchid *Callosobruchus maculatus* was able to develop in only one of 63 nonhost seeds to which it was exposed. He established that seed coats have a different effect on this species than do the seed contents. Therefore, he cautioned against treating seed coats the same as seed contents in studies of bruchids and interactions with seeds. Other studies have reenforced Janzen's findings (see below). Janzen et al. (1977) incorporated secondary compounds in the normal diets of *C. maculatus*. Their results were that alkaloids, amino acids and other secondary compounds were likely to be toxic to at least some animal and are likely to be responsible to some extent for host specificity of seed-eating insects.

Six species of the non-economic bruchid genus *Stator* were offered a variety of nonhost seeds to oviposit upon by Johnson (1981b). His results were that some species with a broad host range in nature could not feed in many nonhost seeds in the laboratory. The antithesis was a species with a very narrow, specialized range of hosts in nature was found to be able to feed in a very broad range of nonhosts. Generally, though, those with very narrow host ranges are not able to feed in nonhost seeds. In addition, some species with few natural hosts were very selective in the seeds that they would oviposit upon in the laboratory. Thus, ovipositional behavior of the adult female has a profound effect on the ability of larvae to utilize seeds because larvae cannot move to an alternate host. Other results were: (1) It is probable that some species of *Stator* have varying degrees of success even when feeding on different species of their natural hosts and (2) Seed coats are formidable barriers to entry of larvae of some species of *Stator*, even in some of their natural hosts.

Data that purportedly showed evidence of coevolution between *Cassia* (s.l.) and bruchids were published by Johnson and Slobodchikoff (1979). Their evidence was that species of bruchids that feed in seeds of species of *Cassia* are naturally divided into specialist and generalist species. Most (82.5%) are specialist species and feed on 1-3 species of host plants. The few generalist species feed on 5-10 species of host plants, are homodynamic, most feed on many of the same species of host plants, all have a broad geographic range, and all feed in a broader range of seed sizes (volume) than the specialist species. Their hypothesis was that seeds of species of *Cassia* that host 5-7 bruchid species have few toxins but the species of *Cassia* that host 1-3 species of bruchids are very toxic.

Toxic seeds and predator satiation are not the only traits that have been suggested to be in a coevolutionary race between bruchids and their hosts. Janzen (1969) emphasized that plants had developed many defenses against seed predation but that "virtually all [defensive traits] seem effective against at least one species of bruchid but only rarely against all bruchids." He suggested that legumes exhibit at least 31 "traits that may be functional in eliminating or lowering bruchid destruction of seeds." Center and Johnson (1974) discovered apparent countermechanisms to eleven of the traits evolved by the Leguminosae. Countermechanisms to defensive traits against bruchids such as gum production by pods, dehiscence, smaller size of seeds, indehiscence, flaking of pod surface, etc., were described and discussed.

This was expanded upon by Johnson (1981a) when he described a series of complex, presumably coevolutionary, interactions between bruchids and their hosts. Several species of bruchids in the genus *Stator* oviposit on exposed seeds only after they have fallen to the ground or are exposed in animal dung. This enabled Johnson to delineate three guilds of bruchid beetles based on observed behavior patterns of many other species of bruchids. He found that *Mimosestes* and *Merobruchus* in guild 1 oviposit *on* the surface of legume pods attached to the plant and species of *Stator* in guild 2 oviposit on seeds while they are still inside a partially dehiscent pod on the plant. Other species of *Stator* in guild 3 oviposit only on seeds on the ground. Pod behavior includes indehiscent pods (fed upon by guilds 1 and 3), partially dehiscent pods (fed upon by guilds 1, 2, and 3), and dehiscent pods (fed upon by guild 3, sometimes by guild 2). Many factors other than the effects of bruchids are responsible for the structure and behavior of pods (e.g. seed dispersal agents, vertebrate seed predators, etc.) but there is the distinct probability that there are some coevolutionary interactions between pod and seed structure and the impact of bruchid beetle depredations. In a similar paper Johnson (1987) documented the preference of species of *Mimosestes* for species of *Acacia* with thick, woody valves. He concluded that bruchids and acacias seem to be in a series of coevolutionary interactions with the structure and chemistry of fruits and seeds of acacias.

2.1. COEVOLUTIONARY RESEARCH DURING THE PAST 25 YEARS

After Ehrlich and Raven (1964) presented their evidence for coevolution, a series of correlative papers appeared that professed to show evidence that coevolution had occurred. The sentiments of most scientists who study coevolution today are probably best summarized by Spencer (1988a) as follows: "Correlation studies are not adequate to detect and describe the coevolution of a system. Experimental and manipulative multivariate studies addressing chemistry, biochemical regulation, and detoxification in both plants and herbivores, edaphic factors, primary metabolic requirements, genetics, ecological factors, and variation of these over space and through time are needed to ultimately define the process of coevolution." See Thompson (1986a,b, 1988) for additional ideas about coevolutionary research.

In recent years the papers that reliably show that coevolution occurs or does not occur in a system use experimental methods. All of the literature on **bruchid-host plant coevolutionary studies** is correlative.

Although there is disagreement between scientists as to what are good examples of coevolution, there seems to be agreement that examples of coevolution are rare, and those most often cited as good examples are fig wasps and figs (Janzen, 1979; Wiebes,

1979), yucca moths and yuccas (Powell and Mackie, 1966), rust and flax (Flor, 1971), ants and acacias (Janzen, 1966), and heliconine butterflies and *Passiflora* (Benson et al., 1976; Spencer, 1988b). The first two examples were termed cospeciation by Thompson (1986a).

Using a variety of experimental and other data, Craig et al. (1988) demonstrated that coevolution probably does not occur in a willow-sawfly system. Using their ideas and those of others (Ehrlich and Raven, 1964; Fox, 1981; Thompson, 1982; Feinsinger, 1983; Futuyma, 1983; Gilbert, 1983; Schemske, 1983; Strong et al., 1984) they proposed that five qualities must be present in interactions for coevolution to occur. These are: (1) **Strong interaction.** Each species must have the potential to significantly affect the fitness of individuals of the other species. Thus to coevolve, each species must exert strong selection on the other. (2) **High specificity.** Each species must tightly interact with the other species. (3) **Widespread.** The interaction must involve a large proportion of each population, otherwise gene flow will swamp evolving traits. (4) **Predictable.** The interaction must be predictable in frequency and intensity to coevolve. (5) **Dominant selective force.** Selection for coevolution must not be compromised by other stronger selective forces.

Many bruchid-plant systems could be studied using the above criteria and much more could be learned about coevolution depending upon the system being investigated.

Thompson (1989) described and discussed what he believed the concepts of coevolution to be as: (1) **Gene-for-gene coevolution.** Parasites and hosts have complementary loci for virulence and resistance. (2) **Specific coevolution.** Coadaptation of two species without specifying the genetic basis is often referred to as specific coevolution. (3) **Guild coevolution.** When reciprocal evolutionary change occurs among groups of species rather than pairs of species. (4) **Diversifying coevolution.** This is reciprocal evolution between species in which the interaction causes at least one of the species to become subdivided into two or more reproductively isolated populations. (5) **Escape-and-radiation coevolution.** A kind of diffuse coevolution which may involve both adaptation and speciation. There are periods of time during which the interaction between the taxa does not occur.

3. Conclusions

Research on bruchid-plant coevolution has evolved in the 20 years since Janzen (1969) introduced the topic.

As explained above, all the research that has been done on "coevolution" of bruchids with their hosts is correlative. Even the paper by Ehrlich and Raven (1964) is correlative. Thus coevolution is equivocal and paradoxical and research on coevolution should follow the more rigorous methods outlined above.

I will briefly discuss some of the problems that have arisen because of our correlative studies. Janzen (1969, p. 16) stated that 13 species of plants had toxic seeds "either by published reports or Central American local folklore". Only one of the 13 species had had its seeds analyzed for toxins at that time. He explained that the evidence that he presented was circumstantial and that bruchids can feed in seeds containing toxins. The problem is that his research has been quoted by others in ensuing papers as showing that toxins are in these seeds although the evidence was only correlative. Moreover, one of the principal examples that he used as a plant that produces large, toxic seeds and is not fed upon by bruchids was *Enterolobium cyclocarpum*. This plant subsequently has been shown to be a host for bruchids in South America (Johnson and Janzen, 1982; Johnson, 1982). In addition, he cited *Leucaena leucocephala* as a small-seeded plant that used predator satiation as a method of escape from predators. As stated above, the seeds of this plant are very toxic and are fed upon by several species of bruchids. The 31 other traits that bruchids "use" to escape from predators (Janzen, 1969) and the 11 that are "countered" by bruchids (Center and Johnson, 1974) are all correlative. The classic biochemical work by Rosenthal on *Caryedes brasiliensis* only demonstrates, in a coevolutionary sense, that these beetles are able to utilize toxins as nutrients and is only one step in clarifying the interaction as coevolution. There are many examples of the inadequacy of correlative studies in coevolution of bruchids and their hosts. Based on data in the literature, some of which has been reviewed here, at this time I think those who study bruchids and plant coevolution should gather all the natural history and observational data possible for possible coevolutionary interactions of bruchids and hosts. Then much rigorous experimental research on bruchids and their interactions with their hosts should be done.

Because the host plants of New World bruchids are relatively well known and their taxonomy using classical techniques is better known than most beetle groups, I contend that the best plan for future research on bruchid coevolution is to study their systematics using cladistics and morphometric methods which can then be used in conjunction with rigorous ecological and biogeographical studies to arrive at reasonable explanations of bruchid-seed interactions. Because most bruchids are host specific, in conjunction with these studies, research on the coevolution of bruchids and their hosts using methods and ideas elucidated by the authors mentioned in this paper should be done.

Acknowledgements.   Funds for this research were provided by NSF Grants DEB81-09995, BSR82-11763 and BSR88-05861.

REFERENCES

Benson, W. W., Brown, K. S., and Gilbert, L. E. (1976) Coevolution of plants and herbivores: Passion flower butterflies, Evolution 29, 659-680.

Bleiler, J. A., Rosenthal, G. A., and Janzen, D. H. (1988) Biochemical ecology of canavanine-eating seed predators, Ecology 69, 427-433.

Bradford, D. F. and Smith C. C. (1977) Seed predation and seed number in *Scheelea* palm fruits, Ecology 58, 667-673.

Center, T. D. and Johnson, C. D. (1974) Coevolution of some seed beetles (Coleoptera: Bruchidae) and their hosts, Ecology 55, 1096-1103.

Craig, T. P., Price, P. W., Clancy, K. M., Waring, G. W., and Sacchi, C. F. (1988) Forces Preventing Coevolution in the Three-Trophic-Level System: Willow, a Gall-forming Herbivore, and Parasitoid, in K. C. Spencer (ed.), Chemical Mediation of Coevolution, Academic Press, New York, pp. 57-80.

Ehrlich, P. R. and Raven, P. H. (1964) Butterflies and plants: a study in coevolution, Evolution 18, 586-608.

Feinsinger, P. (1983) Coevolution and Pollination, in D. J. Futuyma and M. Slatkin (eds.), Coevolution, Sinauer, Sunderland, Massachusetts, pp. 282-310.

Flor, H. H. (1971) Current status of the gene-for-gene concept, Ann. Rev. Phytopathol. 9, 275-295.

Fox, L. R. (1981) Defense and dynamics in plant-herbivore systems, Amer. Zool. 21, 853-864.

Futuyma, D. J. (1983) Evolutionary Interactions among Herbivorous Insects and Plants, in D. J. Futuyma and M. Slatkin (eds.), Coevolution, Sinauer, Sunderland, Massachusetts, pp. 207-231.

Gilbert, L. E. (1983) Coevolution and Mimicry, in D. J. Futuyma and M. Slatkin (Eds.), Coevolution, Sinauer, Sunderland, Massachusetts, pp. 263-281.

Janzen, D. H. (1966) Coevolution of mutualism between ants and acacias in Central America, Evolution 20, 249-275.

Janzen, D. H. (1969) Seed-eaters versus seed size, number, toxicity, and dispersal, Evolution 23, 1-27.

Janzen, D. H. (1977) How southern cowpea weevil larvae (Bruchidae: *Callosobruchus maculatus*) die on nonhost seeds, Ecology 58, 921-927.

Janzen, D. H. (1979) How to be a fig, Ann. Rev. Ecol. Syst. 10, 13-51.

Janzen, D. H. (1980) When is it coevolution?, Evolution 34, 611-612.

Janzen, D. H., Juster, H. B., and Bell, E. A. (1977) Toxicity of secondary compounds to the seed-eating larvae of the bruchid beetle *Callosobruchus maculatus*, Phytochemistry 16, 223-227.

Johnson, C. D. (1970) Biosystematics of the Arizona, California, and Oregon species of the seed beetle genus *Acanthoscelides* Schilsky (Coleoptera: Bruchidae), University Calif. Publ. Entomol. 59, 1-116.

Johnson, C. D. (1981a) Interactions between bruchid (Coleoptera) feeding guilds and behavioral patterns of pods of the Leguminosae, Environ. Entomol. 10, 249-253.

Johnson, C. D. (1981b) Host preferences of *Stator* (Coleoptera: Bruchidae) in non-host seeds, Environ. Entomol. 10, 857-863.

Johnson, C. D. (1982) Survival of *Stator generalis* (Coleoptera: Bruchidae) in host seeds from outside its geographical range, J. Kans. Entomol. Soc. 55, 718-724.

Johnson, C. D. (1983) Ecosystematics of *Acanthoscelides* of southern Mexico and Central America, Misc. Publ. Entomol. Soc. Amer. 56, 1-370.

Johnson, C. D. (1987) Relationships between *Mimosestes* (Coleoptera) and *Acacia* (Leguminosae): is there coevolution between these genera?, in V. Labeyrie et al. (eds.), Insects - Plants, Proceedings of the 6th International Symposium on Insect - Plant Relationships (PAU 1986), Series Entomologica, Vol. 41, W. Junk, The Hague, pp. 347-352.

Johnson, C. D. and Janzen, D. H. (1982) Why are the seeds of the Central American guanacaste tree (*Enterolobium cyclocarpum*) not attacked by bruchids except in Panama?, Environ. Entomol. 11, 373 -377.

Johnson, C. D. and Kingsolver, J. M. (1976) Systematics of *Stator* of North and Central America (Coleoptera: Bruchidae), U. S. D. A. Tech. Bull. 1537.

Johnson, C. D. and Slobodchikoff, C. N. (1979) Coevolution of *Cassia* (Leguminosae) and its seed beetle predators (Bruchidae), Environ. Entomol. 8, 1069-1064.

Mitchell, R. (1977) Bruchid beetles and seed packaging by palo verde, Ecology 58, 644-651.

Nelson, D. M. and Johnson, C. D. (1983a) Stabilizing selection on seed size in *Astragalus* (Leguminosae) due to differential predation and differential germination, J. Kans. Entomol. Soc. 56, 169-174.

Nelson, D. M. and Johnson, C. D. (1983b) Selenium in *Astragalus* (Leguminosae) seeds and its effects on host preferences of bruchid beetles, J. Kans. Entomol. Soc. 56, 267-272.

Powell, J. A. and Mackie, R. A. (1966) Biological interrelationships of moths and *Yucca whipplei*, Univ. Calif. Publ. Entomol. 42, 1-46.

Rosenthal, G. A. and Bell, E. A. (1979) Naturally occurring, toxic nonprotein amino acids, in G. A. Rosenthal and D. H. Janzen (eds.), Herbivores Their Interaction with Secondary Plant Metabolites, Academic Press, London and New York, pp. 353-385.

Rosenthal, G. A. and Janzen, D. H. (1983) Arginase and L-canavanine metabolism by the bruchid beetle, *Caryedes brasiliensis*, Ent. Exp. Appl. 34, 336-337.

Rosenthal, G. A. and Janzen, D. H. (1985) Ammonia utilization by the bruchid beetle, *Caryedes brasiliensis* [Bruchidae], J. Chem. Ecol. 11, 539-544.

Schemske, D. W. (1983) Limits to specialization and coevolution in plant-animal mutualisms, in M. H. Nitecki (ed.), Coevolution, Univ. of Chicago Press, Chicago, pp. 67-109.

Siemens, D. H. and Johnson, C. D. 1990. Host-associated differences in fitness of a seed beetle (Bruchidae): effects of plant variability, Oecologia (in press).

Smith, C. C. (1975) The coevolution of plants and seed predators, in L. E. Gilbert and P. H. Raven (eds.), Coevolution of Animals and Plants, University of Texas Press, Austin and London, pp. 53-77.

Spencer, K. C. (1988a) The Chemistry of Coevolution, in K. C. Spencer (ed.), Chemical Mediation of Coevolution, Academic Press, London and New York, pp. 581-587.

Spencer, K. C. (1988b) Chemical Mediation of Coevolution in the *Passiflora-Heliconius* Interaction, in K. C. Spencer (ed.), Chemical Mediation of Coevolution, Academic Press, London and New York, pp. 167-240.

Strong, D. R., Lawton, J. H., and Southwood, T. R. E. (1984) Insects on Plants, Harvard University Press, Cambridge.

Thompson, J. N. (1982) Interaction and Coevolution, Wiley, New York.

Thompson, J. N. (1986a) Patterns in Coevolution, in A. R. Stone and D. L. Hawksworth (eds.), Coevolution and Systematics, The Systematics Association, Clarendon Press, Oxford, pp. 119-143.

Thompson, J. N. (1986b) Constraints on arms races in coevolution, TREE 1, 105-107.

Thompson, J. N. (1988) Coevolution and alternative hypotheses on insect/plant interactions, Ecology 69, 893-895.

Thompson, J. N. (1989) Concepts of coevolution, TREE 4, 179-183.

Trelease, S. F. and Trelease, H. M. (1937) Toxicity to insects and mammals of foods containing selenium, Amer. J. Botany 24, 448-451.

Wiebes, J. T. (1979) Co-evolution of figs and their insect pollinators, Ann. Rev. Ecol. Syst. 10, 1-12.

SYSTEMATICS OF THE AZUKI BEAN GROUP IN THE GENUS *VIGNA*

Y. TATEISHI AND H. OHASHI
*Biological Institute, Faculty of Science, Tohoku University,
Sendai, Miyagi 980 Japan*

ABSTRACT. The Azuki bean group, subgenus *Ceratotropis* of the genus *Vigna*, is characteristic in having peltate stipule, standard-protuberance, incurved keel-rostrum, keel-pocket, and style-beak. From these features the group is considered to be monophyletic derived by a high degree of specialization in floral structures. Morphological variations were observed in each of these characters in the mature stage and during their ontogenetic development. Correlations in these variations were found. Evolutionary trends of these characters were assumed. Based on these trends a comparison of characters is made amongst subgenera *Ceratotropis*, *Plectrotropis* and *Vigna*. The three subgenera are monophyletic and their phylogenetic relationships are very close.

1. Introduction

Azuki bean, mung bean, black gram, moth bean and rice bean are traditionally cultivated pulses in Asia (Table 1). These five bean species comprise a natural group together with over ten wild species. They occur in subtropical and warm-temperate regions in Asia (Fig. 1). This Azuki bean group is homogeneous with highly specialized floral structures.

Systematic position of the group is historically summarized in Table 2. De Candolle (1825) and Bentham (1837, 1865) placed the Azuki bean group in the section *Strophostyles* of the genus *Phaseolus*. Bentham (1865) characterized the genus *Phaseolus* as having spirally coiled keel-petals, while the nearest genus *Vigna* as having erostrate or rostrate but not spirally coiled keel-petals. Piper (1914, 1926) segregated the Azuki bean group as *Ceratotropis* from *Strophostyles*.

Ohwi (1953) validated *Azukia*, which was proposed by Takahashi (1909) from an agricultural point of view, for seven East Asian species. He defined the genus by peltate stipules, yellow flowers, and the characters of keel-petals, style and pollen grains. Maekawa (1955) divided *Azukia* into *Azukia* s. str. and *Rudua* based on seedling characters. On the basis of an analysis of low molecular weight carbohydrates in the seeds, Yasui et al. (1985) recognized two subgroups in the Azuki bean group that correspond to Maekawa's two genera. Wilczek (1954) redefined *Vigna* on the basis of peltate stipules and the beaked style. This was adopted by Hutchinson (1964) and Ohwi and Ohashi (1969).

Verdcourt (1970) proposed a very restricted concept of *Phaseolus*, limiting it exclusively to those American species with a tightly coiled style and pollen grains lacking coarse reticulation, significantly increasing the concept of *Vigna* containing *Ceratotropis* as a subgenus. Maréchal et al. (1978, 1981) followed Verdcourt and considered two evolutionary tendencies in Old World *Vigna*; (1) specialization of floral morphology which led to the homogeneous subgenus *Ceratotropis* in Asia and (2) simplification of

189

K. Fujii et al. (eds.), Bruchids and Legumes: Economics, Ecology and Coevolution, 189–199.
© 1990 *Kluwer Academic Publishers. Printed in the Netherlands.*

Table 1.   Pulses in Azuki bean group (after Tateishi, 1989).

Cultivated form	Wild form
Vigna aconitifolia 1)*Phaseolus aconitifolius* 2)Moth bean 3)Pakistan and India	**Vigna aconitifolia** (Indistingushable from cultivated form) 4)Arabia, Pakistan, India and W.China
Vigna angularis var. angularis 1)*Phaseolus angularis* 2)Azuki bean 3)China, Korea and Japan	**var. nipponensis** 4)Himalaya, N.Burma, China, Taiwan, Korea and Japan
Vigna mungo var. mungo 1)*Phaseolus mungo* 2)Black gram, Black matpe, Urd 3)Pakistan, India and Burma	**var. silvestris** 4)India and Burma
Vigna radiata var. radiata 1)*Phaseolus radiatus* *Phaseolus aureus* 2)Mung bean, Green gram 3)Tropics and subtropics in the old world	**var. sublobata** 1)*Phaseolus sublobatus* 4)E. and W.tropical Africa, India, Burma, Thailand, Indo-China, W.China, Taiwan, Malaysia, New Guinea and Australia
Vigna umbellata 1)*Phaseolus calcaratus* *Phaseolus ricciardianus* 2)Rice bean 3)Subtropics and warm temperates in the old world	**Vigna umbellata (twining form)** 4)E.India, Thailand, Indo-China and China

1)Synonyms; 2)English names; 3)Area of traditional cultivation;
4)Natural distribution.

the floral morphology which led to the species-rich subgenus *Vigna* in Africa.

Recently, Tateishi(1989) accomplished a monograph of the Azuki bean group in which 17 species, 3 subspecies and 5 varieties are recognized as valid in the subgenus *Ceratotropis* of the genus *Vigna*. Phylogenetic relationships between these species were discussed on the basis of morphological variations in ontogenetic development of diagnostic characters of the subgenus. He agreed with Maréchal et al. (1978) to place subgenus *Ceratotropis* next to subgenus *Plectrotropis*.

In the present paper, phylogenetic relationships of the subgenus *Ceratotropis* with subgenera *Plectrotropis* and *Vigna* are discussed on the basis of variations and developmental morphology of their diagnostic characters.

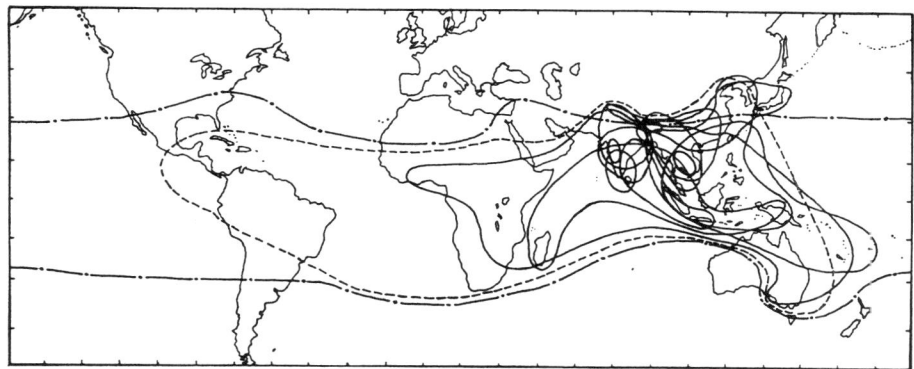

Figure 1.   Distribution of all the species of Azuki bean group (subgenus *Cerototropis*)(——) showing approximate distributional ranges of subgenera *Vigna* (— · —) and *Plectrotropis* (-----).

2.   Morphology of Stipules and Flowers in the Azuki Bean Group

The Azuki bean group is homogeneous and characteristics so far known are summarized as follows: 1) the peltate stipule; 2) the standard with a protuberance near the center of the inner surface of the lamina; 3) the incurved keel-petals in the upper part; 4) the horn-like appendage on the left keel-petal (the keel-pocket); 5) the style prolonged beyond the stigma as a beak; and 6) the pollen grains with coarsely reticulate sculpture.   Of these the pollen character is common with both subgenera *Vigna* and *Plectrotropis*.   Also, the group has a granulate infratectum of pollen grains in common with these two subgenera (Horvat and Stainier 1980; Ohashi and Takahashi 1981). Hence, only the other five characters will be discussed below.

2.1. PELTATE STIPULES

Although the Azuki bean group has peltate stipules in common, there are some varia-

Table 2.   History of the taxonomical treatment of the Azuki bean group to which genera it belonged.

De Candolle (1825) Bentham (1837, 1865) Baker (1876)	*Phaseolus*	(Strophostyles)	*Vigna*
Piper (1914, 1926)	*Phaseolus*	(Ceratotropis)	*Vigna*
Ohwi (1953)	*Phaseolus*	*Azukia*	*Vigna*
Maekawa (1955)	*Phaseolus*	*Azukia* & *Rudua*	*Vigna*
Verdcourt (1970) Maréchal et al. (1978) Tateishi (1989)	*Phaseolus*	(Ceratotropis)	*Vigna*

Names in parenthesis were recognized as a subgenus or section of the genus in the same square.

192

tions in the degree of development of portions below the point of insertion to the stem (Fig. 2). The degree of this development in *Vigna khandalensis* is the least in the group, and the ratio of length of the portions below the point of insertion to the whole length of the stipule is 1/15 to 1/7 (Fig. 2: A). The ratio is 1/8 - 1/4 in *V. stipulacea* (Fig. 2: B), and is more than 1/3 in almost all of the other species (Fig. 2: C - E).

Juvenile plants of the Azuki bean group, however, never have the peltate stipule. Development of the stipule from seedlings to mature plants showed that its insertion mode changes from basifixed with truncate base at the next node of cotyledons, i.e., the second node, through basifixed with auriculate base at several succeeding nodes to peltate at higher nodes (Fig. 3). There are transitional forms between these three stipule types. In *V. stipulacea* (Fig. 3: A) the auricles at the base are very small at the third and fourth nodes, and are larger at the sixth and seventh nodes. At the eighth or ninth node the two auricles fuse into a lower lobe of a peltate stipule. In *V. khandalensis* such stipules first occur at the eighth to tenth nodes, and in *V. aconitifolia* at the sixth node. However, in almost all of the other species, the portion below the insertion to the stem develop well and fuse earlier, usually at the fourth or fifth node as in *V. angularis* var. *nipponensis* (Fig. 3: B). In such species the lobes below the point of insertion are more developed at the higher nodes. These facts suggest a trend of specialization of the stipule morphology in the Azuki bean group from basifixed with truncate base through that with auriculate base, to peltate (Tateishi, 1989). Further, a trend of specialization towards the stipules with well-developed portions below the point of insertion can be recognized amongst the species with peltate stipules.

In the subgenus *Plectrotropis*, stipules are usually auriculate at the base (Fig. 2: F), but in some species such as *V. kirkii* and *V. nuda* they are often peltate. In *V. vexillata* var. *tsushimensis*, which has auriculate stipules at the mature stage, the stipules at the second node are basifixed with truncate base, and at succeeding nodes are basifixed with auricles at the base (Fig. 3: C-D).

2.2. STANDARDS AND PROTUBERANCES ON THE STANDARD

The corolla of the subgenera *Ceratotropis* and *Plectrotropis*, comprising five petals, i.e., one standard, two wings and two keel-petals, is extraordinarily asymmetric (Fig. 4: fl). The standard twists sinistrally to form an obliquely elliptic cymbiform blade (Fig. 4: st).

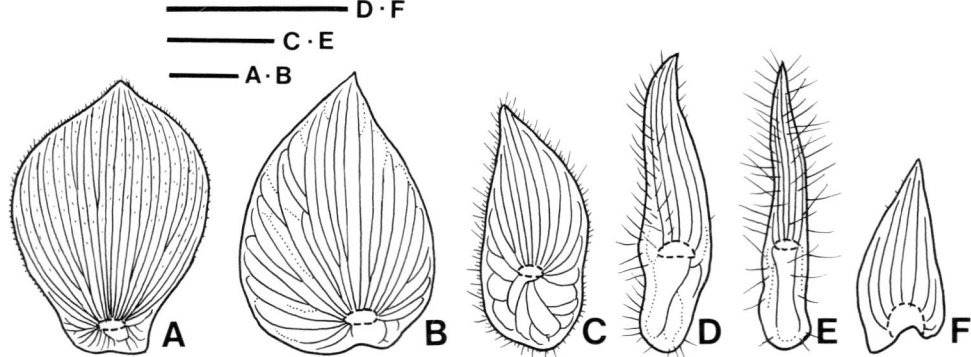

Figure 2. Peltate stipules in *Vigna* subgenus *Ceratotropis* (A-E) and basifixed stipule with auriculate base in subgenus *Plectrotropis* (F). A. *V. khandalensis*. B. *V. stipulacea*. C. *V. radiata*. D. *V. angularis* var. *nipponensis*. E. *V. mungo*. F. *V. vexillata* var. *tsushimensis*. Scale: 5 mm.

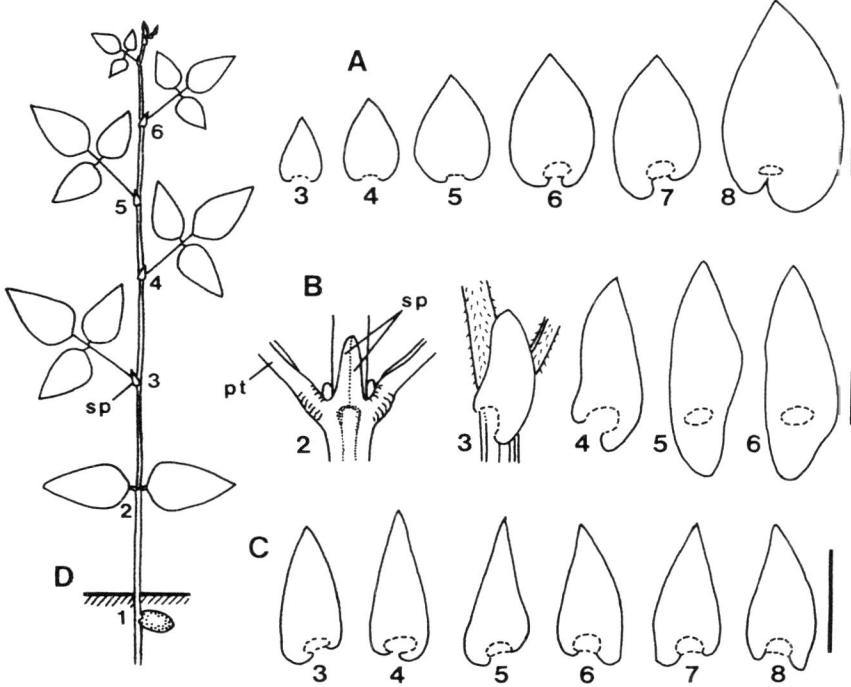

Figure 3. Transition of the insertion mode of stipules to stem in seedlings of *Vigna* subgenera *Ceratotropis* (A-B) and *Plectrotropis* (C-D). A. *V. stipulacea.* B. *V. angularis* var. *nipponensis.* B2. The second node with petioles of opposite juvenile leaves and stipules fused along lateral margin. B3. The third node with stipules basifixed but with small auricles at the base. C-D. *V. vexillata* var. *tsushimensis.* C. Stipules. D. Seedling. Numbers indicate the numerical order of nodes. pt. Petioles. sp. Stipules. Scale: 2 mm.

The subtribe *Phaseolinae* of tribe *Phaseoleae,* to which genera *Vigna* and relatives (the *Phaseolus-Vigna* complex) belong, often has callosities near the center of the inner face of the standard. The callosity is a good diagnostic character of genera or subgenera in the subtribe (Wilczek, 1954; Verdcourt, 1970, 1971; Lackey, 1981). In the subgenus *Vigna* some species have one or two pairs of callosities of diverse shape. In the subgenus *Plectrotropis* a pair of callosities is found on the face of the standard in some species, but in *V. vexillata* var. *tsushimensis* they are obscure (Fig. 4: F, st). In the subgenus *Ceratotropis* the standard has only one callosity rising on the inner face as a hump (Fig. 4: C-E, st and st'). This is quite different in shape from those found in subgenera *Vigna* and *Plectrotropis* and also probably different in function. It is a unique feature among floral characters of *Ceratotropis,* because such callosity is found only in this subgenus in the tribe *Phaseoleae.* Therefore, it is distinguished from the callosities found in the tribe and is called a protuberance (Tateishi, 1989).

In the Azuki bean group, however, some species lack the protuberance. Degrees of development among the standard-protuberance and the lower portions of stipules correlate with each other (Tateishi, 1989). In some species, for example, *V. stipulacea* and *V. khandalensis,* having stipules with a less developed portion below the point of insertion

194

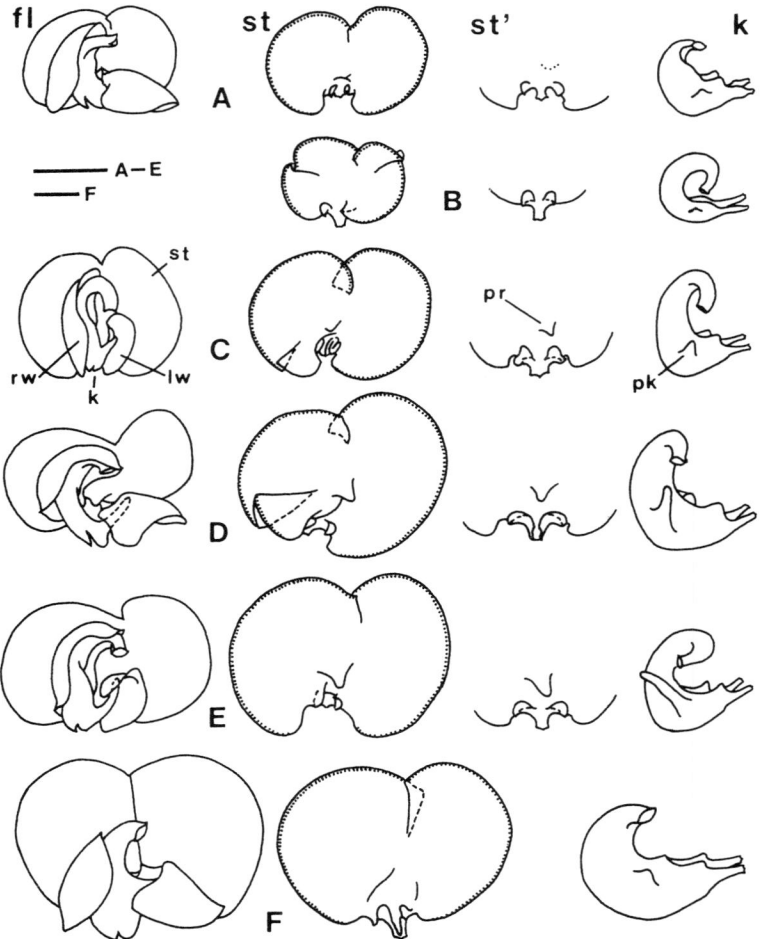

Figure 4. Variation of flowers in *Vigna* subgenera *Ceratotropis* (A-E) and *Plectrotropis* (F). A. *V. stipulacea*. B. *V. khandalensis*. C. *V. radiata*. D. *V. angularis* var. *nipponensis*. E. *V. mungo*. F. *V. vexillata* var. *tsushimensis*. fl. Flowers in front view. st. Standards showing inner face. st'. Basal part of standard with or without a protuberance. k. Keel-petals with a convex or horn-like pocket. lw. Left wings. rw. Right wings. pk. Keel-pockets. pr. Protuberances. Scale: 5 mm.

and with a delayed appearance, the protuberance is obscure (Fig. 4: A and B, st and st'). On the other hand, *V. angularis* var. *nipponensis* and *V. mungo* have a large and prominent protuberance (Fig. 4: D and E, st and st'). They have peltate stipules with well-developed portion below the point of insertion and appear early in the growing stage.

2.3. KEEL-ROSTRUM AND KEEL-POCKET

In the *Phaseolus-Vigna* complex the two keel-petals are fused along both margins except near the top and below the base of the lamina to form a falcate cylinder. In the subgenera *Ceratotropis* and *Plectrotropis* the upper portion of the keel-petals is prolonged and forms an incurved rostrum more or less twisting to the left (Fig. 4: k). The degree of incurving is loosely correlated with that of development of the standard-protuberance with the exception of *Vigna khandalensis*. This species has rather well incurved keel-petals but has no protuberance on the standard (Fig. 4: B, k, st, and st'). The degree of incurving of *Plectrotropis* is less than that in subgenus *Ceratotropis*. The keel-petals of *V. vexillata* var. *tsushimensis* (Fig. 4: F, k) are similar to those of *V. stipulacea* in shape, of which the degree of incurving is least amongst the species of *Ceratotropis*.

In *Ceratotropis* and *Plectrotropis* the left keel-petal usually has one convex or horn-like pocket (keel-pocket), the apex of which faces to the tip of the rostrum of the keel-petals incurved to the left. The keel-pocket seems to play an important role in pollination together with the incurved rostrum. It supports the left wing which extends over the pocket and offers a landing platform for insect pollinators. The insect pollinator lands on the left wing extending horizontally and inserts its proboscis between the standard and the claws of the keel-petals, sliding it down to reach the nectar at the base of the stamen tube. When the pollinator forces its way into the flower the wings and keel-petals are pressed down, and the stamens and style which are enclosed in the keel-rostrum act together as a piston, forcing out pollen masses from the mouth of the keel-rostrum onto the visiting insect.

The occurrence of a keel-pocket is confined to *Ceratotropis* and *Plectrotropis*. Nevertheless, the keel-pocket exhibits a wide range of variation in length and direction (Fig. 4: k) which reflects differences in the position and direction of the landing platform (Fig. 4: fl and k). For example, in *Vigna radiata* the keel-pocket is rather short and ascending (Fig. 4: C). The left wing covering it is compressed laterally, hence, the landing platform of this species is rather narrow. The mouth of the keel-rostrum opens downwardly towards the left wing. On the other hand, the keel-petals of *V. angularis* var. *nipponensis* are incurved left through 210° - 300° (Fig. 4: D). The mouth of the keel-rostrum opens in a half left direction. The keel-pocket projects considerably and spreads horizontally to the left. A wide landing platform is offered at the left side of the flower. The keel-pocket of *V. mungo* is longest in these two subgenera (Fig. 4: E). It projects forwards and supports the left wing spreading forward too, so the landing platform spreads in the same direction. The keel-petals are incurved through 250° - 280° in this species. The mouth of the keel-rostrum opens half-forwardly.

Elongation of the keel-pocket shows considerable correlation with the degree of development of the lower portion of the stipules and the standard-protuberance (Tateishi, 1989). The keel-pocket in subgenus *Plectrotropis* is small but prominent; that of *V. vexillata* var. *tsushimensis* is convex in shape (Fig. 4: F, k).

2.4. STYLES AND STYLE-BEAKS

The pistil is completely concealed in the cylinder of the keel-petals. The style twists left abruptly at the middle but in the upper portion it gradually twists right up to the stigma. This feature seems to be one of the characteristics of the *Phaseolus-Vigna* complex. In the subgenus *Ceratotropis* the upper half of the style shows much in curving (Fig. 5: A), while it is less so in the subgenus *Plectrotropis* (Fig. 5: B) and straight in the subgenus *Vigna* (Fig. 5: C).

Styles are often prolonged beyond the stigma into a beak in these three subgenera. The degree of elongation of the style-beak varies considerably among the species (Fig. 5: D-H). In *Ceratotropis* variation in the length of the style-beak is correlated with those of stipules, standard-protuberances, keel-rostra and keel-pockets mentioned above.

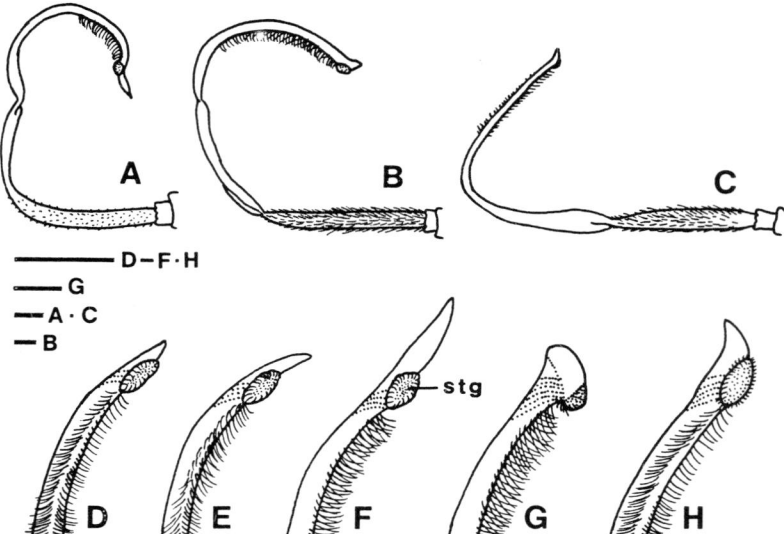

Figure 5. Variation of pistil morphology in subgenera *Ceratotropis* (A and D-F), *Plectrotropis* (B and G) and *Vigna* (C and H). A. Pistil of *V. angularis* var. *nipponensis* with a style much incurved in the upper portion. B. Pistil of *V. vexillata* var. *tsushimensis* with a style incurved above the middle. C. Pistil of *V. marina* with a style straight in the upper portion. D-H. Upper portion of style. D. *V. stipulacea*. E. *V. radiata*. F. *V. angularis* var. *nipponensis*. G. *V. vexillata* var. *tsushimensis*. H. *V. marina*. stg. Stigma. Scale: 1 mm.

3. Evolutionary Trends of Diversification in the Azuki Bean Group

Among the five characteristic features of the subgenus *Ceratotropis*, the standard-protuberance is confined in some species of this subgenus in the tribe *Phaseoleae*. The keel-pocket is restricted in the subgenera *Plectrotropis* and *Ceratotropis*. These two characters are, therefore, thought to be perfections of specialization. The peltate stipule, the keel-rostrum, and the style-beak of the Azuki bean group correlate with these

Table 3. Evolutionary trends of five diagnostic characters of the Azuki bean group.

	Subgenus	
Character	*Plectrotropis* (out group)	*Ceratotropis*
Stipule:	auriculate ---------------- > peltate (lower portion: short ------ > long)	
Standard-protuberance:	absent -------------------- > absent -------- > small -------------- > large	
Keel-rostrum:	incurved ---------------- > incurved ---------------- > much incurved	
Keel-pocket:	short -------------------- > short -------------------------------- > long	
Style-beak:	short -------------------- > short -------------------------------- > long	

two characters. Evolutionary trends of the five diagnostic characters of the Azuki bean group are assumed as given in Table 3. Based on these trends the group must be derived from such plants as having more or less auriculate or peltate stipules, a standard without protuberance, keel-petals with a more or less incurved or straight rostrum and a short pocket, and a style with or without a short beak. The subgenus *Plectrotropis* is most similar to such plants in the genus *Vigna*. This subgenus comprises six African species and one pantropic species, *Vigna vexillata* (Maréchal et al., 1978). In this subgenus the stipule is usually more or less auriculate but in *V. kirkii* and *V. nuda* it is often peltate. The keel-petals are more or less incurved in the upper portion and have a short but prominent pocket. The style is prolonged beyond the stigma into a very short beak.

The standard of *Plectrotropis* has a pair of linear callosities. In the early stage of development of the standard of *Vigna vexillata* var. *tsushimensis* a pair of callosities can be observed clearly, but they do not develop further and become obscure in mature flowers. A trend towards disappearance of the callosity on the inner face of the standard can be assumed in *Plectrotropis*. The protuberance of *Ceratotropis* was supposed by Maréchal et al. (1978) to have been derived by the fusion of two callosities. The developmental morphology, however, showed that only one primordium of protuberance is produced on the center of the standard in *V. radiata* and *V. angularis* var. *nipponensis* (Tateishi, 1989). There is no evidence to combine the protuberance of *Ceratotropis* with the two callosities of *Plectrotropis*. Therefore, the standard without callosities in *Plectrotropis* seems to be continuous to that without protuberance in *Ceratotropis*.

Relationships of the subgenus *Vigna* to *Plectrotropis* and *Ceratotropis* will be discussed below. These three subgenera are compared by the diagnostic characters of *Ceratotropis* as shown in Table 4.

Table 4. Comparison of *Vigna*, *Plectrotropis*, and *Ceratotropis*.

Character	Vigna	Subgenus Plectrotropis	Ceratotropis
Stipule: auriculate(a)/peltate(p)	a (rarely p)	a,p	p
Corolla color: yellow(y)/not(—)	—,y	—	y
Standard: symmetric(+)/not(—)	+ ,—	—	—
Standard callosity: number	0,2,4	0,2	0,1
Keel-rostrum: short & straight(s)/short & incurved(c)/long & much incurved(l)	s	c	l
Keel-pocket: short(s)/long(l)/absent(—)	—	s	s,l
Upper half of style: straight(r)/incurved(c)	r	c	c
Style beak: short(s)/long(l)/absent(—)	—,s	s	s,l
Center of distribution: Asia(As)/Africa(Af)	Af	Af	As

The subgenus *Vigna* comprises thirty or more African and a few pantropic species. In this subgenus stipules vary from basifixed with auricles at the base, to peltate with rather a long spur below the point of insertion. Standards are usually symmetric, but sometimes asymmetric. One or two pairs of callosities are found on the inner face of the standard in the subgenus, but in some species they are absent. Two linear callosities of *Plectrotropis* relate to those of *Vigna*. Differences in *Vigna* from *Plectrotropis* and *Ceratotropis* are found in the keel-petals. The keel-petals of *Vigna* are not elongated sufficiently so as to form a distinct rostrum nor incurved in the upper portion. They have no pocket on the left petal. The rostrum and pocket appear later during growth of the flowers of *V. angularis* var. *nipponensis* and *V. radiata*. These features seem to have less phylogenetic importance between the subgenera *Vigna* and *Ceratotropis*. The other characters of *Vigna* are not so different from *Plectrotropis* and *Ceratotropis* (Table 4). Therefore, these three subgenera are monophyletic and have been diversified on the same evolutionary line. Phylogenetic relationships among them are shown in Figure 6.

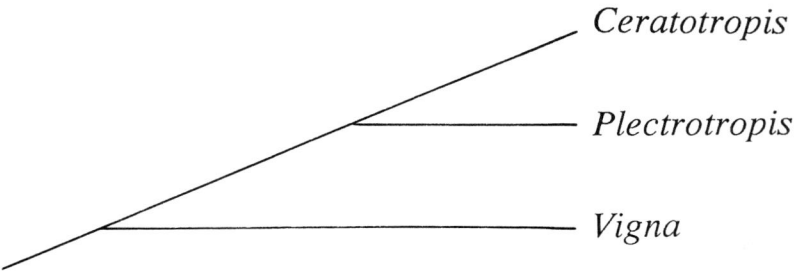

Figure 6.   Supposed phylogenetic relationships between subgenera *Vigna*, *Plectrotropis* and *Ceratotropis*.

REFERENCES

Baker, J. G. (1876-78) Leguminosae, in J. D. Hooker (ed.), The flora of British India 2, (*Phaseolus* and *Vigna* 200-207, 1876), Kent.

Bentham, G. (1837) Commentationes de leguminosarum generibus, Wien.

Bentham, G. (1865) Leguminosae, in G. Bentham and J. D. Hooker (eds.), Genera Plantarum 1, London, pp. 434-600.

De Candolle, A. P. (1825) Leguminosae, in Prodromus Systematis Naturalis Regni Vegetabilis 2, (*Phaseolus* 390-396), Paris, pp. 93-524.

Horvat, F. and Stainier, F. (1980) L'étude de l'exine dans le complexe *Phaseolus-Vigna* et dans des genres apparentés. IV, Pollen et Spores, 22, 139-172.

Hutchinson, J. (1964) Phaseoleae, in The Genera of Flowering Plants 1, Oxford Univ. Press, London, pp. 434-443.

Lackey, J. A. (1981) Phaseoleae, in R. M. Polhill and P. H. Raven (eds.), Advances in Legume Systematics pt. 1, Royal Botanic Gardens, Kew, pp. 301-327.

Maekawa, F. (1955) Topo-morphological and taxonomical studies in *Phaseoleae*, Leguminosae, Jpn. J. Bot. 15, 103-116.

Maréchal, R., Mascherpa, J. M., and Stainier, F. (1978) Etude taxonomique d'un groupe complexe d'espèces des genres *Phaseolus* et *Vigna* (Papilionaceae) sur la base de données morphologiques et pollinique, traitées par l'analyse informatique, Boissiera 28, 1-273.

Maréchal, R., Mascherpa, J. M., and Stainier, F. (1981) Taxonometric study on the *Phaseolus-Vigna* complex and related genera, in R. M. Polhill, and P. H. Raven, (eds.), Advances in Legume

Systematics pt. 1, Royal Botanic Gardens, Kew, pp. 329-335.

Ohashi, H. and Takahashi, H. (1981) Pollen morphology of *Vigna angularis* (Leguminosae), Bot. Mag. Tokyo 94, 177-180.

Ohwi, J. (1953) Flora of Japan, (*Azukia* in 690-691), Shibundo, Tokyo (in Japanese).

Ohwi, J. and Ohashi, H. (1969) Adzuki beans of Asia, J. Jpn. Bot. 44, 29-31 (in Japanese).

Piper, C. V. (1914) Five oriental species of beans, Bull. U.S.D.A. 119, 1-32.

Piper, C. V. (1926) Studies in American *Phaseoleae*, Contr. U. S. Nat. Herb. 22, 663-701.

Takahashi, Y. (1909) Botanical studies on adzuki plant, J. Soc. Agr. For. Sapporo 1, 140-163 (in Japanese).

Tateishi, Y. (1989) A revision of the Azuki bean group, the subgenus *Ceratotropis* of the genus *Vigna* (Leguminosae), Ginkgoana 7 (in press).

Verdcourt, B. (1970) Studies in the Leguminosae-*Papilionoideae* for the 'Flora of Tropical East Africa', IV, Kew Bull. 24, 507-569.

Verdcourt, B. (1971) Phaseoleae, in E. Milne-Redhead, and R. M. Polhill, (eds.), Flora of Tropical East Africa, Leguminosae pt. 4, *Papilionoideae* 2, Crown Agents, London, pp. 503-807.

Wilczek, R. (1954) Phaseolinae, in Flore du Congo Belge et du Ruanda-Urundi 6, INEAC, Bruxelles, pp. 260-409.

Yasui, T., Tateishi, Y., and Ohashi, H. (1985) Distribution of low molecular weight carbohydrate in the subgenus *Ceratotropis* of the genus *Vigna* (Leguminosae), Bot. Mag. Tokyo 98, 75-87.

CROSS COMPATIBILITY AND CYTOGENETICAL RELATIONSHIPS AMONG ASIAN *VIGNA* SPECIES

Y. EGAWA[1], M. NAKAGAWARA[1], AND G. C. J. FERNANDEZ[2]
[1] *National Institute of Agrobiological Resources*
Kannondai, Tsukuba Ibaraki 305 Japan
[2] *Plant Science Department*
University of Nevada, Reno NV 89557 U.S.A.

ABSTRACT. To elucidate interspecific genome relationships among Asian *Vigna* species, *V. radiata*, *V. mungo*, *V. angularis*, *V. umbellata* and *V. glabrescens*, compatibility, pollen fertility and the pattern of meiotic chromosome pairing of their F_1 hybrids were investigated.

Percentage pod set was high when *V. mungo* was crossed as a female parent with *V. radiata* or *V. umbellata*. However, none of the seeds sown germinated in the cross, *V. mungo* x *V. radiata*. Germination rate of the hybrid seeds of *V. mungo* x *V. umbellata* or *V. mungo* x *V. angularis* was relatively high, although the hybrid seedlings died at the initial growth stage. The success of interspecific hybridization between a tetraploid species, *V. glabrescens* and diploid *Vigna* species was high when *V. glabrescens* was used as the female parent.

The hybrids, *V. radiata* x *V. mungo* and *V. radiata* x *V. umbellata*, exhibited irregular meiosis and low pollen fertility. The degree of genome homology between them was low. These three species are clearly differentiated from each other and *V. radiata* is phylogenetically more closely related to *V. mungo* than to *V. umbellata*. Hybrids from the cross *V. glabrescens* x *V. umbellata* showed 11 univalents and 11 bivalents at meiosis, indicating that *V. glabrescens* shares a homologous genome with *V. umbellata*. Hybrids from the crosses *V. glabrescens* x *V. radiata* and *V. glabrescens* x *V. mungo* showed irregular chromosome pairing with considerably lower frequency of bivalent formation, indicating that the genomes of *V. radiata* and *V. mungo* are not included in *V. glabrescens*.

1. Introduction

Vigna radiata (mungbean), *V. mungo* (blackgram), *V. angularis* (adzuki bean) and *V. umbellata* (rice bean) are considered to have originated in Asia and are now cultivated traditionally by small farmers throughout the tropical, subtropical and temperate Asian countries. They are very important sources of protein in these countries and consumed in various forms, such as bean sprouts, boiled dry beans, cakes, paste and so on. These legumes are diploid (2n=22) and predominantly self fertilized. Taxonomic treatment of them has been confusing. They were originally classified into genera *Phaseolus*, *Adzukia* or *Rudua*. Verdecourt(1970) and Maréchal et al. (1978) proposed that they should be put under the genus *Vigna*, based on comparative studies of stipule and flower morphology. Asian *Vigna* species belong to the subgenus *Ceratotropis* and form a different group from African *Vigna* species, with *V. unguiculata* belonging to the subgenus *Vigna*. These two groups are genetically isolated from each other and interspecific hybrids between them have not been successfully produced (Smartt, 1985). Interspecific relation-

201

K. Fujii et al. (eds.), Bruchids and Legumes: Economics, Ecology and Coevolution, 201–208.

ships among the Asian *Vigna* species have been poorly documented so far (Ahn and Hartmann, 1978; Miyazaki, 1982; Chen et al., 1983). We produced intra- and interspecific hybrids among them to investigate their phylogenetical relationships and to promote an effective breeding program through interspecific hybridization (Egawa et al., 1986). Here we describe phylogenetical relationships between *V. radiata*, *V. mungo* and *V. umbellata* by means of studies on meiotic chromosome pairing of the hybrids.

A natural tetraploid *Vigna* species, *V. glabrescens* (2n=44) (AVRDC No. V1160, P.I. No.207655) possesses resistance to major mungbean pests and diseases and is currently utilized in the AVRDC mungbean improvement program (Fernandez and Shanmugasundarum, 1988). However, introduction of these valuable resistance genes from this tetraploid species via conventional breeding methods was unsuccessful due to the difference in ploidy level. This accession regularly formed 22 bivalents with no multivalents at meiosis and is considered to be an amphidiploid but not an autotetraploid (Swindell et al., 1973). The cytogenetical relationship between *V. glabrescens* and diploid *Vigna* species has not yet been analyzed. To clarify the genome constitution of *V. glabrescens*, interspecific hybrids between this species and the Asian diploid species, *V. radiata*, *V. mungo*, *V. umbellata* and *V. angularis* were derived through embryo culture technique and the pattern of meiotic chromosome pairing of the hybrids were analyzed.

2. Materials and Methods

The materials examined cytologically are listed in Table 1. *V. glabrescens* seed was obtained from the Philippines through the Regional Plant Introduction Station at Experiment, Georgia (USDA PI 207655) (Swindell et al., 1973). MR51 is an amphidiploid

Table 1. The strains of *Vigna* species involved in interspecific hybridization.

Strain No.	Locality	Original source
1. *Vigna radiata* (L.) R. Wilczek.		
7085	Nepal	collected by the IBPGR[1] Nepal mission
7124	Iran	supplied by the AVRDC[2]
7125	India	as above
V1856	India	as above
2. *V. mungo* (L.) Hepper		
7080	Nepal	collected by the IBPGR Nepal mission
Tc2210	IARI	
I-92	Nepal	collected by the IBPGR Nepal mission
3. *V. umbellata* (Thunb.) Ohwi & Ohashi		
6090	Japan	supplied by the Kyoto Univ.
6091	Japan	as above
I-275	Nepal	collected by the IBPGR Nepal mission
S-91	Nepal	supplied by the AVRDC
4. *V. glabrescens* M. M. & S.		
V1160	Philippines	PI 207665
5. amphidiploid between *V. radiata* and *V. umbellata*		
MR51		obtained at the AVRDC

[1] IBPGR: International Board for Plant Genetic Resources
[2] AVRDC: Asian Vegetable Research and Development Center

($2n=44$) detected in the F_2 generation of *V. radiata* (V1856) x *V. umbellata* (S-91) at the AVRDC.

All the strains were grown in pots in a green house at the National Institute of Agrobiological Resources, Tsukuba, Japan. For the crossing experiments, flowers were emasculated just before the buds opened and were immediately pollinated in the greenhouse. Thereafter, the flowers were covered with paraffin-paper bags to prevent contamination.

Hybrid seeds between diploid species were aseptically placed on the White's agar medium (White, 1963) in test tubes after sterilization with 70% ethyl alcohol for 30 sec and 5% sodium hypochlorite for 20 min. In order to rescue hybrids between *V. glabrescens* and diploid *Vigna* species, immature embryos were excised from the pods at 2 to 3 weeks after pollination and cultured on White's medium supplemented with 200 mg/l of yeast extract. The resulting seedlings were transplanted to plastic pots filled with vermiculite. After the roots had developed sufficiently, they were transplanted into pots filled with soil. For the cytological examination of meiosis, young anthers were fixed in acetic acid : ethanol (1:3). Chromosome pairing was observed at the first metaphase of pollen mother cells (PMCs) using the acetocarmine squash method. For observation of the mitotic cells, root tips were pretreated in a refrigerator (about 6°C) for 16 to 24 hr and fixed in acetic acid : ethanol (1:3). They were then macerated with the enzymatic mixture at 37°C for 30 min according to the methods described by Nishibayashi (1985). They were stained with 1% of Giemsa in 0.07 M Soerensen phosphate buffer solution (pH 6.8). Pollen fertility was assessed by the percentage of well stained pollen grains with an acetocarmine solution.

3. Results and Discussion

3.1. INTERSPECIFIC CROSS COMPATIBILITY

The results of the interspecific crosses are presented in Table 2. Percentages of pod setting were high when *V. mungo* was crossed as a female parent with *V. radiata*, *V. umbellata* or *V. angularis*. However, none of the seeds germinated in the cross, *V.*

Table 2. Results of interspecific crosses and germination test.

Cross combination	No. of flowers pollinated	No. of pods set (%)	No. of seeds sown	No. of seeds germinated(%)	No. of plants obtained
V. radiata X *V. mungo*	40	9 (22.5)	31	16 (51.6)	7
V. mungo X *V. radiata*	24	15 (62.5)	56	0 (0.0)	0
V. radiata X *V. umbellata*	60	17 (28.3)	63	19 (30.2)	13
V. umbellata X *V. radiata*	27	2 (7.4)	5	0 (0.0)	0
V. radiata X *V. angularis*	15	1 (6.6)	0	-	-
V. angularis X *V. radiata*	14	2 (14.3)	23	0 (0.0)	0
V. mungo X *V. umbellata*	41	17 (41.5)	41	8 (19.5)	0
V. umbellata X *V. mungo*	28	3 (10.7)	12	1 (8.3)	0
V. mungo X *V. angularis*	4	2 (50.0)	4	3 (75.0)	0
V. angularis X *V. mungo*	16	0 (0.0)	0	-	-
V. umbellata X *V. angularis*	25	1 (4.0)	3	1 (33.3)	0
V. angularis X *V. umbellata*	45	17 (37.8)	45	1 (2.2)	0

mungo x *V. radiata*. Germination rates of the hybrid seeds of *V. mungo* x *V. umbellata* and *V. mungo* x *V. angularis* were relatively high, although the hybrid seedlings died at the initial growth stage. The hybrids between *V. mungo* and *V. umbellata* and between *V. mungo* and *V. angularis* were not obtained in the present study. The germination rate of hybrid seeds of *V. radiata* x *V. mungo* and *V. radiata* x *V. umbellata* was 51.6 and 30.2%, respectively, and hybrid plants were obtained with relatively high frequency.

The results of embryo culture and hybrid plants obtained from the crosses *V. glabrescens* x diploid species are summarized in Table 3. The success of embryo rescue and interspecific hybridization between *V. glabrescens* and diploid *Vigna* was high when *V. glabrescens* was used as the female parent. From the cross MR51 x *V. glabrescens*, we were able to obtain hybrid plants relatively easily. However, the success rate was nil when *V. glabrescens* was crossed as the female parent with MR51. These results indicate that generation of interspecific hybrids between *V. glabrescens* and diploid *Vigna* species are also influenced by cytoplasmic factors.

3.2. CYTOGENETICAL RELATIONSHIP BETWEEN *VIGNA RADIATA* AND *V. MUNGO*

The hybrids between *V. radiata* and *V. mungo* grew vigorously and produced numerous flowers continuously. The color of the standard petal in *V. radiata* is greenish yellow. F_1 plants possessed a bright yellow standard petal like the pollen parent, *V. mungo*. The percentages of pollen fertility of the hybrids were very low (13.6%). They set pods rarely and only one seed was set in the pod. Fifty one F_2 seeds were obtained by open pollination without bagging throughout the flowering period.

The hybrids exhibited irregular meiosis with a high frequency of univalent formation. Out of 21 PMCs examined cytologically, 8 (38%) exhibited the configuration of 8II + 6I. The configuration of 11 II was not observed. The number of univalents per cell ranged from 2 to 10. The average chromosome pairing at meiosis was 5.1 I + 8.5 II (3.9 rings + 4.6 rods) as shown in Table 4.

On the basis of low pollen fertility and high frequency of univalent formation at meiosis, the genomes of *V. radiata* and *V. mungo* are thought to be partially homologous. *V. radiata* and *V. mungo* were considered to compromise a single species (Verdecourt, 1970). However, the present results suggested that *V. radiata* and *V. mungo* form separate taxonomic species. Miyazaki (1982) also regarded *V. radiata* and *V. mungo* as a distinct species since they exhibit clear differences in the chemical composition of anthocyanin in the hypocotyle and in the esterase banding pattern in the cotyledon.

Table 3. Results of embryo culture of interspecific hybrids among *Vigna* species.

Cross combination	No. of embryos cultured	No. of embryos germinated		No. of seedlings obtained		No. of plants obtained	
V. glabrescens X *V. radiata*	12	7	(58%)	1	(8%)	1	(8%)
V. radiata X *V. glabrescens*	31	4	(13%)	0	(0%)	-	
V. glabrescens X *V. mungo*	16	11	(69%)	10	(63%)	9	(56%)
MR51 X *V. umbellata*	10	3	(30%)	1	(10%)	1	(10%)
V. glabrescens X *V. umbellata*	39	36	(92%)	35	(89%)	25	(64%)
V. glabrescens X *V. angularis*	4	4	(100%)	4	(100%)	2	(50%)
MR51 X *V. glabrescens*	32	28	(88%)	28	(87%)	21	(66%)
V. glabrescens X MR51	3	0	(0%)	0	(0%)	-	

Table 4. Average frequency of chromosome pairing of the interspecific hybrids at the first metaphase and percentage of pollen fertility.

Cross combination	2n =	No. of cells observed	I	II			III	Pollen fertility
				ring	rod	total		
1. *V. radiata* X *V. mungo*								
7124 X 7080	22	21	5.1	3.9	4.6	8.5	-	13.6%
2. *V. radiata* X *V. umbellata*								
7085 X 6091	22	32	12.7	0.6	4.1	4.7	-	0.2%
7125 X 6090	22	50	10.0	0.7	5.3	6.0	-	0.0%
3. *V. glabrescens* X *V. radiata*								
V1160 X V1856	33	41	18.6	1.8	4.8	6.6	0.4	-
4. *V. glabrescens* X *V. mungo*								
V1160 X Tc2210	33	34	18.3	1.1	5.8	6.9	0.3	-
V1160 X I-92	33	13	15.8	1.3	5.8	7.1	1.0	-
5. *V. glabrescens* X *V. umbellata*								
V1160 X I-275	33	31	11.1	7.5	3.4	10.9	0.0	-
6. amphidiploid between *V. radiata* and *V. umbellata*								
MR51	44	29	0.9	9.8	11.8	21.6	0.0	80.0%
7. (amphidiploid between *V. radiata* and *V. umbellata*) X *V. umbellata*								
MR51 X 6090	33	30	11.0	7.3	3.7	11.0	0.0	-
8. (amphidiploid between *V. radiata* and *V. umbellata*) X *V. glabrescens*								
MR51 X V1160	44	42	11.0	9.0	7.4	16.4	0.1	-

Note: I, II and III denote univalent, bivalent and trivalent, respectively.

Out of 51 F_2 seeds obtained by open pollination without bagging, 36 (70.6%) germinated. Thirty F_2 plants (83.3%) possessed 22 chromosomes and six had 23 chromosomes. The fertility of the progeny plants of *V. radiata* x *V. mungo* was not restored in the F_2 generation (Table 5). In the F_3 generation, however, the percentage of pollen with strong staining became higher.

It is well known that *V. mungo* exhibits a high level of resistance to bean weevil (*Callosobruchus chinensis*), although *V. radiata* is affected by serious damage (Sawa and Tan, 1976; Fujii and Miyazaki, 1987). Moreover, *V. mungo* is considered to be a good source for improving methionine content (AVRDC, 1987). Recovery of fertility in the F_3 generation suggests that there is a high possibility of transferring the useful genes from *V. mungo* to *V. radiata*.

Table 5. Pollen stainability of the F_2 and F_3 progenies derived from the hybrids between *V. radiata* and *V. mungo*.

	Number of plants exhibiting the pollen fertility (%) between										total
	0-10	11-20	21-30	31-40	41-50	51-60	61-70	71-80	81-90	91-100	
F2	4	5	3	3	2	0	0	0	0	0	17
F3	0	0	0	1	0	1	1	1	3	1	8

3.3. CYTOGENETICAL RELATIONSHIP BETWEEN *V. RADIATA* AND *V. UMBELLATA*

The hybrids between *V. radiata* and *V. umbellata* grew vigorously and flowered profusely. The color of the standard petal of the F_1 was bright yellow and was similar to that of the pollen parent, *V. umbellata*. The percentage of pollen with good stainability of the hybrids was 0 or 0.2% (Table 4). They were completely male sterile and no F_2 seeds could be obtained by open pollination. Although more than 300 flowers were backcrossed with *V. umbellata*, no seeds could be obtained. These hybrid plants are thus female sterile as well. *V. radiata* and *V. umbellata* are thought to be strongly isolated reproductively from each other.

Meiosis of the hybrids between *V. radiata* and *V. umbellata* was irregular. The number of univalents per cell ranged from 2 to 18. The configuration of 11 bivalents was not observed. As presented in Table 4, the average frequency was 12.7 univalents + 4.7 bivalents (0.6 rings + 4.1 rods) or 10.0 univalents + 6.0 bivalents (0.7 rings + 5.3 rods). The genome homology between *V. radiata* and *V. umbellata* is very low judging from the hybrid sterility and the irregular meiosis. It is thus difficult to transfer genes between *V. radiata* and *V. umbellata*. At the AVRDC, however, the F_2 plants from *V. radiata* x *V. umbellata* were successfully obtained through spontaneous chromosome doubling. This accession (AVRDC No. MR51) exhibited normal chromosome pairing at meiosis with 22 bivalents and relatively high pollen fertility (75 to 85%). The average meiotic chromosome pairing was 0.9 univalents + 21.6 bivalents (9.8 rings + 11.8 rods). Such normal chromosome pairing was expected from the amphidiploid. MR51 was vigorous in growth and free from pests and diseases (Fernandez and Shanmugasundarum, 1988). Since such resistance to pests and diseases probably comes from the genotype of *V. umbellata* (Duke, 1981), we can expect *V. umbellata* to be a useful gene donor in a mungbean breeding program. We have succeeded in producing triploids from the cross MR51 x *V. umbellata*, but we could not obtain hybrids from the cross MR51 x *V. radiata*.

3.4. PHYLOGENETIC DIFFERENTIATION BETWEEN *V. RADIATA*, *V. MUNGO* AND *V. UMBELLATA*

From the results of cross compatibility, sterility and irregular meiosis of the F_1 hybrids, it is concluded that the genomes of *V. radiata*, *V. mungo* and *V. umbellata* are highly differentiated from each other, and that they form distinct taxonomic species. The analysis of total seed storage protein also supported this taxonomical treatment. The three species exhibited species-specific electrophoretic banding patterns (Egawa et al., 1988).

In meiosis of the hybrids between *V. radiata* and *V. umbellata*, univalents were formed with higher frequency than in the hybrids between *V. radiata* and *V. mungo* (Table 4). Moreover, in hybrids of *V. radiata* x *V. umbellata*, most of the bivalents were loosely paired rod types. Closely associated ring bivalents were rarely observed. These results suggest that *V. radiata* is more closely related to *V. mungo* phylogenetically than to *V. umbellata*.

3.5. CYTOGENETICAL RELATIONSHIPS BETWEEN *V. GLABRESCENS* AND DIPLOID ASIAN *VIGNA* SPECIES

As presented in Table 4, the average meiotic chromosome pairing in *V. glabrescens* x *V. radiata* hybrid (2n=33) was 18.6 univalents + 6.6 bivalents (1.8 rings + 4.8 rods) + 0.4 trivalents. A low frequency of bivalent formation was observed and most of the bivalents were loosely-paired rod types. Such a result indicates that the genome of *V. radiata* is not included in *V. glabrescens*. The average meiotic chromosome pairing of hybrids (2n=33) between *V. glabrescens* and *V. mungo* was 18.3 univalents + 6.9 bivalents (1.1 rings + 5.8 rods) + 0.3 trivalents or 15.8 univalents + 7.1 bivalents (1.3 rings + 5.8 rods) + 1.0 trivalent. These results indicate that *V. glabrescens* does not possess a homologous genome with *V. mungo*.

In contrast, hybrids (2n=33) of *V. glabrescens* x *V. umbellata* showed an average configuration of 11.1 univalents + 10.9 bivalents (7.5 rings + 3.4 rods). Moreover, most of the bivalents observed were closely associated. This result indicates that *V. glabrescens* has a homologous genome with *V. umbellata*.

Hybrids (2n=44) between MR51 and *V. glabrescens* showed, on average, the occurrence of 11.0 univalents + 16.4 bivalents (9.0 rings + 7.4 rods) + 0.1 trivalents. This result is consistent with the average frequency of meiotic chromosome pairing in *V. glabrescens* x *V. radiata* and *V. glabrescens* x *V. umbellata*.

V. umbellata has lanceolate primary leaves and a protruding hilum. However, *V. glabrescens* has cordate primary leaves and a nonprotruding hilum like *V. angularis*. Thus, *V. glabrescens* resembles *V. angularis* more closely than *V. umbellata* in the shape of its primary leaves, hilum and seed. According to Ahn and Hartmann (1977), *V. umbellata* and *V. angularis* hybrids regularly formed 11 bivalents at MI and pollen stainability was 76%. Chen et al. (1983) also reported relatively high fertility of the hybrids between *V. angularis* and *V. umbellata*. This evidence suggests that *V. umbellata* and *V. angularis* share a homologous genome with each other.

Judging from the meiotic data (Table 4), the morphology of *V. glabrescens*, and the fact that chromosome pairing is regular with 11 bivalents in the hybrids between *V. umbellata* and *V. angularis*, we consider that *V. angularis*, probably wild var. *nipponensis*, donates one of the two genomes of a natural tetraploid *V. glabrescens*. Additional cytological examination of the hybrids between *V. glabrescens* x *V. angularis* var. *nipponensis* should be carried out to support the above hypothesis.

If *V. glabrescens* is an autotetraploid of *V. angularis* or an amphidiploid between *V. angularis* and *V. umbellata*, the configuration of 11 univalents and 11 bivalents is expected to occur in the triploid hybrids, *V. glabrescens* x *V. radiata* and *V. glabrescens* x *V. mungo* as a result of autosyndetical chromosome pairing. However, these hybrids showed considerably low frequency of bivalent formation. Based on the present meiotic data, *V. glabrescens* is not considered to be an autotetraploid of *V. angularis*, nor an amphidiploid between *V. angularis* and *V. umbellata*.

Another donor species for the two genomes of *V. glabrescens* has not yet been identified. The candidate for a donor species of another genome, which is different from *V. angularis* and *V. umbellata* but with sufficiently high cross compatiblity to produce a hybrid that can be doubled to produce a fertile amphidiploid, needs to be located from among other species not examined here. As described previously, *V. glabrescens* exhibits resistance against pests and diseases. The genome donor species to *V. glabrescens* is thought to carry such resistance genes. Therefore, such diploid species are very useful as potential breeding materials with cultivated species.

In addition to cultivated species, *Ceratotropis* includes some wild relatives, *V. nakashimae*, *V. riukiuensis* and *V. reflexo-pilosa* (Baudoin and Maréchal, 1988). According to Maréchal et al. (1978), *Ceratotropis* forms an extremely homogenous taxonomic group. It is therefore considered that Asian *Vigna* legumes belonging to *Ceratotropis* can be improved by exchanging useful genes among the species within the group. However, these wild relatives have not yet been utilized in hybridization experiments. To further promote the breeding program of Asian *Vigna* legumes, relationships among the whole species of *Ceratotropis* must be clarified.

REFERENCES

Ahn, C. S. and Hartmann, R. W. (1978) Interspecific hybridization among four species of the genus *Vigna* Savi., Proceedings of the First International Mungbean Symposium, AVRDC, Shanhua, Tainan, pp. 240-246.

AVRDC (1986) 1984 Progress Report, AVRDC, Shanhua, Tainan.

AVRDC (1987) 1985 Progressive Report, AVRDC, Shanhua, Tainan, pp.145-146.

Baudoin J. P. and Maréchal, R. (1988) Taxonomy and evolution of the genus *Vigna*' in S. Shanmugasundarum and B. T. McLean (eds.), Mungbean: Proceedings of the Second International Symposium, AVRDC, Shanhua, Tainan, pp. 2-12.

Chen, N. C., Baker, L. R., and Honma, S. (1983) Interspecific crossability among four species of *Vigna* food legumes, Euphytica 32, 925-937.

Duke, J. A. (1981) Handbook of Legumes of World Economic Importance, Prenum Press, New York.

Egawa, Y., Miyazaki, S., and Nakagahara, M. (1986) Crosscompatibility between four *Vigna* leguminous species, Jpn. J. Breed. 36(Suppl. 1), 260-261.

Egawa, Y., Yamashita, M., Tomooka, N., Kitamura, K., and Nakagahra, N. (1988) Interspecific variation of seed storage protein in Asian *Vigna* species by SDS-polyacrylamide gel electrophoresis, Jpn. J. Breed. 38(Suppl. 2), 442-443.

Fernandez, G. C. J. and Shanmugasundarum, S. (1988) The AVRDC mungbean improvement program: the past, present and future' in S. Shanmugasundarum and B. T. McLean (eds.), Mungbean: Proceedings of the Second International Symposium, AVRDC, Shanhua, Tainan, pp. 58-70.

Fujii, K. and Miyazaki, S. (1987) Infestation resistance of wild legumes (*Vigna sublobata*) to adzuki bean weevil, *Callosobruchus chinensis* (L.) (Coleoptera: Bruchidae) and its relationship with cytogenetic classification, Appl. Entomol. Zool. 22, 229-230.

Maréchal, R., Mascherpa J. M., and Stainier F. (1978) Etude taxonomique d'un groupe complexe d'espèces des generes Phaseolus et Vigna (Papilionaceae) sur la base de données morphologiques et polliniques, traitees par l'analyse informatique, Boissiera (Geneve) 28, 1-273.

Miyazaki, S. (1982) Classification and phylogenetic relationships of the *Vigna radiata-mungo sublobata* complex, Bull. Natl. Inst. Agr. Sci. Ser. D 33, 1-61.

Nishibayashi, S. (1985) Chromosomal analysis in rice. Conditions to observe chromosomes, Jpn. J. Breed. 35(Suppl. 1), 318-319.

Sawa, M. and Tan, S. (1976) Specific difference of resistance to adzuki bean weevil (*Callosobruchus chinensis*) in genus *Phaseolus*, Tohoku Branch Crop Sci. Soc. Jpn. 18, 79-81.

Smartt, J. (1985) Evolution of grain legumes. III. Pulses in the genus *Vigna*, Exper. Agric. 21, 87-100.

Swindell, R. E., Watts E. E., and Evans G. M. (1973) A natural tetraploid mungbean of suspected amphidiploid origin, J. Heredity 64, 107.

Verdecourt, B. (1970) Studies in the Leguminosae-Papilionoideae for the "Flora of Tropical East Africa", IV. Kew Bull. 24, 507-569.

White, R. P. (1963) The Cultivation of Animal and Plant Cells, Ronald Press, New York.

GENETICS AND BREEDING FOR BRUCHID RESISTANCE IN ASIATIC *VIGNA* SPECIES

G. C. J. FERNANDEZ[1] AND N. S. TALEKAR[2]
[1] *Plant Science Department, University of Nevada,
Reno NV 89557 U.S.A.*
[2] *AVRDC, P.O. Box 42, Shanhua, Tainan, Taipei 74199 China*

ABSTRACT. A bruchid [*Callosobruchus chinensis* (L)] is the most common destructive pest of stored grain legumes such as mungbean [*Vigna radiata* (L)], black gram [*V. mungo* (L)], rice bean [*V. umbellata* (L)], and adzuki bean [*V. angularis* (L)]. The initial infestation starts in the field at pod maturity; a secondary infestation under poor storage conditions destroys the seeds within 3 to 4 months. The use of insecticides or fumigants is not practical in bruchid control since the seeds are used for human consumption. Therefore, developing bruchid resistant cultivars could be an effective way to minimize the losses due to bruchid attack. Many sources of resistance to bruchid infestation have been reported in black gram, VM2011 and VM2164 (AVRDC); 'Sindh Khed'(India) and in mungbean, V 2802 (AVRDC); EG-Glabrous, EG-MG4, EG-MG7 (Philippines) and in wild *Vigna* species, *V. glabrescens* and *V. mungo* sub. sp. *silvestris* (AVRDC). Blackgram accession VM2011 was least damaged when the bruchid infestation occurred in the field, while in laboratory screening, VM2164 was highly resistant. Dense pubescence in VM2011 and hard seed coat or possible antibiosis in VM2164 were probably the causes of the resistance. The physical characteristics (seed size, seed coat thickness, seed color) appeared to be related to genotypic susceptibility. The inheritance of bruchid resistance is controlled by the genotype of the maternal parent and appeared to be controlled by recessive genes which complicates the selection and breeding for bruchid resistance. With these constraints in mind, a breeding method is proposed to introduce bruchid resistance from multiple genetic sources to improved mungbean cultivars.

1. Introduction

The Asiatic *Vigna* group consists of several related grain legume species which have been grown for many years in central, southern, and eastern Asia. These annual species are short duration plants and are grown mainly for edible dry seeds, which provide a significant portion of the dietary protein for many low income people in these regions. The immature pods are occasionally used as a green vegetable. The high lysine level in the protein makes it an ideal supplement to cereals. The *Vigna* species can be grown successfully over a wide range of environmental conditions and improve soil fertility by their ability to symbiotically fix atmospheric N_2. The cultivated *Vigna* species are also used as fodder, cover and green manure crops.

Mungbean [*Vigna radiata* (L)] and black gram [*V. mungo* (L)] are the most widely utilized species of this group. Mungbean accounted for about 12% of the total pulse production in the south and south-east Asia and it's production increased from 1.1 million mt in 1976 to 1.8 million mt in 1986, registering a growth rate of only 0.5% for total pulses (Singh, 1988). The adzuki bean [*V. angularis* (L)] is popular in Japan, China, and

K. Fujii et al. (eds.), Bruchids and Legumes: Economics, Ecology and Coevolution, 209–217.
© 1990 *Kluwer Academic Publishers. Printed in the Netherlands.*

Korea. It is the sixth most widely cultivated crop in Japan where 120,000 h are grown annually (Sacks, 1977). Ricebean [*V. umbellata* (L)] is cultivated in India, Burma, Malaysia, and China.

Bruchids are one of the most important insect groups attacking both grain and tree legumes in the arid tropics (Birch et al., 1985; Redden et al., 1983). The most common and destructive bruchid pest of Asiatic *Vigna* species is *Callosobruchus chinensis* (L.) (Fernandez and Shanmugasundaram, 1988; Talekar, 1988). Bruchid infestation begins in the field when the adult beetle lays eggs on immature pods; the larvae bore through the pod wall and feed concealed within the developing seeds (Talekar, 1988). The insects continue to feed on the seeds in storage, and eventually emerge as adults. Secondary infestation results in total destruction within a period of 3 to 4 months. Because of storage and international trade of grain legume seeds, bruchid pests are distributed throughout the world. Bruchid damage results in significant reduction in the quantity and quality of food resources; the quality and viability of seed saved for planting is also reduced (Birch et al., 1985).

The use of insecticides or fumigants in developing countries is not practical in bruchid control because of the primitive nature of the storage facilities, low volume productions, and the use of seeds as human food (Talekar, 1988; Talekar and Lin, 1981). Food legumes have been primarily selected for high yield, large seed, and nutritional value together with low anti-nutritional factors (Birch et al.,1985). This has severely disrupted the co-evolutionary process between plants and insects such that very few cultivated species have retained the insect resistance of their wild relatives (Birch et al.,1985). Partial resistance which reduces the insect's population growth rate in storage could be useful as one of several components in integrated control of bruchids. However, high levels of seed resistance are required to protect seed from bruchids over several months of poor storage conditions. The purpose should therefore be the transfer of multiple component resistance which would be more stable than that based on a single resistance source. The objective of this paper is to review screening methods, identify resistant germplasm, and propose a breeding method to introduce bruchid resistance from multiple genetic sources into cultivated Asiatic *Vigna* species.

2. Screening for Bruchid Resistance

Any resistance screening procedure should simply and reliably identify resistant sources from many genotypes. An adequate population of 20-40 bruchids per sampling unit is recommended to measure resistance to bruchid by the effect on bruchid emergence in cowpea, [*V. unguiculata* (L.)] (Redden and McGuire, 1983). The number of eggs/seed should not be excessively large and should be uniformly distributed over the seeds. A ratio of one adult bruchid pair per 10 seeds for 12-15 hr resulted in about 2 eggs per seed (Redden and McGuire, 1983). Resistant and susceptible checks should be included in screening.

Because adult emergence is influenced by the atmospheric temperature and humidity, both these factors should be controlled to standardize the screening processes. A temperature between 28-32°C and RH of 65-75% are used for bruchid screening in *Vigna* species (Birch et al., 1985; Talekar and Lin, 1981).

Various criteria for assessment of resistance such as number of eggs laid/experimental unit, percentage of adult emergence over a given period, mean number of days to emergence of adults, mean number of emergence holes per seed, percentage weight loss per seed, and percentage of undamaged seed are reported in the literature (Redden and McGuire, 1983). The simplest and most reliable screening method to assess bruchid resistance was percentage of undamaged seeds. However, greater control and manipulation of oviposition on seed would be required to obtain uniform infestation.

3. Genotypes Resistant to Bruchid Damage

Two sources with different modes of resistance to *C. chinensis* were identified in field and laboratory screening of more than 500 mungbean and blackgram accessions at the Asian Vegetable Research and Development Center (AVRDC) (Talekar, 1988; Talekar and Lin, 1981). *C. chinensis* oviposition and adult emergence from pods and seeds from three blackgram accessions are shown in Fig. 1. Blackgram accession VM2011 was relatively less damaged when the insect infestation occurred in the field, whereas in laboratory screening with seeds, accession VM2164 was highly resistant. Dense pubescence in VM2011 and thick seed coat and/or possibly antibiosis in VM2164 were believed to be the cause of the resistance (Talekar and Lin, 1981). Number of eggs laid/50 seeds and the number of adults emerged/50 seeds were used as the screening criteria. Recently at AVRDC, one mungbean accession, V2802, an accession V1160 belonging to *V. glabrescens*, and some wild blackgram accessions belonging to *V. mungo* sub. sp. *silvestris* were also identified as resistant genotypes (Talekar, 1988).

In India, using number of eggs laid on the seed and the mean number of emergence holes per seed, a black gram accession 'Sindh khed' with medium sized seed showed a higher degree of resistance to *C. chinensis* than the two local cultivars (Shivashankar et al., 1972). Both mungbean varieties and accessions were screened for bruchid attack in the Philippines (Doria and Raros, 1975). None of the mungbean varieties were resistant to oviposition, but resistance to larval survival was evident in EG-Glabrous, EG-MG4, and EG-MG7. The blackgram accessions, UPCA-23, UPCA-25, and UPCA-325 were resistant to oviposition and larval survival.

The physical characteristics and chemical components of mungbean seeds appeared to be correlated with varietal susceptibility to bruchid infestation (Epino and Regesus, 1983). Large and heavy seeds were more preferred by the bruchids than small seeds. The resistant accessions had a lower percentage of fats and starch but a higher percentage of protein than the susceptible accessions. However, based on germplasm screening at AVRDC, the associations between physical or chemical characteristics and bruchid resistance were not significant. Thus, these reported relationships might hold only to those specific genotypes used in that investigation.

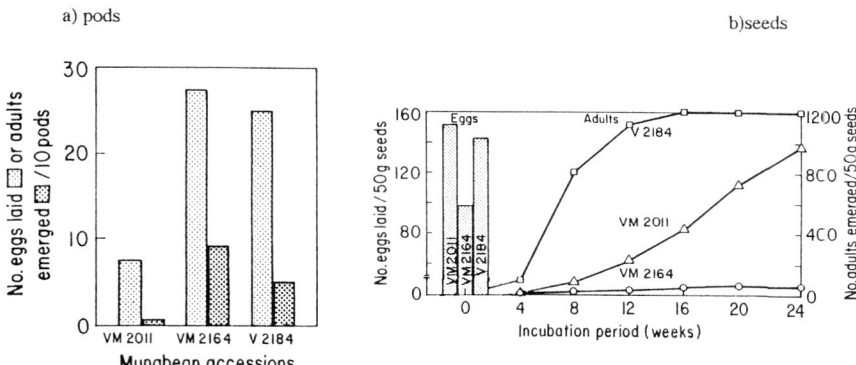

Figure 1. *C. chinensis* oviposition and adult emergence from a) pods and b) seeds of three blackgram accessions.

Trypsin inhibitor, an anti-nutritional component found in the seed, was found to be partially associated with the expression of bruchid resistance in many legume species. However, a correlation analysis of 56 *Vigna* genotypes failed to find any significant association between the level of trypsin inhibitor and the observed seed resistance to *C. chinensis* (Birch et al., 1985). Although isolated trypsin inhibitor from cowpea and common beans [*Phaseolus vulgaris* (L.)] was found to be toxic to *C. maculatus* when incorporated in the artificial diet, their contribution to resistance in the intact seeds of wild species requires further investigation (Birch et al., 1985). Therefore, the contribution of trypsin inhibitor to bruchid resistance in the intact seed of resistant genotypes requires further investigations. Chemical screening alone would probably have little predictive value for resistance breeding since seed resistance to bruchids in many wild legume genotypes depends on the combined effects of several components.

4. Breeding for Resistance to *C. chinensis*

A preliminary breeding effort to incorporate the bruchid resistance found in VM2011 and VM2164 into popular mungbean cultivars was initiated at AVRDC (Talekar and Lin, 1981). Mungbeans and blackgrams are closely related species with the same chromosome number (2n = 2x = 22) and are partially cross compatible. The bruchid resistance found in the two blackgram accessions appeared to be monogenic and recessive. Another noticeable feature was the highly consistent maternal control of seed resistance, i.e. the phenotype of the seed follows the genotype of the maternal parent. Thus, the observed seed resistance in any generation appeared to be governed by the maternal plant rather than by the seed embryo. The seed coat, originating from the maternal tissue, was not totally responsible for the observed resistance since in a no-choice situation considerable seed damage was observed in accessions with hard seed coat. Similar findings were also reported in cowpea (Redden et al., 1983). The maternal control of seed resistance complicates the direct application of backcross or pedigree breeding techniques to incorporate bruchid resistance into popular cultivars. A thorough understanding of why maternal genome expression is favored over the hybrid genome in a endosperm-less legume seed would facilitate bruchid resistance breeding efforts.

4.1. SELECTIVE EXPRESSION OF MATERNAL GENOME IN SEED DEVELOPMENT

Seed development in legumes begins with double fertilization which gives rise to the zygote and the triploid cell that will give rise to endosperm. These events take place within the embryo sac which is embedded in maternal tissue comprised of nucellus tissues and one or more integuments. At first seed development results from the rapid development of the integument tissue. The outer integument cells differentiate into the seed testa or seed coat. The growth of the embryo is very slow at first but accelerates during the second phase. The final cell number in the embryo is reached rather early in its ontogeny and its subsequent increase in mass is the result of cell expansion and concomitant deposition of starch and starch products. The developing embryo totally absorbs the nucellus and endosperm. Thus, the endosperm functions as a transient reservoir of sugar and amino acids. At some point, the cells of the funiculus connecting the ovule to the ovary degenerate making the ovule a nutritionally closed system. Finally, during seed maturation, desiccation sets in as the ovule losses water to the surrounding environment and the seed coat tissue sclarifys and dies. The seed developmental stages in a legume are illustrated in Fig. 2. The duration of individual phases might vary among different species. A detailed account of legume seed development is reviewed

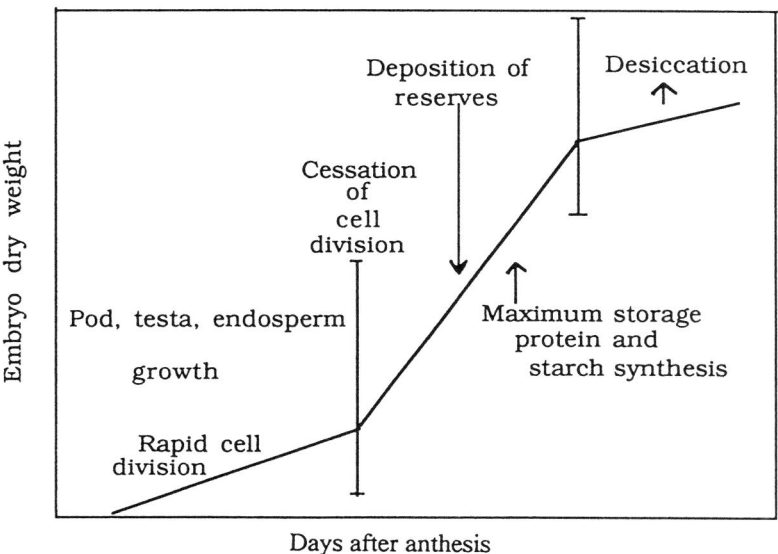

Figure 2. Stages of legume embryogenesis, Dure(1975).

elsewhere (Dure, 1975).

4.1.1. *Seed Texture Layer.* In mungbean, two types of seed texture, dull or shiny, are found (Fernandez and Shanmugasundaram, 1988). During the ripening and drying of the pod wall, the inner membrane is fractured and a part of it protrudes inwards which on contact with the testa adheres to it, and forms the seed texture layer. At seed maturity, the texture layer dehydrates and forms a reticulate deposit of cell wall remnants on the surface of the seed coat. If this deposit is sufficiently thick to mask the shiny seed testa, the seed will appear dull. If the deposit is thin, so that the seed testa is visible through it, the seed will appear shiny. Dull seed texture is genetically dominant over shiny and is monogenic (Fernandez and Shanmugasundaram, 1988). The dull seed has 3 to 4 times more monosaccarides such as glucose, xylose, arabinose, and galactose than the shiny mungbean (Watt et al., 1977). However, an effect of seed texture on bruchid infestation was not apparent.

4.1.2. *Selective Expression of Maternal Genome.* In legume embryogenesis, alleles derived from the maternal genome are selectively expressed for certain traits. The hybrid seed between two diverse legume genotypes closely resemble the seed of the maternal parent. As a result, a pronounced difference may be observed in the reciprocal cross of the parental types such as in the expression of bruchid resistance, seed size, and globulin protein synthesis (Davis, 1973). One possible explanation is a selective repression or activation of loci in the two genomes of the hybrid embryo. The maternal parent might produce an inducer which is recognized only by the repressor locus on the maternal genome. This selective expression of the maternal structural gene would then result in the production of the maternal type in the hybrid embryo (Davis, 1973). However, all loci in the legume embryo genome do not behave in the same way (Davis, 1973). Therefore, the cellular and molecular aspects of the differential expression of the bruchid resistance factors needs further investigation.

4.2. BREEDING PROCEDURES TO IMPROVE BRUCHID RESISTANCE

Selection and breeding for bruchid resistance is complicated by the fact that the inheritance of bruchid resistance in *Vigna* species is apparently controlled by the maternal genome and recessive genes. With these constraints in mind, a breeding method is proposed herein to incorporate resistance from two sources, A and B, into a popular cultivar. The first step in the breeding procedure involves combining the two sources of resistance into one genotype (Fig. 3a).

The resistant genotype-A as the female is crossed with an adapted, high yielding cultivar. Since the resistant genotype is used as the female parent, the F_1 hybrid seeds express bruchid resistance from the maternal parent. A second cross is then made between the resistant genotype-B as the female and the F_1 hybrid as male. The hybrid seeds also express resistance from the maternal parent, the resistant genotype-B. All the hybrid seeds are grown and selfed. These selfed seeds would be susceptible since they express the hybrid genome and the resistance genes are recessive. These seeds are grown, selfed, and the resulting seeds are collected separately from individual plants. The first screening for bruchid resistance is carried out on these selfed, **within single plant seeds**. Both resistance sources, A and B and the susceptible parents are used as checks. The true breeding genotype with high level of resistance is identified by comparing with the two resistant checks and by identifying single plants with expression of 100% resistance in the progeny.

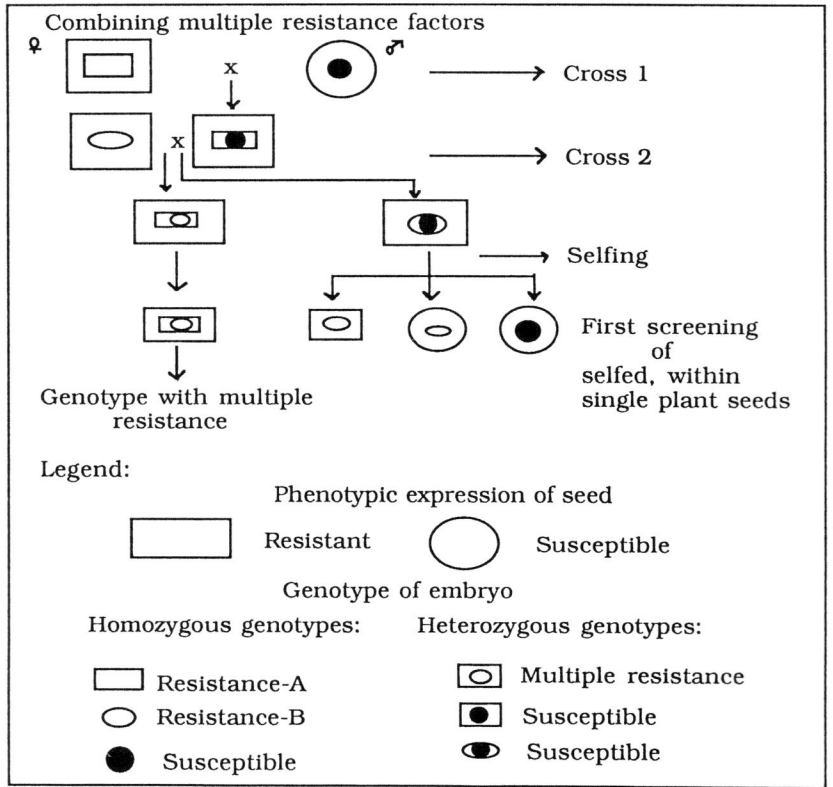

Figure 3a. Breeding procedures to incorporate Bruchid resistance - **combining multiple resistance.**

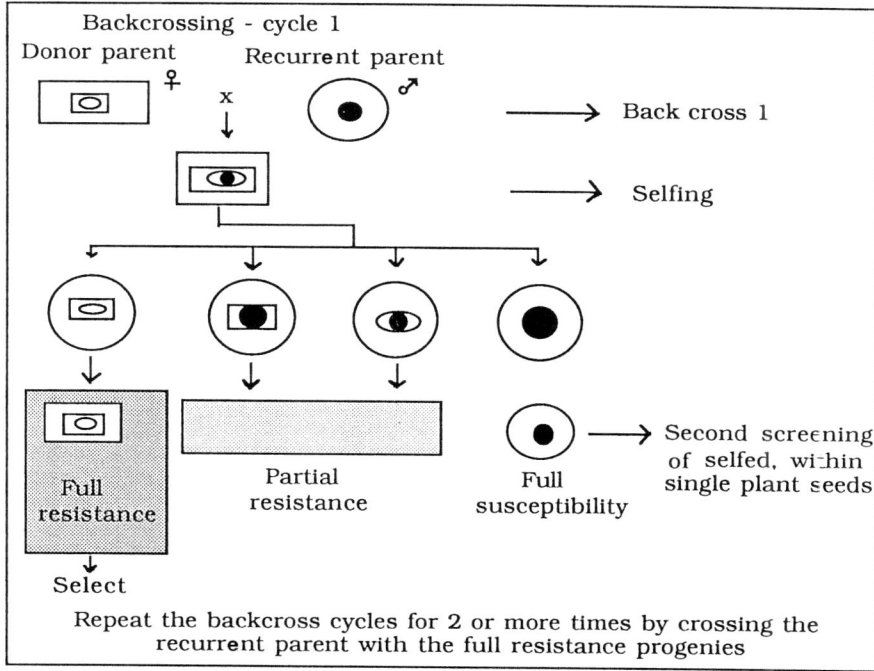

Figure 3b. Breeding procedures to incorporate Bruchid resistance - **backcrossing.**

The second step involves three cycles of backcrossing to introduce the multiple resistance to the popular cultivar (recurrent genotype) (Fig 3b). In the first backcrossing cycle, the multiple resistant genotype is used as the female parent and the hybrid seeds are grown and subsequently selfed. These selfed seeds would be susceptible for bruchid infestation. These seeds are then grown in pots and selfed seeds are harvested separately from individual plants. The second screening for bruchid resistance is carried out on these selfed seeds from individual plants. The true breeding genotype with high levels of resistance is selected and the genotypes with partial resistance and susceptible progeny are discarded. The same backcross procedures are repeated for at least two more cycles. After three backcross cycles, genotypes possessing multiple resistance and high frequency of desirable alleles derived from the recurrent parent would be selected.

The third step in the breeding procedure (Fig. 3c) involves inter-crossing among the selected genotypes possessing high levels of bruchid resistance and desirable alleles from

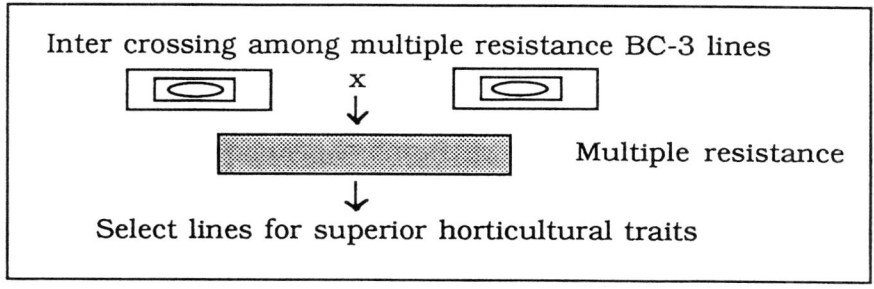

Figure 3c. Breeding procedures to incorporate Bruchid resistance - **intercrossing.**

the popular cultivar. Superior horticultural genotypes are selected from the progenies and the bruchid resistance can be confirmed by a final screening.

This breeding procedure offers several advantages over direct backcross breeding for the introduction of bruchid resistance form multiple sources:

The number of bruchid screenings is kept to a minimum. The use of resistant genotypes as the maternal parents reduces the need for bruchid screening in each generation and protects the susceptible embryo from bruchid damage.

This procedure allows the incorporation of resistance from multiple sources, reduces the number of backcrosses needed, and allows continual selection for horticultural traits.

Detailed analysis and confirmation of the inheritance of bruchid resistance and gene interaction can be successfully carried out in conjunction with this breeding method.

4.3. INTRODUCTION OF BRUCHID RESISTANCE FROM WILD SPECIES

The transfer of resistance factors from wild species outside the primary gene pool to cultivated species depends on the bio-technological skills of making inter-specific crosses, but the progress in this field is encouraging. Modern plant breeding methods using somatic hybridization and embryo rescue methods are making it possible to achieve difficult crosses. Developments in protoplast fusion and plant generation now permit not only the crossing of incompatible species but also the transfer of only a few chromosomes or genes, thus creating many opportunities for crop improvements. Many inter-specific crosses among the wild and cultivated Asian *Vigna* species were recently made (Chen et al., 1983; Egawa, 1988; Gossal and Bajaj, 1983). This should enhance the progress of introduction of high levels of bruchid resistance into popular cultivars.

5. Conclusion

Genotypes with high levels of resistance to bruchids have been identified in both *Vigna* cultivars and accessions of wild *Vigna* species. However, breeding efforts to improve bruchid resistance have been complicated by the fact that resistance is controlled by the maternal genome and recessive genes. The breeding procedure suggested herein consisting of combining multiple resistance, back crossing, and inter-crossing offers several advantages over the direct backcross method for improving resistance to bruchids.

REFERENCES

Birch, N., Southgate, B. J., and Fellows, L. E. (1985) Wild and semi-cultivated legumes as potential sources of resistance to bruchid beetles for crop breeder: a study of *Vigna/Phaseolus*, in G. E. Wickens et al. (eds.), Plants for Arid Lands, George Allen & Unwin, London, pp. 303-318.

Chen, N. C., Baker, L. R., and Honma, S. (1983) Inter-specific crossability among four species of *Vigna* food legumes, Euphytica 32, 925-937.

Davis, D. R. (1973) Differential activation of maternal and paternal loci in seed development, Nature New Biol. 245, 30-32.

Doria, R. C. and Raros, R. S. (1975) Varietal resistance of mungo to the been weevil, *Callosobruchus chinensis* and some characteristics of field infestation, Philipp. Entomol. 2, 399-408.

Dure , L. S. III (1975) Seed formation, Ann. Rev. Plant Physiology 26, 259-278.

Egawa, Y., Nakagahra, M., and Fernandez, G. C. J. (1988) Cytogenetical analysis of tetraploid *Vigna glabrescens* by inter-specific hybridization involving diploid Asian *Vigna* species, in S. Shanmugasundaram and B. T. McLean (eds.), Mungbean: Proceedings of the Second International Symposium, AVRDC, Shanhua, Tainan, pp. 200-204.

Epino, P. B. and Regesus, M. B. (1983) Physico-chemical properties of mungbean [*Vigna radiata* (L.) Wilczek] seeds in relation to weevil resistance, Philipp. Entomol. 5, 607-620.

Fernandez, G. C. J. and Shanmugasundaram, S. (1988) The AVRDC mungbean improvement program: the past, present and future, in S. Shanmugasundaram and B. T. McLean (eds.), Mungbean: Proceedings of the Second International Symposium, AVRDC, Shanhua, Tainan, pp. 58-70.

Gossal, S. S. and Bajaj, Y. P. S. (1983) Inter-specific hybridization between *Vigna mungo* and *Vigna radiata* through embryo culture, Euphytica 32, 129-137.

Redden, R. J., Dobie, P., and Gatehouse, A. M. R. (1983) The inheritance of seed resistance to *Callosobruchus maculatus* F. in cowpea (*Vigna unguiculata* L. Walp.) I. analysis of parental, F_1, F_2, F_3, and backcross seed generations, Aust. J. Agric. Res. 34, 681-695.

Redden, R. J. and McGuire, J. (1983) The genetic evaluation of bruchid resistance in seed of cowpea, Aust. J. Agric. Res. 34, 707-715.

Sacks, F. M. (1977) A literature review of *Phaseolus angularis* - the adzuki bean, Econ. Bot. 31, 9-15.

Shivashankar, G., Urs, K. C. D., Bhat, M. S., and Vishwanatha, S. R. (1972) Study cf varietal resist ANCE TO *Callosobruchus chinensis* L. in black gram (*Vigna mungo*), Mysore J. Agric. Sci. 3, 360-362.

Singh, R. B. (1988) Trends and prospects for mungbean production in south and southeast Asia, in S. Shanmugasundaram and B. T. McLean (eds.), Mungbean: Proceedings of the Second International Symposium, AVRDC, Shanhua, Tainan, pp. 552-559.

Talekar, N. S. (1988) Biology, damage, and control of bruchid pests of mungbean' in S Shanmugasundaram and B. T. McLean (eds.), Mungbean: Proceedings of the Second International symposium, AVRDC, Shanhua, Tainan, pp. 329-342.

Talekar, N. S., and Lin, Y. H. (1981) Two sources with differing modes of resistance to *Callosobruchus chinensis* in mungbean, J. Econ. Entomol. 74, 639-642.

Watt, E. E., Poehlman, J. M., and Cumbie, B. G. (1977) Origin and composition of a texture layer on seeds of mungbean, Crop Sci. 17, 121-125.

BREEDING FOR BRUCHID RESISTANCE IN COWPEA

B. B. SINGH AND S. R. SINGH
International Institute of Tropical Agriculture (IITA)
Oyo Road, PMB 5320, Ibadan Nigeria

ABSTRACT. One of the most important constraints in cowpea storage is widespread infestation by the seed beetle, *Callosobruchus maculatus* (F.), commonly known as bruchid. In addition to causing direct seed weight loss, it reduces seed quality and also affects germination. The International Institute of Tropical Agriculture (IITA) initiated a systematic program in 1974 to develop bruchid resistant cowpea varieties and TVu 2027 was identified as moderately resistant. During the last several years, over 8,000 germplasm accessions have been screened and few additional sources have been identified. The resistance is characterized by delayed, staggered and lower emergence levels of bruchids. Therefore resistant accessions suffer less damage than susceptible ones. On average, within a storage period of 100 days, the resistant accessions show about 25% damaged seeds compared to 95% for the susceptible varieties. The genetics of resistance has been studied and resistance is incorporated in advance breeding lines at IITA. IT84S-2246-4 is one of the superior lines which apart from resistance to bruchid, combines resistance to other pests; aphids, thrips along with resistance to 10 diseases. This line appears to perform well in Nigeria and is being used as an important parent. Because bruchid resistance in cowpea is controlled by 2 pairs of recessive genes, any outcrossing reduces the proportion of resistant plants in succeeding generations. Its implication in germplasm collection and varietal maintenance is discussed.

1. Introduction

Grain legumes, also termed as pulse crops, are a major source of dietary protein in many parts of the world, particularly in the countries situated along the tropical and subtropical belt where the availability and consumption of animal protein is rather low because of social and/or economic constraints. Pulses are much cheaper compared to meat, fish and egg and contain about 25% protein rich in lysine and tryptophane. This makes them a good supplement to cereal and root crop based diets which are usually very low in protein and high in carbohydrates. Different grain legumes are grown in different regions because of differential agro-climatic requirements. For example, pigeon pea (*Cajanus cajan*), chickpea (*Cicer arietinum*), field peas (*Pisum sativum*), green gram (*Vigna radiata*) and black gram (*Vigna mungo*) are more widely grown in Asia; beans (*Phaseolus vulgaris*) in South America and East Africa whereas cowpea (*Vigna unguiculata*) is the principal grain legume in West Africa and a secondary legume in parts of East Africa, Central and Southern America and Asia.

Cowpea, being a warm weather crop and drought tolerant, is grown throughout the tropics, particularly in the semi-arid and low rainfall regions. It provides food and fodder as well as improving the soil fertility. It is cultivated in about 7.7 million ha on a world-wide basis of which about 6 million ha are restricted to Africa alone. However, the average yield of cowpea ranges from 100 to 300 kg/ha due to numerous constraints

K. Fujii et al. (eds.), Bruchids and Legumes: Economics, Ecology and Coevolution, 219–228.
© 1990 *Kluwer Academic Publishers. Printed in the Netherlands.*

particularly insects and diseases. In addition to yield losses in the field, cowpea also suffers considerable loss in storage due to bruchids. Booker (1967) identified 3 species of bruchid on cowpea in Nigeria. These were, *Callosobruchus maculatus* (Fabricius), *Callosobruchus rhodescianus* Pic, and *Bruchidius atrolineatus* Pic. However, the most damaging species among these is *C. maculatus*. The other two are seldom found in cowpea that have been in store for more than a month (Booker, 1967). The cowpea bruchid, *C. maculatus* not only causes direct reduction in dry weight but it also reduces grain quality and seed viability making it unfit for human consumption as well as for planting and thus, causing substantial reduction in market value. The initial infestation occurs in the field and from there, it is carried over to stores (Prevett, 1961) where the population can rapidly build up. The adult females lay eggs on cowpea seeds which hatch within 5 to 7 days. The young larvae bore into seeds and complete their development inside, feeding on the seed tissues. At the end of their development, adults emerge from the seeds leaving a hole at the exit points (Fig. 1). The life cycle varies from 20 - 30 days depending upon the temperature (Williams, 1980). Thus, on average, one cycle is completed every month under tropical conditions. Caswell (1973) estimated that 4.5% of the annual production of cowpea valued over at 30 million dollars is lost each year in Nigeria alone due to bruchids. However, this is thought to be a conservative estimate.

Because of the serious yield losses in field as well as in storage, the International Institute of Tropical Agriculture (IITA) which has the global mandate for cowpea improvement, initiated a systematic screening program in 1974 to identify cowpea lines resistant to cowpea bruchid *C. maculatus*.

A breeding program to incorporate resistance was initiated in 1976. Through the cumulative contribution of research over the years, a great deal of progress has been made and a number of improved varieties have been developed which combine resistance to bruchid and other pests; aphids, thrips along with several diseases. This paper

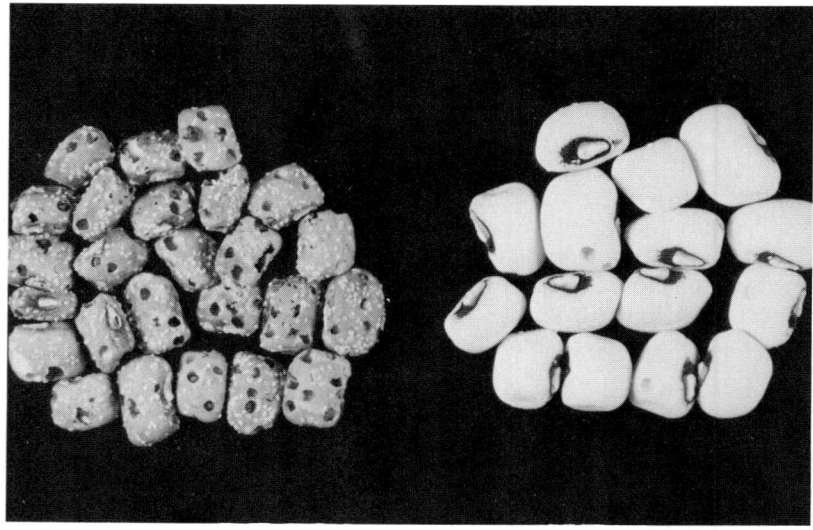

IFE BROWN   TVu 2027

Figure 1. Level of bruchid damage in susceptible and resistant cowpea varieties after about 3 months of storage.

summarizes the work done on various aspects of bruchid resistance in cowpea.

2. Method of Screening for Bruchid Resistance

A rapid and reliable screening method is essential for initiating a breeding program so that the resistant plants/progenies could be easily identified from segregating populations. Singh and Jackai (1985) described the bioassay method used for screening for bruchid resistance in cowpea. The method utilizes infestation of the cowpea seeds by bruchids, and measurement of adult emergence and emergence pattern is computed. The emergence pattern is also taken into consideration along with percent adult emergence for classifying materials into resistant and susceptible classes.

Recently a biomonitor has been developed at Purdue University, W. Lafayette, Indiana, U.S.A. which may offer an extremely fast method of screening not only for bruchid resistance but other seed weevils as well (Murdock, 1988, Personal communication). The seeds with only one egg are incubated until the eggs have hatched. The seeds are then kept on a sound sensitive plate which is connected to a computerized monitor. Each feeding activity (a bite) of the larva inside generates a sound wave which is picked up as a signal and recorded by the computer as a dot, and thus, the number of bites taken by the larvae can be continuously recorded throughout the developmental period of the insect inside the seed. The feeding behavior of the larvae in a resistant seed is different than in a susceptible seed and thus, within 30 minutes to 1 hour of monitoring, resistant seeds can be distinguished from susceptible seeds. IITA is collaborating with Purdue University to develop this device for routine screening of a large number of plants/progenies.

Table 1. Bruchid emergence pattern in resistant and susceptible varieties of cowpea (Singh et al., 1985).

Varieties	Percent adult emergence 50 DAI	Number of days after infestation																							
		27	28	29	30	31	32	33	34	35	36	37	38	39	40	41	42	43	44	45	46	47	48	49	50
		Number of adults emerged																							
TVu2027	26.6	0	0	0	0	0	0	1	0	0	2	0	0	0	0	0	1	1	0	0	0	1	1	1	0
TVu11952	32.8	0	0	0	1	1	0	3	0	1	1	0	1	0	1	1	0	0	1	0	0	0	1	0	0
TVu11953	22.1	0	0	0	0	1	2	1	0	1	0	0	0	1	0	0	0	0	0	0	0	0	0	0	0
IfeBrown	86.4	3	11	7	5	0	0	0	0	0	1	0	0	0	0	0	0	0	0	0	0	0	0	0	0
LSD0.05	11.4																								

222

3. Sources and Level of Bruchid Resistance

Through systematic screening of over 6000 cowpea germplasm lines, Singh (1977) identified TVu 2027, a local line from Nigeria, as moderately resistant to bruchid. Singh et al. (1985) reported two additional sources of resistance, TVu 11952 and TVu 11953 and described in detail the level of resistance in these lines. The emergence pattern and percent adult emergence is given in Table 1. The resistant lines are characterized by delayed, staggered and slow adult emergence while in susceptible lines like Ife Brown, the adult emergence is relatively early and extremely rapid. Thus, the resistant lines are not immune to bruchids but suffer considerably less damage compared to the susceptible line. Singh et al. (1985) reported that a 200 g sample infested with 2 pairs of bruchid showed 25 - 26% damaged seeds in the case of resistant lines, but 95% damaged seeds in the susceptible variety after 103 days of storage. The level of resistance in three lines is not significantly different but it is good enough to provide reasonable protection

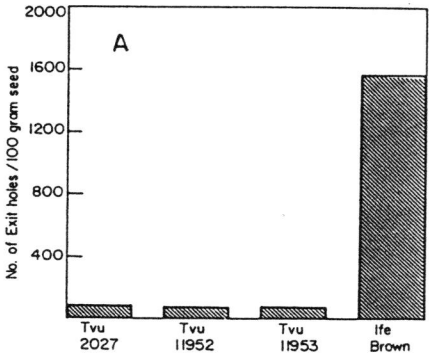

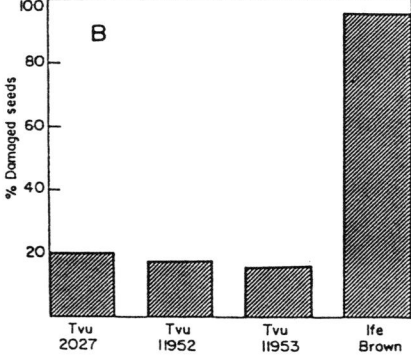
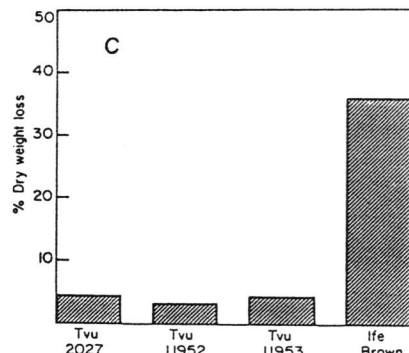

Figure 2. Quantitative differences between bruchid susceptible and resistant lines.
A: Cumulative adult emergence in bruchid resistant and susceptible lines of cowpea 89 days after infestation.
B: Percent damaged seeds in bruchid resistant and susceptible lines of cowpea 89 days after infestation.
C: Percent dry weight loss due to bruchid damage in resistant and susceptible lines of cowpea 89 days after infestation.

against bruchids during storage (Fig. 2).

The resistance in TVu 2027 seems to hold against several geographical strains of *Callosobruchus maculatus* (F.). Dick and Credland (1986) observed resistance of TVu 2027 against *Callosobruchus maculatus* (F.) strains from Brazil, Nigeria and Yemen Arab Republic. Similarly, Ndlovu and Giga (1988) found TVu 2027 derived breeding lines such as IT81D-1032, IT81D-1064 and others to be moderately resistant to *Callosobruchus rhodesianus* (PIC).

4. Nature of Bruchid Resistance

There is no difference between the resistant line TVu 2027 and the susceptible line Ife Brown with respect to oviposition. Also, the rough and smooth seed coats have no effect on bruchid infestation. Therefore, the cause of resistance may be in the seed. Gatehouse et al. (1979) reported a higher level of trypsin inhibitor (about 2 fold increase) in TVu 2027 compared to the susceptible varieties and attributed the bruchid resistance in cowpea to this factor. They also showed that trypsin inhibitor isolated from cowpea and mixed in ground cotyledons of a susceptible cowpea variety TVu 57 reduced the survival of bruchid eggs. However, recent studies do not show correlation between trypsin inhibitor content and bruchid resistance in cowpea. Baker et al. (1989) analyzed trypsin inhibitor activity in ten TVu 2027 derived bruchid resistant breeding lines including TVu 2027 and 5 susceptible lines. The trypsin inhibitor activity ranged from 32.2 T.I.U./mg to 50.0 T.I.U./mg with a mean of 43.2 T.I.U./mg in resistant breeding lines compared to a range of 32.8 T.I.U./mg to 44.5 T.I.U./mg with a mean of 38.8 T.I.U./mg in susceptible lines. The differences were non-significant. It has also been observed that artificial seeds made out of TVu 2027 or its derivatives are as susceptible to bruchids as susceptible lines (L. Murdock and R.E. Shade, personal communication, 1988) indicating that trypsin inhibitor alone may not account for bruchid resistance in cowpea. Osborn et al. (1988) have identified 'arcelin', a major seed protein in wild *Phaseolus vulgaris* as the factor responsible for resistance to bean bruchid *Zabrotes subfasciatus*. Similarly, para-aminophenylalanine in several wild *Vigna* species was shown to be toxic to *Zabrotes subfasciatus* as well as to *Callosobruchus maculatus* (Birch et al. 1986). Ishimoto and Kitamura (1988) showed that a water soluble substance present in kidney beans strongly inhibits the larval growth of *Callosobruchus chinensis*. All these studies indicate a chemical factor to be responsible for bruchid resistance. The only report indicating the involvement of a physical factor is in chickpea where the rough seed coat acts as a deterrent for oviposition (Saxena and Raina, 1970). However, as mentioned earlier, seed coat texture in cowpea has no correlation with bruchid resistance. Apparently, additional research is needed to elucidate the value of bruchid resistance in cowpea.

5. Genetics of Bruchid Resistance

Adjadi et al. (1985) reported a detailed study on the genetics of bruchid resistance in cowpea. They observed that 2 recessive genes (rcm1 rcm1 rcm2 rcm2) are required in the homozygous condition to confer resistance to bruchid. Thus, the F_1 hybrid was susceptible and F_2 segregated into 15 susceptible :1 resistant ratio. The level of resistance in resistant plants selected from segregating populations was similar to TVu 2027 indicating major gene effect. They also observed a maternal effect so that the F_1 seed was similar to the maternal parent in responce to bruchid. The true hybrid genotype was manifested in F_2 seeds produced on F_1 plants. Subsequent allelic tests have shown that the genes for resistance in TVu 2027, TVu 11952 and TVu 11953 are the same.

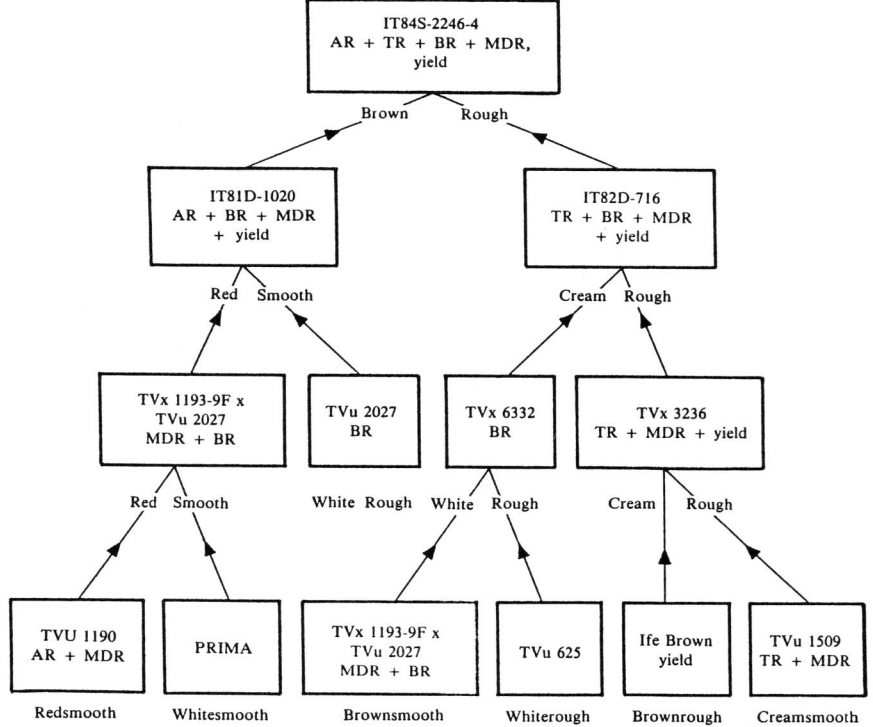

AR = Aphid Res., BR = Bruchid Res., TR = Thrips Res., MDR = Multiple Diseases Res.

Figure 3. Pedigree of IT84S-2246-4 and sequence of selection. This forms part of the work done by K. O. Rachie and B. Smithson in the early phases of the cowpea breeding program.

6. Breeding Strategy for Bruchid Resistance

Since cowpea is attacked by several diseases as well as insect pests in the field before it reaches the storage stage, a strategy involving multiple cycles of crossing and selection was adopted to develop high yielding cowpea varieties combining multiple disease and insect resistance including resistance to bruchid. The segregating populations were handled using pedigree method. Initially individual plants were selected from F_2 populations based on resistance to diseases, plant type, seed type and maturity and then F_3 seeds from individual F_2 plants were tested for aphid and bruchid resistance. The selected progenies were further advanced to F_4-F_5-F_6 generations and retested for insect and disease resistance as well as evaluated for yield potential. The selected lines were then used in another cycle of crosses. Cowpea being a short duration crop, we are able to grow 4 generations each year and thus, within 2 years one cycle of crossing and selection is completed (Fig. 3). Among the newest bruchid resistant lines, IT84S-2246-4 combines resistance to 10 diseases along with resistance to aphid, bruchid and thrips. Its pedigree and sequence of selections is presented in Fig. 3. This variety along with others has been tested in International Trials from 1986 onwards. Its performance is as good or better than best varieties at most of the locations (Table 2). IT84S-2246-4 is a single F_6 plant selected from the cross between IT81D-1020 and IT82D-716 which was made in 1982. It matures in 65 days and combines moderate to high level of resistance to 10 diseases and three insects - aphid, bruchid, and thrips. It has brown rough medium size seeds that are preferred in West Africa. It contains about 30% protein and cooks between 35 - 40 minutes compared to 25% protein and 60 minutes cooking for most of the varieties.

Table 2. Performance of bruchid resistant lines in international trials in 1986/87.

Line	Number of Locations	Yield Range kg/ha		Overall Mean Yield kg/ha	Resistance*
		Lowest	Highest		
IT84S-2246-4	42	394	3208	1200	BR + AR + TR + MDR
IT84D-460	42	405	3232	1150	BR + MDR
IT82D-716	42	371	2899	1100	TR + BR + MDR
IT81D-1137	42	250	2656	1100	BR + MDR
Best check variety	42	270	2984	1150	MDR

* BR = Bruchid resistant; AR = Aphid resistant; TR = Thrips resistant; MDR = Multiple disease resistant

Table 3. Bruchid emergence pattern in IT84S-2246-4 compared to TVu 2027 and Ife Brown.

Varieties	% Adult emergence 50 DAI	Emergence pattern (no. of adults/day after infestation)																				
		23	24	25	26	27	28	29	30	31	32	33	34	35	36	37	38	39	40	41	42	43
IT84S-2246-4	16.2	0	0	0	2	2	0	1	0	1	0	0	0	0	0	0	0	0	0	0	0	0
TVu 2027 (RC)	25.0	0	1	1	1	2	1	1	0	0	0	1	0	0	1	0	0	0	0	0	0	0
Ife Brown (SC)	80.0	6	12	6	4	0	0	0	0	0	0	0	0	0	0	0	0	0	0	0	0	0

RC = Resistant check; SC = Susceptible check.

This variety is becoming very popular in Nigeria. It is also being used as one of the parents in crossing programs as a source of multiple disease and insect resistance.

The level of bruchid resistance in IT84S-2246-4 is similar to TVu 2027. It is not immune but suffers very little damage compared to susceptible varieties (Table 3).

7.   Search for Better Sources of Resistance

As indicated above, the level of bruchid resistance in TVu 2027 which has been used in breeding programs so far is not very high and therefore search for additional and better sources of resistance is being continued at IITA. Several bruchid resistant lines have been selected from the cultivated *Vigna* but they all are similar in resistance to TVu 2027 and have the same genes for resistance as crosses among them do not segregate for susceptibility. Efforts are now being made to screen wild species of *Vigna*. About 160 accessions of different wild *Vigna* have been screened for bruchid resistance so far and the preliminary results are quite encouraging. About 48 accessions have as good or higher levels of resistance than TVu 2027. These will be further tested and their progenies evaluated for stability of resistance over generations. Ofuya (1987) also screened a number of *Vigna* species for their reaction to *Callosobruchus maculatus*. He observed *V. luteola* and *V. adenantha* to be immune and *V. oblongifolia* and *V. racemosa* to be moderately resistant. Thus, there is a possibility to breed for higher levels of resistance to bruchids. However, it may be pointed out that most of these *Vigna* species are difficult to cross with cultivated *Vigna* at present.

8. Germplasm Collection and Maintenance of Bruchid Resistant Varieties

The genetic studies have revealed that 2 recessive gene pairs (rcm1 rcm1 rcm2 rcm2) must be present in the homozygous condition for manifestation of bruchid resistance. Therefore, any outcross plant involving a susceptible variety will be susceptible and seeds produced on this would segregate in a ratio of 15 susceptible: 1 resistant. Even though cowpea is a self pollinated crop, the extent of outcrossing ranges from 1 to 5% depending upon the pollinating insect population. Thus, most of the land races are genetically heterogeneous and a bulk sample consisting of susceptible and resistant plants will appear susceptible if the proportion of resistant plants is less. In fact this seems to be the case. IITA maintains a germplasm collection of over 14000 and only 1 line, TVu 2027 was initially found to be resistant to bruchid. However, recently several individual plant selections from farmers' fields in northern Nigeria have been found resistant, whereas the bulk samples from the same fields are susceptible. The gene for resistance in these lines is the same as in TVu 2027 which also comes from northern Nigeria. Apparently, in looking for additional sources of bruchid resistance, seed from individual plants should be tested rather than bulk field samples. Also, to maintain the bruchid resistance in an improved variety, breeders' seeds should be produced using single plant progenies and constantly testing them for bruchid resistance.

9. Bruchid Resistance in Other Grain Legumes

Moderate to high level of resistance to bruchids have been identified in chickpea, beans, green gram and its related species *Vigna sublobata*. Saxena and Raina (1970) reported G-109-1 strain of chickpea to be resistant to *Callosobruchus chinensis*. Subsequently Raina (1971) observed that this line was also resistant to *C. analis* and *C. maculatus*. The resistance was due to rough seed coat which acted as a deterrent to oviposition. The mean number of eggs on G-109 ranged from 0.7 to 3.0 for *C. analis*; 0 to 0.3 for *C. maculatus* and 1.3 to 4.7 for *C. chinensis* compared to 28 - 73 on other varieties. Schalk (1973) also observed 3 chickpea varieties with a rough seed coat which were less damaged by *C. maculatus*. No further work has been published on the use of this line in breeding programs. Schoonhaven et al. (1983) identified high levels of resistance to bean weevil (*Acanthoscelides obtectus* (Say)) and Mexican bean weevil, *Zabrotes subfasciatus* (Boheman)) in non cultivated wild forms of beans (*Phaseolus vulgaris* (L.). The number of Mexican bean weevil adults that emerged in the most resistant line ranged from 16 to 31 compared to 247 to 349 in the susceptible check. In the case of bean weevil, the number of adults ranged from 0 to 5.6 in resistant lines compared to 19.8 to 44.6 in the susceptible check. Osborn et al. (1988) reported that 'arcelin', a major seed protein in the wild bean, has a toxic effect on the bean bruchid, *Zabrotes subfasciatus*. Transfer of an 'arcelin' allele to bean cultivars results in resistance to bean bruchid.

Khattack et al. (1987) observed varietal differences in green gram with respect to damage by *Callosobruchus maculatus*. Some varieties suffered only 14 to 17% damage compared to 46% damage for others. However, the level of resistance was not very high. Fujii and Miyazaki (1987) and Fujii et al. (1989) identified resistance to *Callosobruchus chinensis* in TC 1966 strain of *Vigna sublobata*, a wild relative of green gram which is considered to be the direct progenitor of *Vigna radiata*. The level of resistance was very high. The newly hatched larvae are unable to penetrate into the resistant seeds. Kitamura et al.(1988) elucidated the genetics of resistance to *C. chinensis*. Their studies indicated a monogenic dominant inheritance. This gene is being incorporated into cultivated green gram varieties.

This review of various studies conducted on bruchid resistance indicates a good deal of genetic variability in several grain legumes. The causes of resistance may be physical and/or chemical but the inheritance seems to be moderately simple. Therefore, with concerted and sustained efforts it would be possible to develop improved varieties of most of the grain legumes with moderate to high levels of bruchid resistance.

REFERENCES

Adjadi, O., Singh, B. B., and Singh, S. R. (1985) Inheritance of bruchid resistance in cowpea, Crop Sci. 25, 740-742.

Baker, T. A., Nielsen, S. S., Shade, R. E.,and Singh, B. B. (1989) Physical and chemical attributes of cowpea lines resistant and susceptible to *Callosobruchus maculatus* (F.) (Coleoptera: Bruchidae, J. stored Prod. Res. 25, 1-8.

Birch, A. N. E., Fellows, Linda E., Evans, S. V., and Doharty, Katherine (1986) Para-amino phenylalanine in *Vigna*: Possible taxonomic and ecological significance as a seed defense against bruchids, Phytochemistry 25, 2745-2749.

Booker, R. H. (1967) Observation on three bruchids associated with cowpea in northern Nigeria, J. stored Prod. Res. 3, 1-15.

Caswell, G. W. (1961) The infestation of cowpeas in the Western region of Nigeria, Tropical Sci. 3, 154-158.

Caswell, G. H. (1973) The impact of infestation on commodities, Trop. Stored Prod. Inf. 25, 19.

Dick, K. M. and Credland, P. F. (1986) Changes in the response of *Callosobruchus maculatus* (Coleoptera: Bruchidae) to a resistant variety of cowpea, J. Stored Prod. Res. 22, 227-233.

Dobie, P. (1981) The use of resistant varieties of cowpea (*Vigna unguiculata*) to reduce losses due to post-harvest attack by *Callosobruchus maculatus*, in V. Labeyrie (Ed.), The Ecology of Bruchids Attacking Legumes (pulses), Dr. W. Junk Publishers, pp. 185-192.

Fatunla, T. and Badaru, K. (1983) Resistance of cowpea pods to *Callosobruchus maculatus* (F), J. Agric. Sci. (U.K.) 110, 205-209.

Fujii, K. and Miyazaki, S. (1987) Infestation resistance of wild legume (*Vigna sublobata*) to azuki bean weevil *Callosobruchus chinensis* (L.) (Coleoptera: Bruchidae) and its relationship with cytogenetic classification, Appl. Entomol. Zool. 22, 229-230.

Fujii, K., Ishimoto, M., and Kitamura, K. (1988) Patterns of resistance to bean weevils (Bruchidae) in *Vigna radiata-mungo-sublobata* complex inform the breeding of new resistant varieties, Appl. Entomol. Zool. 24, 126-132.

Gatehouse, A. M. R., Gatehouse, J. A., Dobie, P., Kilminster, A. M., and Boulter, D. (1979) Biochemical basis of insect resistance in *Vigna unguiculata*, J. Sci. Food Agric. 30, 948-953.

228

Ishimoto, M., and K. Kitamura (1988) Identification of growth inhibitor on Azuki Bean Weevil in Kidney Bean (*Phaseolus vulgaris* L.), Jpn. J. Bred. 38, 367-370.

Khattak, S. U. K., Hamid, M., Khatoon, R., and Mohammad, Tila (1987) Relative susceptibility of different mungbean varieties to *Callosobruchus maculatus* (F.), J. stored Prod. Res. 23, 139-142.

Kitamura, K., Ishimoto, M., and Sawa, M. (1988) Inheritance of resistance to infestation with azuki bean weevil in *Vigna sublobata* and successful incorporation to *V. radiata*, Jpn. J. Breed. 38, 459-464.

Ndlovu, T. M. and Giga, D. P. (1988) Studies on varietal resistance of cowpeas to the cowpea weevil, *Callosobruchus rhodesianus* (PIC), Insect Sci. Applic. 9, 123-128.

Ofuya, T. I. (1987) Susceptibility of some *Vigna* species to infestation and damage by *Callosobruchus maculatus* (F.), J. stored Prod. Res. 23, 137-138.

Osborn, T. C., Alexander, D. C., Sun, S. S. M., Cardona, C., and Bliss, F. A. (1988) Insecticidal activity and lectin homology of arcelin seed protein, Science 240, 207-210.

Prevett, P. F. (1961) Field infestation of cowpea (*Vigna unguiculata*) pods by beetles of the families Bruchidae and Curculionidae in Northern Nigeria, Bull. Entomol. Res. 52, 635-646.

Raina, A. K. (1971) Comparative resistance to three species of *Callosobruchus* in a strain of chick pea (*Cicer arietinum* (L.)), J. stored Prod. Res. 7, 213-216.

Saxena, H. P. and Raina, A. K. (1970) A bruchid strain of Bengal gram, Current Sci. 39, 189-190.

Schalk, J. M. (1973) Chickpea resistance to C. maculatus in Iran, J. Econ. Entomol. 66, 528-579.

Schoonhoven, A. V., Cardona, C., and Valor, J. (1983) Resistance to the bean weevil and Mexican bean weevil (Coleoptera: Bruchidae) in non cultivated common bean accessions, J. Econ. Entomol. 76, 1255-1259.

Singh, S. R. (1977) Cowpea cultivars resistant to insect pests in world germplasm collection, Trop. Grain Legume Bull., 9, 3-7.

Singh, S. R. and Jackai, L. E. N. (1985) Insect pests of cowpeas in Africa: Their life cycle, economic importance and potential for control, in S. R. Singh and K. O. Rachie (eds.), Cowpea Research, Production and Utilization, Chichester, John Wiley & Sons, England, 459pp.

Singh, B. B., Singh, S. R., and Adjadi, O. (1985) Bruchid of resistance in cowpea, Crop Sci. 25, 736-739.

Williams, J. O. (1980) Note on Bruchidae associated with stored products in Nigeria, Trop. Grain Legume Bull. 21, 5-10.

BRUCHID RESISTANCE FACTORS IN *PHASEOLUS* AND *VIGNA* LEGUMES

KEISUKE KITAMURA, MASAO ISHIMOTO, AND SHOZIRO ISHII[1]
Legume Breeding Laboratory, National Agriculture
Research Center, Tsukuba, Ibaraki 305 Japan

[1]*Emeritus Professor of Kyoto University*

ABSTRACT. An investigation was undertaken to characterize bruchid resistance factors present in kidney bean, *Phaseolus vulgaris* and the wild mung bean, *Vigna sublobata*. Kidney beans contain water-soluble substances which strongly inhibit the larval growth of the Azuki bean weevil, *Callosobruchus chinensis*. We have purified and identified a proteineous α-amylase inhibitor as a major inhibitory substance from the seeds. *C. chinensis* larvae could not develop by feeding on artificial beans containing 0.2-0.5% of the purified inhibitor, and died before the second instar. Contrary to other reports, a lectin preparation from kidney beans hardly inhibited the larval growth at this concentration. The α-amylase inhibitor was also just as lethal for *C. maculatus* larvae. The inhibitor markedly suppressed α-amylase enzymatic activities of the larval midgut homogenates of both *C. chinensis* and *C. maculatus* larvae. On the other hand, larvae of *Zabrotes subfasciatus*, a well known storage pest of kidney beans, fed and developed well on artificial beans containing 0.5-2.0% of the purified inhibitor. The larval midgut amylase activity of this weevil species was not appreciably affected by the inhibitor. The results suggest that the resistance of kidney beans to the three weevil species is closely related to the inhibition of the larval-amylase activity by the α-amylase inhibitor.

One strain (TC 1966) of *V. sublobata*, the direct progenitor of mung bean, *V. radiata*, has complete resistance to infestation by *C. chinensis*. By a genetic study of the progeny seeds from the crosses between two mung bean cultivars (susceptible) and TC 1966, the weevil resistance of TC 1966 was shown to be controlled by a single dominant gene designated tentatively as *R*. The TC 1966 strain has very poor seed production; seeds from the productive BC_3F_3 lines whose genotypes are homozygous dominant (*R/R*) and are morphologically similar to the recurrent parent Osaka-ryokutou were therefore used for fractionation of the resistance factor (substance) in TC 1966. The substance showed water-soluble, high molecular weight-, and heat- and protease- stable characteristics. These characteristics suggest that it may not be a protein but a polysaccharide.

1. Introduction

Bruchids are the most serious pests attacking food legume seeds during storage. Leguminous plants, especially wild legumes, have evolved to produce antibruchid chemicals such as alkaloids (Janzen et al., 1977; Evans et al., 1985; Nash et al., 1986), non-protein amino acids (Rosenthal et al., 1976; Janzen et al., 1977) and saponins (Ishii, 1952; Applebaum et al., 1969) in their seeds, all of which have been shown to be detrimental to the larval growth of bruchids. However, these compounds are not actually responsible for the pest resistance in the food legume seeds, since these compounds have been consciously or unconsciously reduced or eliminated from the seeds by breeding selection because of their toxicity and tastelessness to humans and animals. So far,

K. Fujii et al. (eds.), Bruchids and Legumes: Economics, Ecology and Coevolution, 229–239.

proteins such as lectin (Janzen et al., 1976; Gatehouse et al., 1984), trypsin inhibitors (Gatehouse et al., 1983), amylase inhibitor (Ishimoto and Kitamura, 1988) and arcelin (Osborn et al., 1988), and high molecular weight heteropolysaccharides (Applebaum et al., 1970; Gatehouse et al., 1987) have been reported as bruchid resistance factors in food legumes.

In the present investigation, a search was made for bruchid resistance factors present in Kidney bean and in the TC 1966 strain of *Vigna sublobata*. Ishii (1952) reported that kidney beans contain water-soluble substances which strongly inhibit the larval growth of the Azuki bean weevil, *Callosobruchus chinensis*. Ishimoto and Kitamura (1988) have purified and identified a proteinous α-amylase inhibitor as one of the major inhibitory substances. The α-amylase inhibitor at levels of 0.2-0.5% was highly toxic to the larvae. The inhibitor was also equally lethal to the cowpea weevil, *C. maculatus*. On the contrary, the two species of weevils, *Acanthoscelides obtectus* and *Zabrotes subfasciatus* infest not only azuki beans but also kidney beans. Effects of the α-amylase inhibitor on the the larval growth of *Z. subfasciatus* was studied in comparison with that of *C. chinensis* and *C. maculatus*. The α-amylase activities of larval midgut homogenates of the latter two species were strongly inhibited by the inhibitor, while that of *Z. subfasciatus* was hardly affected. It is thus likely that the α-amylase inhibitor in kidney beans is responsible for the protection against attack by the azuki and cowpea weevils.

Recently, one strain (TC 1966) of *Vigna sublobata*, the direct progenitor of mung bean, *V. radiata*, has been shown to confer complete resistance against the azuki bean weevil (Fujii and Miyazaki, 1987; Kitamura et al., 1988): most larvae dug their head into the beans of TC 1966 and the rest of the body was still in the egg shell as observed in the case of *Phaseolus vulgaris* (e.g., Ishii, 1952) and *V. mungo* (Singh, 1976; Sawa and Tan, 1976). We report here the results of a genetic analysis of the weevil resistance in TC 1966, and investigate the possibility of incorporating the resistance to mung bean and the substance(s) responsible for the resistance.

2. Materials and Methods

2.1. BEANS AND INSECTS

Seeds of kidney bean (cv. Taishou Kintoki) and azuki bean (cv. Dainagon) were purchased from a local market. The F_2 seeds were obtained from the selfed F_1 plants between two susceptible mungbean cultivars, Osaka-ryokutou and No.3-ryokutou, and the resistant TC 1966 in the green house. The F_3 seeds from each F_2 plant derived from the crosses Osaka-ryokutou X TC 1966 and No.3-ryokutou X TC 1966 were harvested in the field. Three backcrosses were made to Osaka-ryokutou with selection for the azuki bean weevil resistance. Since *C. maculatus* and *Z. subfasciatus* are not native to Japan, these weevil species were reared on azuki beans at 30°C, about 70% R.H. under a special plant Quarantine licence.

2.2. PURIFICATION OF THE PROTEINEOUS α-AMYLASE INHIBITOR

Kidney bean meal was extracted with 0.02M sodium phosphate buffer, pH 6.7 solution (PBS) at 4°C. The fraction precipitating with 20-60% ammonium sulfate saturation was dissolved and dialyzed against PBS. After dialysis, the crude inhibitor preparation was heated at 70°C for 15 min and then centrifuged. The supernatant was applied to a DEAE-Sephacel column equilibrated with PBS. The adsorbed fraction was eluted with 0.25 M sodium phosphate buffer, pH 6.7 and then applied to a Con A-Sepharose column. The Con A-Sepharose-adsorbed fraction was applied again to a DEAE-Sepha-cel column. The adsorbed proteins were recovered by an elution gradient (0.02-0.25M) of sodium phosphate buffer, pH 6.7. Porcine pancreatic α-amylase (Sigma Type I-a) was used for the detection of the α-amylase inhibitor. The hemagglutination (lectin) activity was tested by serial dilution of rabbit erythrocytes.

2.3. FEEDING TEST

The effect of the α-amylase inhibitor on insect development was examined in feeding traials using artificial beans made of azuki bean meal containing various concentrations of the inhibitor. Each artificial bean was made into a columnar shape of 0.5g (8 mm in diameter) by pressing with a hand compressor, and covered with a collodion film. The artificial beans were placed in a plastic dish into which the adults of the weevil were introduced for oviposition. After 24 hr, the adults were removed from the dishes which were kept in a chamber at 30°C and about 70% R.H. Seven days after the initial oviposition the number of eggs hatched on the surface of the artificial beans was counted. After 30 days, in the cases of *C. chinensis* and *C. maculatus*, and 40 days in the case of *Z. subfasciatus*, the artificial beans were dissected and the number of adults, living larvae and pupae were recorded. Each of the treatments was carried out in at least 4 replications.

2.4. TESTS FOR RESISTANCE

For the tests of the parental and progeny seeds, four to fifty seeds were placed in a plastic dish into which adults of the azuki bean weevil were introduced for oviposition. After 15 hr, the adults were removed from the dishes which were kept in a chamber at 30°C and 70% R.H. After incubation for 30 days, the number of susceptible and resistant seeds was counted. The seeds showing exit holes for adult emergence as well as those easily crushed with the tip of a finger were designated as susceptible seeds, and those showing no damage were designated as resistant seeds. Segregation for the resistance in the F_2 population was examined by testing bulked F_3 seeds obtained from individual F_2 plants.

2.5. PREPARATION OF α-AMYLASE

The midguts were dissected from late forth (last) instar larvae of the three weevil species. They were homogenized with PBS containing 20 mM NaCl and 0.1 mM $CaCl_2$ (100 μl/midgut), and centrifuged at 10,000g for 20 min at 4°C. The supernatant was used as a larval α-amylase preparation.

2.6. ASSAY FOR α-AMYLASE AND α-AMYLASE INHIBITOR ACTIVITIES

The activity of the crude larval amylase was measured using a modification of the Bernfeld method (Bernfeld, 1955). Namely, the amylase preparation (250 μl) was incubated with 250 μl of 1% soluble starch in 0.1 M sodium phosphate buffer, pH 5.5 at 30°C. After 5 min, the reaction was stopped by the addition of 500 μl of 3,5-dinitrosalicylic acid reagent and heated in boiling water for 10 min. Five ml of water was then added to the solution, mixed and left to stand for 15 min. The absorbance of the solution was read at 546 nm, and the α-amylase activity was expressed in μg maltose liberated/min.

The effect of the α-amylase inhibitor on the larval α-amylase preparations was determined by pre-incubating 250 μl of the enzyme preparations with varying amounts

232

of the inhibitor in 250 μl of PBS at 30°C for 15 min before the addition of the starch solution. The activity of the larval midgut α-amylase preparations from the three weevil species was adjusted by diluting with PBS to liberate 70 μg maltose/min in 250 μl.

2.7. FRACTIONATION OF THE WEEVIL RESISTANCE FACTOR IN TC 1966

Because the wild mung bean, TC 1966 has very poor seed production, resistant seeds obtained from productive BC_3F_3 lines derived from BC_3F_2 were therefore used for fractionation of the resistance factor (substance) in TC 1966; these seeds are resistant and morphologically similar to the recurrent parent Osaka-ryokutou. The seed meal was extracted with 20-fold (v/w) of PBS at 4°C. The inhibitory effects of the fractions on the larval growth of the azuki bean weevil were examined using the artificial beans incorporating the fractionated lyophilized materials at various concentrations. Thirty days after initial oviposition the artificial beans were dissected and examined.

3. Results

3.1. PURIFICATION OF THE α-AMYLASE INHIBITOR IN KIDNEY BEAN

The typical elution profile of DEAE-Sephacel chromatography of the Con A-Sepharose-adsorbed fraction is shown in Fig. 1. The first peak contained hemagglutination (lectin) activity, as tested by serial dilution of rabbit erythrocytes, and the second peak had inhibitory activity against the porcine pancreatic α-amylase. It was found that the

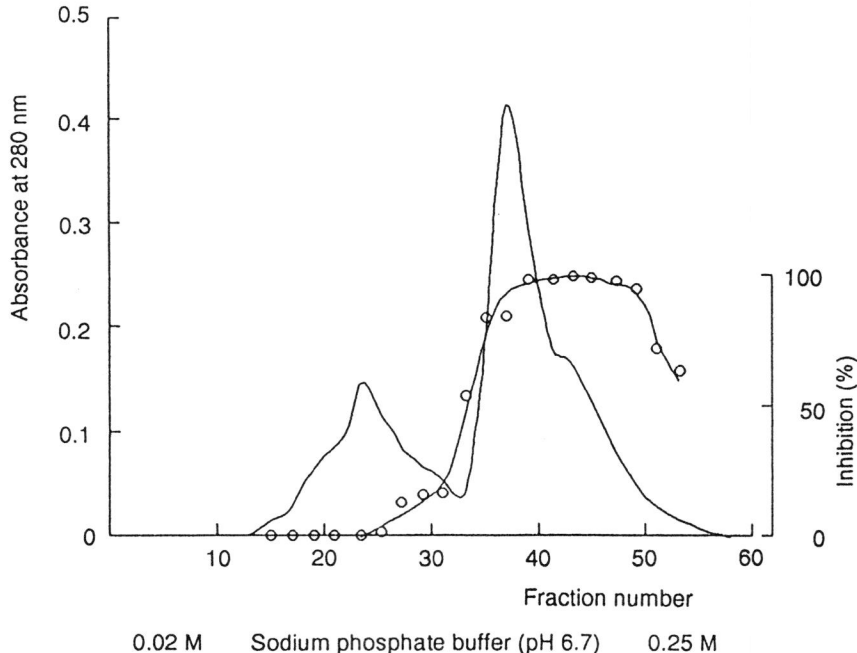

Figure 1. DEAE-Sephacel chromatography of the Con A-Sepharose-adsorbed fraction of the kidney bean. o —— o : inhibitory activities against α-amylase from porcine pancreas.

second is much more toxic to the weevil larvae than the first. The second peak fraction was rechromatographed on the DEAE-Sephacel column. The single major peak containing the α-amylase inhibitory activity was recovered and used as the purified kidney bean α-amylase inhibitor. The purified inhibitor gave a single protein band on a native polyacrylamide electrophoresis (PAGE) gel, and one major and several minor bands with molecular weights lower than 17,000 on a sodium dodecyl sulfate (SDS)-PAGE gel (Fig. 2). No hemmagglutination activity was detected in the purified α-amylase inhibitor.

3.2. EFFECTS OF α-AMYLASE INHIBITOR ON INSECT DEVELOPMENT *IN VIVO*

For the feeding test artificial beans were prepared by adding the purified inhibitor into the azuki bean meal at various concentrations ranging from 0.1 to 2.0 (w/w)%. Such a range of inhibitor concentrations were selected because the α-amylase inhibitor is present at levels of 0.4-0.5% in kidney beans seeds (Marshall and Lauda, 1975). The result of the feeding test is expressed as the total number of adults, living pupae and larvae per gram of artificial beans (Fig. 3).

C. *chinensis* females laid the largest number of eggs among the three weevil species on the surface of the artificial beans. The average number of eggs hatched was 36.9 (S.D. = 3.5) per gram of the artificial beans. One gram of artificial beans made of the azuki bean meal produced, on the average, 12.3 adults. The addition of the inhibitor to the azuki bean meal resulted in a decreased number of total adults, surviving pupae and larvae. No adult weevil emerged from the artificial beans containing the inhibitor at a concentration of 0.2% within 30 days. In the artificial beans containing the inhibitor at concentrations higher than 0.5%, all the larvae were unable to develop and died at the first instar.

In *C. maculatus* the average number of eggs hatched was 16.7 (S.D = 3.5) per gram of the artificial beans. One adult weevil emerged from 1g of artificial beans containing the inhibitor at 0.2%. However, the addition of the inhibitor at levels higher than 0.5% resulted in the larval death at the first instar, as in *C. chinensis*.

In *Z. subfasciatus* the average number of eggs hatched was 31.9 (S.D. = 5.2) per gram of the artificial beans. Adults emerged from all of the treatments within 40 days. The addition of the inhibitor did not appreciably affect the development of *Z. subfasciatus* unlike the cases of *C. chinensis* and *C. maculatus*. Adults emerged even from the artificial beans containing the inhibitor at the concentration of 2.0%.

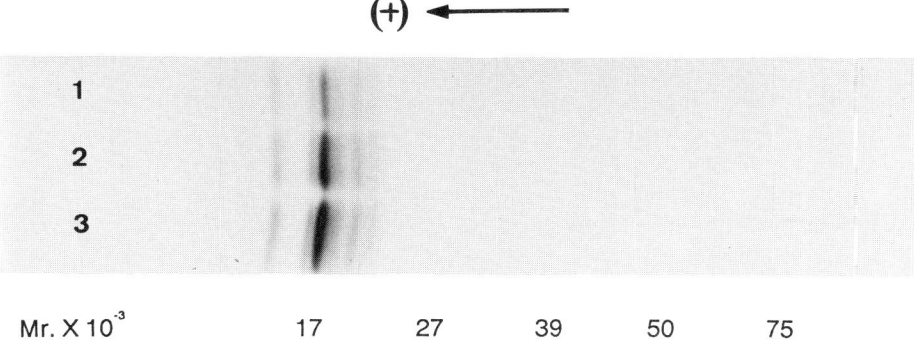

Figure 2. SDS-PAGE analysis of the α-amylase inhibitor in the kidney bean on 13.5% gel slabs. Lanes 1, 2, and 3 are 30, 50, and 70 μg of the purified inhibitor, respectively.

234

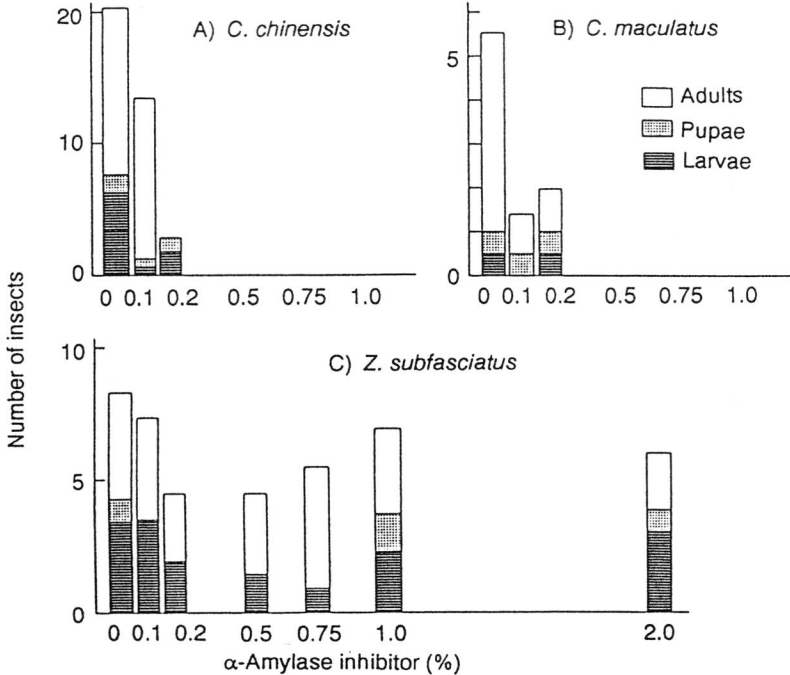

Figure 3. Effects of the α-amylase inhibitor on the development of the three weevil species.

3.3. INHIBITORY ACTIVITY OF THE α-AMYLASE INHIBITOR ON LARVAL AMYLASE PREPARATIONS *IN VITRO*

The supernatant obtained by homogenization of larval midguts dissected from the 4 th instar larvae of each weevil species was used as an enzyme preparation. The enzyme preparation exhibited an α-amylase activity ranging from 40 to 70 μg maltose liberated /min/midgut under the given conditions.

The larval midgut α-amylase activity in the crude enzyme preparations of both *C. chinensis* and *C. maculatus* which liberate 70 μg maltose/min almost completely disappeared when preincubated with 3 to 5 μg of the inhibitor, while the α-amylase activity of *Z. subfasciatus* was not significantly affected by preincubation even with 100 μg of the inhibitor (Fig. 4).

3.4. INHERITANCE OF RESISTANCE AGAINST *C. CHINENSIS* IN *VIGNA SUBLOBATA*

All the F_1 seeds tested, five from the cross Osaka-ryokutou X TC 1966 and four from the cross No.3-ryokutou X TC 1966, were resistant. A total of 301 F_2 seeds from both crosses were tested. Segregation for the weevil resistance was observed in the F_2 seeds. The genetic segregation of the F_2 seeds gave a close fit to 3 resistant : 1 susceptible ratio (Table 1), suggesting that the resistance is dominant over the susceptibility to the weevil.

The mode of inheritance of the resistance was further investigated by testing separately at least 40 F_3 seeds obtained from individual F_2 plants. The segregation patterns in the two crosses are shown in Table 2 and 3. Three phenotypes were observed in the

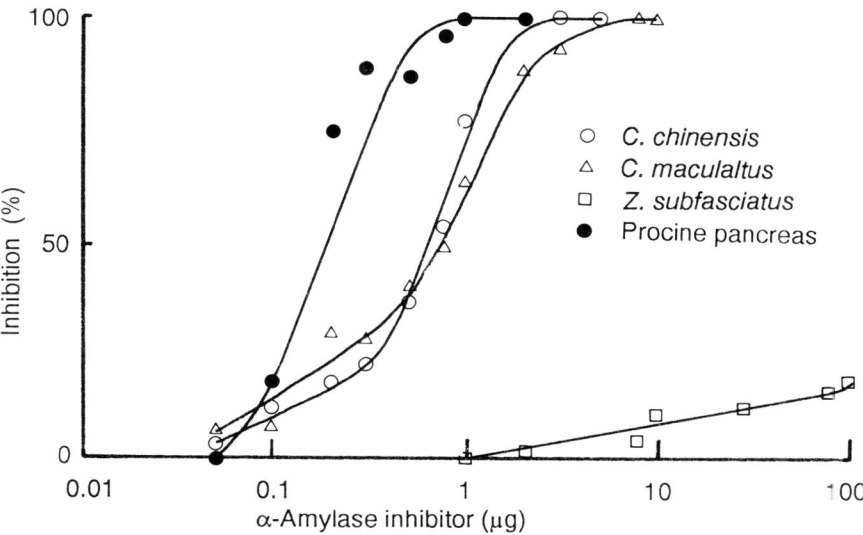

Figure 4. Effects of the α-amylase inhibitor on larval amylase preparations of the three weevil species, and porcine pancreatic amylase. Inhibition shows relative values against each amylase activity without the pre-incubation with the inhibitor.

Table 1. Segregation for the weevil resistance in the F_2 seeds from the crosses Osaka-ryokutou X TC 1966 and No. 3-ryokutou X TC 1966.

Crosses	No. of F_2 seeds subjected to test	Frequency of seed Resistant/Susceptible		χ^2	Probability
Osaka-ryokutou X TC 1966	137	100	37	0.295	0.5 - 0.7
No. 3-ryokutou X TC 1966	164	116	47	1.59	0.2 - 0.3
Total	301	216	85	1.68	0.1 - 0.2

Table 2. Segregation for the weevil resistance in the F_2 population from the cross Osaka-ryokutou X TC 1966 (based on F_3 seeds harvested in 1986).

F_2 population		F_3 seeds	
Genotypes estimated	No. of plants subjected to test	Resistant	Susceptible
R/R	8 [1]	338	5
R/r	27 [1]	831 [2]	285 [2]
r/r	16 [1]	0	675

[1] χ^2 for 1 R/R : 2 R/r : 1 r/r = 2.69 $0.2 < P < 0.3$
[2] χ^2 for 3 $R/-$: 1 r/r from R/r = 0.172 $0.5 < P < 0.7$

Table 3. Segregation for the weevil resistance in the F_2 population from the cross No.3-ryokutou X TC 1966 (based on F_3 seeds harvested in 1986)

F_2 population		F_3 seeds	
Genotypes estimated	No. of plants subjected to test	Resistant	Susceptible
R/R	20 [1]	369	11
R/r	31 [1]	1,006 [2]	352 [2]
r/r	16 [1]	7	693

[1] χ^2 for 1 R/R : 2 R/r : 1 r/r = 0.85 $0.5 < P < 0.7$
[2] χ^2 for 3 $R/-$: 1 r/r from R/r = 0.613 $0.4 < P < 0.5$

F_2 populations: the resistant plants which produced nearly all resistant F_3 seeds, the susceptible plants which produced all or nearly all susceptible F_3 seeds, and the intermediate plants whose F_3 seeds segregated according to the 3 resistant : 1 susceptible ratio. When the F_2 populations in Table 2 and 3 were pooled, there were 28 resistant, 32 susceptible and 58 intermediate plants which gave a close fit to 1 resistant (R/R) : 1 susceptible (r/r) : 2 intermediate (R/r) ratio $(\chi^2 = 0.24)$. These results suggested that the weevil resistance of TC 1966 is controlled by a single dominant gene designated as R.

3.5. FRACTIONATION OF THE GROWTH INHIBITORY SUBSTANCE IN TC 1966

The inhibitory activity of resistant BC_2F_4 seeds derived from TC 1966 was recovered in the PBS extract. The growth inhibitory substance did not pass through a dialyzing cellulose tube, and was concentrated by precipitation with 20-40% ammonium sulfate saturation. The precipitate was dialyzed against PBS. The dialysate was applied to a DEAE-Sephacel column equilibrated with PBS. The inhibitory activity was recovered in the non-adsorbed fraction. The non-adsorbed fraction was dialyzed against water and lyophilized. Larval growth was completely inhibited on the artificial beans containing 5% of the lyophilized material.

The inhibitory activity of the material was little affected after treatments of either heating at 70°C for 15 min, or by incubating with 1/50 (w/w) protease (Sigma protease from bovine pancreas Type I). Judging from these characteristics, it is thought that the inhibitory substance must be a type of polysaccharide.

4. Discussion

4.1. EFFECTS OF THE KIDNEY BEAN α-AMYLASE INHIBITOR ON BRUCHID GROWTH AND DEVELOPMENT *IN VIVO* AND ON LARVAL AMYLASE PREPARATIONS *IN VITRO*

The feeding tests showed that the kidney bean α-amylase inhibitor strongly inhibited the larval growth of both *C. chinensis* and *C. maculatus* which can not be reared on kidney bean seeds. Janzen et al. (1976) suggested that the lectin in kidney beans was responsible for protection from attack by *C. maculatus*, because the larvae could not grow in artificial beans containing lectins at 5%. However, the lectin content in kidney bean is less than 3% (Osborn et al., 1985), and the lectin preparations from kidney beans at the levels of 3 to 5% scarcely inhibits the larval growth of *C. chinensis* (Ishimoto and Kitamura, 1988). In contrast, the present study revealed that the α-amylase inhibitor is sufficiently effective to completely inhibit the larval growth of *C. chinensis* and *C. maculatus* at a concentration of 0.4-0.5%, at which it exists in kidney beans seeds (Marshall and Lauda, 1975). Thus, it is likely that the inhibitory activity of the α-amylase inhibitor in kidney beans on the development of *C. chinensis* and *C. maculatus* is much more effective than that of lectin. The mechanism underlying the growth inhibition seems likely to be ascribed to the direct inhibition of starch digestion by the inhibitor causing a large reduction in carbohydrate assimilation in the larvae.

On the other hand, the larval midgut α-amylase activity from *Z. subfasciatus* was not suppressed by the α-amylase inhibitor, and the larvae could grow well on artificial beans containing the inhibitor at a concentration of 2.0%, a level which was sufficient to cause the death of the larvae of *C. chinensis* and *C. maculatus* at the first instar. Silano et al. (1975) suggested that insects feeding on wheat grains may become adapted to α-amylase inhibitors contained in the grains due to the presence of a high α-amylase activity in the midgut. However, this is not the case for the ineffectiveness of the α-amylase inhibitor on the larval α-amylase from *Z. subfasciatus*, because the α-amylase activity of the larval midgut homogenate in *Z. subfasciatus* was not significantly higher than that of the other two weevils. The results indicate that the *Z. subfasciatus* larvae may be able to detoxify the kidney bean α-amylase inhibitor in the midgut and/or the larvae may have α-amylase enzyme(s) unaffected by the inhibitor.

4.2. THE BRUCHID RESISTANCE FACTOR(S) IN ONE STRAIN (TC 1966) OF *V. SUBLOBATA*

The genetic study suggested that the azuki bean weevil resistance of TC 1966 is controlled by a single dominant gene, *R*. However, the occurrence of susceptible F_3 seeds and resistant F_3 seeds, from the F_2 plants whose genotypes were estimated as homozygous dominant (*R/R*) and homozygous recessive (*r/r*), respectively, can not be explained because the resistant (*R/R*) and susceptible (*r/r*) parents showed complete resistance and susceptibility to the weevil, respectively (Fujii and Miyazaki, 1987). The discrepancy can be explained as follows. First: modifier gene(s) in addtion to the *R-r* genes could be involved in the weevil resistance; second: outcrossing among the F_2 plants and/or seed contamination may have occurred when the F_3 seeds were harvested from individual F_2 plants; third: mutant individuals which had broken down the resistance of TC 1966 may have emerged in the azuki bean weevil populations. In order to determine the cause of the observed phenomenon, we have reexamined the inheritance of the weevil resistance of TC 1966. The preliminary results using the progeny seeds from the cross Osaka-ryokutou X a BC_2F_3 line harboring the resistance showed that the weevil resistance and susceptibility can be explained by the *R-r* alleles.

It may be easy to incorporate the resistance to mungbean cultivars because some resistant and productive progeny plants whose seeds are morphologically similar to Osaka-ryokutou were obtained in the BC_2F_3 populations (Fig. 5). Since the resistant lines are also completely resistant to *C. maculatus* which is a major bruchid species infesting mungbean during storage in many producing areas in the world (Fujii et al., 1989), it is very worthwhile to breed mungbean cultivars harboring the resistance.

Recently, Fujii et al. (1989) showed that the azuki bean weevil resistant lines derived from TC 1966 and TC 1966 itself are resistant to *C. maculatus*, *C. phaseoli* and *Z. subfasciatus*. We observed that in addition to the resistance against these weevil species the resistant lines strongly inhibit the growth of *Riptortus clavatus* (Coreidae: Hemiptera). It is likely that an identical substance whose production is controlled by a domi-

238

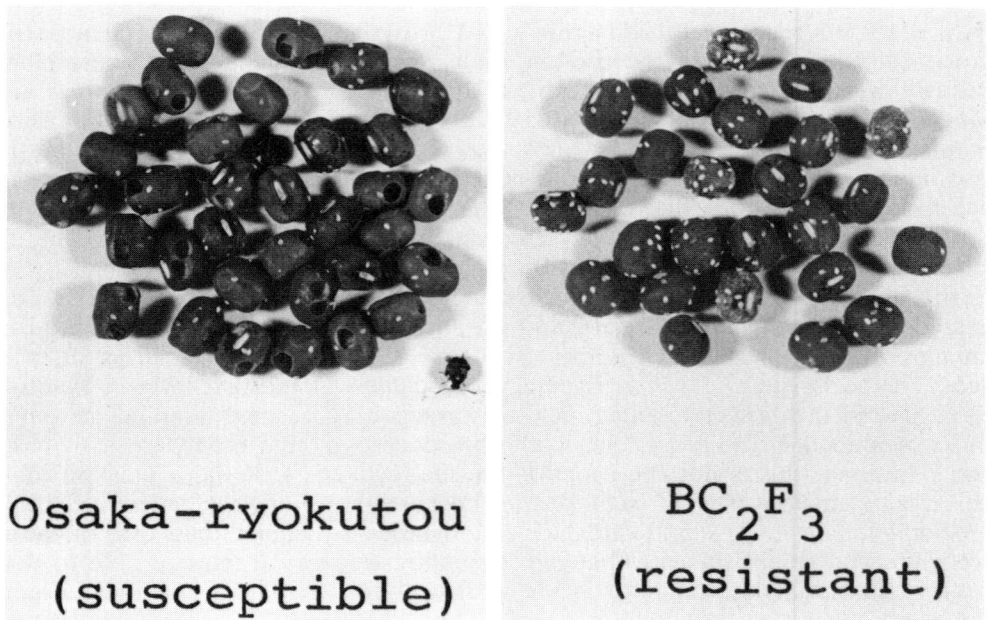

Osaka-ryokutou
(susceptible)

BC$_2$F$_3$
(resistant)

Figure 5.   Extent of seed damage by the azuki bean weevil in a susceptible mung bean cultivar (Osaka-ryokutou) and a resistant BC$_2$ line at 30 days after oviposition.

nant gene from TC 1966 may be responsible for both inhibitory activities to the weevil species and *R. clavatus*. It seems to be very interesting to identify the resistant substance of TC 1966 because such a substance having broad and potent inhibiting activity to insects has not been identified in food legume seeds, so far.

The inhibitory activity of TC 1966 was estimated to have been concentrated to 16 fold of the original activity in the seed meal, since the azuki bean weevil larvae died at the first instar on artificial beans containing 5% of the inhibitory substance concentrated from the resistant seeds derived from TC 1966 and on artificial beans incorporating 80% of the resistant seed meal. The substance has been shown to have water-soluble, high molecular weight-, and heat- and protease-stable characteristics. These characteristics suggest that it may not be a protein but a polysaccharide.

Acknowledgements.   We express our sincere thanks to Dr. K. Fujii, Professor of Tsukuba University, for his helpful advice and supplying the insects.

REFERENCES

Applebaum, S. W., Marco, S., and Birk, Y. (1969) Saponins as possible factors of resistance of legume seeds to the attack of insects, J. Agric. Food Chem. 17(3), 618-622.

Applebaum, S. W., Tadmor, U., and Podoler, H. (1970) The effect of starch and of a heteropoly-saccharide fraction from *Phaseolus vulgaris* on development and fecundity of *Callosobruchus chinensis*, Entomol. Exp. Appl. 13, 61-70.

Bernferd, P. (1955) Amylases, α- and β-, in S.P. Colowick and N. O. Kaplan (eds), Methods in Enzymology, Academic press, New York, pp. 149-158.

Evans, S. V., Gatehouse, A. M. R., and Fellows, L. A. (1985) Deterimental effects of 2,5-Dihy-droxymethyl-3,4-dihydroxy-pyrroidine in some tropical legume seeds on larvae of the bruchid

Callosobruchus maculatus, Entomol. Exp. Appl. 37, 257-261.

Fujii, K. and Miyazaki, S. (1987) Infestation resistance of wild legumes (*Vigna sublobata*) to azuki bean weevil, *Callosobruchus chinensis* (L.) and its relationship with cytogenetic classification, Appl. Entomol. Zool. 22, 229-230.

Fujii, K., Ishimoto, M., and Kitamura, K. (1989) Patterns of resistance to bean weevils (Bruchidae) in *Vigna radiata - sublobata* complex inform the breeding of new resistant varieties, Appl. Entomol. Zool. 24, 126-132.

Gatehouse, A. M. R. and Boulter, D. (1983) Assessment of the anti-metabolic effects of trypsin inhibitors from cowpea (*Vigna unguiculata*) and other legumes on development of the bruchid beetle (*Callosobruchus maculatus*), J. Sci. Food Agric. 34, 345-350.

Gatehouse, A. M. R., Dewey, F. M., Dove, J., Fenton, K. A., and Pusztai, A. (1984) Effect of seed lectins from *Phaseolus vulgaris* on the development of larvae of *Callosobruchus maculatus*; mechanizm of toxicity, J. Sci. Food Agric. 35, 373-380.

Gatehouse, A. M. R., Dobie P., Hodges, R. J., Meik, J., Pusztai, A., and Boulter, D. (1987) Role of carbohydrates in insect resistance in *Phaseolus vulgaris*, J. Insect Physiol. 33, 843-850.

Ishii, S. (1952) Studies on the host preference on the cowpea weevil (*Callosobruchus chinensis* L.), Bull. Nat. Inst. Agr. Sci. Japan Ser. C 1, 185-256.

Ishimoto, M. and Kitamura, K. (1988) Identification of the growth inhibitor on azuki bean weevil in kidney bean (*Phaseolus vulgaris* L.), Jpn. J. Breed. 38, 367-370.

Janzen, D. H., Juster, H. B., and Liener, I. E. (1976) Insecticidal action of the phytohemagglutinin in black beans on a bruchid beetle, Science 192, 795-796.

Janzen, D. H., Juster, H. B., and Bell, E. A. (1977) Toxicity of secondary compounds to the seed-eating larvae of the bruchid beetle *Callosobruchus maculatus*, Phytochemistry. 16, 223-227.

Kitamura, K., Ishimoto, M., and Sawa, M. (1988) Inheritance of resistance to infestation with azuki bean weevil in *Vigna sublobata* and successful incorporation to *V. radiata*, Jpn. J. Breed. 38, 459-464.

Marshall, J. J. and Lauda, C. (1975) Purification and properties of phaseolamin, an inhibitor of α-amylase, from the kidney bean, *Phaseolus vulgaris*, J. Biol. Chem. 250, 8030-8037.

Nash, R. J., Fenton, A. K., Gatehouse, A. M. R., and Bell, E. A. (1986) Effects of the plant alkaloid castanospermine as an anti-metabolite of storage pests, Entomol. Exp. Appl. 42, 71-77.

Osborn, T. C., Brown, J. W. S., and Bliss, F. A. (1985) Bean lectins. 5. Quantitative genetic variation in seed lectins of *Phaseolus vulgaris* L. and its relationship to quatitative lectin variation, Theor. Appl. Genet. 70, 22-31.

Osborn, T. C., Alexander, D. C., Sun, S. S. M., Cardona, C., and Bliss, F. A. (1988) Insecticidal activity and lectin homology of arcelin seed protein, Science 240, 207-210.

Rosenthal, G. A., Dahlman, D. L., and Janzen, D. H. (1976) A novel means for dealing with L-canavanine, a toxic metabolite, Science 192, 256-258.

Sawa, M. and Tan, T. S. (1976) Specific difference in resistance to adzuki bean weevil (*Callosobruchus chinensis*) in genus *Phascolus*, Tohoku Br. Crop Sci. Soc. Japan 18, 79-81.

Singh, Y. (1976) Studies on relative resistance of imported pulses to *Callosobruchus maculatus* (Fabricius) and *C. chinensis* (Linnaeus), Entomol. Newsl. 6, 8-19.

Silano, V., Furia, M., Gianfreda, L., Macri, A., Palescandola, R., Rab, A., Scardi, V., Stella, E., and Valfre, F. (1975) Inhibition of amylases from different origins by albumins from the wheat kernel, Biochim. Biophys. Acta 391, 170-178.

BIOCHEMICAL RESISTANCE TO BRUCHID ATTACK IN LEGUME SEEDS; INVESTIGATION AND EXPLOITATION

A. M. R. GATEHOUSE, B. H. MINNEY, P. DOBIE[1], AND
V. HILDER
*Department of Biological Sciences, Durham University,
Durham, U.K.*
[1] *ODNRI, Central Avenue, Chatham Maritime, Chatham, Kent, U.K.*

ABSTRACT. In attempting to explain the relatively low susceptibility of legume seeds to phytophagous insects in general, a correlation between the presence of secondary metabolites and resistance to attack has been made in many cases. The seeds of legumes are a rich and varied source of secondary plant compounds, many of which are known to be antimetabolic or toxic towards vertebrates. However, direct evidence of the protective role of these compounds against insects is only available in a limited number of examples. These involve alkaloids, saponins, non-protein amino acids, polysaccharides, and proteins such as lectins and enzyme inhibitors. Insect attack on mature seeds of legumes is primarily limited to a specialized family of insects, the Bruchidae, and different species within this family show varying degrees of specialization with respect to host species.

In considering the biochemical defenses employed in legume seeds, two levels of resistance mechanisms can be identified. First, general defensive substances are present, which confer protection against the non-pest species. Secondly, targeted resistance mechanisms often showing marked varietal differences within a host species, which give resistance to the host's specific pests also, may operate. As an example of the first type of resistance mechanism, the cowpea (*Vigna unguiculata*) trypsin inhibitor (CpTI) has been demonstrated to confer resistance to both Lepidopteran and Coleopteran pests in transgenic plants and artificial diets. The second type of resistance mechanism is illustrated by a novel protein (LLP: arcelin) present in some wild lines of *Phaseolus vulgaris* which are resistant to attack by a common pest of cultivated varieties, *Zabrotes subfasciatus*. This protein has been shown to be toxic to *Z. subfasciatus* in artificial diet. Both these examples will be considered in detail.

The exploitation of various resistance mechanisms in crop breeding can be aided by identification of the active component(s), and understanding of its mode of action. This allows conventional breeding programmes to be planned, executed and followed without the necessity for extensive bio-assay trials at every stage. The possibilities for genetic engineering of crop plants with foreign genes to confer insect resistance will be discussed.

1. Introduction

Recent years have shown an increased awareness in the necessity of elucidating the mechanisms of seed resistance to insect attack. A greater knowledge of the bases of resistance enables the plant breeder to assess their potential, and possibly exploit such traits in breeding programmes. Although crop diseases and insect pests can be controlled by a combination of practices which include pesticides, crop rotation, field sanitation and the use of pest-free seeds, the safest, most economical and most satisfactory method is to use plant resistance, particularly when it is a component of an integrated

K. Fujii et al. (eds.), Bruchids and Legumes: Economics, Ecology and Coevolution, 241–256.
© 1990 *Kluwer Academic Publishers. Printed in the Netherlands.*

pest control programme (Meiners and Elden, 1978). Crop plants have been primarily selected for high yields, nutritional value, and where necessary, adaptation to certain environmental conditions, together with low anti-mammalian chemical factors. This "selection pressure" by plant breeders has severely disrupted the co-evolutionary relationships between plants and insects, such that very few cultivated species have retained the degree of resistance exhibited by their wild relatives (Feeny, 1976). In an attempt to redress this balance, and particularly as a consequence of the increased concern of heavy reliance on chemical insecticides in recent years, some breeders and biotechnologists are now attempting to exploit inherent resistance either by conventional plant breeding (Redden et al., 1983; Harmsen et al., 1988) or by recombinant DNA technology (Hilder et al., 1987). Of prime concern, however, is the potential toxicity of such compounds to the 'intended' consumer of the crop in question (Gatehouse, 1984). Examples of both of these different approaches will be discussed in detail later.

The seeds of legumes are a rich and varied source of secondary plant compounds, many of which are known to be antimetabolic or toxic towards vertebrates (Pusztai et al., 1975; 1979; Liener, 1980; Ressler et al., 1961). However, direct evidence for the protective role of these compounds against insects is only available in a relatively limited, albeit increasing, number of examples. The aim of this presentation is two-fold. Firstly to discuss specific examples of biochemical resistance to bruchid attack in legume seeds, and secondly to discuss two successful examples of their exploitation.

In considering the biochemical defenses employed in legume seeds, two levels of resistance mechanisms can be identified. First general defensive substances are present which confer protection against the non-pest species and secondly, there are targeted resistance mechanisms, often showing marked varietal differences within a host species, which give resistance to the host's specific pests. The former constitutes by far the largest category of resistance mechanisms so far investigated.

Although various studies have been carried out indicating the presence of toxic or antifeedant chemicals in the testa of certain leguminous seeds (Janzen, 1977; Birch et al., 1989) this aspect will only be considered in brief. Despite the tissue itself having been shown to be toxic, their role as an effective defense mechanism remains unproven since many bruchid larvae are thought to tunnel through the intact testa without actually ingesting any tissue (Southgate, 1984). Stamopoulos and Huignard (1980) demonstrated that when milled testa from the seeds of *Phaseolus vulgaris* was incorporated into a diet at a level of 10% and fed to *Acanthoscelides obtectus*, larval mortality of 98% occurred; they subsequently demonstrated toxicity of the lignin fraction towards these larvae (1988). In certain legume species, such as *Vicia faba* (Griffiths, 1981) there is evidence for the localized accumulation of polyphenolic compounds, including condensed tannins, in the seed testa. It has been shown that the condensed tannins from these seeds have an adverse effect upon development of *Callosobruchus maculatus* (Boughdad et al., 1986). On the basis of these results the authors suggest that use of cultivars rich in weakly condensed tannins could cause a decrease of infestation rate of *V. faba* by *C. maculatus*. The protective role of tannins themselves, despite the fact that they tend to be concentrated in the testa, still remains a matter of debate (Bernays, 1978; Martin and Martin, 1984).

2. Non-protein Antimetabolites

The range of secondary compounds present in the cotyledons of legume seeds which have been implicated in plant protection is both varied and extensive (Birch et al., 1985). However, evidence for this role is often only circumstantial and impure fractions have been tested. One of the earliest citations on the possible protective role of secondary plant compounds was in 1918 when Bridwell suggested that the presence of natural insecticides, such as alkaloids, protected the seeds of certain legumes in the Hawaiian Islands from attack by members of the Bruchidae (Bridwell, 1918). Toxicity of alkaloids towards several different Bruchidae has subsequently been demonstrated, forming the basis of extensive studies. These compounds, together with non-protein amino acids and rotenoids, have been shown to be involved in the biochemical resistance of many legume seeds, but they are more generally associated with non-food crops, many being extremely toxic to mammals (Bell, 1976).

2.1. ALKALOIDS

Polyhydroxy alkaloids are particularly interesting since several are structural analogues of sugar molecules. One, 2,5-dihydroxymethyl 3,4-dihydroxypyrrolidine (DMDP), an analogue of β-D-fructofuranose, has been detected in the seeds of several species of *Lonchocarpus* at levels of up to 6%. Feeding trials carried out against a non-pest species, *C. maculatus*, showed that the compound reduced larval survival in a dose-dependent manner, being lethal at 0.1% (Evans et al., 1985). This study also found that the larval α-D-glucosidase was strongly inhibited by DMDP in a competitive manner. Other polyhydroxy alkaloids with activity against bruchid glycosidase enzymes have also been isolated from legume seeds, e.g., castanospermine which is present in seeds of *Castanospermum australe* at about 0.1%. At a concentration of 0.03% in artificial diets it completely inhibited larval development, and even at levels as low as 0.005% it significantly reduced the rate of development (Nash et al., 1986). Clearly, the physiological levels present in the seeds are sufficient to afford complete protection against attack by *C. maculatus*. As with DMDP, it was found to be a strong inhibitor of larval gut carbohydrases, inhibiting both α-D- and β-D-glucosidase activities.

2.2. NON-PROTEIN AMINO ACIDS

Non-protein amino acids have often been associated with a protective role within the seed and Rehr et al. (1973) found that some legume seeds that were relatively free from insect attack contained high concentrations of non-protein amino acids. In an attempt to ascertain whether 'the large array of secondary compounds found in seeds is of adaptive significance in preventing a very large portion (but not all) of the potential seed-predators in the habitat from feeding on this or that species of food' Janzen et al. (1977) carried out a detailed set of feeding trials using *C. maculatus* as the test organism. The results showed that several non-protein amino acids were lethal at concentrations of 0.1%; however, in the same study they found that alkaloids were generally the most toxic of the compounds tested. From the data, they concluded that many of the compounds tested were likely to be responsible, at least in part, for the extreme host-specificity shown by seed-eating insects. Another non-protein amino acid shown to be toxic to bruchids *in vitro* is p-aminophenylalanine which is present in the seeds of a few species from the genus Vigna, in particular the subgenus *Plectotropis*, at levels of 0.1% to 1.3%. This compound was found to completely inhibit larval development of *Zabrotes subfasciatus* and *Callosobruchus maculatus* at a concentration of 0.3% and 0.7% respectively (Birch et al., 1986). Perhaps of more interest are the few examples where specialist pest species are able to detoxify such compounds. The classic example is the ability of the larvae of the bruchid beetle *Caryedes brasiliensis* to detoxify, and exploit, the non-protein amino acid L-canavanine (Rosenthal et al., 1976; 1978). This highly toxic compound is an analogue of the amino acid L-arginine. On the basis that only *C.*

brasiliensis is able to attack mature seeds of *Dioclea megacarpa*, which contains approximately 13% dry wt of the compound, the authors consider that it acts as an effective barrier against predation established by this leguminous plant. It has been shown to be toxic not only to bruchid species (Rosenthal et al., 1976) but also to other insects (Rosenthal, 1986).

2.3. ROTENOIDS

Although a few specialist bruchids do develop in host seeds containing rotenoids (Johnson, 1981) several of these complex isoflavonoids are known to exhibit insecticidal properties, and therefore may play a role in legume seed protection against the non-pest species; their major effect is to cause a dramatic decrease in oxygen uptake. Birch et al. (1985) have recently isolated four rotenoids from the seeds of *Lonchocarpus salvadorensis* which appears to be attacked by very few indigenous bruchids in Costa Rica. When tested at the physiological concentrations, they were either found to be lethal or detrimental to development of *C. maculatus* (Birch et al., 1989).

2.4. SAPONINS

A comparative study on the toxicity of certain legume saponin fractions to *Callosobruchus* larvae and the relative resistance of different legume seeds to damage by this beetle suggests that there is a correlation between saponin content and resistance (Applebaum et al., 1969). Various saponin fractions from soybean have been demonstrated to be toxic to this bruchid species (Applebaum et al., 1965). This class of compound is thought to operate at several different levels including acting as feeding deterrents, causing hormonal imbalance and by altering the membrane permeability of the insect gut (Applebaum and Birk, 1972). Ishii (1952) suggested that those present in the seeds of *P. vulgaris* might be detrimental to the larval development of *C. chinensis* on the basis of their antagonism to essential steroids, although he was unable to confirm this experimentally.

2.5. POLYSACCHARIDES

One of the first detailed systematic investigations into host preference studies of bruchids on leguminous food crops was carried out in 1952 when Ishii was able to attribute the failure of larvae of *Callosobruchus chinensis*, a pest of mature seeds of *Phaseolus angularis*, to develop in the seeds of *P. vulgaris* to the presence of pectosans. The mature seeds of *P. vulgaris* are economically very important since they are grown widely as a staple food throughout South America and also in parts of Central Africa where they form a major source of dietary protein. Possibly, because of their economic importance, the resistance of the mature seeds to non-pest species and, in a few cases to pest species, has been extensively studied. There is good evidence to show that many different classes of compound are involved in seed resistance of *P. vulgaris* to non-pest species. Most studies concerning biochemical resistance to non-pest species have primarily been confined to the genus *Callosobruchus*, although there have been a few limited observations of *C. maculatus* surviving on some varieties of *P. vulgaris* (Johnson, 1981). As mentioned previously, the inability of *C. chinensis* to develop and survive on seeds of *P. vulgaris* was attributed to the presence of pectosans (Ishii, 1952). Subsequently Applebaum and Guez (1972) isolated a soluble heteropolysaccharide fraction which when fed to *C. chinensis* at the physiological concentrations present in the seed, i.e., at a level of 1% w/w, prevented larval development, but at this concentration had no deleterious effects on development of *Acanthoscelides obtectus*, for which *P. vulgaris* is a host. However, a concentration of 2% w/w was found to be deleterious to *A. obtectus*. Incorporation of starch granules of *P. vulgaris* into artificial beans was also found to increase larval mortality and decrease the rate of larval development of *C. chinensis* (Applebaum et al., 1970). Not only have certain heteropolysaccharides been shown to be involved in seed resistance to a non-pest species, *C. chinensis*, but recent studies have demonstrated that they are involved in resistance towards a pest species. Due to the severe post-harvest loss of *P. vulgaris* as a result of infestation by *Acanthoscelides obtectus* and *Zabrotes subfasciatus* a breeding programme was established at the Centre International de Agriculture Tropical (CIAT) in Colombia to select for resistance against these insect pests. Out of the lines of *P. vulgaris* screened, resistance to these bruchids was found in several wild lines, one of which (G12953) was resistant to both species (Schoonhoven et al., 1983). Gatehouse et al. (1987) found that at a concentration of 4% w/w, the approximate physiological concentration present within the seed, the heteropolysaccharide fraction from the resistant line was toxic, resulting in 80-85% larval mortality of·*A. obtectus* (LC50 of 2.5%); furthermore, surviving larvae showed a marked increase in their development period. The corresponding fraction from a susceptible line, on the other hand, even at twice the physiological concentration, had a negligible effect upon larval development. This therefore is an example of a varietal form of resistance targeted against a specific pest of the host. Interestingly, the deleterious effects of this component on the other pest species, *Z. subfasciatus*, was less marked (Minney et al., submitted).

3. Protein Antimetabolites

3.1. LECTINS

Another class of compound which is present in many legume seeds and which has been demonstrated to be involved in resistance of *Phaseolus vulgaris* is the lectins (phytohaemagglutinins); these are carbohydrate binding proteins. Toxicity of the purified lectin from *P. vulgaris* towards mammals (Jaffé and Vega-Lette, 1968; Evans et al., 1973; Pusztai et al., 1975; 1979) and birds (Jayne-Williams and Burgess, 1974) is well documented. The first report of toxicity towards insects, indeed a bruchid, was by Janzen et al. (1976). In this study the purified lectin was added to an artificial diet for *C. maculatus* at a range of concentrations from 0.1% to 5.0%. (In seeds of different varieties of *P. vulgaris* the lectin levels vary from approximately 1% to 3% w/w (Pusztai and Watt, 1974.) At 5% there was no survival and even at 0.1% the lectin had caused a significant reduction in the number of larvae developing into adults. Gatehouse et al. (1984) subsequently confirmed the toxicity of the seed lectins of *P. vulgaris* towards the developing larvae and on the basis of indirect immunofluorescence investigations using monospecific antisera for globulin lectins showed that the molecules, when ingested, bound to the midgut epithelial cells. On the basis of this observation it was suggested that the mechanism of lectin toxicity is analogous to that known to occur in rat, namely that the ingested lectin causes disruption of the epithelial cells of the larval midgut leading to a breakdown of the transport of nutrients into these cells, and facilitating the absorption of potentially harmful substances. Pest species such as *A. obtectus* are able to avoid the harmful effects of the lectin demonstrated by the non-pest species, *C. maculatus*, since the intact lectin molecules are unable to bind to the midgut epithelial cell surfaces (Gatehouse et al., 1989).

3.2. AMYLASE INHIBITORS

Another class of secondary compound which is thought to be involved in seed resistance of *P. vulgaris*, and other economically important legume seeds, to non-pest species is α-amylase inhibitors. In a very recent study Ishimoto and Kitamura (1988) claimed that an α-amylase inhibitor was responsible for the protection of *P. vulgaris* seeds against attack by the azuki bean weevil (*Callosobruchus chinensis*). They found this inhibitor to be extremely toxic to the larvae, all of which died before the second instar when fed artificial beans containing 0.2% - 0.5% of the protein. In contrast to earlier studies by Janzen et al. (1976) and Gatehouse et al. (1984) they found lectin preparations to have no detrimental effects upon larval development, although a different species of *Callosobruchus* was used. Birch et al. (1989) also found a commercially available preparation of *P. vulgaris* α-amylase inhibitor to have some detrimental effects upon larval development at concentrations occurring naturally in seeds. Numbers of emerging adults of a non-pest species, *C. maculatus*, were reduced by 30% whilst those of a pest species, *Zabrotes subfasciatus*, were reduced by 10%. However there is no indication as to how pure this preparation was. A very specific α-amylase inhibitor has recently been isolated from the seeds of a wild line of *P. vulgaris* (G12953) (Minney et al., in prep). Although the purified inhibitor has not as yet been tested in feeding trials against *Z. subfasciatus*, its presence is correlated with seed resistance to this particular pest species. It has been shown to be a very potent inhibitor of *Z. subfasciatus* larval α-amylase activity *in vitro* (Gatehouse et al., 1987), but has so far failed to inhibit either other sources of insect α-amylases, or mammalian or bacterial α-amylases tested to date. It therefore may represent another example of a targeted resistance mechanism operative within a host species against the host's specific pests.

It is not clear whether the different mechanisms for seed resistance mentioned above, particularly for examples operating in *Phaseolus vulgaris*, illustrate varietal differences or whether, which seems more likely, they are all part of a complex multiple mechanism required for a broad spectrum of protection.

3.3. PROTEASE INHIBITORS

Having considered some of the plant secondary compounds which have either been implicated or demonstrated to be involved in seed resistance, details of two examples of resistance which have subsequently been exploited are now given. The first example relates to what can be considered as a general defensive substance which confers protection against non-pest species, and when present at sufficiently high concentrations is involved in protection of the seed against a pest species. This is exemplified by a protease inhibitor present in the cowpea, *Vigna unguiculata*, a principal grain legume of West Africa and the north-east of South America (FAO, 1970). Post-harvest loss due to the bruchid beetle *Callosobruchus maculatus* is unacceptably high, with up to 100% seed damage after 5 months storage (Singh, 1978). In view of these serious losses a breeding programme was established at the International Institute of Tropical Agriculture (IITA) in Nigeria to select for resistance against this pest. Out of some 5000 accessions only one, TVu 2027, showed significant levels of resistance towards the larvae of this pest (Singh, 1978). After establishing that resistance did not have a physical basis, seeds of the resistant variety TVu 2027 and seeds of several different susceptible varieties were screened for a range of secondary compounds including endopeptidase inhibitors, exopeptidase inhibitors, lectins (phytohaemagglutinins), saponins, alkaloids and the major non-protein amino acids. Of all these antimetabolic secondary compounds screened for, only inhibitory activity against trypsin and, to a much lesser extent chymotrypsin, could be detected. The resistant variety of cowpea contained significantly higher levels of inhibitors, at least twice as much as any other variety and in the order of 3 to 4 times more than the majority of the varieties tested (Table 1). Differences in the inhibitor content of the resistant variety compared to susceptible varieties were shown to be quantitative and not qualitative (Gatehouse et al., 1979).

The antimetabolic properties of the purified cowpea trypsin inhibitor were demonstrated in feeding trials with the larvae of *C. maculatus*, a pest species (Gatehouse et al., 1979). Initial feeding trials using protein fractions extracted from the resistant cowpea line showed that the albumin fraction (containing the protease inhibitors) was toxic whereas the globulin fraction was not. Removal of trypsin inhibitors from the albumin fraction rendered it non-toxic. To confirm the toxicity of the trypsin inhibitors further feeding trials were carried out by adding the purified inhibitor at a range of concentrations. Since the differences in trypsin inhibitors between different cowpea lines was considered to be primarily in the amounts of inhibitors present, rather than the resistant one containing a unique "effective" inhibitor, trypsin inhibitor was purified from commercially available non-resistant cowpea lines, rather than from the resistant line TVu 2027. Addition of inhibitor to diet at a level of 0.8%, which is marginally lower than the physiological concentration found in the resistant seeds, resulted in complete larval mortality. These results confirm that the trypsin inhibitors play a major role in conferring seed resistance in this particular example of 'field' resistance. However subsequent studies on the mechanism of inheritance of bruchid resistance suggested that resistance was not solely attributed to protease inhibitors but in fact was very complex (Redden et al., 1983).

Not only has this particular trait been utilized in a conventional breeding programme at IITA, but since these protease inhibitors are proteins, i.e., primary gene products, they were potentially useful candidates for genetic engineering. A question of prime importance to the biotechnologist before exploiting this form of insect resistance is how broad a spectrum of insects will this protein be effective against, and will it provide the required levels of protection in the chosen crops? The purified cowpea trypsin inhibitor (CpTI) was therefore tested in artificial diets against a wide range of economically important field and storage insect pests, including members of the Lepidoptera such as *Heliothis* and *Spodoptera*, Coleoptera such as *Diabrotica* and *Anthonomus* and Orthoptera such as *Locusta*, all of which cause crop losses of major economic importance. In

Table 1. Performance of *Callosobruchus maculatus* and physiological concentration of trypsin inhibitors in seeds of different accessions of *Vigna unguiculata*.

Accession	Total number of eggs laid	% adult emergence + 1 S.D.	Trypsin inhibitor content (% w.w)
TVu 2027	105 b,c	0.0 ± 0.0 b	0.92
TVu 4557	85 c	95.1 ± 6.2 a	0.44
TVu 76	116 a,b	90.0 ± 2.5 a	0.34
TVu 3629	70 c	90.6 ± 9.5 a	0.30
TVu 37	116 a,b	86.6 ± 4.3 a	0.26
TVu 57	105 b,c	91.7 ± 6.7 a	0.25
TVu 1190 E	168 a	89.0 ± 6.7 a	0.23
TVu 1502-1D	151 a	92.0 ± 6.0 a	0.19

All figures followed by the same letter are not significantly different at the 5% level of significance (Duncan's Multiple Range Test).

all cases CpTI was found to be an effective insecticide. Preliminary studies have shown that not only does CpTI inhibit a prime metabolic enzyme at the catalytic site in the insect gut but it may also interfere with the endocrine system and water balance. A list of insect pests against which CpTI was found to be toxic is given in Table 2, and includes pests of major economic importance, i.e., pests of cereal and cotton where insecticide costs account for the majority of the world insecticide expenditure. Despite the insecticidal properties of CpTI, there is no suggestion that cowpea seeds are toxic to humans, and although normally cooked, they can be eaten raw (Peterson, 1984). Recent feeding trials showed that the inhibitor in untreated cowpea seed meal did not adversely affect the growth of rats (unpublished data).

4. Production of Insect Resistant Crops by Genetic Engineering

Cowpea trypsin inhibitors form ideal candidates for genetic engineering of crops for insect resistance, for not only are they effective against a broad spectrum of insect pests, but also they exhibit very low or no mammalian toxicity. Chemical studies showed them to be small polypeptides of around 80 amino acids belonging to the Bowman-Birk type of double-headed serine protease inhibitors (Gatehouse et al., 1980). They are products of a middle repetitive gene family (Hilder et al., in press). The CpTI gene which was used was derived from plasmid pUSSRc3/2, a member of a complementary DNA library prepared from cowpea cotyledon polyadenylated RNA (Fig. 1). The cDNA library was, like CpTI, prepared from commercially available cowpeas. A 550-base-pair long Alu 1 - Sca 1 restriction fragment containing the entire coding sequence for the mature protein, a long leader sequence and the majority of the 3'-non-translated sequence was transferred to the Sma 1 site of *Agrobacterium tumefaciens* Ti plasmid binary vector, pROK 2 (Baulcombe et al., 1986). This placed the cowpea sequence under the control of a strong constitutive promoter derived from cauliflower mosaic virus (Guilley et al., 1982) and the nopaline synthase gene-transcription termination sequence (Bevan et al., 1983). Constructs were identified which contained an insert in the correct orientation relative to the CaMV promoter to produce CpTI (pROK/CpTI+5) and in the 'reverse' orientation (pROK/CpTI-2), which has six short open reading frames with no identifiable features.

Table 2. Insecticidal efficacy of cowpea trypsin inhibitors (CpTI).

Insects killed by CpTI in artificial diets		CpTI transgenic tobacco plants resistant to following insects
LEPIDOPTERA	*Heliothis virescens*	*Heliothis virescens*
	Heliothis zea	*Heliothis zea*
	Spodoptera littoralis	*Spodoptera littoralis*
	Chilo partellus	*Manduca sexta*
COLEOPTERA	*Callosobruchus maculatus*	*
	Anthonomus grandis	*
	Diabrotica undecimpunctata	*
	Tribolium confusum	*
	Costelytra zealandica	
ORTHOPTERA	*Locusta migratoria*	*

* insects unable to attack control tobacco plants.

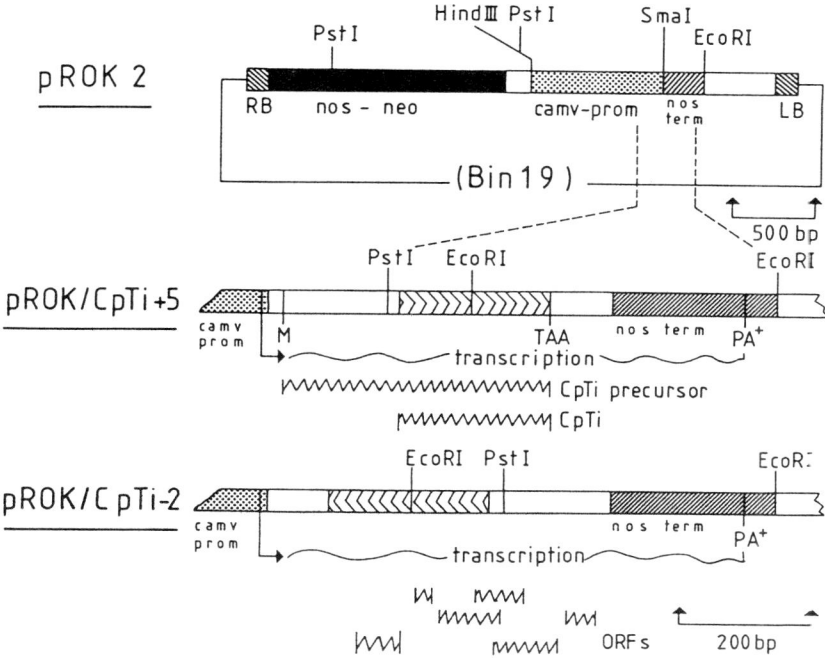

Figure 1. Structure of the binary vector pRok2 and of the chimaeric CpTI-gene constructs in pRok CpTI + 5 and pRok CpTI-2 resulting from insertion of the CpTI sequence into the *Sma* 1 site of pRok2.

This 'reversed' construct was used to produce control transformants. In the present study the crop plant chosen for transformation was tobacco (*Nicotiana tabacum* c.v. Samsun N. N.). The constructs were mobilized into *A. tumefaciens* (Bevan, 1984) and used to transform leaf discs of *N. tabacum*. The transformants were selected by their antibiotic resistance to kanamycin, and the transformed plants were regenerated from shootlets by transfer to a root-inducing, kanamycin-containing agar medium (Horsch et al., 1985). Rooted plants were grown on in soil-based compost.

The presence and levels of CpTI production in the original transformants were measured by dot-immunobinding assays (Jahn et al., 1984) using polyclonal antibodies raised in rabbits against total CpTI. The level of expression in young leaves from different individual pROK/CpTI+5 transformants ranged from below the limit cf detection to ~ 1% of total soluble protein. No CpTI expression could be detected in the pROK/CpTI-2 transformants, i.e. when the gene was inserted in the incorrect orientation no inhibitor was produced (Hilder et al., 1987). Western blotting of soluble leaf proteins from CpTI expressing transformants showed that polypeptides produced and processed in the transformants corresponded to one of the isoinhibitors present in the cowpea seed; no corresponding polypeptides were produced in the control transformants. The functional integrity of the CpTI produced in these transformed tobacco plants was demonstrated by *in vitro* trypsin inhibitor activity assay. Thus the transformed tobacco plants were able to express the foreign CpTI gene and produce an active trypsin inhibitor whose levels of expression in the leaves of the highest expressing plants were similar to those present in the mature seeds of the resistant variety of cowpea, TVu 2027.

The critical test on these CpTI expressing transformed tobacco plants was the bioas-

250

say to test their respective levels of insect resistance or tolerance. This was carried out by infesting the young plants with newly emerged larvae of the Lepidopteran *Heliothis virescens*. The infested plants were sealed into individual plantaria and kept under controlled light and temperature regimes within a growth cabinet. *H. virescens*, the tobacco budworm, was tested in the first instance as it is classified as a serious economic pest, one of whose primary hosts is tobacco. After a trial period of seven days all larvae (both dead and surviving) were removed, their size recorded and the extent of leaf damage measured by computer aided image analysis. The results clearly showed that those CpTI transformants which expressed the foreign protein at approximately 1% were relatively resistant to attack compared to control plants. Some of the transformants showing enhanced levels of insect resistance and some of the control plants were replicated as stem cuttings (Baulcombe et al., 1986) to provide sets of genetically identical plants on which statistically sound insect feeding trials could be run. These further trials provided convincing evidence that the CpTI-producing plants were much more resistant to insect attack. Control plants were devastated by this level of infestation; in trials which we ran beyond seven days, i.e., to 'termination' these control plants were reduced to a stalk. However, on the CpTI-producing plants, although the larvae begin to feed and do some very limited damage to the leaves, they either die or fail to develop as they would on control plants (Fig. 2). This observation is consistent with the mechanism of CpTI toxicity proposed by Gatehouse and Boulter (1983) relying upon a finely controlled balance within the host plant which has to make sufficient nutrients for itself but insufficient to maintain predation, thus the larvae die of starvation at a very early stage. Although the initial bio-assays of the transformed plants were carried out using *H. virescens*, trials were subsequently carried out using *H. zea* (corn earworm) *Spodoptera littoralis* (armyworm) and *Manduca sexta* (tomato and tobacco hornworm). In all cases the CpTI transformants were resistant to attack compared to control plants; routinely ~ 75% larvae die within 2 - 3 days. Unfortunately, it was not possible to carry out trials on these plants using any members of the Coleoptera which we are interested in since they would not attack the control plants; in these instances the only information available on the toxicity of CpTI is from artificial diets.

Figure 2.   Effect of *M. sexta* larvae on *N. tabacum* transformed with CpTI + 5 and CpTI-2 constructs. The plant on the left is a clonal replicate of a CpTI-2 line (control); on the right, a CpTI-expressing CpTI + 5 line.

The CpTI gene was shown to be stably inherited through subsequent generations, and these plants at each generation were also screened for insect resistance. As with the original clonal CpTI transformants, these seed derived plants were resistant to insect attack. Details relating to the genetics and inheritance of CpTI in the transformants are given elsewhere (Hilder and Gatehouse, 1989). The engineering of the CpTI gene into tobacco is the first successful example of the genetic engineering of insect resistance using genes of plant origin. Bio-assays carried out on the CpTI transgenic plants clearly demonstrate that the presence of CpTI in the cowpea seeds is also an example of biochemical resistance operative against non-pest species.

The cowpea trypsin inhibitor gene (obtained from a commercial variety, California Black-eye beans) has been licensed by the Agriculture Genetics Company Ltd. to many of the world's leading seed companies for incorporation into targeted crops.

5. Production of Insect Resistant Crops by Conventional Breeding

The second level of resistance mechanism which can be identified in legume seeds is a targeted resistance directed towards the host's specific pests; this form of resistance often shows marked varietal differences. An example of this is exemplified by a novel protein (Mr 35000 - 38000) designated arcelin (Romero Andreas et al., 1986; Osborn et al., 1988) or LLP (Minney et al., submitted) which has been found to be present in several resistant wild lines of *Phaseolus vulgaris* (Cardona et al., 1989). Since there was a high level of correlation between the presence of this protein and *Z. subfasciatus* seed resistance (Fig. 3) (Osborn et al., 1986; Minney et al., submitted), there was therefore a strong possibility of its involvement in bruchid resistance. Furthermore, seeds of backcross lines containing arcelin (cultivar Sanilac) showed high levels of resistance to *Z. subfasciatus*, whereas lines derived from back crossing experiments lacking arcelin were fully susceptible compared to the control cultivar; lines segregating for arcelin had

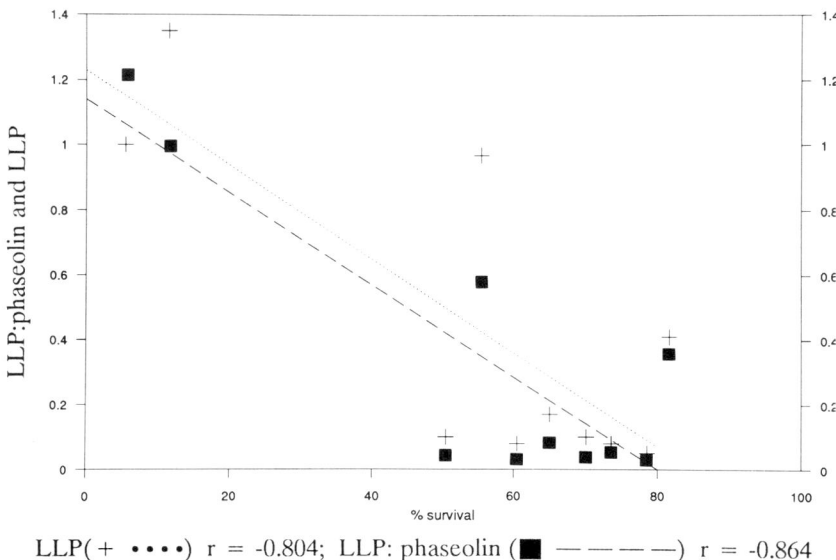

LLP(+ • • • •) r = -0.804; LLP: phaseolin (■ — — — —) r = -0.864

Figure 3. Comparison of the regressions of LLP, and the ratio of LLP: phaseolin against percent survival of Zabrotes subfasciatus *for 10 wild accessions of* Phaseolus vulgaris.

intermediate levels of resistance (Osborn et al., 1989). From these crossing experiments and further backcrosses using different cultivated bean types as recurrent parents, the authors concluded that arcelin was associated with high levels of resistance to *Z. subfasciatus*. Minney et al. (submitted) found this protein to be an effective antimetabolite of *Z. subfasciatus* since incorporation of the purified protein into artificial seeds at a level of 2.5% (i.e., at levels significantly lower than the physiological concentration present in wild resistant lines) caused a 52% reduction in adult survival. Arcelin (LLP), when present in resistant wild accessions, constitutes the major seed storage protein at the expense of phaseolin (also known as glycoprotein II). *In vitro* hydrolysis experiments using purified proteins showed that whereas the larval digestive proteases of *Z. subfasciatus* are able to partially digest phaseolin, LLP is virtually indigestible (Minney et al., submitted). It would, therefore, appear that one possible mechanism for the antimetabolic properties of this protein may be its non-digestibility to the developing larvae, thus depriving them of essential amino acids. The effects of arcelin upon development of another pest species, *Acanthoscelides obtectus*, was negligible, neither was there any correlation between its presence and resistance to this bruchid in backcrossing experiments.

Although arcelin has been shown to be associated with resistance, and to be antimetabolic when fed as purified protein, seed resistance in the wild lines is more complex since the ratio of arcelin (LLP) to phaseolin appears to be more important than the level of arcelin *per se* (Minney et al. submitted); furthermore, as mentioned previously, there is a correlation between resistance and a specific α-amylase inhibitor (Table 3).

Arcelin has now been exploited by scientists at CIAT where it has been successfully and stably incorporated into commercial varieties. Pending further mammalian feeding trials these improved *Zabrotes* resistant cultivars should soon by ready for release into the main breeding programme. This, therefore, is an example of the identification of an insecticidal protein which has been successfully exploited by conventional plant breeding.

In conclusion we would like to suggest that the seeds of legumes not only provide a rich and varied source of plant secondary compounds, but since many of these are insecticidal, they therefore provide a valuable pool for insect resistance genes. However,

Table 3. Relationship of seed components of 10 wild accessions of *Phaseolus vulgaris* to survival of *Zabrotes subfasciatus*.

Accession	Adult emergence (%)	Median dev. period (days)	α-amylase inhibitory activity (%)	LLP content	Phaseolin content	LLP: Phaseolin ratio
GO9989B	60.5	38.1	24.84	0.08	2.11	0.04
G10000	78.6	38.1	57.32	0.05	1.19	0.04
G10007	50.4	39.1	26.75	0.10	2.04	0.05
G10019	55.5	37.1	21.42	0.97	1.68	0.58
G12882	11.5	48.9	80.00	1.35	1.35	1.00
G12888	73.6	38.1	12.74	0.08	1.41	0.06
G12922	81.7	42.3	40.00	0.41	1.14	0.36
G12871	70.1	39.1	35.67	0.10	2.25	0.04
G12880	65.1	38.2	22.93	0.17	1.94	0.09
G12949	5.4	70.1	95.00	1.00	0.82	1.22

LLP and Phaseolin content were determined by laser densitometric scanning, and are given in arbitrary units.

caution should be taken before exploiting such genes as many of the gene products will be toxic or undesirable to consumers other than insects.

Acknowledgements. The authors would like to thank Dr. J. A. Gatehouse for critically reviewing the manuscript, and Mrs. Ethne Ellis and Mrs. Jean Mather for typing it. Financial assistance from ODA and AGC Ltd. is gratefully acknowledged.

REFERENCES

Applebaum, S. W., Gestetner, B., and Birk, Y. (1965) Physiological aspects of host specificity in the Bruchidae IV. Developmental incompatibility of soybeans for *Callosobruchus*, J. Insect Physiol. 11, 611-616.

Applebaum, S. W., Marco, S., and Birk, Y. (1969) Saponins as possible factors of resistance of legume seeds to the attack of insects, Agric. Food Chem. 17(3), 618-622.

Applebaum, S. W., Tadmor, U., and Podoler, H. (1970) The effect of starch and of a heteropolysaccharide fraction from *Phaseolus vulgaris* on development and fecundity of *Callosobruchus maculatus*, Entomol. Exp. Appl. 13, 61-70.

Applebaum, S. W. and Guez, M. (1972) Comparative resistance of *Phaseolus vulgaris* beans to *Callosobruchus chinensis* and *Acanthoscelides obtectus* (Col. Bruchidae): the differential digestion of soluble heteropolysaccharide, Entomol. Exp. Appl. 25, 64-74.

Applebaum, S. W. and Birk, Y. (1972) Natural mechanisms of resistance to insects in legume seeds, in J. G. Rodriguez (ed), Insect and Mite Nutrition, North-Holland Publishers, Amsterdam, pp. 629-639.

Baulcombe, D. C., Saunders, G. R., Bevan, M., Mayo, M. A., and Harrison, B. D. (1986) Expression of biologically active viral satellite RNA from the nuclear genome of transformed plants, Nature 321, 446-449.

Bell, E. A. (1976) 'Uncommon' amino acids in plants, FEBS Letters 64(1), 29-35.

Bevan, M., Barnes, W. M., and Chilton, M. D. (1983) Structure and transcription of the nopaline synthase gene region of T-DNA, Nucleic Acids Res. 11, 369-385.

Bevan, M. (1984) Binary *Agrobacterium* vectors for plant transformation, Nucleic Acids Res. 12, 8711-8721.

Bernays, E. A. (1978) Tannins: an alternative viewpoint, Entomol. Exp. Appl. 24, 44-53.

Birch, A. N. E., Crombie, L., and Crombie, W. M. (1985) Rotenoids of *Lonchocasrpus salvadorensis*: their effectiveness in protecting seeds against bruchid predation, Phytochem. 24(12), 2881-2883.

Birch, A. N. E., Fellows, L. E., Evans, S. V., and Docherty, S. V. (1986) Para-aminophenylalanine in *Vigna*: possible taxonomic and ecological significance as a seed defense against bruchids, Phytochem. 25(12), 2745-2751.

Birch, A. N. E., Simmonds, M. S. J., and Blaney, W. M. (1989) Chemical interactions between bruchids and legumes, in Advances in Legume Biology, Monographs in Systematic Botany, Missouri Botanic Gardens (in press).

Boughdad, A., Gillon, Y., and Gagnepain, C. (1986) Influence des tannins condenses du tegument de feves (*Vicia faba*) sur le developpement larvaire de *Callosobruchus maculatus*, Entomol. Exp. Appl. 42, 125-132.

Bridwell, J. C. (1918) Notes on the Bruchidae and their parasites in the Hawaiian Islands, Proc. Hawaiian Entomol. Soc. 3, 4365-4505.

Cardona, C., Posso, C. E., Kornegay, J., Valor, J., and Serrano, M. (1989) Antibiosis effects of wild dry bean accessions on the Mexican bean weevil and the bean weevil (Coleoptera: Bruchidae), J. Econ. Entomol. 82(1), 310-315.

Dobie, P., Dendy, J., Sherman, C., Padgham, J., Wood, A., and Gatehouse, A. M. R. (1989) New sources of resistance to *Acanthoscelides obtectus* (Say) and *Zabrotes subfasciatus* Boheman (Coleoptera, Bruchidae) in mature seeds of five species of *Phaseolus*, J. stored Prod. Res. (in press).

Evans, R. J., Pusztai, A., Watt, W. B., and Bauer, D. H. (1973) Isolation and properties of protein fractions from navy beans (*Phaseolus vulgaris*) which inhibit growth of rats, Biochim. Biophys. Acta. 303, 175-184.

Evans, S. V., Gatehouse, A. M. R., and Fellows, L. E. (1985) Detrimental effects of 2,5-Dihydrozy-methyl-3,4-dihydroxypyrrolidine in some tropical legume seeds on larvae of the bruchid *Callosobruchus maculatus*, Entomol. Exp. Appl. 37, 257-261.

FAO (1970) The State of Food and Agriculture, Rome, pp.274.

Feeny, P. P. (1976) Plant apparency and chemical defense, Rec. Adv. Phytochem. 10, 1-40.

Gatehouse, A. M. R. (1984) Antinutritional proteins in plants, in B. J. F. Hudson (ed.), Development in Food Proteins, vol. 3., Elsevier Applied Science Publishers, London, pp. 245-294.

Gatehouse, A. M. R., Gatehouse, J. A., Dobie, P., Kilminster, A. M., and Boulter, D. (1979) Biochemical basis of insect resistance in *Vigna unguiculata*, J. Sci. Food Agric. 30, 949-958.

Gatehouse, A. M. R., Gatehouse, J. A., and Boulter, D. (1980) Isolation and characterization of trypsin inhibitors from cowpea, Phytochem. 19, 751-756.

Gatehouse, A. M. R. and Boulter, D. (1983) Assessment of the anti-metabolic effects of trypsin inhibitors from cowpea (*Vigna unguiculata*) and other legumes on development of the bruchid beetle *Callosobruchus maculatus*, J. Sci. Food Agric. 34, 345-350.

Gatehouse, A. M. R., Dewey, F. M., Dove, J., Fenton, K. A., and Pusztai, A. (1984) Effect of seed lectin from *Phaeolus vulgaris* on the development of larvae of *Callosobruchus maculatus*; mechanism of toxicity, J. Sci. Food Agric. 35, 373-380.

Gatehouse, A. M. R., Dobie, P., Hodges, R. J., Meik, J., Pusztai, A., and Boulter, D. (1987) Role of carbohydrates in insect resistance in *Phaseolus vulgaris*, J. Insect Physiol. 33, 843-850.

Gatehouse, A. M. R. and Hilder, V. A. (1988) Introduction of genes conferring insect resistance, in Proceedings of Brighton Crop Protection Conference, Vol. 3., Lavenham Press Ltd., Suffolk, UK., pp. 1245-1254.

Gatehouse, A. M. R., Shackley, S. J., Fenton, K. A., Bryden, J., and Pusztai, A. (1989) Mechanism of seed lectin tolerance by a major insect storage pest of *Phaseolus vulgaris*, *Acanthoscelides obtectus*, J. Sci. Food Agric. 47, 269-280.

Griffiths, D. W. (1981) The polyphenolic content and enzyme inhibitory activity of testas from bean (*Vicia faba*) and pea (*Pisum* spp) varieties, J. Sci. Food Agric. 32, 797-804.

Guilley, H., Dudley, R. K., Jonard, G., Balazs, E., and Richards, K. E. (1982) Transcription of cauliflower mosaic virus DNA: detection of promoter sequences and characterization of transcripts, Cell 30, 763-773.

Harmsen, R., Bliss, F. A., Cardona, C., Posso, C. E., and Osborn, T. C. (1988) Transferring genes for arcelin protein from wild to cultivated beans: implications for bruchid resistance, Ann. Report Bean Improvement Cooperative 31, 54-55.

Hilder, V. A., Gatehouse, A. M. R., Sheerman, S. E., Barker, R. F., and Boulter, D. (1987) A novel mechanism of insect resistance engineered into tobacco, Nature 330, 160-163.

Hilder, V. A., Barker, R. F., Sammour, R. A., Gatehouse, A. M. R., Gatehouse, J. A., and Boulter, D. (1989) Protein and cDNA sequences of Bowman-Birk protease inhibitors from cowpea, Plant Molec. Biol. (in press).

Hilder, V. A. and Gatehouse, A. M. R. (1989) Genetic engineering of crops for insect resistance using genes of plant origin, in D. Grierson and G. Lycett (eds), Genetic Engineering of Crop Plants, Butterworths (in press).

Horsch, R. B., Fry, J. E., Hoffman, N. L., Eichholtz, D., Rogers, S. G., and Fraley, R. T. (1985) A simple and general method for transferring genes into plants, Science 227, 1229-1231.

Ishii, S. (1952) Studies on the host preference of cowpea weevil (*Callosobruchus chinensis*), Bull. Nat. Inst. Agric. Sci. Japan 1(c), 185-256.

Ishimoto, M. and Kitamura, K. (1988) Identification of the growth inhibitor on azuki bean weevil in kidney bean (*Phaseolus vulgaris* L.), Jpn. J. Breed. 38, 367-370.

Jaffé, W. G. and Vega-Lette, C. L. (1968) Heat-labile growth inhibiting factors in beans (*Phaseolus vulgaris*), J. Nutr. 94, 203-210.

Jahn, R., Schiebler, W., and Greengard, P. (1984) A quantitative dot-immunobinding assay for proteins using nitrocellulose membrane filters, Proc. Nat. Acad. Sci. USA 81, 1584-1687.

Janzen, D. H., Juster, H. B., and Liener, I. E. (1976) Insecticidal action of the phytohemagglutinin in black beans on a bruchid beetle, Science 192, 795-796.

Janzen, D. H. (1977) The interaction of seed predators and seed chemistry, Coll. Int. C.N.R.S. 265, 414-418.

Janzen, D. H., Juster, H. B., and Bell, E. A. (1977) Toxicity of secondary compounds to the seed-eating larvae of the bruchid beetle *Callosobruchus maculatus*, Phytochem. 16, 223-227.

Jayne-Williams, D. J. and Burgess, C. D. (1974) Further observations on the toxicity of navy beans (*Phaseolus vulgaris*) for Japanese quail (*Coturnix coturnix* japonica), J. Appl. Bacteriol. 37, 149-169.

Johnson, C. D. (1981) Relations of *Acanthoscelides* with their plant hosts, in V. Labeyrie (ed), The Ecology of Bruchids Attacking Legumes (pulses), Dr. W. Junk, The Hague, The Netherlands, pp. 73-81.

Liener, I. E. (1980) Toxic constituents of Plant Foodstuffs, 2nd edition, Academic Press, New York.

Martin, M. M. and Martin, S. J. (1984) Surfactants: their role in preventing the precipitation of proteins by tannins in insect guts, Oecologia 61, 342-345.

Meiners, J. P. and Elden, T. C. (1978) Resistance to insects and diseases in *Phaseolus*, in R. S. Summerfield and A. H. Buting (eds), Advances in Legume Science, International Legume Conference, Kew, pp. 359-364.

Nash, R. J., Fenton, K. A., Gatehouse, A. M. R., and Bell, E. A. (1986) Effects of the plant alkaloid castanospermine as an antimetabolite of storage pests, Entomol. Exp. Appl. 42, 71-77.

Osborn, T. C., Blake, T., Gepts, P., and Bliss, F. A. (1986) Bean arcelin 2. Genetic variation, inheritance and linkage relationships of a novel seed protein of *Phaseolus vulgaris* L., Theor. Appl. Genet. 71, 847-855.

Osborn, T. C., Burow, M., and Bliss, F. A. (1988) Purification and characterization of arcelin seed protein from common bean, Plant Physiol. 86, 399-405.

Osborn, T. C., Alexander, D. C., Sun, S. S. M., Cardona, C., and Bliss, F. A. (1989) Insecticidal activity and lectin homology of arcelin seed protein, Science 240, 207-210.

Peterson, V. (1984) The Natural Food Catalogue 69, McDonald, London.

Pusztai, A. and Watt, W. B. (1974) Isolectins of *Phaseolus vulgaris*. A comprehensive study of fractionation, Biochem. Biophys. Acta 365, 57-71.

Pusztai, A., Grant, G., and Palmer, R. (1975) Nutritional evaluation of kidney beans (*Phaseolus vulgaris*): the isolation and partial characterization of toxic constituents, J. Sci. Food Agric. 26, 149-156.

Pusztai, A., Clarke, E. M. W., and King, T. P. (1979) The nutritional toxicity of *Phaseolus vulgaris* lectins, Nutr. Soc. 38, 115-120.

Redden, R. J., Dobie, P., and Gatehouse, A. M. R. (1983) The inheritance of seed resistance to *Callosobruchus maculatus* F. in cowpea (*Vigna unguiculata* L. Walp.). I. Analysis of parental, F1, F2, F3 and backcross seed generations, Aust. J. Agric. Res. 34, 681-695.

Rehr, S. S., Bell, E. A., Janzen, D. H., and Feeny, P. P. (1973) Insecticidal amino acids in legume seeds, Biochem. Systematics 1, 63-67.

Ressler, C., Redstone, P. A., and Erenberg, R. H. (1961) Isolation and identification of a neuroactive factor from *Lathyrus latifolius*, Science 134, 188-190.

Romero Andreas, J., Yandell, B. S., and Bliss, F. A. (1986) Bean arcelin 1. Inheritance of a novel seed protein of *Phaseolus vulgaris* L. and its effect on seed composition, Theor. Appl. Genet. 72, 123-128.

Rosenthal, G. A., Dahlman, D. L., and Janzen, D. H. (1976) A novel means for dealing with L-canavanine, a toxic metabolite, Science 192, 256-258.

Rosenthal, G. A., Dahlman, D. L., and Janzen, D. H. (1978) L-canaline detoxification: a seed preda-

256

tors's biochemical mechanism, Science 202, 528-529.

Rosenthal, G. A. (1986) Biochemical insight into insecticidal properties of L-canavanine, a higher plant protective allelochemical, J. Chem. Ecol. 12, 1145-1156.

Schoonhoven, A. van, Cardona, C., and Valor, J. (1983) Resistance to the bean weevil and the Mexican bean weevil (Coleoptera: Bruchidae) in non-cultivated common bean accessions, J. Econ. Entomol. 76, 1255-1259.

Singh, S. R. (1978) Resistance to insect pests of cowpea in Nigeria, in S. R. Singh et al. (eds), Pests of Grain Legumes: Ecology and Control, Academic Press, pp. 267-297.

Southgate, B. J. (1984) Observations on the larval emergence of *Callosobruchs chinensis* (L.) (Coleoptera: Bruchidae), Entomol. Gen. 9, 177-180.

Stamopoulos, D. and Huignard, J. (1980) L'influence des diverses parties de la graine de haricot (*Phaseolus vulgaris*) sur le development des larves d'*Acanthoscelides obtectus*, Entomol. Exp. Appl. 28, 38-46.

Stamopoulos, D. C. (1988) Toxic effect of lignin extracted from the tegument of *Phaseolus vulgaris* seeds on the larvae of *Acanthoscelides obtectus* (Say) (Coleoptera, Bruchidae), J. Appl. Entomol. 105(3), 317-320.

LIFE HISTORIES OF STORED-PRODUCT INSECTS

OSAMU IMURA
Stored-Product Entomology Laboratory
National Food Research Institute
MAFF, Tsukuba, Ibaraki 305 Japan

ABSTRACT. Stored-product insects have been recorded from various families and orders. The life-histories of important pest species fall into two major types; short-lived type and long-lived type. Irrespective of the types, the insects developed fast and reached sexual maturity early, resulting in the high intrinsic rate of increase r_m. The reproductive lives of the two types were distinctly different. Adults of the short-lived type were semelparous and usually died in 1-2 weeks. The long-lived species were iteroparous and their adult longevity was more than 100 days. Life history types follow phylogenetic lines. All lepidopterous species were short-lived type and coleopterous species were mostly long-lived type. Life history variations were present in a species. The variations were genetic and phenotypic. Reproductive allocations were compared between related species. The pattern of population dynamics depended on the types of life history of the species. The two types of insects tend to exploit stored-product environments in different ways.

1. Introduction

The more than 1,000 insect species which infest stored products and reduce their quality are members of the orders of Thysanura, Dictyoptera, Orthoptera, Dermaptera, Psocoptera, Hemiptera, Coleoptera, Lepidoptera, Hymenoptera, and Diptera. However, most of the important stored-product pest insects are derived from two orders, Coleoptera and Lepidoptera. About 600 stored-product pest species are from more than 30 families of the Coleoptera and Lepidoptera includes about 75 pest species from more than 10 families.

These pests have come from various natural habitats. Linsley (1944) defined eight insect groups according to their natural habitats: seed-infesting species, fungus feeders, scavengers on dead plant materials, scavengers on dead animal materials, scavengers or semipredators living under bark, wood borers and wood scavengers, scavengers or depredators in the nests of other insects and spiders, parasites and predators. Among innumerable insect species in these natural habitats, only species which were preadapted to stored-product environments are considered to have become stored-product pests. Each stored-product insect, therefore, must carry a life history which has originally evolved under its natural habitat. The life history of these insects also must have been modified by domestic life in stored-product environments after they started to exploit the new habitat. By analyzing the life histories of stored-product insects, we may gain a better understanding of the biology of these insects which may enable us to establish improved management of these pest insects. This review of the life histories of stored-product was strongly influenced by important studies by Japanese entomologists, Kiritani (1961) and Yoshida (1981).

K. Fujii et al. (eds.), Bruchids and Legumes: Economics, Ecology and Coevolution, 257–269.
© 1990 *Kluwer Academic Publishers. Printed in the Netherlands.*

2. Nature of Stored-Product Environments

Stored-product environments are usually isolated from outdoors and their physical environmental factors such as temperature and humidity are relatively stable. The large quantities of dried products, such as fruits, seeds, roots and animal products are stored by humans. These stored products are energy rich but never grow by themselves, unlike living plants. Stored products are only the primary resource on which all consumers including insect pests directly depend on the energy for their life. Thus, the primary resource is consumed irreversibly by the consumers. All insect pests basically compete for the same resource as food as well as habitat. The products may be stored for fairly long periods, or replaced and removed frequently, leading the environmental uncertainties.

3. Life-History Traits and Types of Life Histories

The principal traits of life histories are related to development, reproduction and mortality, because these are crucial components of the fitness of organisms. The basic life-history traits of 22 species from 10 families which were measured under optimal laboratory conditions are listed in Table 1.

Although the optimal conditions differ for each species, all species listed complete their development less than 50 days except for Ptinid species. Other species whose developmental period is more than two months (not included in Table 1) are those of *Tenebrio* and *Neatus* of Tenebrionidae and *Dermestes*, *Attagenus* and *Anthrenus* of Dermestidae and these are rather minor pests (Cotton and Wilber, 1974). Thus, the major stored-product insects develop rapidly.

The pre-oviposition periods are less than 6 days except for that of *T. freemani* which has not presumably established itself well in stored-product environment (Imura, 1987).

Mean adult longevity of all species of Lepidoptera and those of Bruchidae and Anobiidae are less than 3 weeks, while those of other coleopterous species are longer than 100 days (*C. pusilloides* possibly lives more than 100 days).

Thus, the life histories of the major stored-product insects can be classified into short-lived types with fast development and short imaginal life and long-lived types with fast development and long imaginal life.

Adults of the short-lived type usually do not feed and lay most of their eggs within few days after adult emergence. While adults of the long-lived type feed on the same food as their larvae do and lay eggs continuously for a long period. The short-lived species are semelparous and the long-lived species are iteroparous (Cole, 1954). Generally, the long-lived species lay more eggs than the short-lived species in terms of lifetime egg production, although the daily egg production of the former is lower than that of the latter (the 5th and 6th columns in Table 1).

Since these species were examined under optimal laboratory conditions, their juvenile survival rate is high (Hutchinson, 1978). The patterns of survivorship curves, therefore, correspond to the type A or B_1 defined by Pearl and Miner (1935): these types have convex survivorship curves and adult mortality occurs more abruptly in the type A than in the type B_1.

Yoshida (1981) classified the life histories of stored-product insects into four types according to the length of developmental period and reproductive period. The short-lived and long-lived types correspond to Yoshida's type 1 and type 3, respectively, and also correspond to the Dick's (1937) and Kiritani's (1961) the first type and the second type, respectively, although their classification based only on the reproductive life of stored-product insects.

All these species listed in Table 1 have relatively a high intrinsic rate of increase, irrespective of their life history. The precocity due to fast development and early maturity accounts for the high r_m's (Cole, 1954; Lewontin, 1965). Additionally, the longer mean generation times were compensated with the larger net reproduction rates in the long-lived species and *vice versa* in the short-lived species, resulted in the equivalently high r_m's (c.f. $r_m = \ln R_0/T$).

4. Life-History Variations

Life-history traits of each species may vary. Imura (1986) selected eight strains of *E. kühniella* from a wild stock depending on their larval color. The strains differed in their age-specific survival and fecundity rates. The juvenile mortality ranged from 1.5 to 34.0% and the gross rate of reproduction (Σm_x) from 84.8 to 129.8. The coefficient of variation (C.V.) for the mean generation time (T) and that for the net reproduction rate (R_0) were 16.1 and 7.7%, respectively. These two parameters were highly correlated with the fitness of the strains.

Evans (1977) measured life-history parameters at 15°C for nine Australian strains of *S. oryzae* collected from different regions. The C.V.'s for R_0 and T were 36 and 3.3%, respectively, and the former had major effect on r_m of *S. oryzae* strains. Lewontin (1965) predicted the larger genetic variability in fecundity, a component of R_0, than in developmental period, a major component of T, in colonizing species.

Variations in life-history traits such as developmental period, juvenile and imaginal survival, fecundity and fertility, are also reported in *T. castaneum* and *T. confusum* strains (Park et al., 1961; Dawson, 1965; Mertz et al., 1965; Englert and Bell, 1969; White, 1984). The variations in developmental period and fecundity are due to the additive genetic effects (Englert and Bell, 1970; Orozco and Bell, 1974; Rich et al., 1984) on which natural selection operates.

Stored-product insects show plasticity in their life-history under different environmental conditions. Mertz (1971) observed prolonged development, increased adult survivorship and modification of age-dependent fecundity rates in *T. castaneum* and *T. confusum* under crowded conditions.

The adult longevity of *C. chinensis* is prolonged and the lifetime egg production increased when the weevils feed on fungi (Yoshida et al., 1986). Similar effects of feeding are observed in the Phycitid moths (Norris, 1934) and the Dermestid beetles (Kiritani and Kawahara, 1963).

A polymorphism, flightless and flight forms (Utida, 1954) or normal and active forms (Caswell, 1960), has been reported in *Callosobruchus* species. Population density, changes in temperature, water content of the host beans and photoperiod are the factors responsible for the determination of the forms (Utida, 1976). These two forms differ in life-history traits such as developmental period, pre-oviposition period, fecundity, hatchability and adult longevity (Caswell, 1960; Utida, 1972).

Density dependent diapause or dormancy is reported in *P. interpunctella* (Tsuji, 1963), *E. cautella* (Hagstrum and Sharp, 1975), *Trogoderma granarium* (Burges, 1963), and *T. freemani* (Nakakita, 1982). Genotype-environment interactions in the polymorphism and variations are expected in *C. chinensis* (Utida, 1972), *P. interpunctella* (Tsuji, 1963), *E. cautella* (Hagstrum and Silhacek, 1980), and *T. freemani* (Imura, unpublished).

Both genetic variation and phenotypic plasticity in the life-history traits allow stored-product insects to adapt to the temporal and spatial changes in these environments (Emlen, 1984).

Table 1. Life-history attributes of laboratory reared stored-product insects measured under optimal conditions.

Species	Develop. period[a]	Adult longevity[b]	Pre-ovi. period[c]	Fecun-dity[d]	Reprod. effort[e]	Juvenile survival[f]	Survival curve[g]	R_0	T^e	r_m/day	References
LEPIDOPTERA											
Phycitidae											
Ephestia cautella	48.8	8.3	1.	170.	23.3	47.3	B_1	264.4	51.8	0.1008	1
E. kühniella	35.1	6.8	1.	461.5	79.6	86.	A	127.	37.5	0.1292	2
E. elutella	50.	6.1	1.	159.	31.2	93.	A	26.8	54.1	0.0607	3
Plodia interpunctella	37.1	11.7	1.	305.	30.5	99.	A	95.2	40.5	0.1125	3
Spectrobates ceratoniae	39.3	7.6	2.	113.	20.2	61.8	A	33.1	42.4	0.0826	4
Gelechiidae											
Sitotroga cerealella	30.0	7.5	2.5	145.8	29.2	67.	A	112.3	35.8	0.1318	5,6
COLEOPTERA											
Tenebrionidae											
Tribolium castaneum	32.9	134.	4.5	1459.	11.3	83.7	A	641.	50.9	0.1271	7,8,9
T. confusum	28.1	250.	5.0	742.	3.0	95.	A	423.5	70.4	0.0859	9,10
T. freemani	25.8	303.3	12.5	1009.	3.5	87.3	B_1	351.	80.6	0.0727	11
Gnathocerus cornutus	41.8	140.	5.6	479.9	3.1	97.	A	167.9	72.2	0.0710	12
Curculionidae											
Sitophilus oryzae	28.0	119.4	3.5	344.	3.0	90.	B_1	134.2	44.8	0.1103	13,14
S. zeamais	30.8	126.9	3.5	208.	1.7	90.	B_1	89.4	56.0	0.0805	13,14

							B_1				
Bostrychidae											
Rhyzopertha dominica	24.5	120.	6.	415.	3.6	78.		141.7	45.5	0.1089	13
Prostephanus truncatus	24.	100.	5.	430.	4.5	60.				0.1047	15
Ptinidae											
Ptinus ocelus	63.4	ca. 1 year	1.	276.		83.		122.7	98.0	0.0491	16
Gibbium psylloides	77.	ca. 1 year		283.		55.		76.4	129.5	0.0336	16
Silvanidae											
Oryzaephilus surinamensis	19.4	133.	5.5	375.	2.9	82.	A	107.3	44.8	0.1047	17,18
O. mercator	20.4	126.	5.5	200.	1.7	87.	A	94.3	51.1	0.0894	17,18
Cucujidae											
Cryptolestes trucicus	29.	73.	1.>	102.<		70.	A	32.8	59.2	0.0590	19,24
C. pusilloides	29.	76.<		141.<		80.		51.0	52.9	0.0740	20
Bruchidae											
Callosobruchus chinensis	22.	6.	1.	83.2	16.6	81.	A	28.9	24.5	0.1372	21
Anobiidae											
Lasioderma serricorne	20.	20.	0-3	110.	5.9		A	13.2	35.5	0.0730	22,23

[a] Mean developmental period in day; [b] mean adult longevity in day; [c] pre-oviposition period in day; [d] lifetime egg production per female; [e] reproductive effort per day calculated by d/(b-c); [f] % survival rate in immature stages; [g] type of survival curve labeled by Pearl & Miner (1935); [h] mean generation time.

References: 1.Nawrot(1979), 2. Imura(1986), 3. Imura (1985), 4. Navarro et al.(1986), 5. Teotia and Singh (1976), 6. Ayertey (1975), 7. Sonleitner(1978), 8. Leslie & Park(1949), 9. Dawson(1964), 10. Young(1970), 11. Imura(1987), 12. Tsuda & Yoshida (1984), 13. Birch (1953a), 14. Satomi (1955), 15. Bell & Watters (1982), 16. Howe (1953), 17. Howe (1956), 18. Arbogast (1976), 19. Lefkovitch (1962), 20. Lefkovitch (1964), 21. Shimada (unpublished), 22. Lefkovitch (1963), 23. Lefkovitch & Currie (1967), 24. Bishop (1959).

262

5. Reproductive Allocation

Fisher (1958) implicitly stated that organisms which allocate the available energy or resources towards reproduction and other demands in the way that maximizes future offspring will be selected for under the particular environment. The rule was termed "principle of allocation" (Cody, 1966). The reproductive efforts (Table 1), defined as the mean number of eggs laid per female per day, of the short-lived species are larger than those of the long-lived species, indicating the former spend more energy for reproduction per unit time than the latter. The reproductive effort used here is a convenient but rough measure.

Fig. 1 shows the reproductive allocations, expressed as the ratio of egg weight produced daily per female to the weight of the female in%, of two related species of the short-lived species and those of the long-lived species. A female of *E. kühniella* laid about twice as many as that of *E. elutella* (Fig 1, left panels). However, the total reproductive allocation, the ratio of the weight of eggs produced in lifetime to the weight of newly emerged adult female, were 34.9 ± 2.0 (s.e.)% and 44.3 ± 1.7% in *E. kühniella* and *E. cautella*, respectively, due to the fact that the smaller *E. elutella* females laid ca. 23% heavier eggs than the *E. kühniella* females. *T. castaneum* laid more eggs than *T. confusum*, but *T. confusum* eggs were 17% heavier than *T. castaneum* eggs and the mean weight of adult females of *T. confusum* was 16% lighter than that of *T. castaneum*. As the result, these two species had almost equivalent daily reproductive allocations (Fig. 1, right panels): the mean daily reproductive allocation during the first 30-day-reproductive-period of *T. castaneum* and *T. confusum* were 24.3 ± 1.0% and 24.2 1.8%, respectively.

E. elutella and *T. confusum* compensate in part for their smaller fecundity by laying relatively heavier eggs. This may be an adaptation to harsh environments (Ito, 1978). In fact, *E. elutella* infests leguminous seeds and tobacco which contain toxic substances such as saponins, inhibitors of digestive enzymes and nicotine which usually inhibit the

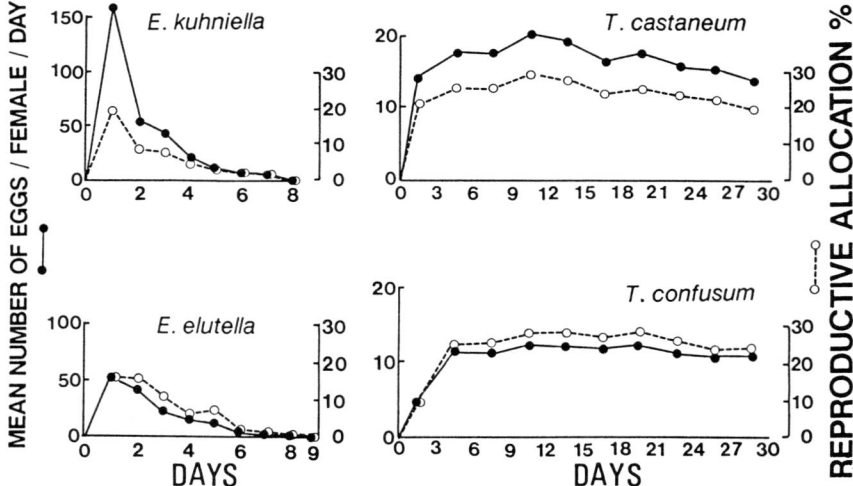

Figure 1. Mean daily fecundity per female and reproductive allocation in % (ratio of weight of eggs to weight of adult female which produced the eggs) of *Ephestia kühniella*, *Ephestia elutella*, *Tribolium castaneum*, and *Tribolium confusum*. The fecundity data during first 30-day-reproductive-period of the *Tribolium* species derived from Park and Frank (1948).

development of *E. kühniella*. Similarly, *T. confusum* can develop on more conditioned medium than *T. castaneum*. The results also support the statement that small species generally show greater reproductive allocation than large ones (Williams, 1966).

6. Reproduction-Survival Trade-off

According to "the principle of allocation" (Cody, 1966), the allocation of more energy to reproduction would decrease the allocation to maintenance (e.g. Williams, 1966; Pianka and Parker, 1975).

Ayertey (1975) reported that females of *S. cerealella* which mated normally and laid eggs lived a shorter time than females which were allowed to mate on the 5th day after emergence and delayed reproduction. Sonleitner (1978) also reported that *T. castaneum* females reared on fresh medium which laid more eggs in her early life lived shorter than those reared on conditioned medium which delayed reproduction but had the same lifetime fecundity as the former. Females of *T. castaneum* strains which were selected for an increase in early fecundity decreased in longevity, but males responded to the selection was heterogeneous (Mertz, 1975). Sokal (1970) reported a similar result in the same species.

In contrast, there were no significant negative correlations between lifetime fecundity and female adult longevity among the strains of a short-lived species *E. kühniella* (Imura, 1986) and long-lived species of *T. castaneum* and *T. confusum* (Park et al., 1961).

These results indicate a trade-off between age-specific fecundity and age-specific survival rate. Rose and Charlesworth (1981) reported that there was a negative genetic correlation between early fecundity and longevity, a genetic trade-off, in *Drosophila melanogaster*.

7. Population Dynamics

The pattern of population dynamics of an organism is determined by life-history traits (e.g. Southwood et al., 1974; Tsuda, 1982, 1988). Fig. 2 illustrates a typical population dynamics of a short-lived species of *E. cautella* (Takahashi, 1961) and a long-lived species of *S. zeamais* (Birch, 1953b). The equilibrium population number of *E. cautella* shows large fluctuations, while that of *S. zeamais* is rather stable. Similarly, the population numbers of other short-lived species such as *S. cerealella* (Crombie, 1945), *E. kühniella* (Flanders, 1968; Imura unpublished), *L. serricorne* (Lefkovitch, 1966), *C. chinensis* (Utida, 1941a) show large fluctuations. While, the population dynamics of long-lived species such as *T. confusum* (e.g. Stanley, 1932, McDonald, 1963, Takahashi and Yamamoto, 1972), *T. castaneum* (e.g. Mertz, 1969; Fujii, 1974) *T. freemani* (Imura, unpublished), *S. oryzae* (Birch, 1953b), *R. dominica* (Crombie,1945; Birch, 1953b), *O. surinamensis* (Crombie, 1945) are stable or if there is change that is rather gradual.

Fig. 3 shows a reproduction curve for a short-lived species *E. cautella* (Takahashi, 1956) and that of a long-lived species *S. zeamais* (Utida, 1956). The reproduction curve of *E. cautella* is humped and that of *S. zeamais* is monotonic. The reproduction curves of short-lived species *E. kühniella* (Imura, unpublished), *C. chinensis* (Utida, 1941b), *C. maculatus* (Bellows, 1982) and possibly *L. serricorne* (Bellows, 1981) are all humped type. While the reproduction curves of long-lived species *T. confusum* (Chapman, 1928; Takahashi and Yamamoto, 1972), *Cryptolestes sp.* (Varley et al., 1973) and *Cathartus*

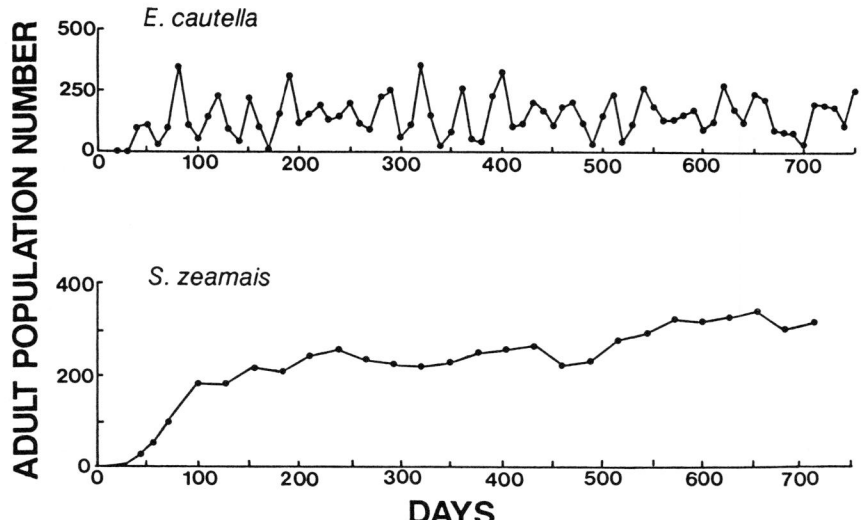

Figure 2. Long-term population dynamics of *Ephestia cautella* and *Sitophilus zeamais* (after Takahashi (1961) and Birch (1953b)).

quadricollis (Varley et al., 1973) are possibly a monotonic type. The humped type curve is liable to cause oscillations in equilibrium population density and the monotonic type stabilizes the densities (Ricker, 1954a, b). Smith and Sibly (1985) also discussed this. Along with the response to population density, the long and iteroparous life probably stabilizes the population dynamics by itself, because it possibly mitigates time-lag effect in the density response.

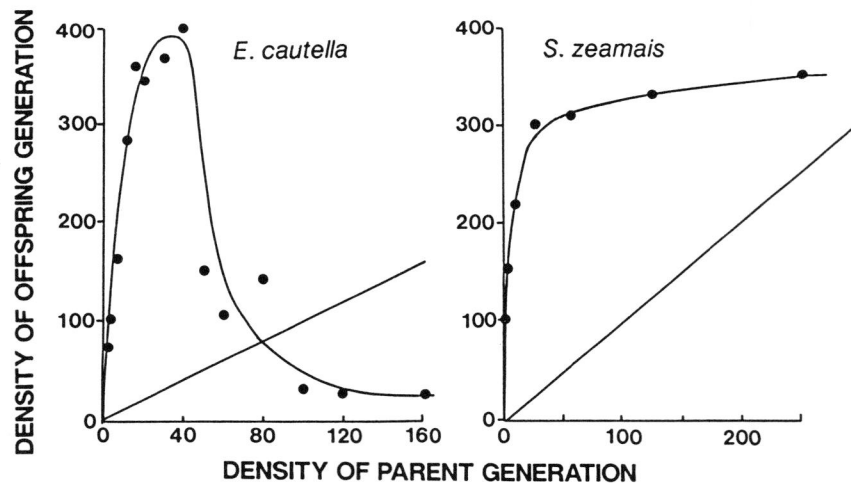

Figure. 3. Reproduction curves of *Ephestia cautella* and *Sitophilus zeamais*. Egg densities and adult densities were plotted in *E. cautella* and *S. zeamais*, respectively, and the offspring density of *S. zeamais* included parents surviving (after Takahashi (1956) and Utida (1956)).

8. Discussion

Major stored-product insects potentially have high r_m. Isolated stored-product environments are obviously an island (MacArthur, 1972). Stored-product insect populations often start with few founders which have been introduced with a large quantity of stored-products into the environment. A species with a high r_m has an advantage when colonizing in such insular environments (MacArthur and Wilson, 1967). The high r_m is mainly realized by fast development and precocity. In fact, the pre-reproductive period (developmental period + pre-oviposition period) was highly correlated with r_m ($r = -0.824^*$) in the short-lived species (Table 1), although there was no significant correlation between these traits ($r = -0.278$) in the long-lived species (Table 1).

The fact in the long-lived species does not support Cole's (1954) statement that species with long pre-reproductive periods could gain more from iteroparity than forms which mature more rapidly. The life-history of the long-lived species typically shows a continuous iteroparity (Begon et al., 1986). Bell (1976) suggested from a theoretical study that early maturity along with high fecundity was likely to evolve in the animals of continuous iteroparity.

Because stored products are a nonrenewable resource, after the establishment of the insect populations, the populations are inevitably subjected to decline in number. When the population declines, selection would favor long-lived and iteroparous individuals (Mertz, 1971). A long-lived iteroparous species is likely to be selected for under environments with low and unpredictable juvenile survival and low adult mortality (Murphy, 1968; Schaffer, 1974) and was termed "bet-hedger" (Stearns, 1976). Mertz (1971) reported that the juvenile mortality was often extremely high (more than 99%) and the adult survivorship increased in crowded and decreasing *Tribolium* populations. The bet-hedging theory explains satisfactory the evolution of life history of *Tribolium* species (Dawson, 1977; Imura, 1987). Subcortical habitats which are the natural habitat of Tenebrionid, Ptinid, Cucujid and Silvanid species must be this kind of environment, since these species exhibit similar life histories.

The short-lived and long-lived species exhibit different responses to population density. The type of competition was scramble in the former and contest in the latter at least in their larval stage. The type of competition in the short-lived species is also possibly exploitative and that in the long-lived species mostly interference (The definition of competition for seed-infesting species such as Bruchids, *Sitophilus* and *Sitotroga* is more complex. Within-seed and between-seeds competition should be distinguished).

We may conclude that the short-lived species are opportunists and the long-lived species are bet-hedgers. Therefore the two types of species must exploit stored-product environments in different way: short-lived species exploit them as ephemeral and long-lived species as uncertain and/or competitive. As the result, the short-lived species tend to exploit the resources quickly and then disperse to find new habitat. While the long-lived species tend to exploit the resources over a long period and persist in the same environment until new resources are added or next colonizing opportunity (*Sitophilus* and *Rhyzopertha* may more possibly bet-hedge in space rather than in time. Thus they may not persist in the same environment but disperse). Diapause and dispersion phenomena should be incorporated for further comprehensive explanations of life histories of stored-product insects (Denno and Dingle, 1981). We should undertake different pest-management strategies depending on the types of life histories of the insect pests.

Acknowledgments. I thank Dr. M. Shimada for allowing me to use his unpublished data and Dr. R. Mitchell for correcting the manuscript.

REFERENCES

Arbogast, R. T. (1976) Population parameters for *Oryzaephilus surinamensis* and *O. mercator.* Effect of relative humidity, Environ. Entomol. 5, 738-742.

Ayertey, J. N. (1975) Egg laying by unmated females of *Sitotroga cerealella* (Lepidoptera: Gelechiidae), J. stored Prod. Res. 11, 211-215.

Begon, M, Harper, J. L., and Townsend, C. R. (1986) Ecology, Blackwell Sci. Pub., Oxford.

Bell, G. (1976) On breeding more than once, Amer. Natur. 110, 57-77.

Bell, R. J. and Watters, F. L. (1982) Environmental factors influencing the development and rate of increase of *Prostephanus truncatus* (Horn) (Coleoptera: Bostrichidae) on stored maize, J. stored Prod. Res. 18, 131-142.

Bellows, T. S. Jr. (1981) The descriptive properties of some models for density dependence, J. Anim. Ecol. 50, 139-156.

Bellows, T. S. Jr. (1982) Analytical models for laboratory populations of *Callosobruchus chinensis* and *C. maculatus* (Coleoptera, Bruchidae), J. Anim. Ecol. 511, 263-287.

Birch, L. C. (1953a) Experimental background to the study of the distribution and abundance of insects. I. The influence of temperature, moisture and food on the innate capacity for increase of three grain beetles, Ecology 34, 698-711.

Birch, L. C. (1953b) Experimental background to the study of the distribution and abundance of insects. II. The relation between innate capacity for increase in numbers and the abundance of three grain beetles in experimental populations, Ecology 34, 712-726.

Bishop, G. W. (1959) The comparative bionomics of American *Cryptolestes* (Coleoptera-Cucujidae) that infest stored grain, Ann. Entomol. Soc. Amer. 52, 657-665.

Burges, H. D. (1963) Studies on the dermestid beetle *Trogoderma granarium* Everts. VI. Factors inducing diapause, Bull. Entomol. Res. 54, 571-587.

Caswell, G. H. (1960) Observations on an abnormal form of *Callosobruchus maculatus* (F.), Bull. Entomol. Res. 50, 671-680.

Chapman, R. N. (1928) The quantitative analysis of environmental factors, Ecology 9, 111-122.

Cody, M. L. (1966) A general theory of clutch size, Evolution 20, 174-184.

Cole, L. C. (1954) The population consequences of life history phenomena, Q. Rev. Biol. 29, 103-137.

Cotton, R. T. and Wilber, D. A. (1974) Insects, in C. M. Christensen (ed.), Storage of Cereal Grains and Their Products, AACC, St. Paul, pp. 193-231.

Crombie, A. C. (1945) On competition between different species of graminivorous insects, Proc. R. Soc. B 132, 362-395.

Dawson, P. S. (1964) Age at sexual maturity in female flour beetles, *Tribolium castaneum* and *T. confusum*, Ann. Entomol. Soc. Amer. 57, 1-3.

Dawson, P. S. (1965) Estimation of components of phenotypic variance for developmental rate in *Tribolium*, Heredity 20, 403-417.

Dawson, P. S. (1977) Life history strategy and evolutionary history of *Tribolium* flour beetles, Evolution 31, 226-228.

Denno, R. F. and Dingle, H. (1981) Considerations for the development of a more general life history theory, in R. F. Denno and H. Dingle (eds.), Insect Life History Patterns, Springer-Verlag, New York, pp. 1-6.

Dick, J. (1937) Oviposition in certain coleoptera, Ann. Appl. Biol. 24, 762-796.

Emlen, J. M. (1984) Population Biology, Macmillan Pub. Co., New York.

Englert, D. C. and Bell, A. E. (1969) Components of growth in genetically diverse populations of *Tribolium castaneum*, Can. J. Genet. Cytol. 11, 896-907.

Englert, D. C. and Bell, A. E. (1970) Selection for time of pupation in *Tribolium castaneum*, Genetics 64, 541-552.

Evans, D. E. (1977) The capacity for increase at a low temperature of several Australian populations of *Sitophilus oryzae* (L.), Aust. J. Ecol. 2, 55-67.

Fisher R. A. (1958) The Genetical Theory of Natural Selection, 2nd ed., Dover Pub. Inc., New York.

Flanders, S. E. (1968) Mechanisms of population homeostasis in *Anagasta* ecosystem, Hilgardia 39, 367-404.

Fujii, K. (1974) Internal and external factors governing the fluctuation of *Tribolium castaneum* populations, Proc. 1st. Intern. Work. Conf. Stored-Prod. Entomol., pp. 597-607.

Hagstrum, D. W. and Sharp, J. E. (1975) Population studies on *Cadra cautella* in a citrus pulp warehouse with particular reference to diapause, J. Econ. Entomol. 68, 11-14

Hagstrum, D. W. and Silhacek, D. L. (1980) Diapause induction in *Ephestia cautella*: an interaction between genotype and crowding, Entomol. exp. appl. 28, 29-37.

Howe, R. W. (1953) Studies on beetles of the family Ptinidae VIII. The intrinsic rate of increase of some ptinid beetles, Ann. Appl. Biol. 40, 121-134.

Howe, R. W. (1956) The biology of the two common storage species of *Oryzaephilus* (Coleoptera, Cucujidae), Ann. Appl. Biol. 44, 341-355.

Hutchinson, G. E. (1978) An Introduction to Population Ecology, Yale Univ. Press, New Haven.

Imura, O. (1985) Biology of phycitid moths associated with stored products, Shokuryo 25, 47-76 (in Japanese).

Imura, O. (1986) Studies on the color variation in larvae of *Ephestia kühniella* Zeller (Lepidoptera: Phycitidae) III. Fitness of different larval color strains, Res. Popul. Ecol. 28, 281-293.

Imura, O. (1987) Demographic attributes of *Tribolium freemani* Hinton (Coleoptera: Tenebrionidae), Appl. Entomol. Zool. 22, 449-455.

Ito, Y. (1978) Comparative Ecology 2nd ed., Iwanami, Tokyo (in Japanese).

Kiritani, K. (1961) The origin of the household pests, and the process of adaptation to indoor conditions, Seitai-Konchyu 9, 22-40 (in Japanese).

Kiritani, K. and Kawahara, S. (1963) Effect of adult diet, on the longevity, fecundity, oviposition period and phototaxis in the black carpet beetle, *Attagenus megatoma* (F.), Jpn. J. Ecol. 13, 21-28.

Lefkovitch, L. P. (1962) The biology of *Cryptolestes trucicus* (Grouvelle)(Coleoptera: Cucujidae), a pest of stored and processed cereals, Proc. Zool. Soc. London 138, 23-35.

Lefkovitch, L. P. (1963) Census studies on unrestricted populations of *Lasioderma serricorne* (F.)(Coleoptera: Anobiidae), J. Anim. Ecol. 32, 221-231.

Lefkovitch, L. P. (1964) The biology of *Cryptolestes pusilloides* (Steel & Howe)(Coleoptera, Cucujidae), a pest of stored cereals in the southern hemisphere, Bull. Entomol. Res. 54, 649-656.

Lefkovitch, L. P. (1966) A population growth model incorporating delayed responses, Math. Biophys. 28, 219-233.

Lefkovitch, L. P. and Currie, J. E. (1967) Factors affecting adult survival and fecundity in *Lasioderma serricorne* (F.)(Coleoptera, Anobiidae), J. stored Prod. Res. 3, 199-212.

Leslie, P. H. and Park, T. (1949) The intrinsic rate of natural increase of *Tribolium castaneum* Herbst, Ecology 30, 469-477.

Lewontin, R. C. (1965) Selection for colonizing ability, in H. G. Baker and G. L. Stebbins (eds.), The Genetics of Colonizing Species, Academic Press, New York, pp. 77-91.

Linsley, E. G. (1944) Natural sources, habitats, and reservoirs of insects associated with stored products, Hilgardia 16, 187-224.

MacArthur, R. H. (1972) Geographical Ecology, Harper and Row Pub., New York.

MacArthur, R. H. and Wilson, E. O. (1967) The Theory of Island Biogeography, Princeton Univ. Press, Princeton.

McDonald, D. J. (1963) Natural selection in experimental populations of *Tribolium*. III. Characteristics of *Tribolium confusum* populations in equilibrium, Amer. Natur. 97, 383-396.

Mertz, D. B. (1969) Age-distribution and abundance in populations of flour beetles, I. Experimental studies, Ecol. Monog. 38, 1-31.

Mertz, D. B. (1971) Life history phenomena in increasing and decreasing populations, in G. P. Patil, E. C. Pielou and W. E. Waters (eds.), Statistical Ecology, Vol. 2, Pennsylvania State Univ. Press, University Park, pp. 361-396.

Mertz, D. B. (1975) Senescent decline in flour beetle strains selected for early adult fitness, Physiol. Zool. 48, 1-23.

Mertz, D. B., Park, T. and Youden, W. J. (1965) Mortality patterns in eight strains of flour beetles, Biometrics 21, 99-114.

Murphy, G. I. (1968) Pattern in life history and the environment, Amer. Natur. 102, 391-403.

Nakakita, H. (1982) Effect of larval density on population of *Tribolium freemani* Hinton (Coleoptera: Tenebrionidae), Appl. Entomol. Zool. 17, 269-276.

Navarro, S., Donahaye, E. and Calderon, M. (1986) Development of the carob moth, *Spectrobates ceratoniae*, on stored almonds, Phytoparasitica 14, 177-186.

Nawrot, J. (1979) Effect of temperature and relative humidity on population parameters for almond moth (*Cadra cautella* Wlk.)(Lepid. Phycitidae), Prace Naukowe IOR Poznan 21, 41-52.

Norris, M. J. (1934) Contributions towards the study of insect fertility III. Adult nutrition, fecundity, and longevity in the genus *Ephestia* (Lepidoptera, Phycitidae), Proc. Zool. Soc. London 2, 333-360.

Orozco, F. and Bell, A. E. (1974) Reciprocal recurrent selection compared to within-strain selection for increasing rate of egg lay of *Tribolium* under optimal and stress conditions, Genetics 77, 143-161.

Park, T. and Frank, M. B. (1948) The fecundity and development of the flour beetles, *Tribolium confusum* and *Tribolium castaneum*, at three constant temperatures, Ecology 29, 368-374.

Park, T., Mertz, D. B., and Peterusewicz, K. (1961) Genetic strains of *Tribolium*: their primary characteristics, Physiol. Zool. 34, 62-80.

Pearl. R. and Miner, J. R. (1935) Experimental studies on the duration of life. XIV. The comparative mortality of certain lower organisms, Q. Rev. Biol. 10, 60-79.

Pianka, E. R. and Parker, W. S. (1975) Age-specific reproductive tactics, Amer. Natur. 109, 453-464.

Rich, S. S., Bell, A. E., Miles, D. A. and Wilson, S. P. (1984) An Experimental study of genetic drift for two quantitative traits in *Tribolium*, J. Heredity 75, 191-195.

Ricker, W. E. (1954a) Effects of compensatory mortality upon population abundance, J. Wild. Manag. 18, 45-51.

Ricker, W. E. (1954b) Stock and recruitment, J. Fish. Res. Bd. Canada 11, 559-623.

Rose, M. R. and Charlesworth, B. (1981) Genetics of life history in *Drosophila melanogaster*. I. Sib analysis of adult females, Genetics 97, 173-186.

Satomi, H. (1955) A comparative study of some physiological and ecological characters of the rice weevils, *Carandra oryzae* L. and *C. sasakii* Takahashi collected from different districts of the world, Botyu-Kagaku 20, 55-61 (in Japanese).

Schaffer, W. M. (1974) Optimal reproductive effort in fluctuating environment, Amer. Natur. 108, 783-790.

Smith, R. H. and Sibly, R. (1985) Behavioural ecology and population dynamics: towards a synthesis, in R. M. Sibly and R. H. Smith (eds.), Behavioural Ecology, Blackwell Sci. Pub., Oxford, pp. 577-591.

Sokal, R. R. (1970) Senescent and genetic load: evidence from *Tribolium*, Science 167, 1733-1734.

Sonleitner, F. J. (1978) Longevity and oviposition rate in *Tribolium castaneum*: Rubner's hypothesis and natural selection, Tribolium Inf. Bull. 21, 150-154.

Southwood, T. R. E., May, R. M., Hassel, M. P. and Conway, G. R. (1974) Ecological strategies and population parameters, Amer. Natur. 108, 791-804.

Stanley, J. (1932) A mathematical theory of the growth of populations of the flour beetle, *Tribolium confusum* Duv., Can. J. Res. 6, 632-671.

Stearns, S. C. (1976) Life-history tactics: a review of the ideas, Q. Rev. Biol. 51, 3-47.

Takahashi, F. (1956) The maximum reproduction of population and the larval density. 3. On the effect of population density on the power of the reproduction of the almond moth, *Ephestia cautella*, Res. Popul. Ecol. 3, 27-35 (in Japanese).

Takahashi, F. (1961) The equilibrium state in the experimental population of the almond moth,

Ephestia cautella Walker, Jpn. J. Appl. Entomol. Zool. 5, 239-244 (in Japanese).

Takahashi, F. and Yamamoto, Y. (1972) Upper limit of the population growth of *Tribolium confusum* Duval, Kontyu 40, 55-64.

Teotia, T. P. S. and Singh, Y. (1976) Studies on the growth of population of *Sitotroga cerealella* Olivier, Bull. Grain Technol. 14, 9-17.

Tsuda, Y. (1982) Reproductive strategy of insects as adaptation to temporally varying environments, Res. Popul. Ecol. 24, 388-404.

Tsuda, Y. (1988) Life-history strategies and domestication, in F. Nakasuji (ed.), Insect Evolutions and Life-History Strategies, Tohjusha, Tokyo, pp. 144-196 (in Japanese).

Tsuda, Y. and Yoshida, T. (1984) Population biology of the broad-horned flour beetle, *Gnathocerus cornutus* (F.) (Coleoptera: Tenebrionidae) I. Life table and population parameters, Appl. Entomol. Zool. 19, 129-131.

Tsuji, H. (1963) Experimental Studies on the Larval Diapause of the Indian-meal Moth, *Plodia interpunctella* Hubner (Lepidoptera: Pyralidae), Kokodo Ltd., Tokyo.

Utida, S. (1941a) Studies on experimental population of the azuki bean weevil, *Callosobruchus chinensis* (L.) V. Trend of population density at the equilibrium position, Mem. Coll. Agr., Kyoto Imp. Univ. 51, 27-34.

Utida, S. (1941b) Studies on experimental population of the azuki bean weevil, *Callosobruchus chinensis* (L.). II. The effect of population density on progeny populations under different conditions of atmospheric moisture, Mem. Coll. Agr., Kyoto Imp. Univ. 49, 1-20.

Utida, S. (1954) "Phase" dimorphism observed in the laboratory population of the cowpea weevil, *Callosobruchus quadrimaculatus*, Oyodobutsu 18, 161-168 (in Japanese).

Utida, S. (1956) Effect of population density upon reproductive rate in relation to adult longevity, Jpn. J. Ecol. 5, 137-140 (in Japanese).

Utida, S. (1972) Density dependent polymorphism in the adult of *Callosobruchus maculatus* (Coleoptera, Bruchidae), J. stored Prod. Res. 8, 111-126.

Utida, S. (1976) Polymorphism in the adult of *Callosobruchus maculatus*, Proc. Joint US-Jpn. Semi. Stored Prod. Insect, pp. 174-185.

Varley, G. C., Gradwell, G. R., and Hassell, M. P. (1973) Insect Population Ecology, Blackwell Sci. Pub., Oxford.

White, G. G. (1984) Variation between field and laboratory population of *Tribolium castaneum* (Herbst) (Coleoptera: Tenebrionidae), Aust. J. Ecol. 9, 153-155.

Williams, G. C. (1966) Adaptation and Natural Selection, Princeton Univ. Press, Princeton.

Yoshida, T. (1981) Recent advances in stored-product entomology, in S. Ishii (ed.), Recent Advances in Entomology, Univ. Tokyo Press, Tokyo, pp. 384-401 (in Japanese).

Yoshida, T., Igarashi, H., and Shinoda (1986) Life history of *Callosobruchus chinensis* (L.) (Coleoptera, Bruchidae), in E. Donahaye and S. Navarro (eds.), Proc. 4th Intern. Work. Conf. Stored-Product Prot., pp. 417-477.

Young, A. M. (1970) Predation and abundance in populations of flour beetles, Ecology 51, 602-619.

BIOTYPE VARIATION AND HOST CHANGE IN BRUCHIDS : CAUSES AND EFFECTS IN THE EVOLUTION OF BRUCHID PESTS

PETER F. CREDLAND
*Department of Biology, Royal Holloway and Bedford New College,
University of London, Egham, Surrey TW20 0EX, U.K.*

ABSTRACT. Results are presented to demonstrate that populations of *Callosobruchus maculatus* may be genetically different with respect to bionomic characteristics and that they also differ in their capacity to utilize unusual hosts. They may, however, rapidly adapt to such hosts and utilize them more efficiently. The multivoltine, polyphagous habits of pest bruchid species are then considered in the context of their origin from oligophagous, wild conspecifics which appear, in many respects, to resemble the typical non-pest bruchids in their life-cycles. It is concluded that the different habits of the pest species are a consequence of their lives as pests of stored legumes and not pre-adaptations for such an existence. The cultivation of their primary wild hosts is suggested to be the key factor that has resulted in their importance. Their spread into additional cultivated hosts can logically be expected but the emergence of new pest species is considered unlikely unless their current wild hosts are brought into cultivation and intensively bred for mammalian consumption.

1. Introduction

There are around 1300 species in the family Bruchidae (Johnson, 1981a) but only around 20 of these, belonging to just 6 genera, attack grain legumes, grown, stored and eaten by man (Southgate, 1979). In this same review, written 10 years ago, Ben Southgate pointed out that "the possible movement of bruchids into new areas of host association must be borne in mind if agricultural development is to continue". Obviously any study of this problem should have regard to both the capacity of known pest species to move into new hosts and the possibility of species which are not currently living on cultivated species evolving into pests. Realistically, it is probably impossible to guess which of several hundred species could become pests although it may theoretically be feasible to try and identify characteristics which may provide some level of pre-adaptation to pest status. It may however be possible to examine species which are currently pests and look for situations or characteristics which may enable them to increase their host range among agricultural crops, thereby providing an early warning of possible future trends. In approaching the problem it should always be remembered that host changes in the laboratory, under intensive selection, can only indicate what may occur in the real world outside, and not what will happen if alternative hosts are present. Nevertheless, laboratory experiments are perhaps the only practical method of approaching such a potentially vast area of study.

In considering bruchids as pests, it would be wise to recall that Labeyrie (1981) in the first Symposium on Bruchids and Legumes in Tours noted that "there can be no such things as stored product insects, but there are insects which can invade and infest

K. Fujii et al. (eds.), Bruchids and Legumes: Economics, Ecology and Coevolution, 271–287.
© 1990 *Kluwer Academic Publishers. Printed in the Netherlands.*

stored products". Nevertheless, there is some suggestion that individual species of bruchid may utilize different hosts which may be "wild", which are not cultivated or harvested and are therefore attacked only in the field, "primary" which are attacked both in the field and store having been cultivated and harvested, or "secondary" in which case they are attacked only in stores (Watanabe and Sugimoto, 1988). There may therefore be cases of bruchid-host associations which are found only under storage conditions and need recognition as such. Labeyrie (1981) further emphasized the importance of examining the biocenotic relationships in wild plant populations and the crops in the areas of domestication of cultivated plants. We had then (Labeyrie, 1981), and still have now, only rudimentary information on the ecology of 4 or 5 harmful bruchid species and, furthermore, "we cannot think of achieving an efficient protection of legume crops without taking into account the remarkable polymorphism of bruchids" (Labeyrie, 1981).

Polymorphism is a term with an established meaning in the context of population or evolutionary genetics. When "two or more discretely different phenotypes are fairly common within a population (and) the rarest of them exceeds some arbitrarily high frequency - say one percent or so - this condition is referred to as polymorphism" (Futuyma, 1986). In the field of bruchid biology, the term has been used in this and other senses (see Pajni, 1986). In the case of *Callosobruchus maculatus* there is a phenotypic and behavioural polymorphism which falls clearly within the general definition. This was reviewed by Utida (1981) and has been further studied by Messina and Renwick (1985) and Messina (1987). In essence there is a genetic and environmental control of the production of active, "flight" forms and "normal" individuals in single populations. Comparable polymorphism may also occur in *C. chinensis* and *C. analis* (see Utida, 1981) and a similar dimorphism has been reported in laboratory cultures of *Zabrotes subfasciatus* (Pajni, 1986). Huignard and Biemont (1979) referred to genetic polymorphism when noting the different reproductive activities of Colombian and European strains of *Acanthoscelides obtectus* although the two strains may not now be considered as part of a single population. The same authors even used the term "reproductive polymorphism" in the title of their paper comparing the reproductive strategies of 2 allopatric populations of the same species from Colombia, although the term was not actually used in the text (Huignard and Biemont, 1981). Garaud (1984) discovered a chromosomal translocation polymorphism in a population of *A. obtectus* from Burundi but this was not apparently associated with any phenotypic polymorphism.

It may therefore be advantageous to distinguish between polymorphism of the kind recognized by Utida (1981) and others in *Callosobruchus*, or the special case of chromosomal polymorphism in *Acanthoscelides*, and the distinctions noted between populations by Huignard and Biemont (1979) in *Acanthoscelides* or the interstrain differences recorded, for example, by Fujii (1968) in *Callosobruchus chinensis*, Dick and Credland (1984) in *C. maculatus*, Robert (1985) in *Caryedon serratus*, or Meik and Dobie (1986) in *Zabrotes subfasciatus*. All these latter cases are not strictly polymorphism but may better be described as biotypical differences with respect to the use of the term biotype proposed by Diehl and Bush (1984) as "a temporary and provisional designation for cases where biological differences have been observed between organisms but where the genetic basis and evolutionary status of the differences have yet to be ascertained".

The use of the biotype concept will bring bruchid studies into line with many other investigations of host association. Saxena and Barrion (1987) referred for example to the concept as being one which is recognized by function and that biotypes represent "evolutionary transients in the process of speciation and develop through natural selection acting upon genetic variation within the pest populations". They considered that biotypes can be sympatric or allopatric but differ in one or more biological attributes such as host or food preferences or reproductive analyses. The assumption was made that there is a similarity in the genetic composition of individuals comprising a biotype. The principles embodied in this concept are in no way at odds with those to which Labeyrie (1981) referred in considering polymorphism; indeed Diehl and Bush (1984) said that "biotypes have obvious implications for pest management and biological control, as the failure to recognize distinct populations can have costly and frustrating consequences".

Two consequences emerge from the previous comments. Firstly, there should be genetic differences between biotypes. These could be recognized by the maintenance of bionomic distinctions between them when they are kept under as near identical conditions as possible, thereby eliminating environmentally induced differences. Such distinctions have already been observed in the studies of reproductive performance in *A. obtectus* to which reference has been made, in various bionomic parameters of *C. chinensis* (Fujii, 1968), and in various attributes of *C. maculatus* (e.g. Dick and Credland, 1984; Credland, 1986; Credland et al., 1986). It should also be possible to demonstrate that parameters of these kinds are inherited and have some kind of genetic basis; results of one such experiment are presented below. Secondly, one might anticipate that there will be differences in the capacity of different biotypes to utilize an unusual host meaning, in this context, one to which the biotypes have not been exposed for a considerable time, if ever, in their history. This would imply that host change under field conditions may be dependent not only on the availability of an alternative and unusual host but also on the biotype of the bruchid species which encounters it. A simple preliminary experiment which may indicate the significance of such a hypothesis is described and some of the observations made are used in a consideration of short-term bruchid evolution.

2. Materials and Methods

2.1. BIOTYPES AND CULTURING METHODS

Biotypes used in the experiments described were all maintained in a constant temperature and humidity (CTH) room at $27 \pm 1°C$ and $70 \pm 10\%$ R.H. Cultures were maintained on Californian black-eye peas, a cultivar of the cowpea, *Vigna unguiculata* (Walp.), and only one generation was allowed to develop in each batch of seeds before subculturing. Further details of the culturing procedure have been given previously (Dick and Credland, 1984).

2.2. INTERBIOTYPE HYBRIDIZATION

The Campinas, IITA and Yemen biotypes, all of which have been described previously (Credland, 1986), were used in these experiments. Newly emerged, virgin adults were collected from subcultures and the two sexes of each biotype were placed separately in different containers. When more than enough for each predetermined cross had been collected, the contents of the requisite containers were added together and mating pairs removed. Each pair was transferred to a tube (7.5 x 2.5 cm), lined with emery paper to inhibit oviposition on the walls, and containing 5 seeds which had previously been conditioned for at least 10 days in the CTH room. The seeds had previously been sieved in a 6.7 mm grain sieve, only those retained and with intact testa being utilized. The number of eggs laid, the number which hatched, and the number of adults (F_1) finally emerging from the eggs in each tube were determined. These results are referred to subsequently as the 'parental generation'. Some newly emerged virgin adults from each tube were collected as before and used to establish second (F_1) generation self- and back-crosses and the whole procedure repeated again with the emerging adults (F_2) for third (F_2) generation crosses. The pairings established with F_2 adults were all either self-crosses or were back-crosses of F_2 adults to individuals of the same sex and biotype as had been mated with the hybrid parent in the previous generation.

Twenty replicates of every cross were initially established in each generation. Detailed data are given below only for crosses between the Yemen and Campinas biotypes.

2.3. RESPONSES OF DIFFERENT BIOTYPES TO HOST CHANGE

Ten biotypes have all been tested in the same way. All those not mentioned previously were initially collected from cowpeas in the country with the name by which the biotype is designated. They had all been in culture under the conditions described for at least 3 years, more than 30 generations, before the experiments were undertaken; during this period none had been exposed to lentils. Newly emerged virgin adults were collected in the usual way, allowed to mate and then each pair was immediately transferred to a 7.5 x 2.5 cm tube lined with emery paper containing either 40 sieved and conditioned cowpeas or 120 conditioned, whole green lentils (*Lens culinaris* Medic.) which have a total surface area approximately the same as 40 cowpeas (Credland, 1987). Twenty replicates were set up for each biotype on each host. Each pair was left until both adults had died after which time the number of eggs laid and the number hatching was recorded. The tubes were then stored in the CTH room and examined at regular intervals until either all the adults which would do so had emerged, or it was impossible to distinguish between the adults emerging from the eggs initially laid and those which were the progeny of adults which had emerged earliest and oviposited before their removal from the tube. Since there were cases of individuals taking up to 259 days to emerge, there was no feasible way of avoiding this problem. It should be noted that the times of emergence recorded below are the number of days which elapsed from the introduction of the adults to the tube until the emergence of their progeny from the host seeds. They are not, therefore, strictly speaking the time from oviposition to adult eclosion. Adults normally oviposit for only about 6 days but the time from adult eclosion to emergence of the adult from a lentil is unknown; it is, however, unlikely to be more than a day or two. The errors ensuing from these unknown factors are most unlikely to affect the interpretation of the results obtained but should be considered carefully before too much reliance is placed on the absolute values for development periods.

Two biotypes, Campinas and Mali, were subcultured onto lentils in January and May, 1988, respectively, and have been subcultured only onto lentils since that time; such subculturing was undertaken when a number of adults were present, not necessarily representing the passage of a generation for the reason expressed previously. Probably between 4 and 6 generations had been passed in lentils before adults were collected in the usual way in April 1989 and put through the same experimental protocol as that described for adults from cowpeas.

3. Results

3.1. INTERBIOTYPE HYBRIDIZATION

A preliminary experiment indicated that males and females of all three biotypes would mate readily, leading to the production of viable and fertile offspring.

3.2. PARENTAL GENERATION

The females of all pairings laid similar numbers of eggs (Fig. 1a) and there was no statistical difference between the intra- and interbiotype crosses ($F_{1,118}=2.54$). Similarly

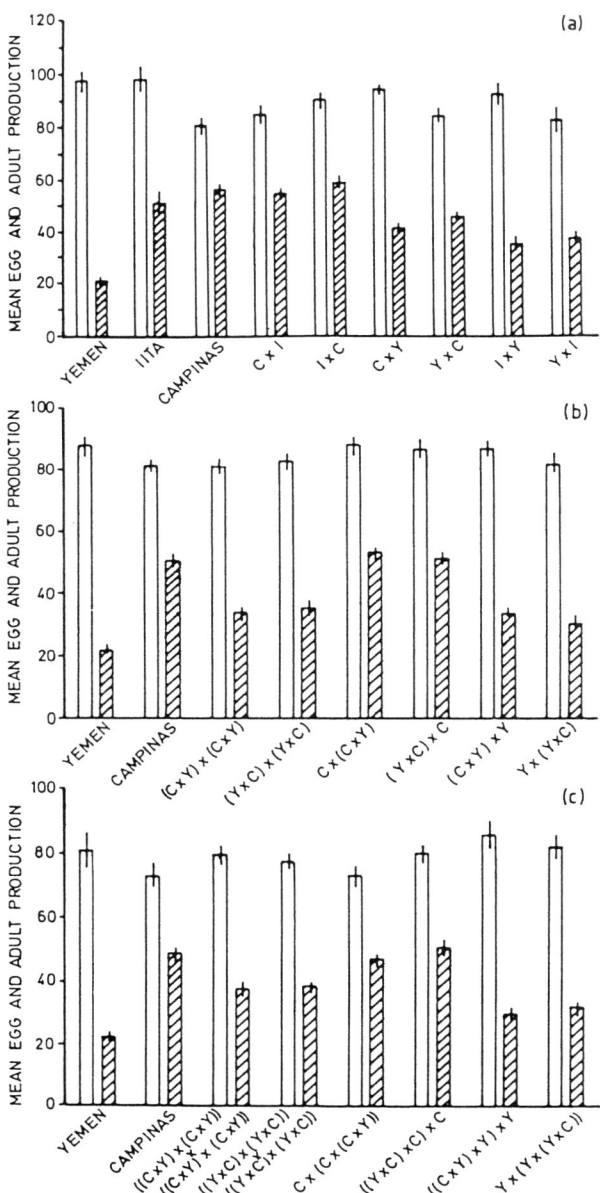

Figure 1. Mean egg production (open columns) and adult progeny (hatched columns) of individual females of the classes of pairings indicated in the (a) parental, (b) F_1 and (c) F_2 generations. Each female had access to 5 seeds.

there were no differences in the numbers of eggs which failed to hatch ($F_{1,118}$=1.26). As these figures include the numbers of first instar larvae which failed to penetrate the testa, having hatched from the egg itself, this would indicate that there are no differences in fertility, embryonic or first instar viability. Throughout the whole experiment, the mean percentage of eggs failing to produce larvae which entered seeds was only 6.5%.

The greatest number of adults were produced from intrabiotype pairings of Campinas and IITA individuals or crosses between them. The pairing of two Yemen individuals produced the smallest number of F_1 adults. Crosses between the other two biotypes and that from the Yemen produced intermediate values in both reciprocal crosses. As a consequence of these results, crosses established in subsequent generations concentrated on pairings between the Yemen and either of the other biotypes.

3.3. F_1 GENERATION

As in the parental generation, the numbers of eggs laid by females were again similar (Fig. 1b) although it was noted that the number of eggs laid by IITA females mated with (IITA x Yemen) males was greater than Campinas females mated with (Campinas x Yemen) males. The data on adult emergence again indicated that there were significant differences in the number of adults emerging in intrabiotype pairings, with Yemen producing the fewest and Campinas the most adult progeny ($F_{2,51}$=60.8, P<0.001). Analysis of the data for the interbiotype crosses was undertaken sequentially. An initial factorial analysis was undertaken omitting the self-crosses to examine 4 effects:
(1) "Biotype", involving a separation of crosses involving Yemen and Campinas from those between Yemen and IITA,
(2) "Yemen input", separating hybrids back-crossed to Yemen from those back-crossed to one of the other biotypes,
(3) "Yemen sex", separation on the basis of the sex of the Yemen parent in the parental generation,
(4) "Reciprocal combinations", comparing male hybrids back-crossed to single biotype females with female hybrids back-crossed to single biotype females.

This analysis (Table 1) revealed that both "biotype" and "Yemen input" had a very significant effect on the number of adults produced, whereas the other two factors had no significant effect. The largest source of variation was caused by "Yemen input"; back-crossing hybrids to either IITA or Campinas produced more F_2 progeny than crossing hybrids to Yemen. The "biotype" effects result from the production of more adults from Yemen and Campinas hybrids than those involving Yemen and IITA.

The data from reciprocal crosses was then combined, and the "Yemen input" factor expanded to include three levels - self-crosses, back-crosses to IITA or Campinas, and back-crosses to Yemen. A two-way analysis of variance was then performed to test simultaneously for the effect of "biotype" and "Yemen input". The latter ($F_{2,218}$=96.54, P<0.001) was the larger source of variation, a significantly higher number of adults being produced by back-crosses to IITA or Campinas than back-crosses to Yemen, or self-crosses. The "biotype" factor was also significant ($F_{1,218}$=9.94, P<0.001) for the reasons already given.

3.4. F_2 GENERATION

Again, the number of eggs laid by individual females was similar in all cases, although IITA females laid more than those of the other biotypes in intrabiotype pairings. Crosses involving Yemen and IITA also laid more than those involving Yemen and

Table 1. Factorial analysis of variance in the number of adult progeny of females in the F_1 interbiotype crosses (excluding self-crosses) showing only the results for the four main factors and significant interactions.

	Source	DF	SS	MS	F	P
A	Biotype	1	738.03	738.03	15.96	<0.001
B	Yemen input	1	9184.03	9184.03	198.63	<0.001
C	Sex	1	78.03	78.03	1.69	N.S.
D	Reciprocal	1	14.69	14.69	0.32	N.S.
	AB	1	455.11	455.11	9.84	<0.005
	AD	1	186.78	186.78	4.04	<0.05
	Error	128	5918.44	46.24		
	Total	143	16954.00	118.56		

Campinas. The numbers of adult progeny (Fig. 1c) were analyzed, as before, in a two-way analysis of variance to test for the effects of "biotype" and "Yemen input", the three levels of "Yemen input" now being - self-crosses among F_2 hybrids, F_2 hybrids back-crossed to IITA or Campinas individuals, and back-crosses of F_2 hybrids to Yemen individuals (Table 2). "Yemen input" had the larger effect on adult production, the highest number arising from back-crosses to IITA or Campinas, and the lowest from back-crosses to Yemen. As in the previous generation Yemen and Campinas crosses produced more offspring than Yemen and IITA crosses.

TABLE 2. Two-way analysis of variance and Newman-Keuls Multiple Range Test on adult production by F_2 females in interbiotype crosses.

Source	DF	SS	MS	F	P
Biotype	1	680.0	680.0	14.57	<0.001
Yemen input	2	11250.0	5265.1	120.56	<0.001
Interaction	2	5.9	3.0		N.S.
Error	266	12411.6	46.7		
Total	271	24379.5			

Newman-Keuls Multiple Range Test

Pairing	Mean	N	Separation
Back-cross to Yemen	28.90	68	a
Self-crosses	36.38	136	b
Back-crosses to IITA or Campinas	46.96	68	c

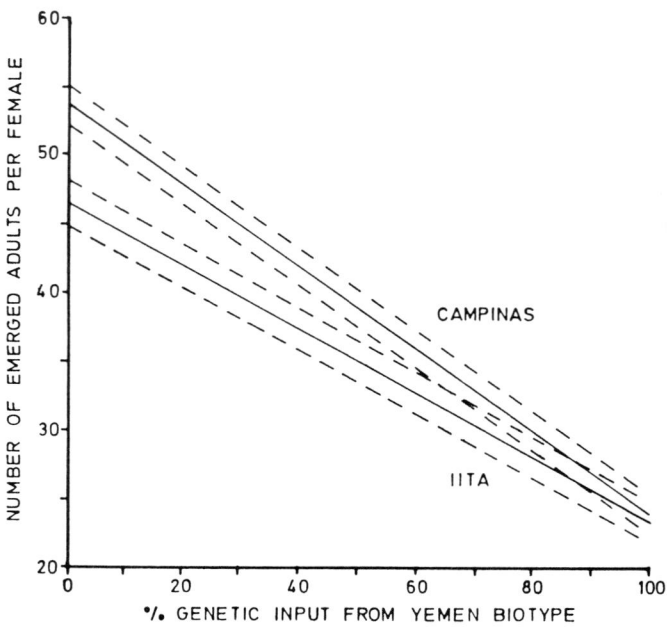

Figure 2. Linear regression of the number of emerged adults per female on percentage genetic input from the Yemen biotype for crosses involving Yemen and IITA (IITA) and Yemen and Campinas (Campinas). Broken lines represent 95% confidence intervals.

The adult emergence data for all the biotypes and hybrids was summarized and related to the mean percentage genetic input to each cross from the Yemen biotype. If the regression of adult production per female on mean percentage genetic input from the Yemen biotype is calculated and plotted separately for hybrids between Yemen and IITA or Campinas (Fig. 2), both regression coefficients are significantly different from zero, additional Yemen input leading to a decline in adult production in both cases. The regression coefficients for IITA and Campinas are significantly different, Yemen input causing a greater reduction in the number of emergences in Campinas than IITA hybrids.

3.5. RESPONSES OF DIFFERENT BIOTYPES TO HOST CHANGE

All ten biotypes that were tested laid eggs on both lentils and cowpeas. The mean number of eggs laid by each female on 40 cowpeas varied from a minimum of 79.10 by females of the Brazil biotype to a maximum of 120.16 by females of the Yemen biotype (Fig. 3). Mean numbers of eggs laid on 120 lentils varied from a minimum of 13.38, laid by Campinas females, to a maximum of 76.42 laid by females of the Nigeria biotype. The percentage of eggs hatching did not show much variation, ranging from 98.5 to 82.46, regardless of the host seed and biotype. There was no evidence that hatching was consistently affected by the type of seed on which the eggs were laid. Between 82.6 and 98.6% of the hatched eggs on cowpeas led to the production of an adult beetle. In most cases emergence was around 90% (Fig. 4). There were, however, many fewer emergences from hatched eggs on lentils.

The most successful biotype was from Mali in which case 33.5% of the hatched eggs produced adults and the least successful was from the Cameroons where only 3.7% of

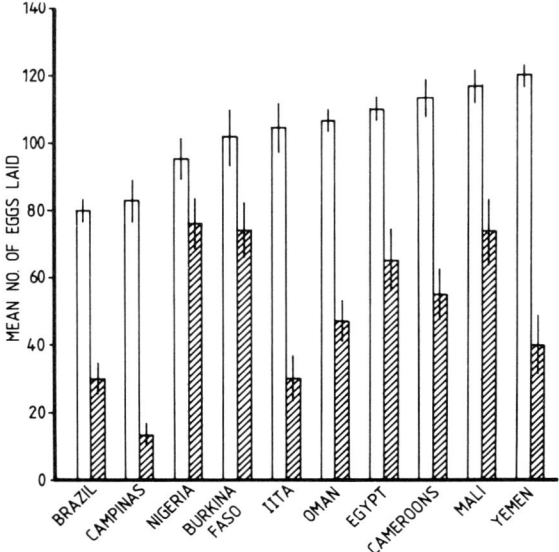

Figure 3. Mean numbers of eggs laid by individual females of different biotypes reared in cowpeas on either cowpeas (open columns) or lentils (hatched columns).

the hatched eggs produced adults. Not only was larval survival suppressed but the duration of development was significantly extended when compared with the rates recorded from cowpeas. The median day of emergence from cowpeas varied from 24 to 30 days after the adults were put into the tubes; the true values are probably two or three days

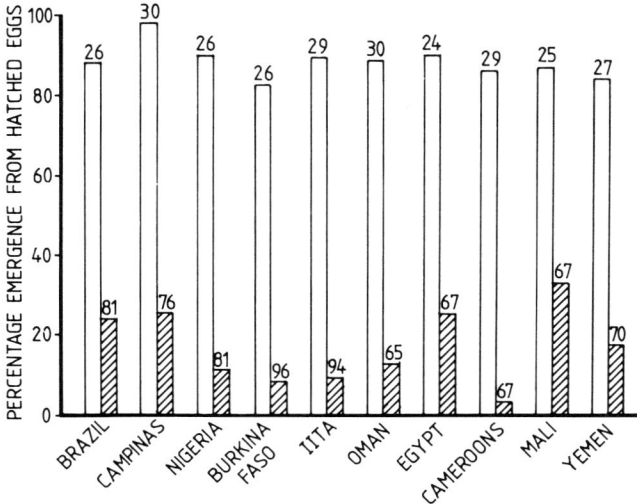

Figure 4. Percentage emergence of adults from hatched eggs laid on either cowpeas (open columns) or lentils (hatched columns) by females of different biotypes reared in cowpeas. The median day of emergence is shown above each column.

less for the reasons explained previously. In lentils, development was invariably much slower; median days of emergence varied from 65 to 96 days, values around 70 days being most common. The first beetles to emerge from lentils usually did so after around 50 days but there was enormous variation in development periods with the maxima recorded usually being in excess of 150 days, and over 250 in some cases.

Maintenance of the Campinas and Mali biotypes on lentils had rather different effects when adults from these stocks were tested. The Campinas females laid a few more eggs on cowpeas than those which had been kept on that host but egg laying on lentils dramatically increased, from a mean of 13.38 to 58.21 eggs per female. Conversely, the Mali females from lentils laid rather fewer eggs on cowpeas than similar females reared in cowpeas, but almost the same number on lentils as those females which had been reared in cowpeas. Emergences from cowpeas showed a small decline in the case of the Campinas biotype, but no change in the case of Mali when the parental generation had been reared in lentils as opposed to cowpeas. Emergences of Campinas adults from lentils rose from 25% to 56% after rearing the parental stock in lentils for less than 18 months and the median development time fell from 76 to 53 days. Similarly, there was a marked increase (33 - 69%) in the percentage of hatched Mali eggs which produced an adult after keeping the stock on lentils for 12 months and, again, the median development time fell, from 67 to 45 days. It can therefore be concluded that both biotypes significantly adapted to using lentils as hosts by an increase in the number of larvae surviving and the mean rate of their development.

4. Discussion

The experiment in which three biotypes were crossed and hybrids bred for two further generations, indicated quite clearly that some bionomic characteristics such as the number of adults which can be produced in each seed are inherited. It is, of course, necessary to remember that recording the mean emergence may conceal a discrepancy in the patterns of larval survival exhibited by progeny of individual pairings and that a preponderance of different survival rates will be sufficient to yield different overall means. The experiment does not provide any insight into how such characteristics are transmitted or the genetic basis of inheritance; indeed, it is probable that a whole complex of factors are involved and that simple Mendelian inheritance of the sophisticated suite of activities which lead to adult emergence is unlikely. Nevertheless, it is evident that despite the complexity of the larval behaviour patterns which may be involved, the interactions of larvae of different genotypes yield evidence that there are genetic differences between biotypes which are heritable and materially affect bionomic characteristics. It is reassuring that Wasserman (1988) has recently, and independently, demonstrated the inheritance of realized fecundity in hybrids between two strains of *C. maculatus.*

In the second experiment all the biotypes, as expected, were extremely successful when reared in cowpeas. Development was rapid and the great majority of larvae survived to produce adults. Conversely, there were fundamental differences in the capacity to utilize lentils, although all the biotypes developed more slowly and with greater mortality than when reared in cowpeas. Differences in egg laying were immense. Campinas, as reported previously (Credland, 1987), laid few eggs and many females did not lay at all. The same was true to a lesser extent of the IITA, Cameroons and the Yemen biotypes, but females of others, such as the Nigeria and Mali biotypes, laid a mean of more than 70 eggs. In no cases could the limited survival of larvae in lentils be attributed to competition as 120 seeds were available and eggs were distributed fairly uniformly in most cases. One must therefore assume that one or more antibiotic factors were responsible for larval mortality. There was great variation among the biotypes with successful emergences varying from 4 to 33% of the hatched eggs. Since all the biotypes had been kept on cowpeas for some years before the experiments, so that they had not been "selected" for utilization of lentils, it seems reasonable to assume that there is some innate variation among them in the capacity to use this unusual host, although it should be noted that a population from Turkey and the Yemen biotype were both initially collected from lentils, so that this host is used by *C. maculatus* under some field conditions. The capacity of biotypes to adapt to a novel host is revealed by the Campinas and Mali biotypes kept on lentils for a year and then retested in the usual experimental protocol. The Mali biotype was among the most successful when initially tested in terms of the numbers of eggs laid and the percentage emergence. No changes were noted in the number of eggs laid on lentils but the speed of development has dramatically increased and the survival of larvae has more than doubled. In the case of the Campinas biotype, unlike Mali, there has been a great increase in the number of eggs laid on lentils (13 - 58 eggs/female) after keeping preceding generations on that host for just over one year. Larval development was also much quicker and the percentage survival from hatched eggs much greater than when the adults were cultured on cowpeas (Fig. 5). Similar differences between biotypes in their response to a resistant cultivar of cowpea and adaptations to it in just three generations have been demonstrated previously (Dick and Credland 1986a,b), indicating that the responses to atypical hosts are not unusual and may be expected to occur quite widely.

It has been established that *Callosobruchus maculatus* does occur in a number of genetically distinct biotypes. Nobody knows how many exist, how stable the situation is, or indeed how it has arisen, but one may assume that a degree of geographical isolation allied to local selection pressures and perhaps an initially limited range of genotypes

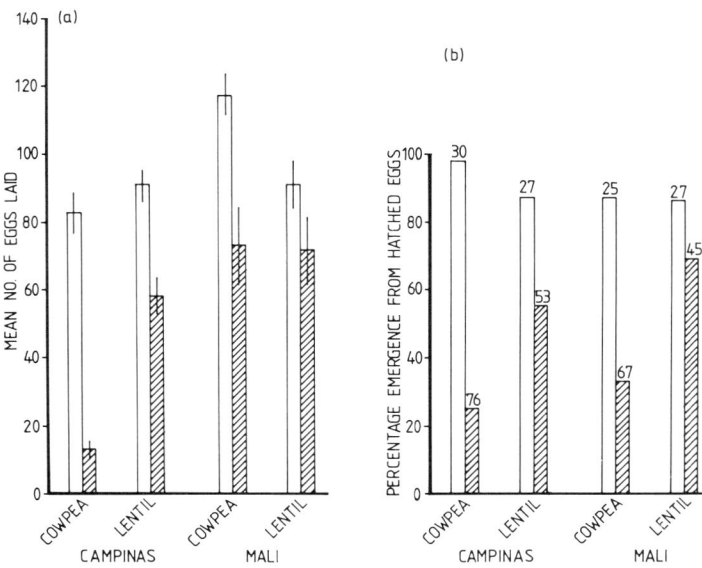

Figure 5. Comparison of egg laying and percentage emergence from hatched eggs on cowpeas (open columns) and lentils (hatched columns) when the Campinas and Mali biotypes had been reared on cowpeas or lentils as indicated.

were involved. There is some evidence to suggest that other bruchid pest species may show a similar if less well defined pattern of occurrence. *C. chinensis* is known to exhibit variation among different cultures of distinct origin (Fujii, 1968), *Acanthoscelides obtectus* is similar (Huignard and Biemont, 1978; Biemont et al., 1987), and *Zabrotes subfasciatus* is also variable (Meik and Dobie, 1986). In all these species there is evidence that individual stocks respond differently in similar experimental situations, as opposed to the numerous cases of the same stock showing a variety of responses when offered different hosts or being kept under different conditions (e.g. Chandrakantha and Mathavan, 1986). This biotypical variation is additional to the polymorphism shown by some species of *Callosobruchus* and *A. obtectus*, which was described previously. The situation in "wild" or "field" species is totally unknown but many have somewhat restricted distributions compared with the pest species and might be expected to show less variation. A relevant question to the evolution of bruchid pests is therefore whether the pest species were initially more variable and could therefore become pests, or whether variation has arisen after their attainment of pest status and subsequent transport to different isolated areas where distinct genotypes have evolved through selective pressure or some drift or founder effect.

Wild species of bruchid which are not pests are usually mono- or oligophagous; Janzen (1974) reported that over 80% of Costa Rican bruchids had only one known host species and Johnson and Slobodchikoff (1979) produced results leading to a similar conclusion. Where accounts exist of the storage pest species living wild, they always indicate that they are essentially mono- or oligophagous, just like the preponderance of other bruchid species. *Acanthoscelides obtectus* is apparently found on two species of *Phaseolus*, *P. vulgaris* and *P. coccineus*, with the latter actually being attacked more frequently (Biemont and Bonet, 1981). Johnson (1981b) suggested a wider current host range but noted that some records needed verification and was prepared to state that "the original and still principal host .... is *Phaseolus vulgaris* and other species of *Phaseolus*". Similarly, *Z. subfasciatus* was only found on one wild host species, *P. lunatus* (Pimbert, 1985) although it may well also infest *P.vulgaris*, and *Callosobruchus maculatus* was only found in cowpeas, *Vigna unguiculata*, out of 34 species of legume examined in Niger (Alzouma, 1981). *Callosobruchus chinensis*, the azuki bean weevil, has been found on only two wild hosts, *Dubaria villosa* and *Vigna trilobata* (*V. angularis* var. nipponensis) (Yoshida et al., 1986), the latter being closely related to mungbean, *V. radiata*, with which it shows partial crossing compatibility, and cultivated azuki bean, *V. angularis* (Fujii and Miyazaki, 1987). *Caryedon serratus* represents a rather different situation from these other species. It is a cosmopolitan bruchid which may be of African (Robert, 1985) or Asian (Davey, 1958) origin; it has 6 known wild hosts in Africa where it is also a pest of groundnuts, *Arachis hypogea*, (Robert, 1985) but in southern Asia it is predominantly found in only a single wild host, *Tamarindus indica*, besides occurring as a pest in locally grown groundnuts (Dick, 1987). It therefore differs from the other pest species in that its primary cultivated host, groundnut, is not infested in the one part of the insect's range, South America, from which that host originated. The story is, however, complicated since one of its wild hosts, *Tamarindus indica*, is also cultivated for its pod contents which are eaten by man and has been widely distributed to Africa and elsewhere, probably with *C. serratus* in its pods or seeds, from its Asian origin. This situation is apparently unique but may actually be informative of the evolutionary processes involved in host change since it demonstrates once again the initial association of a bruchid with a wild host brought into cultivation, although the beetle's greatest importance is in the context of a different crop. Even allowing for a slightly wider spectrum of hosts utilized which may emerge if collection from wild sources were extended into areas from which collections have not been made, it does therefore seem eminently reasonable to consider all these species no more than oligophagous in the wild although they are typically polyphagous in their associations with cultivated legumes (e.g. Pajni, 1986).

Furthermore, like those species which are not pests, wild populations of the pest species may also be univoltine unlike their conspecifics living in stored, cultivated, seeds; Biemont and Bonet (1981) found univoltine *A. obtectus* in wild hosts alongside a multivoltine population infecting a cultivated crop. Similarly, *Callosobruchus chinensis* produces four generations each year in adjacent stores but is univoltine on its wild hosts in Japan, overwintering in the seeds and the adults emerging in spring to feed on pollen, nectar and fungi before ovipositing on pods in the autumn (Yoshida et al., 1986). Prevett (1966) noted reports of *Caryedon serratus* also hibernating in tamarind pods in Pakistan and Japan but recorded continuous breeding in Nigeria. As Southgate (1981) suggested previously, "the potential for multivoltinism is inherent to a great number of bruchid species" but it is actually rare outside 'storage' situations.

What then has inhibited wild species from developing closer associations with cultivated legumes and prevented host changes? Although some wild species such as *Stator vachelliae* which normally develops in *Acacia* can produce the occasional adult from *Parkinsonia aculeata* (Johnson, 1988) this may signal the potential for host change, but it may not be enough to suggest the mechanism for the evolution of a 'new' pest. A requirement for pollen, often though not always from the normal host (Labeyrie and Hossaert, 1985), to initiate further reproduction would necessitate departure from accumulated seeds and, frequently, a return to the normal host. This requirement for food as "pollen and honey" (nectar?), to stimulate reproductive activity is a common factor among 'wild' bruchids (Watanabe, 1985) and field pests (Pajni, 1986) although nutrition may not always be of primary importance. *Bruchidius atrolineatus*, for example, is not restricted to its host for pollen, but food alone was not enough to terminate imaginal diapause; contact with floral scars, the corolla, the terminal part of the pod or extrafloral nectaries were ultimately required to stimulate development and diapause termination, leading Huignard et al. (1987) to deduce that host allelochemicals were the critical factors. The dependence of wild species on contact with the growing host may be enough to militate against their successful transition to storage situations. In the case of *Acanthoscelides obtectus* although some wild populations require hosts for oogenesis and reproduction, others do not and this could have been a key factor in their successful transition to storage situations (Huignard and Biemont, 1978). *Callosobruchus maculatus* has a similar duality in terms of the response of the female reproductive system to host plants since the 'flight' form requires the presence of the host and the 'flightless' form will begin egg production in its absence (Ouedraogo and Huignard, 1981). Although Ouedraogo and Huignard (1981) correctly reject the notion of the 'flight' form as the "nature" form and "flightless" one as that "found in stores" one cannot automatically reject the hypothesis that the latter may have evolved or become more numerous as a consequence of man's activities. Initial association with *V. unguiculata* in the wild state preceded a situation where prolonged storage of seed after cultivation extended the availability of the host to adult beetles. This new evolutionary pressure may have led to polymorphism, widespread dispersal of the species, pest status and subsequent polyphagy. It would be interesting to know if the 'flight' form reproduces in the field in the absence of cowpeas, where that crop is rarely grown and the beetle is found on different hosts in stores. It is known that populations of the beetle differ in their capacity to produce the 'flight' form and in the suggested situation it may be expected to disappear altogether as has happened with the IITA biotype after prolonged culture (Messina and Renwick, 1985). The ultimate condition, where a species was only found in stores, would then have been reached. Polymorphism in *C. maculatus* may therefore be viewed as a consequence of their attainment of pest status or at least their life in seeds stored and available in bulk for prolonged periods, and not as a precondition for the attainment of pest status. The variation in the reproductive physiology of *A. obtectus* could have a similar origin, since it could have evolved in response to different agricultural practices or it may represent variation in wild populations which has evolved under different climatic conditions.

What distinctive features of their biology have therefore limited the evolution of bruchid species into pests of cultivated, stored seeds to just 20 out of the 1300 or more species in the world? Grain legumes which are cultivated differ from wild ones in at least two significant ways. Firstly they are grown in large amounts and the seeds are stored in bulk under conditions designed for their long-term survival so that they can be consumed or planted months after their production, and secondly they must either contain few vertebrate (human) toxins or such toxins as do occur must be easily degraded, usually by some form of cooking. The pest species of bruchid of which we have significant knowledge are each found in one wild host which has been brought into cultivation or is (closely related to) an ancestor of a cultivated species (Fujii and Miyazaki, 1987). After cultivation there would be selection for diminished human toxin content and some form of seed storage would be undertaken. Since the factors toxic to humans or other mammals are the same, in many cases, as those which inhibit beetle development (Bell, 1978; Liener, 1982), the beetles would encounter a progressively more favourable situation for their development as plant breeders deliberately or inadvertently selected cultivars of progressively lower toxicity. Furthermore, trade in seeds would favour movement of the beetle away from its source to other areas where human activities, including agriculture, took place and therefore where alternative legume crops again with diminished toxin levels, may be in cultivation. In some cases, as demonstrated in the experiment described previously, beetles could survive in the alternative host but one might expect that infestation of such abnormal hosts would not be universal since the origin or biotype of the beetles exposed to them will determine whether or not colonization occurs even if the new host is treated as uniform throughout its range, itself an unfounded and patently false assumption (e.g. Simmonds et al., (1989). A patchy and irregular distribution of pest species on abnormal hosts, such as the occurrence of *Caryedon serratus* on groundnuts is therefore likely to occur. Suitable biotypes of *C. serratus* may simply not have reached Latin America where groundnuts are grown. A similar situation may explain the fact that *Callosobruchus subinnotatus* is a pest of *Voandzeia subterranea*, bambarra nuts, in only a few widely separated parts of the host range (Southgate, 1978), although the influence of climatic factors cannot be discounted in this example. *Acanthoscelides obtectus* occasionally occurs in a range of abnormal hosts over only a part of their common distribution, for example on cowpeas in Jalapa in Mexico (Jarry and Bonet, 1982) or *Zabrotes subfasciatus* on cowpeas in Uganda and Burma (Meik and Dobie, 1986) whilst other populations or biotypes have been unable to colonize the same hosts. These and similar observations may be attributed to biotypical variation allied with variation in host quality.

Regular exposure to cultivated plants is an essential requirement for the evolution of pest species and transfer of established pest species between crops is, I believe, a greater cause of concern than bruchid species currently known only from wild legumes. However, should any new legume species be brought into cultivation, as seems quite possible in the foreseeable future, then bruchids living on the ancestors of such species should be viewed as potential pests of other related crops already under cultivation.

I therefore believe that the very small number of bruchid species recognized as pests is not surprising but is a direct consequence of the hosts occupied by their wild ancestors being brought into cultivation. There is, however, a real prospect of the spread of established pests into other cultivated legumes which they do not currently infest but a relatively small chance of new bruchid species attaining pest status unless their current hosts are brought into regular intensive cultivation. Quarantine regulations inhibiting the transmission of established pest species from one area to another still play an important part in control programmes as the dispersal of biotypes as well as species could lead to unforeseen problems. The argument that a pest species is already present in a particular country or region is invalid since, as shown above, biotypes are inherently variable in their capacity to infest different hosts.

Acknowledgements. I am greatly indebted to Dr Kenny Dick who provided most of the information on interbiotype hybridization and to Ros Jones who has spent many tedious hours helping with the experiments on host transfers. I also value the comments of many people with whom I have discussed the subject of this contribution and, on this occasion, I would like to single out the late Ben Southgate whose invaluable experience and knowledge of bruchids was inspirational in the early stages of my work on these fascinating insects. I am indebted to the Overseas Development Natural Resources Institute, the SERC, and the Royal Society for their financial support.

REFERENCES

Alzouma, I. (1981) Observations on the ecology of *Bruchidius atrolineatus* Pic and *Callosobruchus maculatus* F. (Coleoptera, Bruchidae) in Niger, in V. Labeyrie (ed.), The Ecology of Bruchids Attacking Legumes (Pulses), Dr. W. Junk Publishers, The Hague, pp. 205-213.

Bell, E. A. (1978) Toxins in seeds, in J. B. Harbourne (ed.), Biochemical Aspects of Plant and Animal Coevolution, Academic Press, London, pp. 143-161.

Biemont, J. C. and Bonet, A. (1981) The bean weevil populations from the *Acanthoscelides obtectus* Say group, living on wild or subspontaneous *Phaseolus vulgaris* L. and *Phaseolus coccineus* L. and on *Phaseolus vulgaris* L. cultivated in the Tepotzlan region state of Morelos - Mexico, in V. Labeyrie (ed.), The Ecology of Bruchids Attacking Legumes (Pulses), Dr. W. Junk Publishers, The Hague, pp. 23-42.

Biemont, J. C., Butare, I., and Huignard, J. (1987) Grouping and inhibition of oogenesis in two strains of *Acanthoscelides obtectus* Say (Coleoptera Bruchidae) of different geographical origin - Specificity of this inhibitory effect, Int. J. Invert. Reprod. Devel. 12, 185-197.

Chandrakantha J. and Mathavan, S. (1986) Changes in the developmental rates and biomass energy in *Callosobruchus maculatus* (F.) (Coleoptera: Bruchidae) reared on different foods and temperatures, J. stored Prod. Res. 22, 71-75.

Credland, P. F. (1986) Effect of host availability on reproductive performance in *Callosobruchus maculatus* (F.) (Coleoptera: Bruchidae), J. stored Prod. Res. 22, 49-54.

Credland, P. F. (1987) Effects of host change on the fecundity and development of an unusual strain of *Callosobruchus maculatus* (F.) (Coleoptera: Bruchidae), J. stored Prod. Res. 23, 91-98.

Credland, P. F., Dick, K. M. and Wright, A. W. (1986) Relationships between larval density, adult size and egg production in the cowpea seed beetle, *Callosobruchus maculatus*, Ecol. Entomol. 11, 41-50.

Davey, P. M. (1958) The groundnut bruchid, *Caryedon gonagra* (F.), Bull. ent. Res. 49, 385-404.

Dick, K. M. (1987) Pest management in stored groundnuts, Information Bulletin no. 22., Patancheru, A.P. 502 324, India: International Crops Research Institute for the Semi-arid Tropics.

Dick, K. M. and Credland, P. F. (1984) Egg production and development of three strains of *Callosobruchus maculatus* (F.) (Coleoptera: Bruchidae), J. stored Prod. Res. 20, 221-227.

Dick, K. M. and Credland, P. F. (1986a) Variation in the response of *Callosobruchus maculatus* (F.) to a resistant variety of cowpea, J. stored Prod. Res. 22, 43-48.

Dick, K. M. and Credland, P. F. (1986b) Changes in the response of *Callosobruchus maculatus* (Coleoptera Bruchidae) to a resistant variety of cowpea, J. stored Prod. Res. 22, 227-233.

Diehl, S. R. and Bush, G. L. (1984) An evolutionary and applied perspective of insect biotypes, Ann. Rev. Entomol. 29, 471-504.

286

Fujii, K. (1968) Studies on interspecific competition between the azuki bean weevil and the southern cowpea weevil. III. Some characteristics of strains of two species, Res. Popul. Ecol. 10, 87-98.

Fujii, K. and Miyazaki, S. (1987) Infestation resistance of wild legumes (*Vigna sublobata*) to azuki bean weevil, *Callosobruchus chinensis* (L.) (Coleoptera: Bruchidae) and its relationship with cytogenetic classification, Appl. Ent. Zool. 22, 229-230.

Futuyma, D. J. (1986) Evolutionary Biology (Second edition), Sinauer Associates, Sunderland, Massachusetts.

Garaud, P. (1984) Mise en evidence d'un polymorphisme chromosomique de translocation dans une population naturelle d'*Acanthoscelides obtectus* (Coleoptere, Bruchidae) du Burundi, Genetica 63, 85-91.

Huignard, J. and Biemont, J. C. (1978) Comparison of four populations of *Acanthoscelides obtectus* (Coleoptera: Bruchidae) from different Colombian ecosystems, Oecologia 35, 307-318.

Huignard, J. and Biemont, J. C. (1979) Vitellogenesis in *Acanthoscelides obtectus* (Coleoptera: Bruchidae). II. The conditions of vitellogenesis in a strain from Colombia. Comparative study and adaptive significance., Int. J. Invert. Reprod. 1, 233-244.

Huignard, J. and Biemont, J. C. (1981) Reproductive polymorphism of populations of *Acanthoscelides obtectus* from different Colombian ecosystems, in V. Labeyrie (ed.), The Ecology of Bruchids Attacking Legumes (Pulses), Dr. W. Junk Publishers, The Hague, pp. 149-164.

Huignard, J., Germain, J. F., and Monge, J. P. (1987) Influence of the inflorescence and pods of *Vigna unguiculata* Walp. (Phaseolinae) on the termination of the reproductive diapause of *Bruchidius atrolineatus* (Pic) (Coleoptera, Bruchidae), in V. Labeyrie et al. (eds.), Insects - Plants, Dr. W. Junk, Dordrecht, pp. 183-188.

Janzen, D. H. (1974) The role of the seed predator guild in a tropical deciduous forest, with some reflections on tropical biological control, in D. P. Jones and M. E. Soloman (eds.), Biology in Pest and Disease Control, Blackwell, Oxford, pp. 3-14.

Jarry, M. and Bonet, A. (1982) La bruche du haricot, *Acanthoscelides obtectus* Say (Coleoptera, Bruchidae), est-elle un danger pour le cowpea, *Vigna unguiculata* (L.) Walp.?, Agronomie 2, 963-968.

Johnson, C. D. (1981a) Seed beetle host specificity and the systematics of the Leguminosae, in R. M. Polhill and P. H. Raven (eds.), Advances in Legume Systematics, Royal Botanic Garden, Kew, pp. 995-1027.

Johnson, C. D. (1981b) Relations of Acanthoscelides with their plant hosts, in V. Labeyrie (ed.), The Ecology of Bruchids Attacking Legumes (Pulses), Dr. W. Junk Publishers, The Hague, pp. 73-81.

Johnson, C. D. (1988) The possible beginning of adaptation to a new host by bruchid beetles in Venezuela, Biotropica 20, 80-81.

Johnson, C. D. and Slobodchikoff, C. N. (1979) Coevolution of *Cassia* (Leguminosae) and its seed beetle predators (Bruchidae), Environ. Entomol. 8, 1059-1064.

Labeyrie, V. (1981) Ecological problems arising from weevil infestation of food legumes, in V. Labeyrie (ed.), The Ecology of Bruchids Attacking Legumes (Pulses), Dr W Junk Publishers, The Hague, pp. 1-15.

Labeyrie, V. and Hossaert, M. (1985) Ambiguous relations between *Bruchus affinis* and the *Lathyrus* group, Oikos 44, 107-113.

Leiner, I. E. (1982) Toxic constituents in legumes, Oxford and IBH Publishing Co., New Delhi, pp. 217-257.

Meik, J. and Dobie, P. (1986) The ability of *Zabrotes subfasciatus* to attack cowpeas, Entomol. exp. appl. 42, 151-158.

Messina, F. J. (1987) Genetic contribution to the dispersal polymorphism of the cowpea weevil (Coleoptera: Bruchidae), Ann. Entomol. Soc. Am. 80, 12-16.

Messina, F. J. and Renwick, J. A. A. (1985) Dispersal polymorphism of *Callosobruchus maculatus* (Coleoptera: Bruchidae): variation among populations in response to overcrowding. Ann.

Entomol. Soc. Am. 78, 201-206.

Ouedraogo, A. P. and Huignard, J. (1981) Polymorphism and ecological reactions in *Callosobruchus maculatus* F. (Coleoptera, Bruchidae) in Upper Volta, in V. Labeyrie (ed.), The Ecology of Bruchids Attacking Legumes (Pulses), Dr. W. Junk Publishers, The Hague, pp. 175-184.

Pajni, H. R. (1986) Ecological status of host range and polymorphism in Bruchidae, in E. Donahaye and S. Navarro (eds.), Proc. 4th Intern. Work. Conf. Stored-Product Protection, Tel Aviv, Israel, Sept. 1986, pp. 506-516.

Pimbert, M. (1985) A model of host plant change of *Zabrotes subfasciatus* Boh. (Coleoptera: Bruchidae) in a traditional bean cropping system in Costa Rica, Biol. Agric. Hort. 3, 39-54.

Prevett, P. F. (1966) Observations on biology in the genus *Caryedon* Schonherr (Coleoptera: Bruchidae) in Northern Nigeria, with a list of associated parasitic Hymenoptera, Proc. R. ent. Soc. Lond. (A) 41, 9-16.

Robert, P. (1985) A comparative study of some aspects of the reproduction of three *Caryedon serratus* strains in presence of its potential host plants, Oecologia 65, 425-430.

Saxena, R. C. and Barrion, A. A. (1987) Biotypes of insect pest of agricultural crops, Insect Sci. Applic. 8, 453-458.

Simmonds, M. S. J., Blaney, W. M., and Birch, A. N. E. (1989) Legume seeds: the defences of wild and cultivated species of *Phaseolus* against attack by bruchid beetles, Annls. Bot. 63, 177-184.

Southgate, B. J. (1978) The importance of the Bruchidae as pests of grain legumes, their distribution and control, in S. R. Singh et al. (eds.), Pests of Grain Legumes: Ecology and Control, Academic Press, London, pp. 219-229.

Southgate, B. J. (1979) Biology of the Bruchidae, Ann. Rev. Entomol. 24, 449-473.

Southgate, B. J. (1981) Univoltine and multivoltine cycles: their significance, in V. Labeyrie (ed.), The Ecology of Bruchids Attacking Legumes (Pulses), Dr. W. Junk Publishers, The Hague, pp. 17-22.

Utida, S. (1981) Polymorphism and phase dimorphism in *Callosobruchus*, in V. Labeyrie (ed.), The Ecology of Bruchids Attacking Legumes (Pulses), Dr. W. Junk Publishers, The Hague, pp. 143-149.

Wasserman, S.S. (1988) Partial paternal inheritance of realized fecundity in a bruchid beetle, *Callosobruchus maculatus*, Behav. Genet. 18, 193-200.

Watanabe, N. (1985) Oviposition habit of *Sulcobruchus sauteri* (Pic) and its significance in speculation on the pre-agricultural life of seed beetles attacking stored pulses (Coleoptera, Bruchidae), Kontyu (Tokyo) 53, 391-397.

Watanabe, N. and Sugimoto, S (1988) Geographical variation in male antennae of the azuki bean weevil, *Callosobruchus chinensis* (L.) (Coleoptera: Bruchidae), Appl. Entomol. Zool. 23, 282-290.

Yoshida, T., Igarashi, H., and Shinoda, K. (1986) Life history of *Callosobruchus chinensis* (L.) (Coleoptera, Bruchidae), in E. Donahaye and S. Navarro (eds.), Proc. 4th Intern. Work. Conf. Stored-Product Protection, Tel Aviv, Israel, Sept. 1986, pp. 471-477.

EXPERIMENTAL CHARACTER DISPLACEMENT IN THE ADZUKI BEAN WEEVIL, *CALLOSOBRUCHUS CHINENSIS*

MARK L. TAPER*
Division of Entomology,
National Institute of Agro-Environmental Sciences,
Tsukuba, Ibaraki 305 Japan

* present address
Dept. of Biology, University of New Mexico,
Albuquerque, NM 87131 U.S.A.

Dedicated to **Dr. S. Utida**

ABSTRACT. Laboratory populations of *C. chinensis* in the presence of interspecific competition evolve differently from populations in its absence. A strain of *C. chinensis* with high genetic variation (hereafter **hvC**) was reared for nine generations in competition with *C. maculatus* (Fuji lab **aQ** strain). Two types of beans were used as resources, mungbean (*Vigna radiata*) and lentil (*Lens culinaris*). Both beans were novel to both strains. Initially both beetle strains grew extremely well on mung. Lentil, on the other hand, was essentially lethal to **aQ** and severely reduced the emergence of **hvC** and retarded its development. Control treatments included **hvC** alone on both lentil and mung, **hvC** alone on lentil, **hvC** alone on mung, and a hard selection treatment in which **hvC** was allowed to oviposit on both mung and lentil but all mung were discarded before beetle emergence. At the end of the experiment, no difference was detectable in oviposition preference among the **hvC** from any of treatments. On the other hand, in all treatments where it was exposed to lentil, **hvC** increased its physiological adaptation to lentil, as expressed by changes in the probability of emergence, development time, and size at emergence (size is strongly correlated to fecundity). The increase in adaptation in the competition treatment was significantly greater than that in the treatment where **hvC** was raised on lentil and mung without interspecific competition. Furthermore, there was little difference between the competition treatment and the hard selection or lentil alone treatments, indicating the potency of competition as an evolutionary force.

1. Introduction

Despite its theoretical importance and numerous field studies, there appears to be no published direct experimental test of character displacement. The present study was undertaken to fill this gap.

The clearest definition of character displacement is given by Abrams (1986): "...character displacement is defined as any genetically based change in resource utilization that is caused by interspecific competition." As Abrams pointed out, both processing efficiencies and capture rates are aspects of resource utilization that may be subject to character displacement.

In this experiment two species of seed-beetles were reared for nine generations in competition for two resources. To simplify the design, observations, and analysis, an

K. Fujii et al. (eds.), Bruchids and Legumes: Economics, Ecology and Coevolution, 289–301.
© 1990 *Kluwer Academic Publishers. Printed in the Netherlands.*

evolutionary response to competition was only sought in one of the species. This is termed the **target** species while the other is considered the **driver**. The **target** species could utilize both resources, while the **driver** could only use one.

Character displacement would be indicated if at the end of the experiment the ability of the **target** species from the experimental lines to utilize the resources is significantly different from that of the **target** species in the control lines. Because of the initial asymmetry between species in resource utilization, one can predict that competition lines of the **target** species should evolve an increased utilization of the resource not used by the **driver** species.

2. Materials

2.1. SPECIES

I used two species of the genus *Callosobruchus*. The **driver** species was *C. maculatus* (Fujii laboratory strain code a**Q**). The **target** species was *C. chinensis* (**hvC**). The mnemonic stands for high variance Chinensis. I constructed the **hvC** strain for this experiment with repeated reciprocal crossings of 11 laboratory strains of *C. chinensis*. Females of strain *A* were crossed with males of strain *B* at the same time that males of strain *A* were crossed with females of strain *B*. The offspring of both crosses were pooled to form strain *AB*. A constructed strain such as this will initially contain large amounts of linkage disequilibrium (Falconer, 1981). Therefore, before beginning the experiment the **hvC** strain (**target**) was held for 11 generations to allow the most labile portion of disequilibrium to decay. Although some disequilibrium will remain after 11 generations, the rate of change in this portion will be quite slow (Falconer, 1981). Thus linkage disequilibrium is not likely to have a strong effect on the results of this experiment.

2.2. RESOURCES

Lentil (*Lens culinaris*) and mung (*Vigna radiata*) were the resources. When a single egg is laid on a mung bean about 95% of hatched eggs of both species emerged. Lentil, on the other hand, is essentially lethal to the **driver** and was in the beginning a poor host for the **target** (70% mortality). **Driver** and **target** (and all the strains from which the **target** was constructed) have been maintained on Adzuki bean (*V. angularis*) for many generations. Thus both experimental beans were novel to both species. The beans used in this experiment were screened to control size. The lentils used would pass mesh 4 (4.76 mm) but not mesh 5 (4.00 mm), and the mung beans would pass mesh 5 but not 6 (3.36 mm). This gave beans of very similar weight (approximately 50 mg per bean).

Preference measurements made prior to the experiment showed that the **driver** strongly preferred mung. At low egg densities, no eggs were laid on lentil in mixed bean preference trials. As egg density increased, the proportion of eggs on lentil slowly increased towards an asymptotic value of about 40%. In contrast, the **target** showed only a slight preference for mung. Regardless of egg density (0.1 - 8 eggs per bean), about 40% of its eggs were on lentil. This proportion is indistinguishable from my estimates of the relative surface areas of the two beans.

3. Methods

3.1. MAINTENANCE REGIME

The experimental containers were plastic boxes 20 cm X 13 cm X 4 cm on edge with 3 columns of 8 compartments each. The partitions between compartment of a column were removable. All partitions in the center column were removed. Holes in the inner side of each of compartments 2-7 in columns 1 and 3 allowed beetles to freely visit all compartments. To allow ventilation silk plankton netting was glued over three 1 cm X 15 cm slots cut in the lid over the three columns of compartments. Compartments 2-7 in columns 1 and 3 were used for beans during the experiment. The two types of beans were kept in different compartments. Thus the compartments represented distinct habitats.

Every week 200 fresh lentil and 200 fresh mung were added to the system, 100 beans in each of 4 compartments. After 1 week of oviposition these beans were removed from the boxes and sub-sampled for studies on egg distribution and emergence. One hundred each of egg laden lentil and mung were returned to the experimental container. Used beans were removed after five weeks. All adult seed beetles were removed each week. Beetle populations were started with 20 mated females added to the system in each of the first three weeks. The experiment was maintained under conditions of constant light, 30°C, and 70% r.h..

3.2. TREATMENTS

There were six experimental treatments. These treatments were:
1) A no interspecific competition treatment, with **target** reared alone on lentil and mung. This is termed the **control** treatment.
2) A hard-selection treatment, with **target** allowed to oviposit alone on lentil and mung, but with all mung beans removed before beetle emergence. This is called the **selection** treatment.
3) A **competition** treatment, with **target** and **driver** reared together on lentil and mung.
4) An augmented competition treatment designated **competition+**. By the fifth generation the numbers of **driver** in the **competition** lines were severely reduced. To maintain competition as a selective force on the **target** the beans from each **competition** line normally used for emergence studies were used to start new lines in the sixth generation. In the seventh, eighth, and ninth generations twenty-five mated stock **driver** females were added to each container each week.
5) A lentil as a single resource treatment in which **target** was reared on lentil alone without mung, and in the absence of the **driver** species. This treatment is called **lentil**.
6) A single resource treatment in which **target** was reared alone on mung. This will be referred to as the **mung** treatment.

Treatments 1-4 had four replicate lines apiece, while treatments 5 and 6 only had three. The experiment was maintained for nine generations.

Since the proportion of **target** eggs laid on lentil and mung was approximately the same in the **control** and **competition** treatments, the primary effect of competition was to reduce the fitness of the **target** individuals using mung. Therefore, the **selection** treatment is considered a control estimating the maximum possible effect of competition. The character displacement experiment per se is comprised of the **control** treatment and the competition treatments (**competition**, **competition+**, and **selection**).

3.3. BREAKDOWN PROCEDURES AND MEASUREMENTS

To avoid potentially confounding effects of larval environment on adult ovipositional preference or of maternal environment on larval development, all lines were reared for one generation on Adzuki beans after termination of the selection portion of the experiment. Because genetic differentiation might have occurred among weeks within a line, breakdown measurements were made on the beetles from three successive weeks within each line. Beetles emerging from these lines were held overnight without access to beans. The next day the beetles were sexed and preference trials were conducted. In preference trials, 10 females were introduced to 50 mung beans and 50 lentils mixed together in a Petri dish. The beetles were allowed to oviposit for 4 hours, and then were removed from the bean arrays. On the average, 1.5 eggs were laid per bean.

The eggs were counted after a week, allowing all viable eggs to hatch. In scoring preference, both hatched and unhatched eggs were counted. However, only beans with a single hatched egg were used to determine emergence proportions and development rates. The proportion of eggs on lentil was arcsin of square-root transformed before analysis. Wasserman and Futuyma (1981) utilized ANCOVA with realized fecundity as a covariate to estimate preference in aQ. However, this was not done in this study because of the lack of dependence of the **target**'s preference on egg density.

The emergence study was conducted by placing beans with a single hatched egg each in the wells of clear plastic multi-compartment tissue culture dishes (Nunc multidish). The boxes were inspected for daily for 50 days. Development period and sex of emergent beetles were scored. To confirm that all emergence was accounted for, the boxes were scored one last time after a lapse of several months. No new emergence was detected. The emergence and development period of **target** from all treatments were measured in both types of beans. Development periods had a highly skewed distribution not suitable for parametric analysis. Development period was transformed to development rate which was calculated as the inverse of the number of days from oviposition to emergence.

In blocking the emergence study, equal effort was allocated to each treatment. Treatments with 4 replicates had 18 boxes per replicate for a total of 72 per treatment, while treatments with 3 replicates had 24 boxes per replicate also for a total of 72 per treatment. Effort was not equally allocated between beans, however. Each box contained 8 mung beans and 16 lentil. Because of shortages of beans with only one egg, a few boxes were not completely filled. In all, the emergence and development of 10310 eggs were determined.

After the emergence study was complete, the elytra length of selected individuals was measured using a microscope with a camera lucida attachment. This allowed a magnified image of the elytra to be measured with a digitizer. For each treatment, replicate line, and week within replicate line, four boxes were randomly chosen. From each selected box, one male and one female from each bean type was measured. In all 952 individuals were measured.

In an experimental study such as this, the relevant sampling unit for statistical purposes is the line-replicate. In this report I analyze all characters as population averages. Some of the characters could have been analyzed with slightly greater power using nested ANOVA with individuals as the sampling unit. However, emergence is dichotomous at the individual level and should not be analyzed using parametric ANOVA/MANOVA techniques without very pressing reasons. Therefore to make the complete analysis in a common framework all characters were transformed to population averages. Individuals were first pooled within treatment/replicate/bean/week, and then, after checking for the absence of statistical differences among weeks within treatment/replicate/bean, weeks were pooled within treatment/replicate/bean. All analyses were made using the SYSTAT statistical package (version 4.0).

4. Results

4.1. PREFERENCE

The principal analytic technique used in this paper is a Multi-Variate-Analysis of Variance (MANOVA). However, because of an a priori hypothesis that preference for lentil would be greater in the **competition** and **selection** lines than in the **control** line, preference was first analyzed with a univariate ANOVA. The dependent variable was the arcsin(square root) transform of the proportion of eggs laid on lentil in preference trials. No evolution in preference was detectable ($F=1.07$, $df_{ef}=5$, $df_{er}=16$, $P=0.412$).

4.2. CHARACTER DISPLACEMENT:

4.2.1. *Treatments.* The predictor variables used in the MANOVA were treatment and bean type. In testing for character displacement only treatments 1-4 (**control, selection, competition**, and **competition+**) were used. This "a priori" decision was made while designing the experiment.

4.2.2. *Dependent Variables.* The dependent variables used in the MANOVA were the arcsin(square root) transform of the proportion of emergence (ASPEMRG), the development rates of emerging males and females (MLDRT, FMLDRT), the elytron length of emerging males and females (MLELEN, FMLELEN). Because the proportion of eggs laid on each bean type are not independent, no transformation of proportion of eggs on a bean type can be used as a dependent variable when bean types are used predictor variables. Instead the eggs laid on each bean type per preference trial (EGGSPT) was used. A significant interaction between treatment and bean type in eggs per trial (EGGSPT) would indicate a change in preference. The character eggs per trial (EGGSPT) was included in the initial MANOVA, despite its univariate ANOVA performance, so the omnibus test would have legitimate degrees of freedom. Cell means and standard errors are listed in Table 1. See Figure 1 for a graphical display. To show parallelism in the evolution of traits measured in different units all variables have been transformed to 100*(value-minimum value)/(Maximum value-minimum value). Within a variable, treatments not distinguishable at the 5% level by Tukey's honestly significant difference test are joined with a bar. Readers that believe in the validity of univariate tests protected by a significant MANOVA (Barker and Barker, 1984; Hummel and Sligo 1971) may take the bars at face value. Readers who espouse a simultaneous contrasts approach (Harris, 1984; Bird and Hadzi-Pavlovic, 1983) should view the bars as heuristic indications of variance.

4.2.3. *Analysis.* The MANOVA indicated a significant treatment X bean interaction as well as significant main effects for both treatment and bean (see Table 2). The significance criterion chosen on an a priori bases was the Lawley-Hotelling Trace, but all four common tests were significant and are listed in the table (see Morrison, 1976; or Barker and Barker, 1984 for a discussion of the alternative tests). As expected the univariate test for the TREAT X BEAN on eggs per trial (EGGSPT) was insignificant.

Main effects are not easily interpretable in the presence of a significant interaction. It is a standard practice to follow the discovery of a significant interaction with a series

Table 1. Cell Means and Standard Errors

BEAN = MUNG

TREATMENT	ASPEMRG	MLDRT	FMLDRT	MLELEN	FMLELEN	EGGSPT
control	1.40682	0.05111	0.04961	1.70769	1.84658	91.07192
	0.03053	0.00011	0.00012	0.01206	0.01166	1.92569
selection	1.37275	0.05101	0.04976	1.73688	1.86236	93.90278
	0.01806	0.00016	0.00021	0.01406	0.01839	5.25233
competition	1.38948	0.05055	0.04949	1.76170	1.87618	83.85620
	0.05362	0.00032	0.00019	0.01255	0.01056	8.22055
competition+	1.41194	0.05043	0.04960	1.76250	1.88017	101.81250
	0.01532	0.00049	0.00034	0.00784	0.02093	5.76192
lentil	1.40919	0.05135	0.05014	1.75769	1.88917	80.04762
	0.04468	0.00015	0.00015	0.00955	0.00851	4.58647
mung	1.41750	0.05078	0.04994	1.76731	1.88704	85.11204
	0.02609	0.00022	0.00006	0.02016	0.01501	6.66798

BEAN = LENTIL

TREATMENT	ASPEMRG	MLDRT	FMLDRT	MLELEN	FMLELEN	EGGSPT
control	1.17135	0.04661	0.04559	1.70996	1.79308	57.01141
	0.03927	0.00027	0.00028	0.01153	0.01241	1.38973
selection	1.34356	0.04829	0.04723	1.75090	1.84986	60.94524
	0.00935	0.00016	0.00014	0.00749	0.01140	4.92395
competition	1.30580	0.04816	0.04703	1.73281	1.85858	47.68162
	0.03218	0.00035	0.00037	0.01214	0.00766	3.31563
competition+	1.33645	0.04824	0.04700	1.72698	1.85896	52.44097
	0.03116	0.00038	0.00040	0.01432	0.00833	2.30334
lentil	1.35925	0.04861	0.04706	1.76037	1.88046	51.05049
	0.01934	0.00013	0.00011	0.01092	0.01212	3.50714
mung	0.58362	0.03728	0.03759	1.72787	1.80194	47.54259
	0.00817	0.00032	0.00106	0.00666	0.02540	3.06249

of one-way analyses across rows and down columns (Harris 1985). The MANOVAs within treatment across beans, while all significant, are not very interesting showing simply that the beans are different to seed beetles as well as to humans. I confine my

Table 2. Summary Table: MANOVA treatments 1-4, beans lentil and mung

Source	Wilks' lambda	df_{ef}	df_{er}	F	S	M	N	Roy's GCR	Pillai's trace	Lawley Hotelling trace
TREAT	0.235	18	54	2.014*	3	1.0	8.5	0.634*	1.020*	2.247*
BEAN	0.048	6	19	62.95***						
TREAT X BEAN	0.204	18	54	2.276**	3	1.0	8.5	0.699***	1.047*	2.765***

* <0.05, ** <0.01, *** <0.005.

discussion to MANOVAs among treatments within beans. Since eggs per trial (EGGSPT) can only be interpreted in the context of the TREAT X BEAN interaction, the variable was dropped before performing the one-way MANOVAs.

Character displacement was tested with a series of orthogonal contrasts. First the control treatment is contrasted against the competition treatments (**selection**, **competition**, and **competition+**) [-3 1 1 1]. Then the **selection** treatment is contrasted with **competition** and **competition+** [0 -2 1 1] and finally **competition** and **competition+** are compared [0 0 -1 1].

Within lentil the contrast [-3 1 1 1] between the **control** and the competition treatments is significant at the $P=0.013$ level (see Table 3, note that when there is only a single root all four tests are equivalent). Furthermore all five characters are greater in beetles from the competition treatments than from the **control** treatment. Four of these univariate contrasts are significant at or below the $P=0.001$ level. Male elytron length (MLELEN) was the only marginally significant variable with a probability level of 0.07. These results confirm the prediction, discussed in the introduction, of the effect of competition on the evolution of the utilization of lentil. Further contrasts ([0 -2 1 1] and [0 0 -1 1]) show no detectable differences among the competition treatments.

Within mung the primary contrast [-3 1 1 1] had a probability level of $P=0.056$, which can be considered significant given the conservative nature of the analysis with group averaged data (see Table 4). However, the significance of the MANOVA can be

Table 3. Summary Table: MANOVA treatments 1-4, lentils only, contrast [-3 1 1 1]

Source	Lawley Hotelling trace	df_{ef}	df_{er}	F	P
TREAT	3.800	5	8	6.080	0.013
UNIVARIATE F TESTS					
ASPEMRG		1	12	20.405	0.001
MLDRT		1	12	21.243	0.001
FMLDRT		1	12	17.037	0.001
MLELEN		1	12	4.020	0.068
FMLELEN		1	12	28.633	0.000

Table 4. Summary Table: MANOVA treatments 1-4, mung bean only, contrast [-3 1 1 1]

Source	Lawley Hotelling trace	df_{ef}	df_{er}	F	P
TREAT	2.207	5	8	3.531	0.056
	UNIVARIATE TESTS				
ASPEMRG		1	12	0.164	0.693
MLDRT		1	12	1.568	0.234
FMLDRT		1	12	0.000	0.997
MLELEN		1	12	11.293	0.006
FMLELEN		1	12	2.030	0.180

attributed principally to the single character with univariate significance (MLELEN $P=0.006$). The further contrasts ([0 -2 1 1] and [0 0 -1 1]) also indicate a lack of difference among the competition treatments within the mung bean type.

4.3. UNDERLYING TRAITS

To gain insight into the evolution occurring in this experiment, I will try to interpret the discriminant functions for the treatment effect. In order to observe more of the full scope of evolution in this experiment, I will analyze the discriminant functions of the treatment effect of the within lentil one-way MANOVA using all six treatments. This MANOVA is summarized in Table 5. Table 6 lists the coefficients of the first two discriminant functions and variable loadings on these functions. Also listed in Table 6 are two sets of simplified coefficients, which capture virtually all of the variance explained by the first two discriminant functions, and are more easily interpreted.

The variables in parentheses are suppressor variables. The primary action of a suppressor variable in a discriminant function is to suppress noise in another variable. They are indicated by loadings of opposite sign to the corresponding coefficient (see Wilkinson, 1975).

Ignoring the suppressor variables, we can see that the first simplified function is dominated by male development rate (MLDRT) and the second by female elytron length (FMLELEN). Emergence (ASPEMRG) contributes strongly to both simplified functions. Given the strong correlation between male development rate (MLDRT) and female development rate (FMLDRT) (see Figure 1, it seems likely that male development rate (MLDRT) is serving as representative for both characters, and thus it may be best to consider this function simply a development rate axis. On the other hand, the concordance between male and female elytron length (FMLELEN and MLELEN) is nowhere near as strong, and there are several indications throughout the data set that MLELEN and FMLELEN may not be evolving completely in tandem. For these reasons I would name the second function only as a female elytra length axis. It also might be reasonable to consider it a female body size axis, or given the correlation between elytra length and total fecundity ($r=0.72$, $N=54$, $p<<0.001$) as a fecundity axis.

I checked the reality of these simplified functions by performing an ANOVA on the discriminant scores they generate and testing the resultant F ratios using a post hoc

Table 5. Summary Table: MANOVA treatments 1-6, lentil only

Source	Wilks' lambda	df_{ef}	df_{er}	F	S	M	N	Roy's GCR	Pillai's trace	Lawley Hotelling trace
TREAT	0.003	25	46	7.365***	5	-.5	5.0	0.987***	2.085***	78.91***

*** < 0.005

UNIVARIATE TESTS

VARIABLE	df_{ef}	df_{er}	F	P
ASPEMRG	5	16	98.242	0.000
MLDRT	5	16	187.730	0.000
FMLDRT	5	16	62.459	0.000
MLELEN	5	16	2.505	0.074
FMLELEN	5	16	7.017	0.001

Dimension Reduction Table

ROOTS	X^2	df	P
1-5	92.621	25	0.000
2-5	25.251	16	0.06
3-5	7.571	9	0.57

Greatest Critical Root criterion (see Harris, 1984, pp 185). The first simplified factor is significant at a probability far below 0.01, and the second has a P value between 0.01 and 0.05. A comparison of the correlations shows that the first full discriminant function has an r of 0.994 while the first simplified function has an r of 0.981, and the second full function has an r of 0.825 while the second simplified function does slightly better with an r of 0.881.

These axes are relevant to the character displacement experiment per se (treatments 1-4) because the principal axis of that MANOVA is very similar to the second axis of the MANOVA of treatments 1-6. That is, treatments 1-4 are distinguished among themselves primarily on the female elytra length/emergence axis while all of these treatments are differentiated from the **mung** treatment along the development rate/emer-

Table 6. Discriminant Functions: MANOVA treatments 1-6, lentil only

	Discrim Funct 1	Loading Funct 1	Discrim Funct 2	Loading Funct 2	Simple Funct 1	Simple Funct 2
ASPEMRG	0.367	0.631	-0.496	-0.383	+0.5	-0.5
MLDRT	0.970	0.877	0.145	-0.168	+1	0
FMLDRT	-0.084	0.506	0.593	-0.063	0	(+0.5)
MLELEN	0.049	0.028	0.038	-0.459	0	0
FMLELEN	-0.401	0.103	-1.109	-0.802	(-0.5)	-1

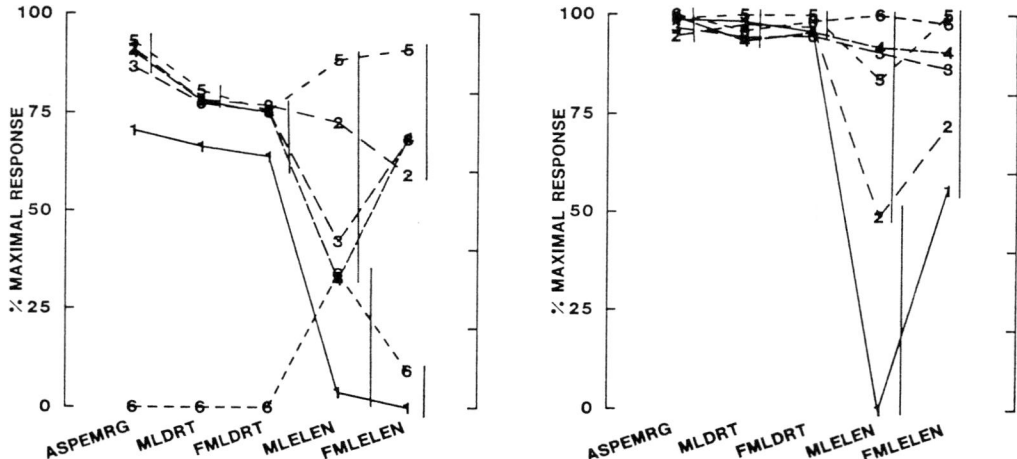

Figure 1. Treatment End States. (a) within lentil, (b) within mung. Treatments indicated by numbers: 1 = **control**, 2 = **selection**, 3 = **competition**, 4 = **competition+**, 5 = **lentil**, 6 = **mung**. Characters: ASPEMRG = arcsin (square root (lentil proportion of total emergence)), MLDRT and FMLDRT = male and female development rates, MLELEN and FMLELEN = male and female elytron lengths. All variables have been transformed to 100 * (value - minimum value) / (Maximum value - minimum value). Within a variable, treatments not distinguishable at the 5% level by Tukey's honestly significant difference test are joined with a bar.

gence axis. Figure 2 shows the six treatments in the space of the three variables emergence, male development rate, and female elytron length (ASPEMRG, MLDRT, and FMLELEN) that principally define the two discriminant functions.

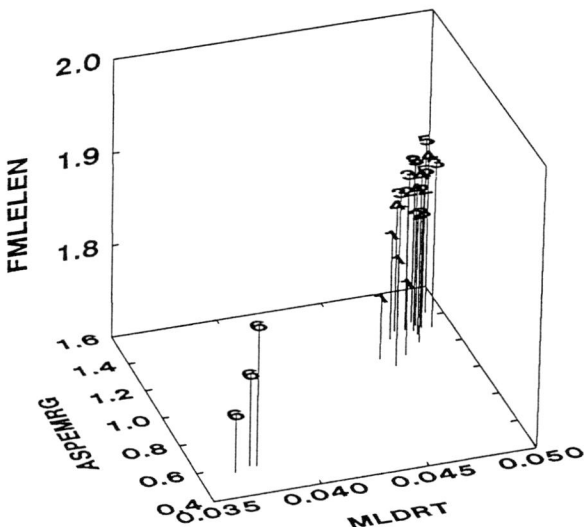

Figure 2. Within lentil, the treatments plotted in the space of ASPEMERG (unitless), MLDRT (days^{-1}) and FMLELEN (mm). See Figure 1 for symbols.

It is likely that the evolution in male elytron length (MLELEN) expressed in mung represents a third underlying trait.

5.   Discussion and Conclusions

The above results may be summarized as follows:

1) There is no indication of any evolution in preference.

2) All lines exposed to lentil show physiological adaptation to the bean.

3) Character displacement has occurred in that the competition and no-competition treatments differ significantly.

4) Predictions of the direction of character displacement are confirmed in that the competition lines do better on lentil.

5) Evolution has been detected in 5 characters representing at least 2 and probably 3 underlying traits.

Character displacement has clearly been demonstrated. But, there are troubling aspects to this study. A character based tradeoff in the ability to utilize different resources is at the heart of all character displacement theory (Abrams, 1986; Brown and Vincent,1987; Case, 1982, Lawlor and Maynard-Smith, 1976; Roughgarden, 1983; Slatkin, 1980; Taper and Case, 1985; Taper, 1988; for a few examples). Without a tradeoff a species should evolve to utilize all resources perfectly despite the presence of competition.

The current study has produced no evidence of a tradeoff between the two bean types. This is not a novel result in the plant/insect interaction literature (Futuyma and Wasserman, 1981; Rausher, 1984; Futuyma and Philippi, 1987; Hare and Kennedy, 1986; James et all, 1988). If the absence of tradeoff is real, then it is likely that the character displacement observed may be a transient phenomenon, in which competition is only facilitating the adaptation to lentil.

Nonetheless, competition has proven to be a potent selective force. Even the un-augmented competition lines are virtually indistinguishable from the hard selection lines.

This study may also shed some light on the evolution of host-choice by insects. From this viewpoint the discrepancy between the results of this experiment and those of Wasserman and Futuyma (1981) is extremely interesting. In that classic study the authors found that a bean beetle (*C. maculatus* Fujii lab aQ strain) was able to respond to hard selection by evolving changes in preference for bean types but not in physiological adaptation to those bean types. This led those authors to conclude "that the diets of phytophagous insects could be more labile at the behavioral level than at the physiological level." Their results have been used to bolster arguments that behavioral adaptation may generally precede physiological adaptation (Futuyma, 1983, 1986). In light of the significance of this proposition it is important to try to elucidate the differences between the experiments.

My hard-selection control treatment is similar to their experimental treatments. The history of the aQ strain is important. It is a very old laboratory strain having been reared in the laboratory since 1946 - over 500 generations at the time of the Wasserman and Futuyma experiment.   This represents strong and unremitting selection for adaptation to a single host, the adzuki bean. According to evolutionary theory (Wright 1977) aQ should be essentially depleted of genetic variation in characters relating to physiological adaptation to beans. On the other hand during aQ's long laboratory history many characters influencing preference would have been neutral since no alternative hosts were available. The strain has been maintained with large effective populations sizes so natural genetic variability in preference is likely not to have been lost to drift and may even have been enhanced by mutation. This may explain why aQ was able to respond to selection by shifting preference but not physiological adaptation. In my experiment, the hvC (target) strain was created with high genetic variation, so it comes as no surprise that it should be able to adapt physiologically.

It is unclear why hvC did not also adapt behaviorally. It is possible that population density is a critical difference between our experiments. The population sizes in the Wasserman and Futuyma experiment were kept quite low with egg densities on the order of 1 per bean, while in my experiment densities were unconstrained except by the natural carrying capacity. It may also be simply that my preference assay was inappropriate. Preference measurements are very fickle things. A preference measured under one set of conditions may be greatly altered or even reversed under another set of conditions (Singer, 1986).

In general, it is likely that both behavior and physiology are labile. Which responds may depend more on selection regime than genetic constraints. This is hardly the place for a review, but at least in seed beetles both behavior (Wasserman and Futuyma, 1981; Mark, 1982; Credland, this volume) and physiology (Credland, 1987 and this volume; this study; and Fujii, ongoing experiments) have repeatedly responded to appropriate selection.

Acknowledgments. This work was supported partly by US-NSF grant #BSR 8600185 and partly by a grant from the Japanese Science and Technology Agency. Perhaps most of all it was supported by the generosity, conviviality, and plain old Bruchid know-how of my host in Japan Dr. Koichi Fujii. Many fruitful hours were spent with Yukihiko Toquenaga in discussions of the psychological pathology of the seed beetle. I would like to thank Ann Thompson for help counting eggs during the critical bottleneck of the breakdown. Also I would like to acknowledge her tolerance throughout the many months when I came home nightly - late, with ether on my breath. This paper has profited from critical reading by Drs. K. Fujii, R. Mitchell, and L. Slobodkin.

References

Abrams, P. A. (1986) Character displacement and niche shift analyzed using consumer-resource models of competition, Theor. pop. Bio. 29, 107-160.

Barker, H. R. and Barker, B. M. (1984) Multivariate Analysis of Variance (MANOVA) A Practical Guide to Its Use in Scientific Decision Making, University of Alabama Press, Alabama.

Bird, K. D. and Hadzi-Pavlovic, D. (1983) Simultaneous test procedures and the choice of a test statistic in MANOVA, Psychol. Bull. 93, 167-178.

Brown, J. S. and Vincent, T. L. (1987) Coevolution as an evolutionary game, Evolution 41, 66-79.

Case, T. J. (1981) Niche packing and coevolution in competition communities, Proc. Nat. Acad. Sci. (USA) 78, 5021-5025.

Credland, P. F. (1987) Effects of host change on the fecundity and development of an unusual strain of *Callosobruchus maculatus* (F.) (Coleoptera: Bruchidae), J. stored Prod. Res. 23, 91-98.

Falconer, D. S. (1981) Introduction to Quantitative Genetics. Second Edition, Longman, New York.

Futuyma, D. J. (1986) The role of behavior in host-associated divergence in herbivorous insects, in M. D. Huettel (ed.), Evolutionary Genetics of Invertebrate Behavior, Plenum Press, New York, pp. 295-302.

Futuyma, D. J. (1983) Selective factors in the evolution of host choice by phytophagous insects, in S. Ahmad (ed.), Herbivorous Insects, Academic Press, New York, pp. 227-244

Futuyma, D. J. and Philippi, T. E. (1987) Genetic variation and covariation in responses to host plants by *Alsophila pometaria* (Lepidoptera: Geometidae), Evolution 41, 269-279.

Futuyma, D. J. and Wasserman, S. S. (1981) Food plant specialization and feeding efficiency in the tent caterpillars *Malacosoma disstria* and *M. americanum*, Entomol. Exp. Appl. 30, 106-110.

Hare, J. D. and Kennedy, D. D. (1986) Genetic variation in plant-insect associations: Survival of

Leptinotarsa decemlineata populations on Solanum carolinese.s, Evolution 40, 1031-1043.

Harris, R. J. (1985) A Primer of Multivariate Statistics: Second Edition, Academic Press, New York.

Hummel, T. J. and Sligo, R. J. (1971) Empirical comparisons of univariate and multivariate analysis of variance procedures, Psychol. Bull. 76, 49-57.

James, A. C., Jakubczak, J., Riley, M. P., and Jaenike, J. (1988) On the causes of monophagy in *Drosophila quinaria*, Evolution 42, 626-630.

Lawlor, L. R. and Maynard-Smith, J. (1976) The coevolution and stability of competing species, Amer. Natur. 110, 79-99.

Mark, G. A. (1982) An experimental study of evolution in heterogeneous environments: phenological adaptation by a bruchid beetle (*Callosobruchus maculatus*), Evolution 36, 984-997.

Messina, F. J. and Renwick, J. A. (1985) Ability of ovipositing seed beetles to discriminate between seeds with differing egg loads, Ecol. Entomol. 10, 225-230.

Mitchell, R. (1975) The evolution of oviposition tactics in the weevil, *Callosobruchus maculatus*, Ecology 56, 696-702.

Morrison, D. F. (1976) Multivariate Statistical Methods. Second Edition, McGraw-Hill, New York.

Rausher, M. D. (1984) Tradeoffs in performance on different hosts: Evidence from within- and between-site variation in the beetle *Deloyala guttata*, Evolution 38, 582-595.

Roughgarden, J. (1983) The theory of coevolution, in D. J. Futuyma and M. Slatkin (eds.), Coevolution, Sinauer, Sunderland, pp. 33-64.

Singer, M. C. (1986) The definition and measurement of oviposition preference in plant-feeding insects, in J. R. Miller and T. A. Miller (eds.), Insect-Plant Interactions, Springer-Verlag New York Inc., pp 65-94.

Slatkin, M. (1980) Ecological character displacement, Ecology 61, 163-177.

Taper, M. L. (1988) The coevolution of resource competition: Appropriate and inappropriate models of character displacement, Bull. Soc. Popul. Ecol. 44, 45-53.

Taper, M. L. and Case, T. J. (1985) Quantititative genetic models for coevolution of character displacement, Ecology 66, 355-371.

Via, S. (1984) The quantitative genetics of polyphagy in an insect herbivore. II. Genetic correlations in larval performance within and among host plants, Evolution 38, 896-905.

Wasserman, S. S. and Futuyma, D. J. (1981) Evolution of host plant utilization in laboratory populations of the southern cowpea weevil, *Callosobruchus maculatus* (Coleoptera: Bruchidae), Evolution 35, 605-617.

Wilkinson, L. (1975) Response variable hypotheses in the multivariate analysis of variance, Psychol. Bull. 82, 408-412.

Wilkinson, L. (1988) Systat, Systat Inc., Evanston.

Wright, S. (1977) Evolution and the Genetics of Populations. Vol. 3: Experimental Results and Evolutionary Deductions, University of Chicago Press, Chicago.

ALTERNATIVE LIFE-HISTORIES IN *CALLOSOBRUCHUS MACULATUS*: ENVIRONMENTAL AND GENETIC BASES

FRANK J. MESSINA
Department of Biology, Utah State University,
Logan, Utah, 84322 U.S.A.

ABSTRACT. This review considers within- and between-population variation in three traits that mediate intraspecific competition in *Callosobruchus maculatus*: 1) the production of the "active" or dispersing morph, 2) the tendency to distribute eggs uniformly among seeds, and 3) the competitiveness of larvae confined to the same seed. The proportion of active progeny in different strains ranged from near zero despite intense crowding to > 30% even in the absence of crowding. Similarly, distributions of eggs varied from nearly uniform to nearly random. A strain producing highly uniform egg distributions also exhibited unusually competitive larvae; if two larvae entered a small seed simultaneously, only one adult emerged. In contrast, > 50% of seeds bearing two larvae from a different strain yielded two adults. Each trait was under polygenic control, but differences between strains were caused by both additive (in the case of morph determination) or non-additive (in the case uniform egg-laying) genetic variation. Quantitative-genetic analysis within an outbred strain produced significant heritabilities for body size, development time, fecundity, and egg dispersion. Comprehensive genetic and demographic analyses are needed to determine if morphological, behavioral, and physiological traits in *C. maculatus* covary in a predictable way to form alternative life-histories.

1. Introduction

Life-history traits can be highly variable within or among natural populations, and much ecological research has attempted to account for this variation (Stearns, 1980; Partridge and Harvey, 1988). One objective has been to distinguish variation that represents adaptive responses to local environments from variation that is the product of random genetic drift. A second goal of life-history studies has been to examine relationships *among* various traits, since most traits cannot be fully understood except in relation to others. If two or more traits each exhibit large variation, are they associated so as to produce alternative "strategies"? Can we identify obvious "trade-offs" between traits (e.g., the oft-cited negative correlation between early reproduction and longevity, Bell, 1980; Imura, this volume)?

Recent studies have documented extensive variation in life-history traits among populations of the cowpea seed beetle, *Callosobruchus maculatus* (F.). Geographic strains show widely different intrinsic rates of increase, dispersal tendencies, responses to competition, oviposition behavior, and host-plant relationships (Bellows, 1982a,b; Dick and Credland, 1984, 1986; Messina and Renwick, 1985a, b; Credland, 1986; Credland and Dick, 1987; Wasserman, 1986a,b; Messina and Mitchell, 1989; see also Credland, Mitchell, and Smith, this volume). This variability may provide a convenient system for examining processes leading to divergent life-histories. Here I consider the genetic and environmental bases of three life-history traits, each of which is actually a composite of

K. Fujii et al. (eds.), Bruchids and Legumes: Economics, Ecology and Coevolution, 303–315.

several behavioral and physiological characters: 1) the production of dispersing or active beetles in response to crowding, 2) the tendency of females to distribute eggs uniformly among seeds and thereby reduce larval competition, and 3) the competitiveness of larvae within seeds.

2. General Considerations Regarding Variation in *C. maculatus*

Several processes influence population differentiation with regard to life-history traits. Genetic drift, local differences in natural selection, and mutation will enhance differentiation, while gene flow serves as the primary force retarding divergence (Slatkin, 1987). Gene flow may promote life-history evolution by allowing the spread of a novel, superior trait, or it may constrain evolution by preventing adaptation to local conditions. Some features of the *C. maculatus* life cycle make differentiation of populations especially likely. Many populations infest stores of grain legumes and have been associated with stored seeds for at least a few thousand years (Southgate, 1978; Smartt, 1985). Storage populations are typically initiated by a few individuals that attack a small percentage of pods in the field (Alzouma, 1981; Hagstrum, 1985; Huignard et al., 1985). Under these conditions, founder effects, inbreeding and genetic drift may be important. *C. maculatus* is a colonizing species that exhibits demographic instability. Large population fluctuations in storage tend to produce small effective population sizes. Finally, separate storage facilities provide ample opportunity for differences in directional or stabilizing selection, as host species, host availability, temperature, and humidity vary considerably among regions.

Gene flow may occur between storage populations in a variety of ways. Human transport of infested seeds can combine individuals from distant populations whose genetic architectures may have become quite different. In regions such as West Africa, the so-called active form of *C. maculatus* may promote gene flow by leaving the storage environment and infesting plants in the field (Utida, 1972, 1981; Ouedraogo and Huignard, 1981). Virtually nothing is known with regard to migration by active beetles, but the exchange of only a few migrants between populations may counter the effects of genetic drift.

Most life-history research on *C. maculatus* has been conducted on long-term, laboratory populations. Although the laboratory mimics the "natural" environment of this stored-product pest, population sizes in laboratory cultures are often small. Increased homozygosity and inbreeding depression may be observed, and the amount of additive genetic variation should be low. An effective population size of 500 has been suggested as needed to maintain heritable variation in selectively neutral, quantitative traits, but this figure depends on several assumptions (Lande, 1988). Surprisingly, some populations of *C. maculatus* have responded strongly to artificial selection (indicating substantial additive genetic variation) despite many years of laboratory culture (Sano-Fujii, 1986). For example, Wasserman and Futuyma (1981) achieved significant alteration of oviposition preferences after 11 generations of selection on a population that had been in the laboratory for > 30 years (see Taper, this volume, for further discussion).

3. Dispersal Polymorphism in Response to Crowding

Utida (1954) and Caswell (1960) first noted that "abnormal" flying adults may appear in crowded laboratory colonies of *C. maculatus*. This active or flight morph is easily separated from the sedentary or normal morph on the basis of a few external characters in most strains (Southgate et al., 1957; Utida, 1972). Experiments by Utida (reviewed in Utida, 1972, 1981) established significant differences in traits of active and normal beetles; active beetles are characterized by slower development, reduced fecundity, delayed reproduction, larger size, lower water content, higher lipid content, and of course a much higher tendency to disperse (Utida, 1972; Nwanze et al., 1976; Messina, 1984; Messina and Renwick, 1985a). Elegant experiments by Sano-Fujii (Sano, 1967; Sano-Fujii 1979, 1984) demonstrated that the proximate cues controlling morph ratios include a rise in temperature and bean water content which results from the metabolic activity of crowded larvae. The second-instar is apparently the sensitive stage.

Utida (1970) and Caswell (1960) suggested that the dispersal polymorphism also has a genetic basis, as laboratory cultures gradually lose the ability to produce active beetles even when crowded. Typical culturing practices favor genes promoting the faster-developing, more fecund, normal morph. An artificial selection experiment by Sano-Fujii (1986) provided evidence of genetic variation among individuals in the tendency to produce active beetles. A strain maintained in the laboratory for > 20 years produced a very low frequency of active beetles. After 7 generations of using only active beetles as parents, the population shifted to an equilibrium where 55% of the beetles were active. Hybrids of selected and control lines yielded intermediate morph ratios, which suggested additive inheritance of the tendency to produce active beetles. Because morph ratios depended more on the *line* from which parents were derived (selected or control) than on whether parents themselves were active or normal, these experiments also indicated polygenic control.

Messina and Renwick (1985a) investigated variation in morph ratios among four geographic strains. A strain that had been maintained in the laboratory for > 50 generations produced virtually no active progeny at all levels of crowding. The proportion of active progeny in two other strains was lowest at the lowest larval density, but further increases in crowding did not produce a simple linear increase in the proportion of active progeny. Instead, the ratio of active : normal individuals was highest at an intermediate level of crowding. The fourth strain was tested after only 2 generations in the laboratory, and the proportion of active beetles was high (> 30%) even without crowding. The response of this "wild" strain was thus opposite to that of the long-term laboratory strain. Reciprocal crosses between the wild strain (M) and the long-term laboratory strain (I) produced morph ratios in the F_1 and F_2 generations that were intermediate to those of the parental strains, with a minor dominance deviation toward producing normal beetles (Messina, 1987). These results, like the within-population study of Sano-Fujii (1986), indicated polygenic, largely additive inheritance of the trait.

Between-strain crosses also indicated a transient maternal effect, as the mean proportion of active beetles in the F_1 generation was 0.0, 12.0, 21.5, and 65.2 in the I x I, I x M, M x I, and M x M progeny, respectively (females listed first). Maternal effects can alter the short- or long-term response to selection on a trait in several ways (Kirkpatrick and Lande, 1989). One type of maternal influence in the dispersal polymorphism of *C. maculatus* was provided by Sano-Fujii (1979; see also Wasserman and Asami, 1985). She showed that eggs laid late in the oviposition period of a female are more likely to develop into active progeny than eggs laid early in the oviposition period. It is tempting to speculate that the increased frequency of dispersing individuals among later progeny represents a maternal "bet-hedging" strategy against future crowding, as the quality of oviposition sites available to a female may be quite different from the quality of sites available to her progeny. In addition, it may be important to produce *non-dispersing* progeny as early as possible in the highly competitive storage environment. Although morph determination is still not fully understood, it appears to depend on a complex genotype x environment interaction involving population density, maternal condition, and larval genotype.

Several questions remain regarding the dispersal polymorphism in *C. maculatus*. What is the role of delayed mating and reproduction by the active form in nature? A majority of active females from two strains did not reproduce at all under conditions sufficient for reproduction by normal beetles (Messina and Renwick, 1985a); does this reproductive failure suggest a requirement of diapause development? What conditions promoted the origin of the polymorphism? One hypothesis is that the highly fecund, non-diapausing, normal form is a neotenic derivative of the active form, and evolved to exploit the bounty of seeds in grain stores (Utida, 1981). The ironically-named "normal" form is thus the aberrant morph with respect to the typical life-history of monomorphic bruchids.

4. Uniformity of Egg Distributions among Seeds

Callosobruchus females have long been known to avoid adding eggs to seeds that already bear eggs, and thus distribute their eggs evenly among seeds (Utida, 1943; Avidov et al., 1965; Umeya, 1966; Mitchell, 1975, 1983; Gokhale and Srivastava, 1975). Because it reduces competition among larvae, a typical uniform distribution of eggs on small hosts increases an individual's fitness by 20 - 60% over the expected fitness if eggs were randomly distributed (Mitchell, 1975). Behaviors that produce uniform egg dispersions in insects have been referred to as "egg-load assessment," "egg-recognition," or "host discrimination", and are usually observed where the sedentary larvae feed on small, discrete resources, such as seeds, fruits, or other insects. Females of *C. maculatus* not only distinguish between egg-laden and egg-free seeds, but also respond to small differences in egg load, i.e., they assess egg loads quantitatively (Messina and Renwick, 1985c). This ability is adaptive in crowded populations, because larval mortality often increases monotonically with each increase in larval density (Messina and Renwick, 1985a; Smith and Lessells, 1985).

The behavioral "rules" that permit quantitative egg-load assessment are unclear; of particular interest is whether or not learning plays a role (Mitchell, 1975; Messina and Renwick, 1985c; Wilson, 1988). Messina and Renwick (1985c) suggested two means by which females prefer hosts with below-average egg loads. First, a female may compare the seed she is currently inspecting with a memory trace for some set of seeds visited previously (a relative rule). Alternatively, she could respond only to the current egg density on a seed, but where each increase in egg load corresponds to a lower probability of acceptance (an absolute rule). Wilson (1988) provided the only attempt to distinguish between these possibilities, and concluded that learning was not involved.

The cues involved in detection of eggs on seeds are also not yet defined. Chemical cues are almost certainly involved, as females prefer to lay eggs on clean seeds than on seeds from which an egg has been scraped off (leaving a chemical residue) (Oshima et al., 1973; Giga and Smith, 1985; Wasserman, 1985). Oshima et al. (1973) isolated an ether-soluble "oviposition marker" from egg-laden artificial hosts (glass beads) (see also Giga and Smith, 1985; Sakai et al., 1986).

The role of ether-soluble chemicals in mediating egg dispersion is uncertain, because similar deterrents can be obtained from egg-free beads exposed to males (Messina and Renwick, 1985d). Additional confounding evidence was provided by Messina et al. (1987a, b), who found that ablation of the maxillary palps transformed a uniform egg-layer into a random one. Palpectomized females were therefore expected to ignore oviposition "markers", but these females still avoided ether-soluble deterrents in bioassays using coated seeds. If palpectomized females respond to these deterrents but cannot distribute eggs evenly, it seems likely that other cues provide the true spacing mechanism. Current evidence indicates that cues involved in egg recognition 1) are associated with the egg itself, and not with other activities on the host, 2) may be physical (tactile) as well as chemical, and 3) persist well after the larva has entered the seed (Messina and Renwick, 1985c,d; Messina et al., 1987).

Messina and Mitchell (1989) found differences among several geographic strains in the tendency to distribute eggs evenly. They developed an index (U) of egg dispersion that was independent of the number of eggs laid. The commonly used variance-to-mean ratio and its derivatives were found to be unsatisfactory in this regard, because the variance of a discrete uniform distribution is zero only when the number of eggs laid is an exact multiple of the number of seeds available. The U-index, which measures the deviation of an observed egg distribution from a uniform one, produces a value of 1 for uniform distributions, 0 for random distributions, and < 0 for clumped distributions.

Egg dispersion varied significantly among strains, and also depended on which of three host species a female was provided (Table 1, Messina and Mitchell, 1989). Females of the S strain frequently achieved perfect uniform distributions on mung bean, while B-strain females frequently distributed eggs at random on black bean. Differences among strains were evident throughout the oviposition period; some strains were relatively "sloppy" even when egg densities were low. Across strains, egg dispersions were generally most uniform on the smallest host, mung bean, and were least uniform on the largest host, black bean. Variation in egg-spacing behavior was in part related to the ancestral host of each strain. The most-uniform egg-layers (S-strain) have been chronically associated with a small host (mung bean), which supports only 1 or 2 larvae/seed. The least uniform egg-layers (B and G) have been chronically associated with a large host (black-eye peas). Host size may thus influence the tendency to distribute eggs uniformly on both ecological and evolutionary time scales.

Realized fecundity also varied significantly among strains. A trade-off (here a negative phenotypic correlation) between egg dispersion and realized fecundity was expected because 1) distributing eggs more uniformly may require greater search costs and 2) the ability to detect differences in egg load should decrease with increasing mean egg load. Under the conditions provided by Messina and Mitchell (1989), however, no evidence of such a trade-off was obtained either within strains, i.e., more fecund individuals did not produce sloppier distributions, or between strains, i.e., a strain that was more fecund on a given host was not necessarily less uniform in its distribution of eggs.

Single-pair crosses between the S and B strains indicated a genetic basis of between-population variation (Table 1, Messina, 1989). F_1 females distributed their eggs among

Table 1. Mean (SE) U-values of 7 geographic strains on two hosts (0 = random egg-laying, 1 = perfect uniformity).

Strain	Mung bean	Black bean
S	0.97 (.01)	0.76 (.04)
F	0.82 (.03)	0.61 (.07)
I	0.72 (.03)	0.34 (.06)
A	0.71 (.03)	0.41 (.04)
C	0.54 (.05)	0.29 (.05)
G	0.60 (.04)	0.10 (.08)
B	0.51 (.05)	0.08 (.05)

seeds nearly as uniformly as S females did; only heterotic females that laid especially high numbers of eggs produced less uniform distributions. F_2 females distributed their eggs as uniformly as S females did, as did the progeny of backcrosses to the S-strain. Females derived from backcrosses to the B-strain produced distributions that were significantly less uniform than those of the S-strain, but even these females (which possess 3/4 of the relevant alleles from the B-strain) displayed a dominance deviation toward uniform egg-laying. These results are perhaps best explained by polygenic control combined with dominance for uniform egg-laying and epistasis. Possession of only a few alleles from the S-strain leads to egg distributions that are nearly as uniform as those of S females.

Two quantitative-genetic experiments (Falconer, 1981) were conducted to examine within-population variability in egg-spacing behavior and to examine the relationship between this trait and others (full methods and results will be presented elsewhere). Parent-offspring regressions were performed to examine heritabilities of egg-laying behavior, body size, and realized fecundity. Forty adult pairs were isolated from a population that was obtained in Florida, USA, and had spent only 2 generations in the laboratory. Pronotal width and elytral length were measured on parent males, parent females, and each of four female offspring, while realized fecundity and egg dispersion were recorded for parent and offspring females.

Although the number of families was too low to produce small standard errors of heritability estimates (Klein, 1974), the results illustrate some important considerations in parent-offspring resemblance. Significant regressions (and narrow-sense heritabilities) were obtained when offspring body size was compared to female parent or mid-parent values, or when offspring fecundity was compared to female parent fecundity (Table 2, the narrow-sense heritability equals the regression coefficient for regressions using mid-parent values, or twice the regression coefficient for regressions using single-parent values). The regression was not significant, however, when offspring body size was compared to male-parent values. Thus, parent-offspring resemblance in size (and perhaps fecundity) was entirely due to non-genetic maternal effects. The regression comparing egg dispersion in parent and offspring was non-significant.

A half-sib breeding design was used to estimate heritabilities of egg dispersion, realized fecundity, body size, and development time. Beetles derived from Alabama, USA, had been maintained at a large population size for eight generations. Twenty-eight males (sires) were each mated to 2 - 5 females (dams), creating a series of half- and full-sib families. Traits were measured in 5 - 13 offspring in 98 families (877 total progeny). Nested analysis of variance was used to calculate the variance components due to sires and to dams within sires, where the former is used to estimate the additive genetic variation, and hence narrow-sense heritability (Falconer, 1981).

Table 2. Narrow sense heritabilities ($\pm$ SE) from parent-offspring regressions of various traits using single-parent or mid-parent values. N = 40 pairs.

Trait	Male parent	Female parent	Mid-parent
Pronotal width	0.10 $\pm$ .40	0.55 $\pm$ .26	0.44 $\pm$ .22
Elytral length	-0.18 $\pm$ .32	0.63 $\pm$ .31	0.45 $\pm$ .28
Fecundity	—	0.70 $\pm$ .17	—
U-value	—	0.08 $\pm$ .25	—

Table 3. Overall means and heritabilities of several traits as determined from a half-sib breeding experiment.

Trait	Mean (SE)	Heritability (SE)
Realized fecundity	75.49 (1.47)	0.30 (.15)
Development time (days)	19.55 (0.15)	0.73 (.35)
Pronotal width (mm)	1.30 (0.01)	0.34 (.17)
Elytral length (mm)	1.89 (0.01)	0.22 (.17)
U-value (dispersion)	0.30 (0.02)	0.29 (.15)

Table 3 presents overall means and heritabilities. Non-zero heritabilities were obtained for realized fecundity, development time, pronotal width, and egg dispersion. Natural populations are expected to harbor little additive genetic variation for traits such as development time and fecundity, which are considered to be important components of overall fitness (Mousseau and Roff, 1987). Several explanations have been proposed for the existence of heritable variation in life-history traits (e.g., frequency-dependent selection, recurrent mutation, etc.). One explanation that can be examined using quantitative-genetic techniques is that of antagonistic pleiotropy, which can arise from a negative genetic correlation between fitness-related traits (Rose, 1982), and would prevent persistent directional selection for any one trait.

Phenotypic and genetic correlations for traits in *C. maculatus* are presented in Table 4. At a phenotypic level, development time was inversely related to body size and fecundity. Body size, as expected, was positively related to fecundity. Surprisingly, there was a positive relationship between fecundity and egg dispersion. Most genetic correlations were clearly non-significant, although there was apparently a negative genetic correlation (but no phenotypic correlation) between egg dispersion and each measure of body size. At the genotypic level, this experiment provided little support for the antagonistic-pleiotropy hypothesis, and did not suggest obvious constraints to the independent evolution of most traits.

5. Larval Interactions within Seeds

A critical determinant to any egg-laying strategy is the nature of interactions between larvae occupying the same host. As Smith and Lessells (1985) noted for insects such as *C. maculatus*, "oviposition ... behavior [affects] the mean and variance of the number of

Table 4. Phenotypic (above the diagonal) and genetic correlations (below the diagonal, SE in parentheses) among traits in Table 3.

	FEC	DT	PN	EL	U
FEC	—	-0.19**	0.18**	0.14**	0.42**
DT	-0.14(.34)	—	-0.12**	-0.16**	-0.05
PN	-0.17(.32)	-0.01(.34)	—	0.73**	-0.03
EL	-0.18(.42)	-0.33(.38)	0.39(.37)	—	-0.02
U	0.32(.32)	-0.04(.35)	-0.56(.25)	-0.77(.18)	—

**P < 0.01.

larvae per bean, which has some effect on what form of larval competition is stable. Conversely, the form of larval competition will determine the shape of the larval competition curve [i.e., the relationship between larval density and fitness] which is a critical determinant of oviposition ... strategies." In parasitic wasps, for example, the degree of larval aggression within a host covaries with the egg-laying strategy of the female (Godfray, 1987). The cost of ovipositing on an occupied host clearly depends on where larval competitiveness lies in the spectrum from pure "contest" to pure "scramble" competition.

Larval interactions of *C. maculatus* are poorly understood, but the outcome of larval competition can be quite variable in different populations. Dick and Credland (1984; see also Credland et al., 1986) allowed multiple larvae from one of three strains to enter cowpea seeds simultaneously. Up to 15 adults could emerge from a heavily attacked seed in two strains, whereas the same initial density of larvae produced only one or two adults emerged from seeds attacked by a third strain. Size differences among strains could not explain the drastically different competitive outcomes (Credland and Dick, 1987). Thanthianga and Mitchell's (1987) ingenious experiments probed the nature of interactions among "competitive" larvae. By inspecting seeds at various times after oviposition, they determined that one larva of a pair exhibited behavioral dominance. Responses of co-occurring larvae were apparently mediated by vibrations during feeding, but which larva would be inhibited was unpredictable.

The highly competitive S-strain used by Thanthianga and Mitchell (1987) and the less competitive I strain used by Dick and Credland (1984) were recently combined in a factorial experiment where the density, composition, and timing of larvae entering azuki bean seeds were each manipulated independently (Messina and Mitchell, unpublished). Only the simplest treatments will reviewed. When two larvae entered seeds simultaneously, nearly all seeds bearing 2 S-strain larvae yielded one adult or no adult, whereas over 50% of seeds bearing 2 I-strain larvae yielded two adults (Table 5). If one I larva and one S larva entered a seed, the outcome was similar to interaction of two S-strain larvae, i.e., most seeds produced one adult. The emerging adult from these "one-winner" seeds was much more likely to be an S-strain individual than an I-strain individual.

The outcome of competition could be altered if one larva obtained a "headstart" in entering a seed. If the I-strain larva had a 2-day headstart, it was more likely to emerge than when the two larvae entered the seed simultaneously, but the probability of emergence was still only ca. 50%. Thus, a headstart was needed simply for I larvae to compete equally with S larvae. An unexpected result was that a 2-day head-start failed to increase the competitive dominance of S larvae. Approximately 75% of emerging adults in single-survivor seeds were from the S-strain in *both* the simultaneous and S-headstart treatments.

Table 5. Survival of larvae in seeds bearing two larvae from the S strain, two from the I strain, or one from each strain. N = 90-110 seeds/treatment.

Treatment	% seeds yielding			Overall
	0	1	2 adults	survival (%)
2 S larvae	15.7	80.6	3.7	44.0
2 I larvae	11.0	36.0	53.0	71.0
1 S + 1 I	15.7	75.9	8.3	46.3*

* 74% of individuals from single-survivor seeds were from the S strain.

Crosses were performed between the S- and I-strains, and the F_1 progeny were allowed to enter the seeds simultaneously. The outcome of larval competition, as indicated by larval survival, was intermediate between the outcomes using S or I larvae. The percentage of seeds yielding two adults was 26 and 34% for seeds bearing hybrid progeny from S x I and I x S crosses, respectively (female listed first). Overall survival of hybrid progeny was 58 and 64% for the S x I and I x S crosses, while overall survival in two-larva seeds in the parental S- and I-strains was 44 and 71%. Differences between reciprocal crosses again suggest a maternal effect.

The competitiveness of S-strain larvae appears to be a consequence of the behavioral inhibition demonstrated by Thanthianga and Mitchell (1987). This sort of behavioral interaction may in turn be related to the large size that these larvae typically attain. In the absence of competition in azuki beans (1 larva/bean), the mean weight of S females was 6.9 mg at emergence. In contrast, I females weighed 4.3 mg. A similar difference was found in males. When two larvae share a medium-sized seed, the weight of emerging adults drops only slightly in both strains, but apparently for different reasons. In the S-strain, the typical lone survivor obtains enough food by dominating (or eliminating) its competitor. In the I strain, a single azuki bean usually provides enough food for two larvae to grow as large as they would without competition. The intermediate survival of the hybrids may be related to their intermediate size (Fig. 1), although the hybrids showed a dominance deviation toward the larger S-strain.

Experiments manipulating larval density and composition in conjunction in host size may provide an explicit test of a simple game-theory model developed by Smith and Lessells (1985) to evaluate alternative interactions of bruchid larvae. The outcomes of larval competition in the S- vs. I-strains conform rather well to the *attack* vs. *avoid* strategies formulated by Smith and Lessells (1985), although *dominance* may be a better description of the S-strain interaction. It should be possible to measure how expected "payoffs" to different strategies change in different sized hosts.

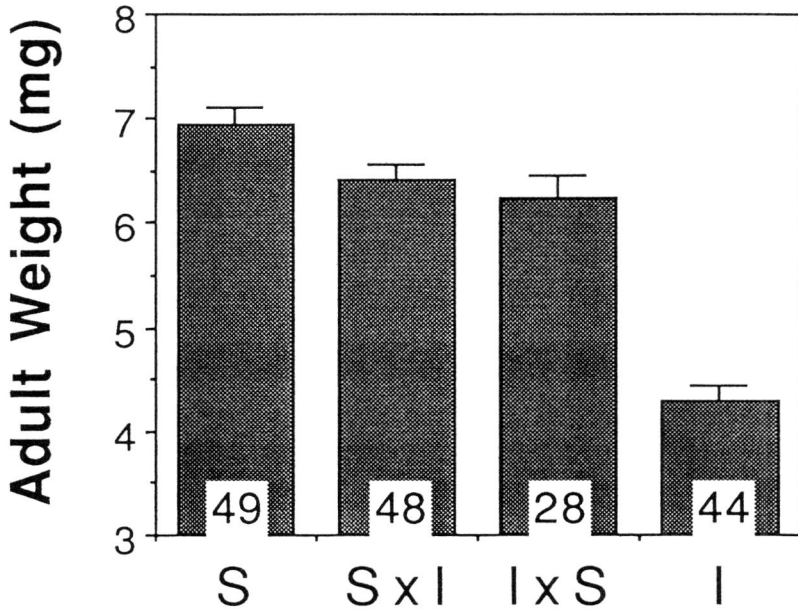

Figure 1. Mean weights (+ 1 SE) at emergence of S, I, or hybrid females.

6. Conclusions

The three traits discussed in this paper can each be considered a response to potential intraspecific competition. Recognition and avoidance of egg-laden seeds tends to prevent multiple occupancy of seeds. Uniform egg-laying cannot prevent strong competition at high densities, however. The nature of larval interactions will then determine the number of larvae that can develop per seed, and the degree to which competition affects the size and fecundity of emerging adults (Mitchell, 1975; Dick and Credland, 1984; Credland et al., 1986; Credland and Dick, 1987). Since the occurrence of multiple larvae per seed is a good indicator of the depletion of local resources, the developmental switch to the dispersing morph produces individuals that search for new sources of seeds. Each of these traits may have other functions in the life-history of *C. maculatus* (e.g., the dispersal polymorphism may serve to synchronize the life-cycle with host availability in the field), but the way in which they mediate intraspecific competition is clear.

Intraspecific competition may be generally rare among natural populations of herbivorous insects (Strong et al., 1984). In some systems, however, certain plant and insect characteristics combine to make competition much more likely. Larvae of bruchid pests are usually confined to developing within a single seed. The number of larvae developing to adulthood in a seed is relatively small and depends on an interaction of seed traits (such as size) and larval traits (such as aggression). Moreover, the number of larvae that can develop per seed without any effect of competition (such as reduced fecundity) will be even smaller. Although the storage environment may provide a bounty of seeds at the start of an infestation, rapid population growth can deplete available seeds in a few generations. Natural enemies, mostly parasitoid wasps, are usually present in storage, but their effectiveness in regulating seed beetle populations is unclear (Utida, 1950; Hassell et al., 1985; Bellows and Hassell, 1988).

If populations of *C. maculatus* frequently experience competition for food or space, it is not surprising to find a trait such as uniform egg-laying. Rather, the remarkable aspect of this species is the extent to which such fitness-related traits vary between and within interfertile populations. Simultaneous comparison of a suite of traits is needed to determine if different selection pressures, such as might be expected from association with small vs. large host seeds, have indeed led to alternative sets of life-history characters. Current evidence, for example, suggests that strains in which females produce less uniform distributions also tend to have tolerant larvae (see Mitchell, this volume). However, such generalizations must await more comprehensive genetic studies of a natural populations. Because the relevant traits are quantitative, either as continuous (e.g., body size), meristic (e.g., fecundity), or threshold (e.g., morph ratios) variables, and because beetle populations are exceptionally well-suited to laboratory study, the *C. maculatus*-legume seed association appears to be an especially promising one for analyzing phenotypic and genetic covariances of traits with obvious ecological significance.

Acknowledgements. I thank R. Mitchell for comments. R. Mitchell, S. S. Wasserman, J. R. McLaughlin, and the Montgomery Seed Co., Alabama, kindly provided strains. Research supported by a Faculty Research Grant from Utah State University and NSF grant BNS-8908541.

REFERENCES

Alzouma, I. (1981) Observations on the ecology of *Bruchidius atrolineatus* Pic. and *Callosobruchus maculatus* F.(Coleoptera, Bruchidae) in Niger, Ser. Entomol. 19, 205-213.

Avidov, Z., Applebaum, S. W., and Berlinger, M. J. (1965) Physiological aspects of host specificity in the Bruhchidae: II. oviposition preference and behaviour of *Callosobruchus chinensis* L., Entomol. Exp. Appl. 8, 96-106.

Bell, G. (1980) The costs of reproduction and their consequences, Amer. Natur. 115, 45-76.

Bellows, T. S., Jr. (1982a) Analytical models for laboratory populations of *Callosobruchus chinensis* and *C. maculatus* (Coleoptera, Bruchidae), J. Anim. Ecol. 51, 263-287.

Bellows, T. S., Jr. (1982b) Simulation models for laboratory populations of *Callosobruchus chinensis* and *C. maculatus*, J. Anim. Ecol. 51, 597-623.

Bellows, T. S., Jr. and Hassell, M. P. (1988) The dynamics of age- structured host-parasitoid interactions, J. Anim. Ecol. 57, 259-268.

Caswell, G. H. (1960) Observations on an abnormal form of *Callosobruchus maculatus* (F.), Bull. Entomol. Res. 50, 671-680.

Credland, P. F. (1986) Effect of host availability on reproductive performance in *Callosobruchus maculatus* (F.) (Coleoptera: Bruchidae), J. stored Prod. Res. 22, 49-54.

Credland, P. F., Dick, K. M., and Wright, A. W. (1986) Relationships between larval density, adult size and egg production in the cowpea seed beetle, *Callosobruchus maculatus*, Ecol. Entomol. 11, 41-50.

Credland, P. F. and Dick, K. M. (1987) Food consumption by larvae of three strains of *Callosobruchus maculatus* (Coleoptera: Bruchidae), J. stored Prod. Res. 23, 31-40.

Dick, K. M. and Credland, P. F. (1984) Egg production and development of three strains of *Callosobruchus maculatus* (F.) (Coleoptera: Bruchidae), J. stored Prod. Res. 20, 221-227.

Dick, K. M. and Credland, P. F. (1986) Variation in the response of *Callosobruchus maculatus* (F.) to a resistant variety of cowpea, J. stored Prod. Res. 22, 43-48.

Falconer, D. S. (1981) Introduction to Quantitative Genetics, 2nd Ed., Longman, New York.

Giga, D. P. and Smith, R. H. (1985) Oviposition markers in *Callosobruchus maculatus* F. and *Callosobruchus rhodesianus* Pic. (Coleoptera, Bruchidae): asymmetry of interspecific responses, Agric. Ecosyst. Environ. 12, 229-223.

Godfray, H. C. J. (1987) The evolution of clutch size in parasitic wasps, Amer. Natur. 129, 221-233.

Gokhale, V. G. and Srivastva, B. K. (1975) Ovipositional behaviour of *Callosobruchus maculatus* (Fabricius) (Coleoptera: Bruchidae). I. Distribution of eggs and relative ovipositional preference on several leguminous seeds, Ind. J. Entomol. 37, 122-128.

Hagstrum, D. W. (1985) Preharvest infestation of cowpeas by the cowpea weevil (Coleoptera: Bruchidae) and population trends during storage in Florida, J. Econ. Entomol. 78, 358-361.

Hassell, M. P., Lessells, C. M., and McGavin, G. C. (1985) Inverse density dependent parasitism in a patchy environment: a laboratory system, Ecol. Entomol. 10, 393-402.

Huignard, J., Leroi, B., Alzouma, I., and Germain, J. F. (1985) Oviposition and development of *Bruchidius atrolineatus* (Pic) and *Callosobruchus maculatus* (F.) (Coleoptera: Bruchidae) in *Vigna unguiculata* (Walp) cultures in Niger, Insect Sci. Appl. 6, 691-699.

Kirkpatrick, M. and Lande, R. (1989) The evolution of maternal characters, Evolution, 43, 485-503.

Klein, T. W. (1974) Heritability and genetic correlation: statistical power, population comparisons, and sample size, Behav. Genet. 4, 171-189.

Lande, R. (1988) Genetics and demography in biological conservation, Science 241, 1455-1459.

Messina, F. J. (1984) Influence of cowpea pod maturity on the oviposition choices and larval survival of a bruchid beetle *Callosobruchus maculatus*, Entomol. Exp. Appl. 35, 241-248.

Messina, F. J. (1987) Genetic contribution to the dispersal polymorphism of the cowpea weevil (Coleoptera: Bruchidae), Ann. Entomol. Soc. Amer. 80, 12-16.

Messina, F. J. (1989) Genetic basis of variable oviposition behavior in *Callosobruchus maculatus* (Coleoptera: Bruchidae), Ann. Entomol. Soc. Amer. 82, 792-796.

Messina, F. J., Barmore, J. L., and Renwick, J. A. A. (1987a) Oviposition deterrent from eggs of *Callosobruchus maculatus*: spacing mechanism or artifact?, J. Chem. Ecol. 13, 219-226.

Messina, F. J., Barmore, J. L., and Renwick, J. A. A. (1987b) Host selection by ovipositing cowpea weevils: patterning of input from separate sense organs, Entomol. Exp. Appl. 43, 169-173.

Messina, F. J. and Renwick, J. A. A. (1985a) Dispersal polymorphism of *Callosobruchus maculatus* (Coleoptera: Bruchidae): variation among populations in response to crowding, Ann. Entomol.

Soc. Amer. 78, 201-206.

Messina, F. J. and Renwick, J. A. A. (1985b) Resistance to *Callosobruchus maculatus* (Coleoptera: Bruchidae) in selected cowpea lines, Environ. Entomol. 14, 868-872.

Messina, F. J. and Renwick, J. A. A. (1985c) Ability of ovipositing seed beetles to discriminate between seeds with differing egg loads, Ecol. Entomol. 10, 225-230.

Messina, F. J. and Renwick, J. A. A. (1985d) Mechanism of egg recognition by the cowpea weevil *Callosobruchus maculatus*., Entomol. Exp. Appl. 37, 241-245.

Messina, F. J. and Mitchell, R. (1989) Intraspecific variation in the egg-spacing behavior of the seed beetle *Callosobruchus maculatus*, J. Insect. Behav. 2, 727-742.

Mitchell, R. (1975) The evolution of oviposition tactics in the bean weevil, *Callosobruchus maculatus* (F.), Ecology 56, 696-702.

Mitchell, R. (1983) Effects of host-plant variability on the fitness of sedentary herbivorous insects, in R. F. Denno and M. S. McClure (eds.), Variable Plants and Herbivores in Natural and Managed Systems, Academic Press , New York, pp. 343-370.

Mousseau, T. A. and Roff D. A. (1987) Natural selection and the heritability of fitness components, Heredity 59, 181-197.

Nwanze, K. F., Maskarinec, J. K., and Hopkins T. L. (1976) Lipid composition of the normal and flight forms of adult cowpea weevils, *Callosobruchus maculatus*, J. Insect Physiol. 22, 897-899.

Oshima, K., Honda, H., and Yamamoto, I. (1973) Isolation of an oviposition marker from azuki bean weevil, *Callosobruchus chinensis* (L.), Agric. Biol. Chem. 37, 2679-2680.

Partridge, L. and Harvey, P. H. (1988) The ecological context of life-history evolution, Science 241, 1449-1455.

Rose, M. R. (1982) Antagonistic pleiotropy, dominance and genetic variation, Heredity 48, 63-78.

Sakai, A., Honda, H., Oshima, K., Yamamoto, I. (1986) Oviposition marking pheromone of two bean weevils, *Callosobruchus chinensis* and *Callosobruchus maculatus*, J. Pestic. Sci. 11, 163-168.

Sano, I. (1967) Density effect and environmental temperature as the factors producing the active form of *Callosobruchus maculatus* (F.) (Coleoptera, Bruchidae), J. stored Prod. Res. 2, 187-195.

Sano-Fujii, I. (1979) Effect of parental age and developmental rate on the production of active form of *Callosobruchus maculatus* (F.) (Coleoptera: Bruchidae), Mech. Aging Develop. 10, 283-293.

Sano-Fujii, I. (1984) Effect of bean water content on the production of the active form of *Callosobruchus maculatus* (F.) (Coleoptera: Bruchidae), J. stored Prod. Res. 20, 153-161.

Sano-Fujii, I. (1986) The genetic basis of the production of the active form of *Callosobruchus maculatus* (F.) (Coleoptera: Bruchidae), J. stored Prod. Res. 22, 115-123.

Slatkin, M. (1987) Gene flow and the geographic structure of natural populations, Science 236, 787-792.

Smartt, J. (1985) Evolution of grain legumes. III. Pulses in the genus *Vigna*, Exp. Agric. 21, 87-100.

Smith, R. H. and Lessells, C. M. (1985) Oviposition, ovicide and larval competition in granivorous insects, in R. M. Sibly and R. H. Smith (eds.), Behavioral Ecology, Blackwell, Oxford, pp. 423-448.

Southgate, B. J. (1978) The importance of the Bruchidae as pests of grain legumes, their distribution and control, in S. R. Singh et al. (eds.), Pests of Grain Legumes: Ecology and Control, Academic Press, London, pp. 219-229.

Southgate, B. J., Howe, R. W., and Brett, G. A. (1957) The specific status of *Callosobruchus maculatus* (F.) and *Callosobruchus analis* (F.), Bull. Entomol. Res. 48, 79-89.

Stearns, S. C. (1980) A new view of life-history evolution, Oikos 35, 266-281.

Strong, D. R., Lawton, J. H., and Southwood, T. R. E. (1984) Insects on Plants: Community Patterns and Mechanisms, Blackwell Scientific, Oxford.

Thanthianga, C. and Mitchell, R. (1987) Vibrations mediate prudent resource exploitation by competing larvae of the bruchid bean weevil *Callosobruchus maculatus*, Entomol. Exp. Appl. 44, 15-21.

Umeya, K. (1966) Studies on the comparative ecology of bean weevils. I. On the egg distribution and the oviposition behaviors of three species of bean weevils infesting azuki bean, Res. Bull. Plant Prot. 3, 1-11.

Utida, S. (1943) Studies on the experimental population of the azuki bean weevil, *Callosobruchus chinensis* (L.). VIII. Statistical analysis of the frequency distribution of the emerging weevils on beans, Mem. Coll. Agric. Kyoto Imperial Univ. 54, 1-22.

Utida, S. (1950) On the equilibrium state of the interacting population of an insect and its parasite, Ecology 31, 165-175.

Utida, S. (1954) "Phase" dimorphism observed in the laboratory population of the cowpea weevil, *Callosobruchus quadrimaculatus*, Oyo Dobutsugaku Zasshi 18, 161-168.

Utida, S. (1970) Secular change of percent emergence of the flight form in the population of the southern cowpea weevil, *Callosobruchus maculatus*, Jpn. J. Appl. Entomol. Zool. 14, 71-78.

Utida, S. (1972) Density dependent polymorphism in the adult of *Callosobruchus maculatus* (Coleoptera, Bruchidae), J. stored Prod. Res. 8, 111-126.

Utida, S. (1981) Polymorphism and phase dimorphism in *Callosobruchus*, Ser. Entomol. 19, 143-147.

Wasserman, S. S. (1985) Oviposition behavior and its disruption in the southern cowpea weevil, *Callosobruchus maculatus* F. (Coleoptera: Bruchidae), J. Econ. Entomol. 78, 89-92.

Wasserman, S. S. (1986a) Genetic variation in adaptation to food plants among populations of the southern cowpea weevil, *Callosobruchus maculatus*: evolution of oviposition preference, Entomol. Exp. Appl. 42, 201-212.

Wasserman, S. S. (1986b) Behavioral analysis of male-induced inter-strain differences in realized fecundity in *Callosobruchus maculatus.*, in M. D. Huettel (ed.), Evolutionary Genetics of Invertebrate Behavior, Plenum, New York, pp. 145-152.

Wasserman, S. S. and Asami, T. (1985) The effect of maternal age upon fitness of progeny in the southern cowpea weevil, *Callosobruchus maculatus*, Oikos 45, 191-196.

Wasserman, S. S. and Futuyma, D. J. (1981) Evolution of host plant utilization in laboratory populations of the southern cowpea weevil, *Callosobruchus maculatus* Fabricius (Coleoptera: Bruchidae), Evolution 35, 605-617.

Wilson, K. (1988) Egg laying decisions by the bean weevil *Callosobruchus maculatus*, Ecol. Entomol. 13, 107-118.

BEHAVIORAL ECOLOGY OF *CALLOSOBRUCHUS MACULATUS*

RODGER MITCHELL
Department of Zoology
Ohio State University
Columbus, Ohio 43210, U.S.A.

ABSTRACT. Each geographic strain of *C. maculatus* appears to have a unique set of physiological and behavioral traits. Some strains disperse eggs uniformly, others place eggs at random, fecundity may or may not be inhibited when few beans are available , and females may select the largest available beans for oviposition. Host preferences and tolerances differ from strain to strain and respond to selection. Before this variability was understood too little attention was given to the need for careful experimental design. A protocol is suggested as providing some standard points of reference for future research. The prospects for exciting research in ecology and evolution in this astonishingly variable species seems unlimited.

1. Introduction

Our view of *C. maculatus* changed when interfertile strains were discovered to be strikingly different in their fecundity and oviposition behavior (Dick and Credland 1984). The catalogue of differences between strains (Table 1) is already astonishingly large and every comparison seems to reveal more intraspecific variation. These revelations have opened up new prospects and this conference comes at an auspicious and critical point. Intraspecific variation in *C. maculatus* is known to be extensive and, perhaps, the excitement of discovery is waning. These findings need to be consolidated and the published data should provide the basis for comparative studies, but the problems of using the literature are apparent from the data on fecundity (Table 2). The inconsistencies are the result of either genetic differences or experimental procedures. A reviwer can not usefully speculate about data which shows the beetle to be more variable and sensitive to experimental conditions than anyone suspected.

Faced with unresolvable data, I designed a protocol to obtain data for comparing four established strains. I wanted to account for the inconsistencies in the literature and to develop a set of standards for comparative studies. Three aspects of the life history will be covered; fecundity, oviposition behavior, and larval behavior.

2. Fecundity

The eggs a female will lay are limited by the reserves she built up during larval feeding. It is the protein in these reserves that presumably limits the number of eggs she can lay. Females may lay fewer eggs if they experience competition as a larva (B-2; the codes refer to Table 1) and more eggs if they can feed on sugar solutions (Larson

K. Fujii et al. (eds.), Bruchids and Legumes: Economics, Ecology and Coevolution, 317–330.

Table 1. The behavioral traits of *C. maculatus* inferred to show heritable variation from strain differences and experiments. Major reviews and new findings on all of these topics will be found in Chapters by Taper, Messina, Toquenaga, R. H. Smith, and Bellows of this volume.

Trait	Evidence[1] and Citation	
A. Fecundity, oviposition behavior		
1. Number of eggs laid.	T	Dick & Credland 1984
2. Oviposition inhibited	E	Credland 1987
with few beans.	T	Dick & Credland 1984, Credland 1986
3. Discriminate number	H	Messina 1988
of eggs on a bean.	H	Messina & Mitchell 1989
4. Discriminate species of host	T	Wasserman 1986
B. Larval responses		
1. Pattern of larval competition	T	Credland et al. 1986 Credland 1986, 1987
2. Competition reduces	M	Utida 1972
growth & fecundity	T	Messina & Renwick 1985 Credland et al. 1986 Credland & Dick 1987
3. Food tolerance	E	Credland 1987
	H	Dick & Credland 1986b
	T	Dick & Credland 1986a
4. Development time	T	Dick & Credland 1984
5. Induction of	H	Messina 1987
dispersal morphs	M	Caswell 1961, Utida 1970
	T	Messina & Renwick 1985

[1] Letters indicate the evidence for genetic differences.

E	- Evolution of traits in mass cultures
H	- Hybrids intermediate between parental stocks
S	- Selection altered traits
T	- True-breeding strains differ in traits

and Fisher 1924). Fecundity may be secondarily reduced if few beans are available (A-2) or an unattractive host presented (A-5).

Three measures of fecundity have been proposed (Thanthianga and Mitchell in prep.). "Standard Fecundity" is the number of eggs laid by unfed females that did not compete as larvae and had a surplus of beans for oviposition. It is the number of eggs produced when the reserves must support both adult activity and egg production. Females that feed on sugar solutions may use those carbohydrates to support all activity so that the reserves can be used for egg production. The number of eggs produced by fed females with ample beans for oviposition is "potential fecundity". "Realized fecundity" is appropriate to use for the eggs laid by females under conditions that could secondarily affect fecundity or when conditions are not specified.

Table 2. Standard fecundity ± s.e. obtained from 100 bean protocols and the published data for four strains. All published data are for females given the number of blackeye peas indicated in parentheses. My data are for generation 102 of South India and generation 29-30 of the other strains.

Protocol on 100 beans			Published records	
Strain	Host	Fecundity	Fecundity	Citation
Campinas	Mung	60.0 ± 6.3		
	Cowpea	57.2 ± 3.6	70.5[*]	
			81.5(40)	Dick & Credland 1984
			113.6(100)	Credland 1986
			112.5(40)	Credland et al. 1986
			72.5(19)	Giga and Smith 1987
Cameroons	Mung	45.8 ± 3.4		
	Cowpea	44.5 ± 5.3	76.4[*]	
			106.0(20)	Messina & Renwick 1985a
			66.6(7)	Messina 1987
IITA	Mung	58.2 ± 3.3		
	Cowpea	51.8 ± 1.2	86.8[*]	
			88.52.6(40)	Dick & Credland 1984
			109.5(140)	Credland 1986
			117.3(40)	Credland et al. 1986
			75.2(20)	Messina & Renwick 1985a
			86.4(7)	Messina 1987
South India	Mung	47.8 ± 2.4	86.2[*]	
	Cowpea	47.4 ± 2.6		

[*] Fecundity recorded on 50 cowpeas for stocks at the time of receiving them from Messina, April 1986.

2.1. STANDARD FECUNDITY

Table 2 lists the maximum number of eggs eggs laid in the study cited, hence, most values are standard fecundity. The protocol for females with 100 beans is standard fecundity. The kind of bean has no effect. Cameroons and South India lay significantly fewer eggs than the other strains. All fecundities are lower than my earlier experiments and published values. Fecundity declined with the use of a culture technique for producing even aged adults (Thanthianga and Mitchell in prep.). Usually cultures are maintained by allowing a beetles to oviposit on fresh beans until they die. I allow females from the first 1-2 days of emergences to oviposit for 1-2 days. Early emerging females of the South India strain lay significantly fewer eggs (Thanthianga and Mitchell in prep). If early emergence is heritably associated with lower fecundity, this regimen will favor less fecund females.

The inconsistencies in Table 2 could be due to sampling errors, variation within the stocks of a strain, or evolved differences between strains. Fecundity varies, even in the reports from one laboratory and differs within strains as much as between strains. Hence, it appears that selection on fecundity is either weak or inconsistent under most laboratory regimens. Because of the genetic variance in fecundity persisting in the laboratory, all studies of fecundity should include standard fecundity controls.

2.2. POTENTIAL FECUNDITY

When fed sugar solutions, fecundity increased by 31 (Larson and Fisher 1924) to 56 eggs (Thanthianga and Mitchell in prep.) from standard fecundities of 88 and 69 respectively. The increase was the egg equivalent of the reserves females had to sacrifice to support their activity when they were unfed.

2.3. INHIBITION BY NUMBERS OF BEANS

The fecundities of three strains given 1, 3, 5, 10, 40, 100, and 140 cowpeas fit one of two response patterns (Dick and Credland 1984, Credland 1986). One strain was indifferent to the number of beans while two strains were inhibited if given 40 or fewer beans. Groups of females given limited numbers of beans responded in the same way (Bellows 1982b, Messina and Renwick 1985). The oviposition responses may be mediated by the average number of eggs/bean in the environment.

Fecundity declines as females are given fewer beans and all strains show a significant reduction of fecundity with the number of mung beans. The inhibition with cowpeas is strongest for Campinas and South India, and weak for Cameroon. IITA is indifferent to the number of cowpeas.

Females with 4 beans would lay some eggs on the dishes, but the response was extremely variable. Most Cameroon and South India females did not oviposit on dishes. IITA and Campinas females laid either more than 10 or zero eggs on dishes.

Surface area has been suggested as the cue inhibiting oviposition. Spray painting estimates of surface area (Wasserman 1986, Messina and Mitchell 1989) are difficult to evaluate. I caculated the suface areas of a cylinder (an overestimate) and a sphere (an underestimate) from the linear dimensions of cowpeas and mung beans. The average of these two approximates of surface area and indicates the surface area of cowpeas to be 83% greater than mung beans (330/180 mm2). If surface area cues regulate oviposition, beetles that significantly reduce fecundity on both beans should lay 83% more eggs on cowpeas. The inconsistencies, Campinas laid 81% more eggs on 4 cowpeas than on 4

Table 3. The number of eggs laid by females given 4, 20, and 100 cowpeas or mung for oviposition. Letters indicate means that are similar under a t-test. Numbers in parentheses are the eggs laid on dishes.

	Number of beans			Per cent reduction on 4 beans
	100	20	4	
On cowpea				
Campinas	$57.2 \pm 3.6a$	$53.4 \pm 3.6a$	$35.4 \pm 5.3b(11.2)$	38
Cameroons	$44.5 \pm 5.3a$	$44.4 \pm 6.6a$	$34.8 \pm 6.2b(2.0)$	22
IITA	$51.8 \pm 1.2a$	$48.4 \pm 2.5a$	$47.4 \pm 3.8a(.4)$	8
South India***	$47.4 \pm 2.6a$	$40.5 \pm 2.1b$	$19.5 \pm 1.0c(1.6)$	59
On Berkin				
Campinas***	$60.0 \pm 6.3a$	$41.2 \pm 1.6b$	$19.5 \pm 0.8c(4.2)$	68
Cameroons***	$45.8 \pm 5.3a$	$43.6 \pm 5.0a$	$25.4 \pm 2.6b$	45
IITA***	$56.2 \pm 3.4a$	$40.2 \pm 9.3ab$	$19.0 \pm 5.3b(15.6)$	66
South India**	$47.8 \pm 2.4a$	$31.4 \pm 1.7b(.6)$	$14.0 \pm 5.0c(1.4)$	71

** $P < .01$, *** $P < .001$ for the regression of fecundity with number of beans.

mung beans, as expected, but South India laid only 32% more eggs on 4 cowpeas, are similar to those Giga and Smith (1987) noted.

The behavior controlling egg dispersion often breaks down when there are several eggs/bean (Thanthianga and Mitchell in prep.). The extremes of inhibition observed in dishes may not occur if females can disperse from areas of high density. The significance and use of data from the extremely high densities used by Bellows (1982a, 1982b) and Smith and Lessels (1984) cannot be decided until it is known whether females remain at such densities in unconfined systems.

3. Oviposition Behavior

A larva hatches and chews through the floor of its egg shell to enter the bean underneath. It must develope in that bean, hence, the cues releasing oviposition by its mother determine the bean it will exploit and whether that larva must compete with older larvae. If the size, kind of bean, or competition affects survival or fecundity of the offspring, then, selective oviposition can evolve. Ovipositing females seem to respond to every aspect of a bean; the species of bean, its size, whether it is cracked, and how many eggs a bean carries (Mitchell 1983, Thanthinaga and Mitchell in prep.). Oviposition decisions are the critical determinant in the life cycle because they set the conditions in which an offspring must develop from egg to adult.

3.1. RESPONSES TO BEANS WITH EGGS

Eggs laid at random should fit a Poisson distribution. Deviations from the Poisson indicate selective oviposition and all strains are biased toward hyperdispersion. Some strains disperse eggs uniformly. Uniform distributions or those close to uniform cannot be discriminated with conventional statistics because the variance of a uniform distributions is a function of its mean. For example, the variance of a uniform distribution of 2 eggs/bean is 0 because every bean carries 2 eggs. Uniform distributions of 1.5 eggs/bean have 1 egg on half the beans and 2 eggs on the other half of the beans. The variance of this, and all distributions with a mean of X.5, is 0.25. Messina and Mitchell (1989) use the number of "errors" to calculate an index of uniformity, "U" explained in Appendix A.

All protocols have positive U-values indicating a consistent deviation toward unformity (Table 4). Eggs are placed more uniformly on mung beans than on cowpeas, except for IITA on 20 mung. The ability to respond to eggs per bean is greater on the smooth mung beans just as Credland and Wright (1988) found beetles dispersing eggs more uniformly on glass beads than on cowpeas. Individual distributions, except for Campinas and one Cameroons replicate, have variances significantly less than the Poisson .

Each strain has a unique response pattern. South India is the most uniform in laying eggs, with 4/10 of the distributions on 100 beans being uniform. Campinas oviposits nearly randomly on cowpeas but is more uniform on mung beans. Cameroon disperses eggs more uniformly than Campinas. IITA hyperdisperses with high U-values but, hyperdispersion does not show the usual decline in U with the number of beans available. The genetically determined ability to hyperdisperse which differs from strain to strain is expressed in a variety of ways (Messina and Mitchell 1989). The trait may be affected by a number of genetic modifiers.

Table 4. The number of eggs that must be moved (errors) to change the observed and the Poisson distributions into uniform distributions. "U" = (Poisson-observed)/Poisson errors for the totals for each set of replicates. Letters indicate errors that are similar under a t-test. There are 5 replicates for each entry.

	Cowpeas			Mung		
	100	20	4	100	20	4
Campinas						
Observed	$13.4 \pm 1.9a$	$8.4 \pm 1.5a$	$4.2 \pm 1.5b$	$8.0 \pm 1.7a$	$5.0 \pm 0.8ab$	$2.5 \pm 0.7b$
Poisson	13.8	10.4	4.0	15.3	9.2	3.7
U	0.02	0.19	0.04	0.48	0.56	0.19
Camerooons						
Observed	$5.2 \pm 1.4a$	$4.2 \pm 0.5a$	$2.8 \pm 0.6a$	$1.4 \pm 0.7a$	$3.2 \pm 0.6a$	$1.8 \pm 0.5a$
Poisson	8.9	8.7	4.3	9.2	8.7	3.8
U	0.42	0.52	0.33	0.85	0.66	0.53
IITA						
Observed	$8.0 \pm 0.8a$	$4.2 \pm 0.6a$	$3.8 \pm 0.6a$	$6.8 \pm 1.9a$	$5.8 \pm 2.3ab$	$1.2 \pm 0.4b$
Poisson	11.4	9.6	5.3	14.2	8.7	2.7
U	0.30	0.56	0.28	0.52	0.33	0.57
South India						
Observed	$3.2 \pm 0.5a$	$4.0 \pm 1.2a$	$2.0 \pm 0.1a$	$0.4 \pm 0.4a$	$0.2 \pm 0.2a$	$0.2 \pm 0.2a$
Poisson	9.8	9.5	3.1	9.9	8.5	2.5
U	0.69	0.58	0.35	0.96	0.98	0.92

3.2. BEHAVIORAL PROCESSES

The cues controling oviposition decisions involve the numbers of egg/bean which are discriminated quite accurately (Messina and Renwick 1985c) with inputs from the palps being the primary source of information (Messina and Renwick 1985a). Wilson (1988) considered how the information might be processed. He distinguished models based on fixed oviposition responses (absolute models) from relative models in which cues based on past experience release oviposition. Absolute model A involves digital processing, an egg is added to beans with no eggs and the probability of laying an egg on a bean with eggs is 0.25. The probability of oviposition in Model B is the exponent of the negative number of eggs/bean. Wilson's relative models (C-E) call for oviposition decisions to be based on relations between two of three measures. The measures are; (1) the number of beans ever visited, (2) the number of beans ever encountered with no eggs on them, and (3) the total number of eggs ever encountered. Females accumulate these totals throughout their life and they are exponentiated to obtain some current index of density which, in turn, determines the probability of ovipositing on a bean.

My simple digital model releases oviposition if a bean carries fewer eggs than the previous bean encountered (Mitchell 1975). A rule to oviposit after a number of rejections is necessary and the simulation is written to release oviposition after 5 rejections.

Oviposition models must satisfy two conditions. It must generate egg dispersions resembling the observed dispersion and, secondly, the behavior of the beetle must correspond to expectations of the the model. I ran simulations for 25 eggs laid on 50 beans under Wilson's models and my comparison model to see how well the outcomes fit the performance of the South India strain.

model	A	B	C	D	E	comparison
errors	1.74	2.48	1.96	2.88	2.94	0.10

South India made 0.025 errors (n = 58 females) in laying 20-30 eggs on 50 beans (Thanthianga and Mitchell in prep.). All of Wilson's (1988) models generate far too many errors. The comparison model is a closer, but not much less preces than South India.

Wilson (1988) presents second step behavioral studies but he did not test his strain for its ability to hyperdisperse eggs. We do not know if the results are for beetles that disperse eggs randomly, uniformly, or in some intermediate fashion.

3.3. CLUTCH SIZE

Charnov and Skinner's (1984) analysis of clutch sizes from egg dispersion data was extended by Smith and Lessels (1985). Groups of 2 or more eggs laid at random would produce variances in eggs/bean greater than the mean. Variances are less than the mean for every known strain (e.g. Table 4), clearly eggs were deposited singly and abandoned, hence, the "clutches" analyzed in these studies were unrelated and unattented groups of eggs laid at different times. I do not understand why such a group should be called a clutch.

Oviposition decisions clearly limit the competition a larva will face from older larvae but females cannot anticipate competition from the larvae of eggs laid at a later time. It would be interesting to model the evolution of oviposition behavior on the basis of what a female percieves when laying an eggs rather than the final number of eggs/bean.

3.4. RESPONSES TO BEAN SIZE

All females add eggs to the larger beans (Table 5). The discrimination of bean size is usually more apparent with cowpeas of 200-300 mg than with 40-80 mg mung beans. When some beans are without eggs, as in the 100 bean tests, the beans with 2-3 eggs rarely weigh 10% more than beans with no eggs. In 20 bean tests, females of all but the South India strain lay 3 or 4 eggs on some beans and usually select beans at least 20% heavier than the beans with one egg. Size cues evidently have a greater effect when females are adding eggs to occupied beans.

Size discrimination by IITA is precise enough that the number of eggs/bean is significantly correlated with bean weight in all tests. When Cameroon females add eggs to beans bearing eggs (20 bean tests), they select cowpeas averaging 248 mg. Since eggs/bean is not correlated with bean weight, it suggests that oviposition is released by any occupied bean above a threshhold size. South India is strongly inhibited if all beans carry an egg and only a few cowpeas carried three eggs. They significantly discriminate size with cowpeas, but not with mung.

If the offspring of larvae from females selecting larger beans must compete with older larvae they will be in an above average sized bean. Selection of large beans would clearly be adaptive if the number of survivors from a bean increases with bean size, but that has not yet been examined.

Table 5. The relation of egg load to the weights of beans with the minimum and maximum number of eggs from the 20 and 100 bean protocols of beans. The percent difference between the maximum and minimum is a measure of the selectivity.

	Campinas		Cameroon		IITA		South India	
Cowpeas	20	100	20	100	20	100	20	100
Minimum (eggs)	174 ±13.6 (0)	236 ±3.3 (0)	162 ±28.4 (0)	246 ±1.7 (0)	222 ±10.1 (1)	255 ±2.6 (0)	230 ±11.6 (1)	248 ±3.8 (0)
Maximum eggs/bean	246 ±10.8 (4)	280 ±9.9 (2)	248 ±17.9 (4)	257 ±11.2 (2)	285 ±9.1 (4)	268 ±6.0 (2)	280 ±11.2 (3)	274 ±8.9 (2)
Per cent difference	41*	19**	53**	4**	28**	5**	22**	10**

	Campinas		Cameroon		IITA		South India	
Mung	20	100	20	100	20	100	20	100
Minimum eggs/bean	42.4 ±2.8 (1)	56.1 ±1.0 (1)	57.2 ±3.9 (1)	59.4 ±0.5 (1)	47.9 ±2.8 (1)	55.1 ±0.6 (1)	49.8 ±3.5 (1)	57.5 ±0.8 (0)
Maximum (eggs)	78.0 ±5.3 (4)	58.8 ±1.2 (3)	68.2 ±3.7 (3)	64.7 ±2.5 (3)	60.8 ±1.6 (3)	63.6 ±2.6 (3)	50.5 ±5.8 (2)	59.7 ±1.0 (1)
Per cent difference	84**	5*	19**	9**	27*	15*	1	4

* P < .05 ** P < .001 of the two weights being similar.

3.5. RESPONSES TO HOST SPECIES

Acetone extracts from whole beans could make glass beads as acceptable for oviposition as the cowpeas from which the extract was taken (Credland and Wright 1988). Presumably the strains respond to the chemical signature of the hosts because (Wasserman 1986) found major differences in the number of eggs various strains laid on 9 different hosts. Each strain may have a unique set of responses. Unfortunately, the number of beans he presented was adjusted for bean size and no controls used for the number of beans. Responses to numbers of beans, if any, cannot be distinguished from responses to the host species.

Preferences may have evolved in Wasserman and Futyma's (1981) experiment. Their measure "realized fecundity/oviposition preference" was obtained by giving 25 females 500 azuki beans and 1000 pigeonpeas for oviposition. The portion of eggs on pigeonpeas, was influenced by fecundity and could be affected if hyperdispersion took precedence over the prefereces for azuki. Preferences could not be measured directly.

Table 6. Comparison of the oviposition traits of four strains of *C. maculatus*.

Host and fecundity (cowpea/mung)	Inhibition with few beans (I)	Disperse eggs uniformly (D)	Select large beans (S)
Campinas (57/60)	Consistent on both hosts	Consistent on both hosts	Consistent for both hosts
Cameroon (44/46)	Weakly expressed	Weakly expressed	Strong on cowpea Weak on mung
IITA (52/58)	None on cowpea Strong on mung	Weak on cowpea Strong on mung	Consistent on both hosts
South India (47/48)	Consistent on both hosts	Consistent on both hosts	Only with cowpea

When given a choice between the beans available to the source population, South India showed preferences corresponding to larval survival in the host (Thanthinanga and Mitchell in prep.)

Preference	cowpea >	mung >>	pigeonpea >>	chickpea
Survival to adult	0.76	0.74	0.39	0.59

Palpal inputs appear to provide the major cues for host discrimination, although inputs from the antennae and fore tarsi may modify responses (Messina et al. 1987a).

3.6. OVIPOSITION STRATEGIES

The differences in the oviposition behavior can be quite bewildering, but the major patterns (Table 6) identify associations. For example, cues for the number of eggs/bean might regulate fecundity as well as release oviposition behavior on individual beans. South India responds sharply to numbers of beans as well as showing a uniform egg dispersion. With few mung beans, IITA is strongly inhibited but the females do not consistently hyperdisperse eggs, hence, egg deposition is not necessarily associated with the the inhibition response.

Females regulate the level of larval competition through three oviposition responses. Two behaviors alter the incidence of competiton. Inhibition reduces competition because fewer eggs are laid and hyperdispersion averages out the level of competition by eliminating the extremes. The effect of competition is altered by placing eggs on larger beans so that competing larvae have more resources. South India virtually excludes competiton through hyperdispersion and extreme inhibition. Larvae exclude competitors, so it is not surprizing that responses to bean size are insignificant. The other strains exhibit some combination of one or two of the strategies, except Campinas that consistently responds to bean number, size, and egg density.

4. Larval Behavior

South India larvae exchange signals between burrows. The signals are presumably vibrations which allow one larva to assume dominance during the first day or two of competition (Thanthianga and Mitchell 1987). That larva will be the sole survivor. The other strains have multiple survivors and the behavior of these larvae can be followed by opening beans to see day to day changes in the beans.

IITA larvae are indifferent to each other unless the burrows happen to intersect. When larval burrows intersect, the larvae retreat to the rear of their burrows. They gather up fass, push it to the intersection and build a wall between the burrows. Beans opened after the intersected burrows are closed usually show larvae burrowing side by side in the same direction. This cooperative use of space in a bean allows more larvae to survive than would be possible if burrows were cut in a haphazard way.

Overall survivorship is density dependent (Table 7) with surviorship of South India falling as 1/(larvae/bean). Larval survival of IITA falls to 57% when there are 5 larvae in a cowpea. Survival is 73% for Cameroon and 82% for Campinas when 5 larvae compete. The Campinas and IITA data fit a general pattern reported by Dick and Credland (1986) and Credland (1986). Mortality of the Yemen strain (Credland et al. 1986) resembles the pattern for South India. These data suggest that the larval interactions vary among the strains.

Campinas was tested in three sizes of *Vigna* to determine how bean size affected competition. One larva usually eliminated competitors in 32 mg moth beans (*V. acontifolius*) so that 85% of the beans with 2 larvae in them had one exit (n=55). Two competitors survived in 50% (n=50) of the mung beans, which provide twice as much food (65 mg). There were two survivors in 74% of the 260 mg cowpeas.

If competitors fed indifferently in the moth beans too small to support two individuals, there would be mutual deaths. Such deaths did not occur suggesting that competitive behavior was modulated in response to bean size. Larvae that tolerated competors in large beans seemed to change their behavior so as to exclude competitors in smaller beans.

5. Summary

Adaptations for exploiting stored beans by *C. maculatus* probably began to evolve while

Table 7. t-tests for the regression of survival of competing larvae in cowpeas tested as a function of the number of larvae competing in a bean.

larvae/bean	Campinas	Cameroon	IITA	South India
1	0.944	0.973	0.896	0.744
2	0.820	0.835	0.826	0.390
3	0.862	0.812	0.770	0.247
4	0.843	0.762	0.590	0.177
5	0.823	0.733	0.567	0.149
6	0.846			0.124
t-values	1.44	4.78	7.21	3.79
	n.s.	$P < .01$	$P < .01$	$P < .05$

cowpeas were being domesticated in Africa nearly 6000 year ago (Ng and Marechal 1985). The beetle is now cosmopolitan and quite catholic in its tastes. All the common old world pulses are attacked by one strain or another. Feeding adaptations which involve physiological and behavioral traits (Credland 1987) seem to evolve readily.

Larval behavior spans the entire range of what students of competition think of as mutually exclusive alternatives. Selective oviposition may ameliorate the intensity of larval competition in three ways: (1) The intensity of competition can be lowered by withholding eggs when few beans are available. (2) The level of competition is averaged as a result of at least a weak tendency to hyperdisperse eggs. (3) When females are forced to add eggs to beans already carrying eggs, they may select the largest beans so that their larvae compete in a bean with a larger than average quantity of food.

The benefits gained by an ovipositing female that reduces larval competiton depends on density dependent increases in larval mortality. The increase in mortality from 1 to 2 eggs is an index of the benefits associated with eliminating competition. The oviposition strategies that are strongly expressed (coded in Table 6) are associated with the increased mortality of 2 eggs as:

Strain	South India	Campinas	IITA	Cameroons
Mortality increase	47%	12%	9%	14%
Oviposition strategies	I,D	I,D,S	S	-

South India can gain the most by reducing larval competition and it virtually eliminates larval competition. The other strains could gain from 9 to 14% in survival from selective oviposition but are quite inconsistent in their adaptations for reducing larval competition.

No other animal that I know offers so many opportunities for research on its ecology, adaptations, and evolution. It can be reared by the most careless and negligent in laboratory environments that resemble the storage conditions where it evolved and thrives. The life cycle is short and a wide range of traits affecting survival or fecundity show heritable variation. The effect of these traits can be measured and expressed in a life tables.

The published work cannot be fit together because of the unexpected variability of the beetle. The genetic variability is a resource that can be exploited, but it is a potential pitfall unless more attention is given to executing well designed experiments with controls based on standard protocols. The protocols explored here and the rationale for their use is presented in Appendix A.

I believe a burst of very exciting work will appear before the next conference on bruchids. In more reckless moments, I think that *C. maculatus* may well allow ecologists and evolutionists to come to grips with some of the most untractable problems in evolution: How can the effects of interacting but independent components of fitness be measured and analysed? Can natural selection, which is driven by reproductive rates alone, account for the evolution of a suite of interacting traits?

REFERENCES

Bellows, T. S. (1982a) Analytical models for laboratory populations of *Callosobruchus chinensis* and *C. maculatus* (Coleoptera, Bruchidae), J. Anim. Ecol. 51, 263-287.

Bellows, T. S. (1982b) Simulation models for laboratory populations of *Callosobruchus chinensis* and *C. maculatus* (Coleoptera, Bruchidae), J. Anim. Ecol. 51, 597-623.

Caswell, G. H. (1961) Observations on an abnormal form of *Callosobruchus maculatus* (F.), Bull.

328

Entomol. Res. 50, 671-680.

Charnov, E. L. and Skinner, S. W. (1984) Evolution of host selection and clutch-size in parasitoid wasps, Florida Entomol. 67, 5-21.

Credland, P. F. (1986) Effect of host availability on reproductive performance in *Callosobruchus maculatus* (Coleoptera: Bruchidae), J. stored Prod. Res. 22, 49-54.

Credland, P. F. (1987) Effects of host change on the fecundity and development of an unusual strain of *Callosobruchus maculatus* (F.) (Coleoptera: Bruchidae), J. stored Prod. Res. 23, 91-98.

Credland, P. F. and Dick, K. M. (1987) Food consumption by larvae of three strains of *Callosobruchus maculatus* (Coleoptera: Bruchidae), J. stored Prod. Res. 23, 31-40.

Credland, P. F., Dick, K. M. and Wright, A. W. (1986) Bionomic variation among three populations of the Southern cowpea weevil *Callosobruchus maculatus*. Ecol. Entomol. 11, 41-50.

Credland, P. F. and Wright, A. W. (1988) The effect of artificial substrates and host extracts on oviposition by *Callosobruchus maculatus* (Coleoptera: Bruchidae), J. stored Prod. Res. 24, 157-164.

Dick, K.M. and Credland, P. F. (1984) Egg production and development of three strains of *Callosobruchus maculatus* (F.) (Coleoptera:Bruchidae), J. stored Prod. Res. 19, 189-198.

Dick, K. M. and Credland, P. F. (1986a) Variation in response of *Callosobruchus maculatus* (F.) to a resistant variety of cowpea, J. stored Prod. Res. 22, 43-48.

Dick, K. M. and Credland, P. F. (1986b) Changes in the response of *Callosobruchus maculatus* (Coleoptera: Bruchidae) to a resistant variety of cowpea, J. stored Prod. Res. 22, 43-48.

Giga, D. P. and Smith, R. H. (1987) Egg production and development of *Callosobruchus rhodesianus* (Pic) and *Callosobruchus maculatus* (F.) (Coleoptera:ruchidae) on several commodities at two different temperaturres, J. stored. Prod. Res. 23, 9-15.

Larson, A. O. and Fisher, C. K. (1924) Longevity and fecundity of *Bruchus quadrimaculatus* Fab. as influenced by different foods, J. Agric. Res. 29, 297-305.

Larson, A. O. and Simmons, P. (1923) Notes on the biology of the four-spotted bean weevil, *Bruchus quadrimaculatus* Fab., J. Agric. Res. 25, 609-616.

Matlock, R. S. and Oswalt, R. M. (1963) Mungbean varieties for Oklahoma. Oklahoma Agric. Exp. Sta. Bull. B-612, 15pp.

Messina, F. J. (1987) Genetic contribution to the dispersal polymorphism of the cowpea weevil (Coleoptera: Bruchidae), Ann. Entomol. Soc. Amer. 80, 12-16.

Messina, F. J. Barmore, J. L. and Renwick, J. A. A. (1987a) Host selection by ovipositing cowpea weevils: patterning of input from separate sense organs, Entomol. exp. appl. 43, 169-173.

Messina, F. J., Barmore, J. L. and Renwick, J. A. A. (1987b) Oviposition deterrent from eggs of *Callosobruchus maculatus*: spacing mechanism or artifact, J. Chem. Ecol. 13, 219-226.

Messina, F. J. and Mitchell, R. (1989) Intraspecific variation in the egg-spacing behavior of *Callosobruchus maculatus*, J. Insect Behav. in press.

Messina, F. J. and Renwick, J. A. A. (1985a) Dispersal polymorphism of *Callosobruchus maculatus* (Coleoptera, Bruchidae): Variation among populations in response to crowding, Ann. Entomol. Soc. Amer. 78, 201-206.

Messina, F. J. and Renwick, J. A. A. (1985b) Mechanism of egg recognition by the cowpea weevil *Callosobruchus maculatus*, Entomol. exp. appl. 37, 241-245.

Messina, F. J. and Renwick, J. A. A. (1985c) Ability of ovipositing seed beetles to discriminate between seeds with differing egg loads, Ecol. Entomol. 10, 225-230.

Mitchell, R. (1975) The evolution of oviposition tactics in the bean weevil, *Callosobruchus maculatus* (F.), Ecology 56, 696-702.

Mitchell, R. (1983) Effects of host-plant variability on the fitness of sedentary herbivorous insects, in R. F. Denno and M. S. McClure (eds.), Variable Plants and Herbivores in Natural and Managed Systems, Academic Press, Inc., New York, U.S.A., pp. 343-369.

Ng, N. Q. and Marichal, R. (1985) Cowpea taxonomy, origin and germ plasm, in S. R. Singh and K. O. Ritchie (eds.), Cowpea Research, Production and Utilization, John Wiley, New York,

U.S.A., pp. 11-21.

Smith, R. H. and Lessels, C. M. (1984) Oviposition, ovicide and larval competition in granivorous insects, in R. M. Silby and R. H. Smith (eds.), Behavioral Ecology; Ecological Consequences of Adaptive Behavior, Symposium British Ecological Society, Blackwell's, Oxford, United Kingdom, pp. 423-448.

Taylor, T. A. (1974) Observations on the effects of initial population densities in culture, and humidity on the production of 'active' females of *Callosobruchus maculatus* (F.) (Coleoptera, Bruchidae), J. stored Prod. Res. 10, 113-122.

Taylor, T. A. and Aludo, J. I. S. (1974) A further note on the incidence of active females of *Callosobruchus maculatus* (F) on mature cowpea in the field in Nigeria, J. stored Prod. Res. 10, 123-125.

Thanthianga, C. and Mitchell, R. (1987) Vibrations mediate prudent resource exploitation by competing larvae of the bruchid bean weevil *Callosobruchus maculatus*, Entomol. exp. appl. 44, 15-21.

Thanthianga, C. and Mitchell, R. The fecundity and oviposition behavior of a South Indian strain of *Callosobruchus maculatus* (Fab.) (Bruchidae:Coleoptera), in prep.

Utida, S. (1972) Density dependent polymorphism in the adult of *Callosobruchus maculatus* (Coleoptera, Bruchidae), J. stored Prod. Res. 8, 111-126.

Wasserman, S. S. (1986) Genetic variation in adaptation to foodplants among populations of the southern cowpea weevil, *Callosobruchus maculatus*: Evolution of oviposition preference, Entomol. exp. appl. 42, 201-212.

Wasserman, S. S. and Futyma, D. J. (1981) Evolution of host plant utilization in laboratory populations of the southern cowpea weevil, *Callosobruchus maculatus* Fabricius (Coleoptera, Bruchidae), Evolution 35, 605-617.

Wilson, K. (1988) Egg laying decisions by a bean weevil *Callosobruchis maculatus*, Ecol. Entomol. 13, 107-118.

APPENDIX A

Three major features of oviposition behavior, fecundity, egg dispersion, and selection of large hosts, can be determined with the protocol for standard fecundity. A second level protocol will reveal responses to the abundances of hosts. The object of the protocols is to measure the traits of a strain, hence, other variables must be controlled. The quality of females must be specified and the hosts presented must be selected to allow comparisons with other studies, and the experimental conditions specified.

Female quality. Females should be virgins from uncrowded cultures, with one egg per bean to exclude larval competition. Virgin females can be readily obtained if adults are removed from an emerging culture in the morning and adults collected as they emerge. Females should be paired and left undisturbed until death (about 10-12 days).

Species of host. A cowpea, the blackeyed pea(*V. unguiculata*), usually used by Europeans and Americans, is a set of varieties and most commercial peas are "California No 5". This may be the ideal candidate for standard tests because it has been widely used and can be obtained from suppliers of seed grains. The disadvantage of blackeyed cowpeas is that females do not fully express inhibition (Table 3) or hyperdispersion (Table 4) on that bean. These traits are expressed on mung beans (*V. radiata*). Mung beans are small smooth surfaced, nearly round beans differing in size. There are no widely used commercial varieties, although Johnson Seed Company (Enid, OK, USA), can provide the variety "Berkin", which is maintained by the Oklahoma Foundation Seed Stocks, Inc. (Matlock and Oswalt 1963).

Numbers of beans. 100 beans are usually sufficient, but 125 beans must be given to strains with unusually high fecundities. It important that the number of beans be greater than the fecundity because some females are inhibited when they do not have a surplus of beans (Credland 1986). Neither responses to bean size nor hyperdispersion can be expressed unless there is an excess of beans.

Experimental conditions. Containers must allow females easy access to all the beans. Behavioral data from controlled conditions, generally temperatures of 29-30° and 70% R.H. are indistiguishable from experments done under ambient laboratory conditions.

Analysis of the standard fecundity protocol. Shortly after the females die, the beans should be sorted according to eggs/bean. Beans with 0 eggs, with 1 egg, etc., should be counted and weighed. Measures for weight discrimination based on the weights of the beans combined will reveal patterns but more information on weight discrimination would require each bean to be weighed.

The average eggs/bean for each replicate can be used to define a uniform and Poisson distributions for each female. With these distributions, the number of eggs that must be moved to change the Poisson distribution into a uniform uniform distribution can be determined by insepection. For example, the uniform distribution for replicate 1 of IITA with 100 mung has 11.5 fewer beans with 0 eggs than the Poisson, thus, 11.5 eggs must be moved from beans with 2 or more eggs to convert the Poisson to a uniform distribution. Poisson errors = 11.5 and it can be seen that there are 8 errors under the observed distribution. Precise statements for the procedures are in Messina and Mitchell (1989). The index for uniformity, "U", is (Poisson-observed)/poisson errors. The results of this are:

No of beans with x egg

	0	1	2	≥3	Errors	
Uniform	48	52				
Random	59.5	30.9	8.0	1.6	11.5	$U = (11.5-8)/11.5 = 0.304$
Observed	56	36	8		8	

The first level protocol is a simple test for all aspects of oviposition known to vary in the species except the degree of inhibition.

Second level protocols. Every aspect of oviposition behavior can change when a small number of beans are available and these responses are so varied that standard protocols are of limited use. Density dependent tests should include an intermediate number of hosts and a very small number. The 20 hosts used here were intermediate in that some strains responded at that density and others did not. The extreme of 4 gave enough hosts to determine whether there was any hyperdispersion and allowed the strains to be differentiated from each other and host effects to be discriminated.

APPENDIX B

Protocols used 4 strains which have been characterized. Three strains are maintained on cowpeas: Campinas and IITA (Dick and Credland 1984), Cameroon (Messina and Renwick 1985c). South India (Thanthianga and Mitchell 1987) was maintained on Berkin. The stocks were kept at ambient temperatures and subcultured by allowing the females from first 1-2 days of emergences to oviposit an abundance of fresh beans for 24-36 h. There were usually about 500 individuals per generation.

Virgin females collected on emergence from lightly infested beans were isolated with males in Petri dishes with 4, 20, or 100 mung beans (Berkin) and the same numbers of cowpeas (California No. 5 blackeyed peas). After the beetles died, the beans were scored and the data analyzed as explained in Appendix A.

SEX-RATIO DETERMINATION IN THREE WASP SPECIES ECTOPARASITIC ON BEAN WEEVIL LARVAE

KOICHI FUJII AND KHIN MAR WAI
Institute of Biological Sciences, University of Tsukuba,
Tsukuba, Ibaraki 305 Japan

ABSTRACT. We investigated the control of the sex ratio in three ectoparasitic solitary wasp species, *Dinarmus basalis* (Rondani), *Anisopteromalus calandrae* (Howard), and *Heterospilus prosopidis* Vier., parasitizing azuki bean weevil (*Callosobruchus chinensis* (L)(Coleoptera: Bruchidae)) larvae. Using host age as a measure of host quality, we ask the question whether these wasps can adjust their sex ratio to different sized hosts.

By either artificially transferring wasp eggs onto larger hosts, or letting eggs develop on the hosts on which these were laid, we measured the primary and emergence sex ratios of offspring wasps on different aged hosts. In all species, sex ratios from older hosts were more female biased. However, there was significant differential mortality between sex during the larval stage, especially on younger hosts.

Further experiments showed that wasps did not lay more female eggs on larger hosts of the same age, suggesting that offspring sex was not controlled by the individual size of the host on which a female oviposited.

There was a strong positive correlation between the size of host on which a male developed and his mating ability and longevity. A similar positive correlation was seen between the size of host on which a female developed and her fecundity and longevity.

We demonstrate 1) that female discrimination of host quality does influence sex allocation and 2) that the allocation of offspring sex by host size significantly increases a female contributions to the next generation (her fitness).

1. Introduction

The way organisms allocate resources to their offspring is a central problem in evolutionary biology (Darwin, 1887; Fisher, 1930; Williams, 1975; Maynard Smith, 1978; Bell, 1982; Charnov, 1982; Goodman, 1982). Fisher (1930) predicted that there should be equal investment in male and female offspring in randomly mating populations.

The haplodiploid sex determination system of most parasitoid wasps allows females to control the sex ratio (ratio of female progeny to total progeny), because they can control whether an egg is female (fertilized) or male (unfertilized). The topic of offspring sex ratios in parasitoid wasps was reviewed extensively by Flanders (1939, 1946), Kochetova (1977), and King (1987).

King (1987) defines four factors that may influence offspring sex ratios in parasitoid wasps; parental characteristics, environmental characteristics, host characteristics, and factors influencing local mate competition.

We shall be primarily interested in how the sex ratio is related to host size. This has been explored mathematically by Charnov (1979) and Charnov et al. (1981). The model predicts that the sex of parasite offspring will vary with host size if host size differentially

K. Fujii et al. (eds.), Bruchids and Legumes: Economics, Ecology and Coevolution, 331–340.
© 1990 *Kluwer Academic Publishers. Printed in the Netherlands.*

affects the fitness gains of sons vs. daughters. Female eggs should be laid in large hosts and sons in small hosts if females gain more by developing in or on large hosts than males.

The relationship between the host size and the sex of the emerging wasp(s) has been extensively examined, and in most species studied, the sex ratio of wasps emerging from large host is higher than from small host, though there are exceptions.

However, the association of sex ratio with host size may result from selective oviposition by mothers or from differential mortality. Few studies address this distinction. In the more than forty studies reviewed by King (1987), differential mortality was ruled out as a causal factor in only eight cases. However, even these studies did not show that there was no sex differential mortality. These studies showed only that sex differential mortality was not the main factor for sex ratio change.

Most studies scored emergence sex-ratio (i.e, the sex ratio at emergence from hosts on or in which eggs had been oviposited) and Charnov's (1982) careful review of the data emphasizes the need to eliminate differential mortality of male and female offspring as a possible cause for the differences in sex-ratio before equating the emergence sex-ratio with the primary sex-ratio (i.e., the sex-ratio at oviposition).

In this paper, we consider 1) how the host size changes with age, 2) how the emergence sex ratio changes with host age (and size), 3) whether there is sex biased mortality of larvae, and 4) the cues female wasps use to determine host quality. We show that the primary sex ratio changes with host age, and that sex-biased mortalities are common on small hosts, and discuss the cues female wasps may use to adjust the sex ratios.

2. Materials

We used three parasitic wasps, *Anisopteromalus calandrae, Heterospilus prosopidis,* and *Dinarmus basalis.* These are ectoparasites on larvae of several grain and bean weevils. We used the azuki bean weevil, *Callosobruchus chinensis* (L.) for their hosts. The stock was maintained as follows. The stock adult beetle density was controlled so that each bean had about 8 hatched eggs. Freshly emerged beetles were allowed to oviposit on about 100 g of fresh azuki beans (*Vigna angularis*) for 6 hours.

The newly emerged adult wasps were transferred to four compartment plastic Petri dishes (Falcon No.1009, 7 cm x 1 cm) with approximately 100 azuki beans from the stock beetle culture. Stock wasp cultures were given 16 or 17 day old larvae (4th instar) throughout the experiment. Stocks and experiments were held in a controlled chamber at 30°C and 70% r.h.

Wasp species *A. calandrae* and *H. prosopidis* have been maintained in the similar manner for more than 40 years. *D. basalis* was introduced from India about 10 years ago and has been maintained in the similar manner as the other wasp species.

All three wasp species can parasitize 10 to 19 day old larvae of the azuki bean weevil. Ten day old larvae are in the late 2nd or early 3rd instar, and prepupae or pupae occur at 19 days.

3. Experiment I. Development and Host Weight

Host age (days since oviposition) was the aspect of host quality controlled in the present study, except as noted otherwise. We first calibrated the way host size changed with age.

Azuki beans with 7 to 8 hatched eggs were drawn from stock azuki bean weevil culture to obtain 10, 11, 12, 13, 14, and 15 day old larvae. The larvae were removed from the beans under a binocular microscope, and individually weighed immediately with Cahn electro balance (larvae less than 1 mg to the nearest 0.002 mg, those between 1 and 5 mg to the nearest 0.01 mg, and larger ones to the nearest 0.02 mg).

The relationship between larval age and fresh weight is shown in Figure 1. The largest change in weight was observed between 12th and 13th days.

4. Experiment 2. The Sex Ratio and Body Weight of Emerging Wasp Adults from Different Host Ages

With the relationship of host weight and size established, we determined the way the sex ratio and body weight of wasp offspring varied with the age (and inferred size) of the host.

The wasps were collected each day from the stock culture and kept in the empty petri dish for 24 hours without access to hosts. This procedure allowed them to oviposit a maximum number of eggs when they were provided with hosts. The wasps were then introduced to a four-compartment Petri dish. The dish contained one layer of azuki beans from stock beetle cultures. Azuki bean weevil larvae at 10, 11, 12, 13, 14, 15, or 17 days old were provided. The wasps were removed from the dish after 24 hours.

Half of the beans were kept in the dish until the wasp adults emerged. The remaining half was used for Experiment 3 (see below). The emerging wasps were removed daily from the dish and kept in the refrigerator until examined. After all wasps emerged, each wasp was sexed and weighed.

The sex ratios of offspring increased with older hosts in all three wasp species. In *A.*

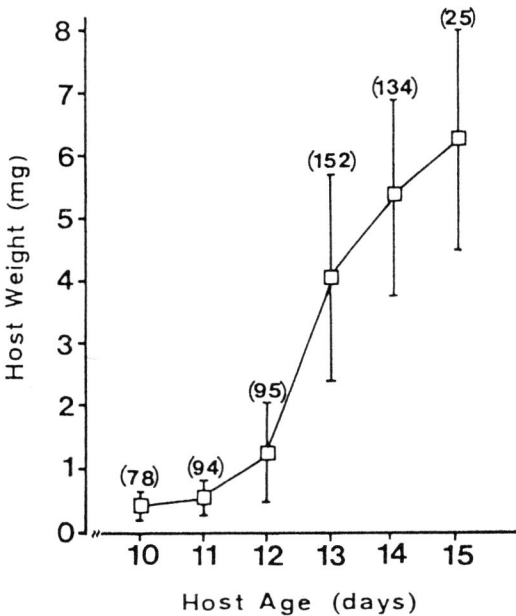

Figure 1. Host age and host weight relationship. Square shows the mean and vertical bar shows one standard deviation. Sample sizes are shown in parenthesis.

calandrae (Fig. 2(a)), no adult wasp emerged from 10 day old hosts. A small number of male wasps emerged from 11 days hosts. The sex ratio increased sharply with 12 day old hosts, and remained between 0.6 and 0.7 in older hosts. The weight of female offspring (Fig. 2(d)) increased on older hosts.

In *D. basalis* (Fig. 2(b) and (e)), no female offspring emerged from both 10 day and 11 day old hosts. The emerging sex ratio from 12 day old hosts rose to 0.1, that from 13 day old hosts increased further to above 0.4, and after that it remained constant at about 0.5.

The results with *H. prosopidis* are shown in Fig. 2(c) and (f). Relatively large numbers of male wasps emerged from 10 day old hosts. However, they were small (less than 0.1 mg, compared with 3.0 mg from 15 days host). The sex ratio increased gradually on hosts 11 days and older. It was around 0.45 on 12 and 13 day old hosts, and 0.65 on 14 and 15 day old hosts. The weight of wasps also increased with host age. It is noteworthy that in *H. prosopidis*, there was no significant difference in weight between males and females on hosts of the same age. The range of female weight is also noteworthy. Females from 15 day old hosts were 6 times heavier than those from 11 day old

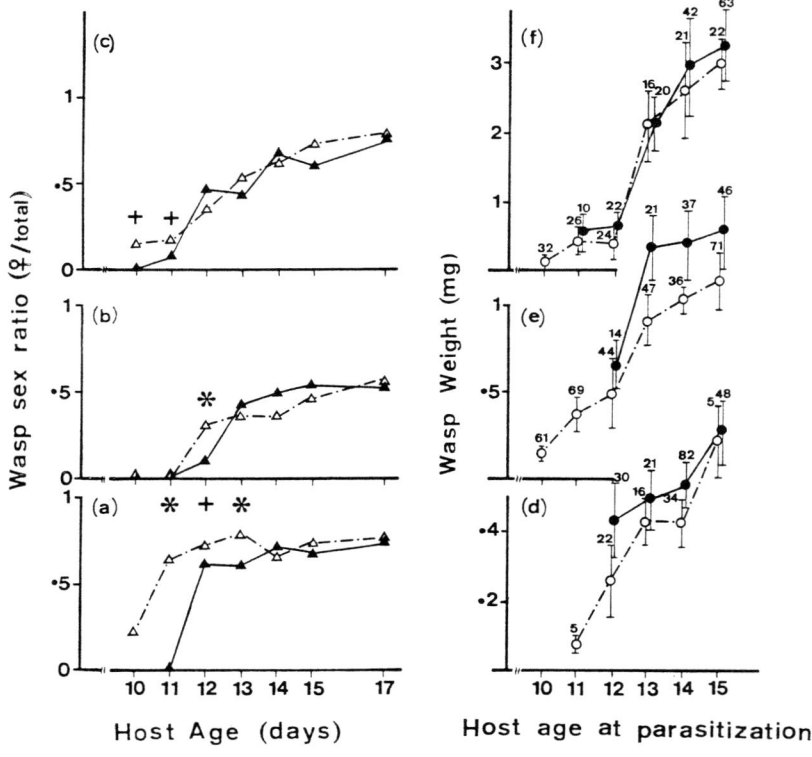

Figure 2. (a)-(c) The relationship between host age and emergence sex ratio (solid mark) (Experiment 2) or primary sex ratio (open mark) (Experiment 3) of wasp offspring. Statistical significance between two ratios is shown by + (0.05 < P < 0.10) or *(P < 0.05). (d)-(f) The relationship between host age and female wasp weight (o) or male wasp weight (o). Vertical bar shows one standard deviation. Number above each bar shows the sample size. (a) and (d) *A. calandrae*, (b) and (e) *D. basalis*, and (c) and (f) *H. prosopidis*.

hosts. The comparable weight increase in *D. basalis* was about 2 times, and in *A. calandrae* it was only 1.5.

5. Experiment 3. Primary Sex Ratios on Hosts of Different Ages

In Experiment 2, the sex ratio increased with host age in all three species. However, the observed changes might have been due to a sex biased mortality of wasp larvae. This can be demonstrated by examining the primary sex ratio (sex ratio at oviposition). The primary sex ratio was determined with the technique of wasp egg transfer established by Khin Mar Wai and Fujii (in manuscript).

We removed 15 or 16 day old larvae from beans of beetle stock culture, and transferred each larvae to a well (4 mm in diameter) in an acrylic plastic plate. The well was about the size of the cavities the 4th instar larvae occupied in the bean.

The other half of the beans prepared in Experiment 2 were dissected immediately after the wasp removal. Each wasp egg on host larvae was individually transferred onto the fresh 15 or 16 day old host in a plastic well. Wasp survival was usually greater than 90% with this treatment (Khin Mar Wai and Fujii, in manuscript). The well in the acrylic plate was covered with slide glass. Emerging wasps were sexed, and counted. The results are superimposed on the results of Experiment 2 in Figures 2(a), (b), and (c).

Upon dissection, we found many *A. calandrae* eggs on 10 day old hosts, although in Experiment 2, no wasps emerged from these hosts (Fig. 2(a)). Most eggs transferred from 10 day old hosts to 15 day old hosts developed into adults, and showed a sex ratio of more than 0.2. The sex ratio of eggs collected from 11 day old hosts was 0.65, significantly higher than that of emerging offspring in Experiment 2. The sex ratio of 0.65 was the same as observed with older hosts in Experiment 2. These results show that female *A. calandrae* wasps laid eggs on 10 day old hosts that were too small to support the wasp larval development. They also show that female wasps laid many female eggs on hosts (11 day old host) which were large enough to support male wasp development but not large enough for females (see Fig. 2(a)). Thus, in *A. calandrae*, there was a drastic primary sex ratio change from 10 to 11 day old hosts, and there also was a large differential mortality between sexes during larval development on young host. Although the differences between the primary and emerging sex ratios were smaller on 12 and 13 day old hosts compared on 11 day old hosts, they were still significantly different, suggesting that sex differential mortality operates even on 12 and 13 day old hosts.

In Experiment 2, only males of *D. basalis* emerged from 10 and 11 day old hosts. No female eggs were found among the transferred eggs. Thus, at least for 10 and 11 day old hosts, exclusive emergence of males in Experiment 2 was not due to sex differential larval mortality. Instead, the mother controlled the primary sex ratio by laying only unfertilized eggs. With 12 day old hosts, however, the primary sex ratio was significantly higher than emerged sex ratio, as with *A. calandrae* on 10 and 11 day old hosts. Female larvae suffered a higher mortality during their larval stage on 12 day old hosts. With 13 or older day hosts, primary sex ratio was not significantly different from the realized sex ratios.

The comparison of *H. prosopidis* sex ratios between control and transferred experiments shows a trend similar to that in *A. calandrae* and *D. basalis*. Wasps tended to lay fewer female eggs on small hosts. Nonetheless, the primary sex ratio on hosts of 10 and 11 days old was significantly higher than the sex ratio of emerging offspring. Thus with these hosts, the sex ratio, which was male biased at oviposition, was further male biased at emergence because of the higher mortality of females. There was no significant difference between primary and emerging sex ratios with hosts 12 or more days old. The primary sex ratio increased gradually from 12 day old host to 15 day old host, as did the sex ratio of emerging offspring.

6. Discussion

6.1. THE EXISTENCE OF DIFFERENTIAL MORTALITY BETWEEN SEXES

When Charnov (1982) discussed the causes of sex ratio change in parasitic wasps with host size, he carefully distinguished between the primary and emergence sex ratios. His model on sex ratio dealt with primary sex ratio, however, he admitted that most of the evidence for a change in the sex ratio with host size (or age) was data for the emergence sex ratio. Several studies were thought to show a change in sex ratio with host size without differential mortality between sexes. However, in most of them, the possibility of biased mortality cannot be discarded.

Charnov cites Kishi(1970)'s study as one of the clearest examples of a change of primary sex ratio change with host size without any differential mortality. A careful examination of Kishi's data shows no direct measure of both the primary and realized sex ratios.

Charnov et al. (1981) showed that the emergence sex ratio of *H. prosopidis* became more female biased with older hosts. Their claim "A series of control experiments showed that this result could not be due to differential, sex-specific mortality" has no supporting data.

Sandlan's (1971) study has been thought to show a sex ratio change with host size without any differential mortality. However, Wellings et al. (1986) questioned the claim of no differential mortality on the statistical grounds.

Our study clearly shows that all three wasp species employed, had a significant sex-specific, differential mortality, especially on small hosts. In general, the already male biased primary sex ratio on small hosts was further biased to males due to differential mortality. However, in *A. calandrae* on 11 day old hosts, the emergence sex ratio was 0 (i.e., males only), whereas the primary sex ratio was more than 0.5. From these results, we strongly urge that all studies on sex ratio take differential mortality into consideration.

When significant differential mortality occurred, females invariably suffered higher mortality than males. This suggests that the threshold host weight below which larvae cannot successfully develop to adulthood is higher for females than for males. This may be related to the reproductive success of emerged wasps of each sex. Detailed study on the reproductive success of each sex, especially for small individuals, are further required.

Our results showed an increase in the sex ratio with older hosts. However, the pattern of the sex ratio change with host age differs among the species. In *A. calandrae*, the emergence sex ratio increased abruptly between 11 and 12 day old hosts, and stayed rather constant thereafter. In *D. basalis*, it increased between 12 and 13 day old hosts, and remained steady thereafter. On the other hand, in *H. prosopidis*, the emerged sex ratio increased gradually from 10 to 15 day old hosts.

One possible explanation for these results can be seen in Figure 3. Figure 3 is a transformation of Figures 2(d), (e) and (f), but here the abscissa is the estimated mean host weight. The association of mean host weight with age from Fig. 1 was used. Figure 3 thus shows the relationship between the mean weight of host on which the wasp developed and mean weight of emerged wasps. The mean body weight of emerged females of *A. calandrae* (Fig. 3(a)) is almost constant (except that from 15 day

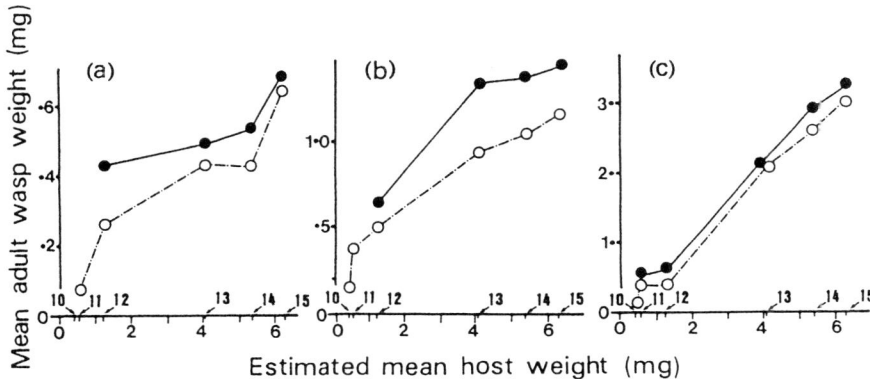

Figure 3. Relationship between estimated mean host weight and emerging female (●) and male (o) wasp weight. The number above abscissa shows the host age corresponding to the mean host weight. (for detail see text)

old host) regardless of host weight. In *D. basalis* (Fig. 3(b)), the female wasps from 12 day old hosts weighed almost half that of females from 13 day old hosts. However, further increases of host weight did not result in substantial increases in emerged female wasp weight. In *H. prosopidis* (Fig. 3(c)), the situation is different. Within the range of the present experiment, the weight of emerged female wasps increased almost linearly with the weight of the host on which the wasp developed.

These three wasp species are all idiophytic (i.e., developing on non-growing host which was paralyzed at oviposition, Haeselbarth, 1979). Thus, the host body weight at oviposition constitutes the whole resource available for wasp development. If females developing on large hosts have a greater increase in reproductive success than the males, it can be argued that females should evolve the capacity to place daughters in large hosts and sons in small hosts. Conversely, if there is a physiological threshold above which wasp weight of both sexes cannot increase on larger host, there should be no more sex ratio change above this host weight. This seems to be the case in these three wasp species.

We must ask what factors determine this equilibrium sex ratio, which is above 0.5 in most cases. This may relate to the relative contribution of each sex or local mate competition. In this regard, it is important to see how the wasp size correlates with his (or her) fitness. Our preliminary experiments showed that the female wasps emerging from the youngest possible host lived only a day or two (compared to 5 or more days from older host), and produced only a few offspring (compared to 30 to 50 in *H. proso-pidis*, and more than 100 in *A. calandrae* and *D. basalis* from older host). Males emerging from the youngest possible host also lived only a day or two (compared 5 days or more from older hosts), and could inseminate only a few females (compared 10 or more from older hosts) during his life-time. However, the estimation of the theoretical equilibrium sex ratio requires further experiments that measure in more detail the relationship of the fitness of each sex with host age.

It should be mentioned here that van den Assem et al. (1984) showed a sex ratio change with host size in *A. calandrae*, but this change was caused mainly by the change in clutch size. A large number of female eggs were deposited on old hosts, and far fewer on young hosts, during the 24 hour egg laying period. However, the number of sons produced per laying period was similar. In our experiment, we did not directly observe the oviposition behavior, thus there is no data comparable to that by van den

338

Assem. However, we usually found fewer eggs on younger hosts when we dissected the beans for transfer experiment. Thus, it is very likely that the change in primary sex ratio with host age observed in the present study was the result of the mechanism shown by van den Assem et al. (1984) at least in *A. calandrae*.

6.2. CUES FOR AND PRECISION OF HOST QUALITY DETERMINATION

The present experiment showed that the sex ratio of a female's eggs was a response to the age or size of host. Host sizes are strongly correlated with host age (see Fig. 1), thus, the observed change in offspring sex ratio with host age may be a response to the size of host. There are many other observations which suggest the relationship between the sex ratio and host size (see King, 1987).

In the present experiment, host age was used to indicate host size, however, as Figure 1 shows, the large size variation exists among individuals of the same age. If the female wasps control the sex of her offspring depending on the size of host, it can be expected that the female wasps may be able to distinguish the different host size independently of host age, and to lay eggs of different sexes based on the individual host size.

To ascertain this point, we conducted another experiment with *H. prosopidis* on 11 day old hosts and with *A. calandrae* on 10 day old hosts. The procedures were exactly the same as in Experiment 3. We transferred 110 *H. prosopidis* and 100 *A. calandrae* eggs to 15 day old host larvae. Additionally, we weighed the host from which each egg was transferred. The emerged wasps were sexed. The hosts from which eggs had been collected were classified into two groups according to the sex of the emerged wasp.

Seventy male and 10 female wasps eventually emerged in *H. prosopidis*, and 58 male and 11 female wasps in *A. calandrae*. The sex ratios (0.125 and 0.159) were comparable to those in Experiment 3. The body weights of hosts oviposited on ranged from 0.228 to 4.680 mg for *H. prosopidis*, and from 0.170 to 1.290 mg for *A. calandrae*. The host weight distributions of two groups are shown in Figure 4(a) and (b). The mean body weights of the groups, indicated in the figure, were not significantly different in either species. Thus, it is concluded that there was no tendency for the female wasps to lay more female eggs on larger hosts of the same age.

It is interesting to compare the above result with that of Jones'(1982) experiment with *H. prosopidis*. When he provided two groups of hosts (e.g., 11 day and 15 day old

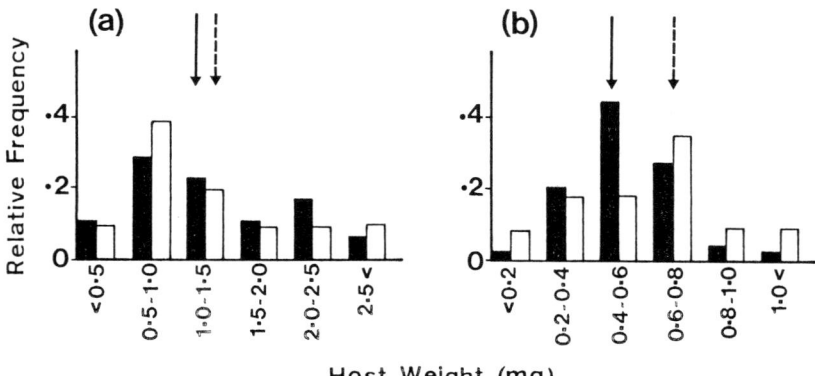

Figure 4. The frequency distribution of host weight on which eggs were oviposited (open bar for female and solid bar for male). (a) by *H. prosopidis* on 11 day old host. (b) by *A. calandrae* on 10 day old host. Vertical arrow (--- for female and for — male) shows the mean of ungrouped data.

hosts) simultaneously, the sex ratio of emerged offspring from 11 day old hosts was more male biased. Here, female wasps might have controlled the offspring sex ratio depending on the host sizes encountered within one arena. Or, the result might have been due mostly to differential mortality during larval stages.

As far as our experiments are concerned, it is certain that female wasps control the primary sex ratio depending on the host age, but do not control the sex ratio in response to the size of the host on which she oviposits. The size variation of the hosts of the same age in the above experiments is still small compared with the situation where larvae of different ages (e.g., 11 and 15 day old hosts together as in Jones' experiment) are provided. Host size variation given in the above experiments might have been too small for female wasps to detect and lay eggs differentially on individual basis.

We can think of two oviposition scenarios. The sex ratio of her batch of eggs (i.e., the ratio of fertilized and unfertilized eggs) is determined by the mean host size in the environment. Wasp adjusts her offspring sex ratio to the mean host size as she encounters, during her successive ovipositions, the hosts of representative size in the environment more often. However, she cannot fine tune her offspring sex according to the size of individual host. Alternatively, the female wasps may control offspring sex ratio by host age, not by host size. An undetermined physiological difference may be used to estimate the mean size of the host and to maximize the offspring fitness.

Either scenario can provide a reasonable explanation of the results shown in Figure 4 and why the female wasps lay more female eggs than emerged sex ratio on small host. Without being able to control the sex of individual eggs directly in response to the host size at oviposition, a female wasp must lay more female eggs than can emerge. Further experiments will distinguish which offspring sex determination mechanism is used by female wasps.

Acknowledgements. We deeply thank Prof. Emeritus Synro Utida, and Drs. Mark Taper and Rodger Mitchell for their valuable comments and suggestions to earlier version. This work was supported in part by grants (62540491 and 01304012) from Ministry of Education, Japan.

REFERENCES

Bell, G. (1982) The Masterpiece of Nature, the Evolution and Genetics of Sexuality, Croom Helm, London.

Charnov, E. L. (1979) The genetical evolution of patterns of sexuality: Darwinian fitness, Am. Nat. 113, 465-480.

Charnov, E. L. (1982) The Theory of Sex Allocation, Princeton University Press, Princeton.

Charnov, E. L., Los-den Hartogh, R. L., Jones, W. T., and van den Assem, J. (1981) Sex ratio evolution in a variable environment, Nature 289, 27-33.

Darwin, C. (1887) The Descent of Man, and Selection in Relation to Sex, 2nd edn., John Murray, London.

Fisher, R. A. (1930) The Genetical Theory of Natural Selection, Oxford University Press, Oxford.

Flanders, S. E. (1939) Environmental control of sex in hymenopterous insects, Ann. Entomol. Soc. Am. 32, 11-26.

Flanders, S. E. (1946) Control of sex and sex-limited polymorphism in the Hymenoptera, Quart. Rev. Biol. 21, 135-143.

Goodman, D. (1982) Optimal life histories, optimal notation, and the value of reproductive value, Amer. Natur. 119, 803-823.

Haeselbarth, E. von (1979) Zur parasitierung der puppen von Forleule (*Panolis flammae* [Schiff.], Kiefernspanner (*Bupalus piniarius* [L.]) und Heidelbeerspanner (*Boarmina bistortata* [Goezel]) in bayerischen Kiefernwaldern, Z. Angew. Entomol. 87, 311-322.

Jones, W. T. (1982) Sex ratio and host size in a parasitoid wasp, Behav. Ecol. Sociobiol. 10,207-210.

Khin Mar Wai and K. Fujii (1989) Intraspecific larval competition among wasps parasitic of bean weevil larvae, Res. Popul. Ecol. (in press).

King, B. H. (1987) Offspring sex ratios in parasitoid wasps, Quart. Rev. Biol. 62, 367-396.

Kishi, Y. (1970) Difference in the sex ratio of the pine bark weevil parasite, *Dolichomitus* sp. (Hymenoptera: Ichneumonidae), emerging from different host species, Appl. Entomol. Zool. 5, 126-132.

Kochetova, N. I. (1977) Factors determining the sex ratio in some entomophagous hymenoptera, Entomol. Rev. (Engl. transl. Entomol. Obor.) 56, 1-5.

Maynard Smith, J. (1978) The Evolution of Sex, Cambridge University Press, Cambridge.

Sandlan, K. (1971) Sex ratio regulation in *Coccygomimus turionella* Linnaeus (Hymenoptera: Ichneumonidae) and its ecological implications, Ecol. Entomol. 4, 365-378.

van den Assem, J., Putters, F. A., and Prins, Th. C. (1984) Host quality effects on sex ratio of the parasitic wasp *Anisopteromalus calandrae* (Chalcidoidea, Pteromalidae), Neth. J. Zool. 34, 33-62.

Wellings, P. W., Morton, R., and Hart, P. T. (1986) Primary sex ratio and differential progeny survivorship in solitary haplo-diploid parasitoids, Ecol. Entomol. 11, 341-348.

Williams, G. C. (1975) Sex and Evolution, Princeton University Press, Princeton.

THE MECHANISMS OF CONTEST AND SCRAMBLE COMPETITION IN BRUCHID SPECIES

Y. TOQUENAGA
Graduate School of Biological Sciences,
University of Tsukuba, Tsukuba, Ibaraki 305 Japan

ABSTRACT. I classified seven strains of three *Callosobruchus* species into contest or scramble strategists on the basis of the larval competition curves and resource intake patterns. The competition curves of *C. maculatus* strains show gradation between contest and scramble strategies. Inside the bean, the competitive mechanism of contest strains is to bite the opposing larvae. The *C. analis* strain used this tactic only at the 3rd and 4th larval stages. In *C. maculatus*, both contest and scramble strains performed this interference at different larval instars. In single generation competition, each strain could overcome an opposing strain when its initial egg density per bean was higher than that of the other. However, the pattern of this density and frequency dependency was more complicated in intraspecific competition of *C. maculatus*. These experiments begin to account for the adaptation of bruchids to stored beans when examined under a mechanism-based concept of competition .

1. Introduction

Bruchid pests of stored beans have played an important role in experimental research on ecological interactions, especially competition (e.g., Utida, 1959; Fujii, 1968; Yoshida, 1966; Mitchell, 1975; Giga and Smith, 1981; Bellows, 1982; Credland et al., 1986). Bruchid competition can be divided into adult and larval phases. Adult bruchids compete for oviposition sites on the most suitable beans as they lay eggs on beans. Larvae, on the other hand, cannot migrate from bean to bean and have to compete for limited resources inside the bean. The larval competition phase is more important when these bruchids compete in a closed environment, such as in stored beans or laboratory cultures.

Nicholson's dichotomy, contest vs. scramble competition (Nicholson, 1954), has been used to interpret results of larval competition (e.g., Bellows, 1982; Fujii, 1968; Smith and Lessells, 1985). The competition curve (the number of survivors after competition plotted against the initial density of competitors) is used to classify competition types in a phenomenological sense. Figure 1 (a) shows a schematic example. In pure scramble competition, no one can survive above the carrying capacity of the environment and the competition curve is like the dotted line. In contest type, the competition curve looks like a saturation curve (the solid line). In many cases, scramble competitors show some temporal or spatial inequality in their access to resources and the competition curve looks like the broken line.

The underlying mechanisms of contest and scramble competition can be characterized by the resource sharing pattern among competitors (Łomnicki, 1988). In contest compe-

K. Fujii et al. (eds.), Bruchids and Legumes: Economics, Ecology and Coevolution, 341–349.
© 1990 *Kluwer Academic Publishers. Printed in the Netherlands.*

342

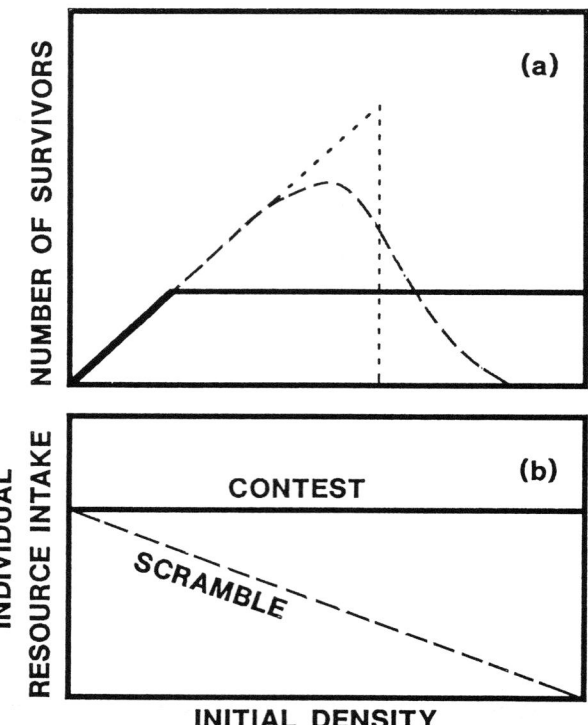

Figure 1. Competition curves (a) and patterns of individual resource intake (b) of contest and scramble strategists. The solid line and the broken line corresponds contest and scramble type respectively. The dotted line represents pure scramble type with no temporal and spatial inequality to access resources among competitors.

tition, the dominant individuals can control resources they need whereas the subordinates can get nothing. Thus, some fitness indicators of the dominants remain constant irrespective of increase of the initial density (the solid line in Fig. 1(b)). In scramble competition, all competitors have to share resources and individual resource intake decreases as the initial density increases (the broken line in Fig. 1(b)).

Unfortunately, only phenomenological criteria have been used to classify bruchid species into contest and scramble strategists (e.g., Fujii, 1968; Bellows and Hassell, 1984). However, there is no simple relationship between competition mechanisms and shapes of competition curves. Several workers have shown that even though the competition type is scramble, some factors such as the insufficient initial density or the aggregated behavior of competitors could result in contest-like, or saturation curves (Kuno, 1983; Ives and May, 1986). My primary aim of this study is to classify bruchid species into contest and scramble type using mechanistic criteria, such as resource intake patterns.

Contest and scramble competition are characteristics at the population level. What mechanisms at the individual level bring these population behaviors? I attacked this question focusing on interference and exploitative competition among competing larvae. I was also interested in the outcome of competition between contest and scramble strategists and tried to interpret these results using the competition mechanisms at the individual level. My competition experiments give us examples comparable to general models of contest vs. scramble competition (Fujii, 1965; Kuno, 1983; Smith and Lessells,

1985; Ives and May, 1986; Łomnicki, 1988). These also help to understand how *Callosobruchus* species from wild beans have evolved adaptations for exploiting stored beans.

2. Materials

There are a few reports of contest competition in a mechanistic sense in bruchids. Only exceptions are those about *Callosobruchus* species (Umeya et al., 1975; Dick and Credland, 1984; Thanthianga and Mitchell, 1988). Thus I chose 7 strains of 3 *Callosobruchus* species as materials for my experiments.

A strain of *Callosobruchus analis*, eA, and the eP strain of *C. phaseoli* were used. *C. analis* is reported as a contest species (Umeya et al., 1975) and *C. phaseoli* as a scramble species (Ike, 1984). These strains have been reared in my laboratory for more than 9 years. I also used 5 strains of *C. maculatus*, aQ, tQ, wQ, yQ, and iQ. Both aQ and tQ have been reared in the laboratory for more than 10 years. All adults of aQ strain were the "active form" (Utida, 1954). wQ was established one year ago (1988) from wild *C. maculatus* in commercial beans. yQ was obtained form Dr. Credland and is referred to as the "Yemen strain" in his papers (Dick and Credland, 1984). Dr. Mitchell provided the iQ strain which he designates the "South India" strain. These last two strains have been reported as contest strains (Dick and Credland, 1986; Thanthianga and Mitchell, 1988). All stock cultures and experiments were maintained at 30°C and 70% R.H. and constant light.

I used mung beans, *Vigna radiata*, in all my experiments. Mung beans are known as the most favorable hosts for many *Callosobruchus* species whose developmental periods are almost equal on mung beans (Gokhale, 1973, Giga and Smith, 1987). Using mung beans eliminates the time advantage in development between competing species. Mung beans were separated into large (larger than No. 5 mesh, 74.2 ± 1.5 mg (Mean ± SE), N=30) and small (smaller than No. 6 mesh, 34.1 ± 1.17 mg, N=30). Adzuki beans, *Vigna angularis*, were used for stock cultures.

3. Larval Competition Curve

I used the elytron length of emerging adults as the indicator of resource sharing patterns among competitors. The elytron length is strongly correlated with the fitness of imagoes in these strains (unpublished data).

Females of a strain were allowed to lay eggs on mung beans in plastic Petri dishes. The number of females of each strain and the beans were adjusted to achieve the assigned number of hatched eggs per bean. After 4 hours, I removed the females and kept the beans for 2 weeks. Beans were segregated by the density of hatched eggs and kept in multiple-well boxes. The maximum number of hatched eggs per bean varied among strains because strains differ in their oviposition responses. Developmental period, sex, and elytron length were measured for each emerging adult. Ten replicates were performed for each initial density for each strain. I performed a linear regression of elytron length against initial densities to classify each strain into contest or scramble type. Negative regression coefficients indicate scramble type competition. If the regression is not significantly different from 0, then contest competition is occuring.

Larval competition curves of 7 strains are shown in Fig. 2. The contest species, eA, had only one emergence from a bean, irrespective of bean size (Fig. 2 (a) and (b)). On the other hand, the scramble species, eP, had maximal emergence at an intermediate initial larval density in both large and small beans. A sudden drop of emergence is clear

344

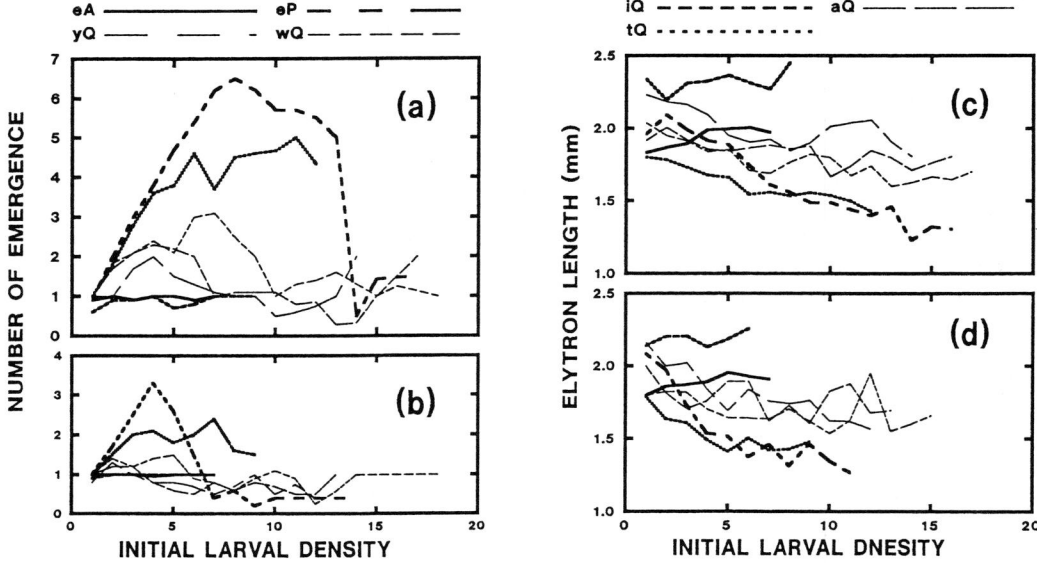

Figure 2. Mean emergence and mean elytron length as a function of initial larval density. (a) Mean emergence in large beans. (b) Mean emergence in small beans. (c) Elytron length in large beans (d) Elytron length in small beans.

in large beans (Fig. 2(a)). The elytron length of eA slightly increased with increase of the initial larval density in both bean sizes (b=.027, P<.000 in small beans: b= .034, P<.000 in large beans: Figs. 2(c), 2(d)). In eP, the elytron length clearly decreased with increase of the initial density (b=-.081, P<.000 in small beans: b=-.059, P<.000 in large beans: Figs. 2(c), 2(d)).

$C.$ $maculatus$ strains show a gradation between contest and scramble strategies. Only one adult of iQ could emerge from a bean as in eA (Fig. 2 (a) and (b)) regardless of bean size and the elytron length did not change with the degree of competition (b=.002, P=.863 in small beans: b=.010, P=.261 in large beans: Figs. 2(c), 2(d)). Like eP, tQ had maximal emergence at an intermediate larval density in both bean sizes (Fig. 2 (a) and (b)). The elytron length of this strain also decreased as the initial larval density increased (b=-.041, P<.000 in small beans: b=-.028, P<.000 in large beans: Figs. 2(c), 2(d)). Thus, we can take iQ as a typical contest strain and tQ as a scramble strain in $C.$ $maculatus.$

Competition in the other 3 strains of $C.$ $maculatus$ seems to be intermediate between that of iQ and tQ in large mung beans. The characteristic of the curves of these three strains is that there is a peak in numbers emerging at some low larval density which falls to some constant level afterward (Fig. 2(a)). Interestingly, this trend disappears in small beans and curves of these 3 strains are similar to that of iQ (Fig. 2(b)). However, the elytron length of these 3 strains declines in both bean sizes (b_{aQ}=-.016, b_{yQ}=-.046, b_{wQ}=-.019 in small beans: b_{aQ}=-.016, b_{yQ}=-.025, b_{wQ}=-.022 in large beans: P<.000 for all these b values: Fig. 2 (c) and (d)) and in this respect these 3 strains behave as scramble types rather than intermediate types between contest and scramble strategies. I used the iQ as the contest strain and the tQ as the scramble strain in $C.$ $maculatus$ in following experiments.

4. Direct Larval Interference

Umeya et al. (1975) showed that *C. analis* larvae bite and kill opponents inside a bean. I wanted to determine when these attacks occurred and whether this mechanism applied in interspecific competition. I also wanted to know whether this mechanism operated in the contest strain of *C. maculatus*. I used eA and eP for interspecific and iQ and tQ for intraspecific contest vs. scramble competition, respectively.

Females of each strain were allowed to lay eggs on mung beans for 4 hours. I dissected larvae from the beans on assigned days. Intra- and inter-strain pairs of larvae without any resources were put into wells drilled on plastic plates. Another set of larva were put into wells individually as control.

Every strain has 4 larval instars and that the duration of each larval stage differs among the strains (Fig. 3) . I assumed that competitors hatched simultaneously and made pairs of larvae of intra- and inter-strain competition on the day marked in Fig. 3. Since I could not distinguish larvae of iQ and tQ, I made only intra-strain combinations for competition for these strains. Although Umeya et al. (1975) reported that they could detect the bite wound, I could not distinguish clearly between a death with or without direct attack. Thus I checked survivorship of each larva every day.

In inter-strain competition, we can detect the existence of interference if the age at mortality of competing larvae is different from that of control larvae. I used Kolmogorov-Smirnov Two Sample Test for this analysis. The analysis of intra-strain competition is more complex. A Monte Carlo simulation was used to compare the actual survival data to the distribution of survival in pairs of individuals randomly drawn from the control data. Interference was considered to occur if both of the following two conditions were fulfilled: (1) competing pairs differ more in longevity than pairs randomly selected from control data; (2) the sum of the longevity of competing pairs is less than that simulated with control data. I ran 10,000 trials to get the following 2 probabilities simultaneously; the probability that the differences of longevity in competing pairs less

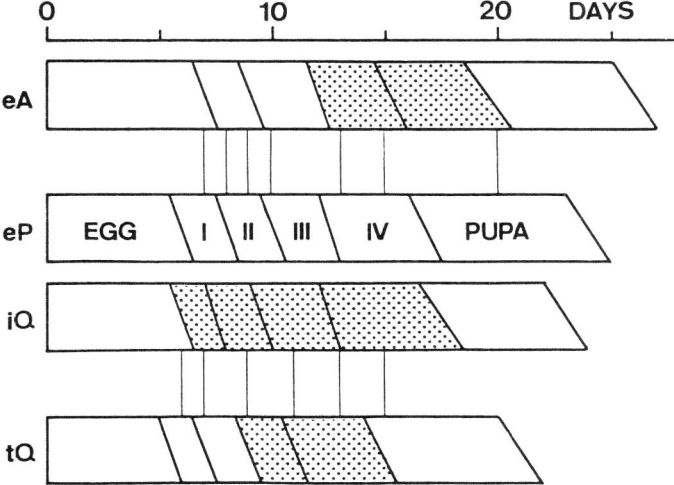

Figure 3. Developmental schedules of 4 strains of *Callosobruchus* species in mung beans. Roman numerals represent the larval stages. Thin lines connecting above and below two bars represent the day when beans were dissected to get larvae for the direct larval interaction experiment. Hatched areas show the period at which eA, iQ, and tQ performed their interference behavior.

than that of simulated pairs and that the sum of actual pairs was larger than that of simulated one. If both these probabilities was small enough (less than .01), we could say that some interference occurred between paired individuals.

The hatched areas in Fig. 3 show at which larval stages each strain performed interference behavior. Interference of eA larvae occurred at only the third and fourth larval stages in both intra- and inter-strain competition. eP showed no interference behavior at any larval stages. In intraspecific competition in *C. maculatus* strains, iQ intra-strain interference behavior was performed throughout the larval stages. tQ, on the contrary, showed intra-strain interference only in the third and the fourth larval stages.

5. Single Generation Competition

Contest and scramble strategies are larval traits. Thus a single generation experiment is adequate to determine the outcome of competition between these two strategies. I used eA and eP for interspecific competition between contest and scramble strategies. I used iQ and tQ for intraspecific competition. Adults of the two *C. maculatus* strains could be easily distinguished by body size (see Figs. 2 (c) and (d)) and the elytoral color pattern. I used two different sized mung beans for this experiment. Females of a contest strains were allowed to lay eggs on mung beans for less than 24 hours to obtain 1 to 5 eggs per bean. Then I allowed scramble strain females to add eggs to beans sorted by the number of contest strain eggs on them for less than 24 hours. I got 5 X 5 densities of hatched eggs of each strain. Each bean was separated and the emergences were recorded daily.

If the contest strain larvae could interfere and dominate in interspecific competition against scramble strains, then only contest strain imagoes would emerge from a bean. However, competition between eA and eP in small beans was density and frequency dependent. At low densities, eA overcame eP and only one eA emerged from a bean. This tendency got stronger at high eA density and low eP density. On the other hand, eP defeated eA when the initial density of eP was much higher than that of eA and more than one eP emerged from a bean in this situation. In some cases, both strains could emerge from a bean at high and low total densities. The responses in large beans resembled those in small beans except that more eP emerged from large beans than from small beans when eP overcame eA.

To indicate the trends in this competition, I used the proportion of a contest strain to total emergence from beans for each combination of initial larval density of each strain. In small beans, the eA proportion increased as the eA initial density increased but decreased as eP initial density increased (Fig. 4 (a)). The minimal proportion dropped far below 0.5 when eP initial density was high and eA initial density was low (at the nearest corner of Fig. 4 (a)). In large beans, this density and frequency dependency was not so strong as in small beans and the minimal proportions of eA were much higher than those in small beans at low eA initial densities and high eP initial densities (Fig. 4(b)).

The general tendency is similar in intraspecific competition between iQ vs. tQ. However, the shape of the plane of the proportion of iQ emergence is much more complicated in small beans (Fig. 4 (c)). This complication is largely due to the fact that the iQ proportion did not decrease monotonically with increase of tQ initial density. Moreover, the iQ proportion reached its maximum at the intermediate initial density of iQ for each tQ initial density. In large beans, the responses are a little simpler than in small beans and the changes of iQ proportions depended more on the initial density of tQ than that of iQ itself (Fig. 4 (d)).

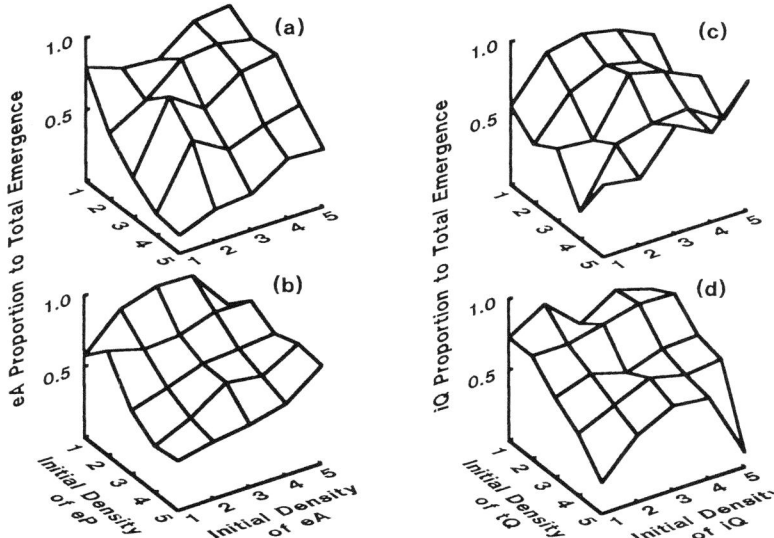

Figure 4. Proportional emergence of a contest strain to the total emergence in the single generation competition experiment. (a) eA vs. eP in small beans. (b) eA vs. eP in large beans. (c) iQ vs. tQ in small beans. (d) iQ vs. tQ in large beans.

6. Discussion

6.1. MECHANISM-BASED INTERPRETATION OF CONTEST AND SCRAMBLE

In the larval competition curve experiment, eA showed a saturated competition curve with a constant elytron length while the response of eP showed a humped curve with decreasing the elytron length. Thus eA and eP showed patterns expected of contest and scramble competitors respectively (Fig. 2). In *C. maculatus* strains, the competition curves do not reveal the mechanisms of competition. It is very hard to distinguish these strains from each other, especially in small beans. The changing pattern of the elytron length indicates that all but iQ strain are scramble type (Fig. 2).

Using small pulses, such as mung beans, might also help to distinguish contest strains from scramble types. In most of previous works, larger pulses, such as cowpeas, *Vigna unguiculata*, or adzuki beans, *Vigna angularis*, were used (Fujii, 1968; Bellows, 1982; Dick and Credland, 1984). It may be difficult to obtain egg densities high enough to show the point at which the competition curve bends downward for scramble species. It is very interesting that yQ strain which Dick and Credland (1984) report to be a contest strain responds as a scramble type in my experiments.

In most previous works, the dichotomy of contest vs. scramble competition has been taken as the same as that of interference and exploitative competition. However, my direct larval competition experiment showed that even scramble strains such as tQ performed interference at some period in larval stages. Especially, the pattern of interference of tQ was almost the same as that of a typical contest stain, eA. Moreover, iQ and eA have different patterns even though they are both classified as contest type. Thus we should consider mechanisms at the individual level (interference or exploitative) as well as the population level (contest or scramble) in discussing the outcome of competition between contest and scramble competitors.

6.2. DENSITY AND FREQUENCY DEPENDENT PHENOMENA

The outcome of the single generation competition depended on the initial density and frequency of each competing strain. Contest strains (eA or iQ) could not overcome scramble strains (eP or tQ) unless their initial density was higher than that of the opponent scramble strain. Scramble strains could overcome contest strains when their initial density was much higher than that of the contest strains. This can be explained by their developmental schedules (Fig. 3) and larval behaviors.

In competition between eA and eP, eP had an advantage due to its short developmental period. If eP used up resources before eA reached the third or fourth larval stage, eP escaped from eA's interference and emerged from the bean. Moreover, eA has tendency to feed at the center and eP feeds at the surface of a bean (Ike, 1984). In small beans, this spatial segregation broke down and the probability of eA meeting eP was high even if the total initial density was low. As a result, eP could emerge only when the initial density of eP was so high that eA could not obtain enough resource to develop to the third instar larva. In large beans, both the spatial segregation and the density and frequency advantage enabled eP to win at lower densities.

The competition between iQ and tQ was more complicated. tQ developed slightly faster than iQ but the difference was not as great as the case of eA and eP. Moreover, both strains exhibited interference behavior. Thus the interference behavior of tQ may have allowed it to act as a contest competitor. If tQ changed its strategy to contest type, then fewer individuals could emerge than when it acted as scramble competitor. This might account for the high proportion of iQ emergences in small beans when tQ initial density was 5 (Fig. 4 (c)). In summary, the contest strain could overcome the scramble strain through interference. The scramble strain could dominate the contest strain through exploitative competition and the outcome was altered by changing the bean size.

6.3. BEAN SIZE AND CONTEST VS. SCRAMBLE COMPETITION

Bean size altered patterns of the density and frequency dependency in competition between eA and eP (Fig. 4). The slope of this dependency is steeper in small beans than large beans. Consequently, there may be a drastic change in competition in small beans which depends on the realized initial larval density of each strain in long-term competition experiments. In multiple generation competition systems started with equal initial adult density of each strain, eA overcame eP in small beans but was defeated by eP in large beans (Toquenaga, unpublished).

Although there are several difficulties, these bean size dependent phenomena also help us to consider the evolution of a bruchid species from wild beans to stored beans. Selective breeding for larger beans has altered the resources so that bruchids must evolve adaptations for exploiting larger beans. Most of *Callosobruchus* species living on stored beans have scramble traits which allow them to fully exploit large beans. On the other hand, many bruchids living on small wild beans exhibit the traits of the contest type (Shimada, personal information). It is not clear yet whether the contest trait observed in *C. maculatus* is established in nature or selected in domesticated beans. However, knowledge of the genetics of this trait and competition experiments with these strains may allow us to understand this adaptation process.

Acknowledgements. I am deeply indebted to Prof. Koichi Fujii, my major professor, for his support and guidance throughout the course of this study. I wish to thank Dr. Credland and Prof. Mitchell for providing me with the contest strains which played such an important roll in my experiments. I thank

Prof. Mitchell and Prof. Slobodkin for their critical reading. Thanks are also due to my Daddy-Long-Legs, Dr. Mark L. Taper, for his highly educational advice and fruitful discussions.

REFERENCES

Bellows, T. S. Jr.(1982) Analytical models for laboratory populations of *Callosobruchus chinensis* and *C. maculatus*, J Anim. Ecol. 51, 263-287.

Bellows, T. S. Jr., and Hassell, M. P. (1984) Models for interspecific competition in laboratory populations of *Callosobruchus* spp., J. Anim. Ecol. 53, 831-848.

Credland, P. F., Dick, K. M., and Wright, A. W. (1986) Relationships between larval density, adult size and egg production in cowpea seed beetle, *Callosobruchus maculatus*, Ecol. Entomol. 11, 41-50.

Dick, K. M. and Credland, P. F. (1984) Egg production and development of three strains of *Callosobruchus maculatus* F. (Coleoptera: Bruchidae), J. stored Prod. Res. 20, 221-227.

Fujii, K. (1965) A statistical model of the competition curve, Res. Popul. Ecol. 7, 118-125.

Fujii, K. (1968) Studies on interspecies competition between the azuki bean weevil and the southern cowpea weevil: III. Some characteristics of strains of two species, Res. Popul. Ecol. 10, 87-98.

Giga, D. P. and Smith, R. H. (1981) Varietal resistance and intraspecific competition in the cowpea weevils *Callosobruchus maculatus* and *C. chinensis* (Coleoptera: Bruchidae), J Appl. Ecol. 18, 755-761.

Giga, D. P. and Smith, R. H. (1987) Egg production and development of *Callosobruchus rhodesianus* Pic. and *C. maculatus* F. (Coleoptera: Bruchidae) on several commodities at two different temperatures, J. stored Prod. Res. 23, 9-16.

Gokhale, V. G. (1973) Developmental compatibility of several pulses in the bruchidae I. Growth and development of *Callosobruchus maculatus* on host seeds, Bull. Grain Technol. 11, 28-31.

Ike, T. (1984) Experimental studies of interspecific competition between bruchid species and system stability in experimental ecosytem, M. A. thesis, Univ. Tsukuba.

Ives, A. R. and May, R. M. (1986) Competition within and between species in a patchy environment relations between microspecific and macrospecific models, J. theor. Biol. 115, 65-92.

Kuno, E. (1983) Factors governing dynamical behaviour of insect populations: A theoretical inquiry, Res. Popul. Ecol. Suppl. 3, 27-45.

Lomnicki, A. (1988) Population Ecology of Individuals, Princeton Univ. Press, Princeton.

Mitchell, R. (1975) The evolution of oviposition Tactics in the bean weevil *Callosobruchus maculatus* (F.), Ecology 56, 696-702.

Nicholson, A. J. (1954) An outline of the dynamics of animal populations, Austral. J. Zool. 2, 9-65.

Smith, R. H. and Lessells, M. (1985) Oviposition, ovicide, and larval competition in granivorous insects, in R. M. Sibly and R. H. Smith (eds.), Behavioural Ecology; Ecological consequences of adaptive behaviour, Blackwell, Oxford, pp 423-448.

Thanthianga, C. and Mitchell, R. (1988) Vibration mediated prudent resource exploitation by competing larvae of the bruchid bean weevil *Callosobruchus maculatus*, Entomol. Exp. Appl. 44, 15-21.

Umeya, K., Kato, T., and Kocha, T. (1975) Studies on the comparative ecology of bean weevils: VI. Intraspecific larval competition in *Callosobruchus analis* (F.), Jpn. J. Appl. Entomol. Zool. 19, 47-53.

Utida, S. (1954) Phase dimorphism observed in the laboratory population of the cowpea weevil, *Callosobruchus quadrimaculatus*, Oyo-Dobuts-Zasshi 18, 161-168.

Utida, S. (1959) The effect of population density on progeny populations observed in the different strains of the Azuki bean weevil. II, Jpn. J. Ecol. 9, 172-178.

Yoshida, T. (1966) Studies on the interspecific competition between bean weevils, Mem. Fac. Lib. Arts Educ., Univ. of Miyazaki 20, 59-98.

ADAPTATIONS OF *CALLOSOBRUCHUS* SPECIES TO COMPETITION

R.H. SMITH
Department of Pure and Applied Zoology
University of Reading
Reading RG6 2AJ, U.K.

ABSTRACT. The discontinuous nature of the stored seed environment poses special problems for bruchids and other granivorous insects. Competition between larvae constrained to develop within a single seed reduces number, size and fecundity of survivors. Evolutionary models have been proposed to suggest theoretical options for larval and adult behaviour. Experimental tests of some predictions of models of oviposition are presented. In general, predictions are supported by data. A genetical model of larval competition is developed incorporating escape of larvae from direct competition. The results are related to seed size and toxin content as major determinants of bruchid life history.

1. Introduction

In a previous paper, Smith and Lessells (1985) considered three problems that arise because larvae are confined within seeds selected for them by their mother: (i) oviposition behaviour, (ii) the possibility of ovicide, (iii) larval competition strategy. In all cases, the optimal evolutionary solution was determined by the effects of larval competition on survival and quality of emergent adults. Such models have value in highlighting the important variables and in pointing towards useful experiments designed to test specific hypotheses. In this paper, I take Smith and Lessells (1985) as a starting point and develop two of the three problems considered there. First, optimal oviposition models are examined in relation to experiments designed to test specific predictions. Second, I develop the game theory model of larval competition as a conventional genetic model which highlights the importance of seed productivity in determining optimal strategy.

2. Competition for Oviposition Sites

If evolution of the oviposition behaviour of female bruchids were not affected by larval competition, females should lay eggs on the first seed encountered; choice of oviposition site might be determined by suitability of the substrate, but not by the presence of eggs or other females. Evidence that competition has moulded adult female behaviour in *Callosobruchus* species is overwhelming; the evidence includes uniform egg distributions (e.g. Mitchell 1975), chemical oviposition deterrent markers (e.g. Giga and Smith 1985) and egg guarding behaviour (e.g. Smith 1987). Quantitative effects of competition on quantity and quality of larvae emerging from seeds were reviewed by Smith and Lessells (1985) and other examples are now in the literature (e.g. Credland et al., 1986). Models of bruchid oviposition behaviour were developed by Mitchell (1975)

K. Fujii et al. (eds.), Bruchids and Legumes: Economics, Ecology and Coevolution, 351–360.
© 1990 *Kluwer Academic Publishers. Printed in the Netherlands.*

and Smith and lessells (1985). Here, I will describe experimental evidence from two unpublished studies designed to test some predictions of optimal oviposition models.

2.1. OPTIMAL CLUTCH SIZE MODELS

Optimal foraging theory (Charnov, 1976) has been used as a basis for predicting optimal clutch size (number of eggs/patch of resource) in insects (Charnov and Skinner, 1984; Parker and Courtney, 1984; Skinner, 1985; Smith and Lessells, 1985). Here, I will use clutch size to mean the number of eggs laid in a patch before another patch is utilised. The basic prediction is that, if a resource is distributed in discrete patches (e.g. individual seeds, or seeds in a pod, or pods on a plant), the extent to which an individual patch is utilised should increase with 'search time' (time and effort required to find and prepare for use another patch). In most bruchids (and other semelparous species with non-feeding adults) the limiting factor which will constrain fitness is probably energetic reserves rather than time, and Smith and Lessells (1985) modified the optimal clutch size model replacing search time by search costs, measured in units of 'egg equivalents' (because energy spent locating a new patch of resource is not available for egg production). Quantitative predictions require detailed information about the relationship between fitness and eggs/seed. However, two qualitative predictions can be made and tested experimentally: 1. for an individual female, as search time (costs) increase, so should time spent ovipositing (hence eggs laid) per patch; 2. for a group of females, as the number of females increases, time spent ovipositing per female in a patch should decrease. These predictions have been tested in *C. maculatus* by reference both to number of eggs laid per patch of resource and the way that females use their time during oviposition.

Search costs for insects that lay eggs on the seed surface in seed stores and laboratory cultures are generally zero or very small (that is, a female can move from one seed to another with little expenditure of time or energy). The prediction from oviposition models based on optimal foraging theory is therefore that a female should lay only one egg before moving on to another seed (cf. Mitchell, 1975). However, it could be argued that models more general than a simple 'one egg per seed' decision rule are superfluous for bruchids and add nothing other than unnecessary mathematics (e.g. Janarden, 1980; Smith and Lessells, 1985). One justification for more general models is that they help put bruchids in context with insects where search costs are significant, because seeds must be tunneled out before oviposition (e.g. *Prostephanus truncatus* Horn). Another justification is that storage bruchids such as *C. maculatus* also infest legumes in the field where travel between seed pods on a plant or between plants does involve significant costs, such that a 'one egg per seed' rule may be sub-optimal. The experiments of Perry (1988) were designed to examine in a simple laboratory system whether *C. maculatus* females retain the ability to respond to search time/costs. Jagoe (1989) followed Perry (1988) and Smith (1987) in recording time budgets of individual female *C. maculatus* to examine how ovipositing females responded to the presence of other females. Both studies used a Brazilian (Campinas) strain of *C. maculatus*.

2.2 VARYING SEARCH COSTS

Perry (1988) examined responses of recently emerged *C. maculatus* females to experimental manipulations of search time/costs. In experiment 1, individual adzuki beans *Phaseolus angularis* were arranged on grids, each with 2, 4, 6, 10, 12 or 16 cm between beans. In experiment 2, beans were placed individually in the compartments of plastic trays. Compartments had a basal area of 2 cm^2 with walls 1 cm high such that beetles had to climb into and out of compartments to search for beans. Basal distance between beans was set at 2, 4, 6, 10 or 12 cm, but travel time was much greater than corresponding distances in experiment 1 because of the added vertical dimension. Individual females (14 in experiment 1, 13 in experiment 2) were placed in the centre of the apparatus and observed continuously for 120 minutes at 25°C. Perry noted when a female discovered a bean and took up the egg-laying position (motionless with abdomen pressed against the bean), and when she moved off after laying the final egg on that bean; the interval between the two times is defined as 'oviposition time', and the time from moving off to taking up egg-laying position on another bean as 'search time'.

When beans were close together, females tended not to return to a bean once they had left it. When beans were more spaced, a female would lay an egg, leave the bean and then quickly return, either to lay another egg or to leave again after examining the bean. If beans were very widely spaced, females would sometimes lay 2 or 3 eggs before eventually finding another bean. Thus the '1 egg per bean' decision rule is not adequate when search costs are significant.

We predicted that search time should increase as spacing between beans increased, and that oviposition time (and number of eggs laid) should correspondingly increase with search time. The predictions were tested by linear regression of oviposition time and number of eggs laid on search time. In both experiments, search time increased with increasing distance between beans; time spent ovipositing correspondingly increased (see Table 1a), as did mean number of eggs laid per bean ($P<0.001$ in both experiments). Thus optimal foraging predictions are supported by data. Results are clearest in experiment 2 where search time/costs are greatest.

TABLE 1. Studies on the behaviour of individual females reveal how individual time-budgets change in response to (a) varying search costs (b) the presence of other females

1a. Null hypothesis: no linear relationship between Oviposition Time and Search Time (data from Perry, 1988)

Experiment	Sample size	Test of null hypothesis
Experiment 1	94	$P < 0.05$
Experiment 2	86	$P < 0.001$
All data	180	$P < 0.001$

1b. Null hypothesis: a female in a group spends as long as a single female examining a bean before oviposition (pre-oviposition time) (data are mean pre-oviposition times in minutes, from Jagoe, 1989).

Focal female	Wild Type	Black	Mean
Single	3.2	3.8	3.5
Group	2.6	3.0	2.8

Compare Single vs. Group means: $t_{32} = 2.78$; $P < 0.01$

2.3 SINGLE *vs.* GROUP OVIPOSITION

Jagoe (1989) manipulated the number of recently emerged *C. maculatus* females ovipositing on 10 adzuki beans in a 10 cm diameter petri dish. A colour mutant was used as a natural marker in order to distinguish between a focal female and other members of a group. A group of females consisted of 4 wild type + 1 black (6 replicates) or 1 wild type + 4 black (8 replicates). Ten individual females of both colour types were also observed in the same apparatus. Position and activity of the focal female were recorded once every minute for 90 minutes at 25°C.

We predicted that a female in a group would spend less time on a bean ovipositing than a single female. Time spent on a bean was divided into pre-oviposition (examination of bean surface) and post-oviposition times and comparisons made between wild type and black as well as between single and grouped females. There were no significant differences between wild type and black females or between mean number of eggs laid by single females or females in a group. Pre-oviposition time on a bean was significantly less for a female in a group (mean = 2.8 min.) than a single female (mean = 3.5 min.; t_{32} = 3.24, P<0.01), though post-oviposition time on a bean was not (t_{32} = 0.45; n.s). Thus time spent ovipositing was less for a female in a group because she spent less time examining a bean before choosing to lay an egg. A consequence of this lack of discrimination is that the distribution of eggs laid by a female tends to be uniform (variance < mean) for a female on her own, but is indistinguishable from a Poisson distribution (variance = mean) when she is in a group of five females (variance/mean ratio = 0.51 for single female, variance/mean ratio = 1.04 for a female in a group; t_{32} = 4.43, P<0.001).

3. Competition between Larvae for Seed Resources

In Smith and Lessells (1985), we considered how parameters defining larval competition strategies could determine which strategy would be evolutionarily stable to invasion by an alternative strategy, using a simple game theory model (Maynard Smith, 1982). For simplicity, two contrasting competition strategies are defined. One is **Attack**, where only one insect emerges from a seed because of active and aggressive competition (e.g. *Callosobruchus analis*; Umeya et al., 1975). The alternative is **Avoid**, where competition is passive exploitation for limited resources and several insects may emerge from a seed (e.g. *Callosobruchus maculatus*).

Two parameters describe the strategies; for **Avoid**, the cost of passive exploitation is reduction of larval survival by a proportion E if another **Avoid** larva is in the same seed; for **Attack**, an **Attack** larva meeting an **Avoid** larva has a probability W (0.5<W<1) of winning that contest, while an **Attack** larva meeting an **Attack** larva has a probability 0.5 of winning (Table 2). With two larvae in a seed, **Attack** is always an evolutionarily stable strategy (ESS) provided W>0.5 but **Avoid** is also an ESS if E<(1-W). Analysis of the more general case of N larvae in a seed shows that **Attack** is always an ESS provided W>0.5 (Smith and Lessells, 1985).

Smith and Lessells (1985) noted that the evolutionary outcome of the game was different if **Avoid** sometimes escaped contact with an **Attack** larva, and that a structured population or IGS model (Harvey, 1985) might explain why **Avoid** is the more common larval strategy even though **Attack** is always an ESS if W>0.5, since the ESS with the higher fitness will give rise to more potential colonisers during migration and colonization. The local population (or trait group) that we had in mind was a temporary seed store colonized by a small number of individuals. Sensible further development of the

TABLE 2. Two larvae in a seed. The table shows the frequency of interaction between the different larval genotypes, the mean numbers of genotypes emerging/seed in the adults, and the mean seed productivity. It is assumed that there is only one survivor when **Avoid** meets **Attack**.

2a. **Avoid** allele D is recessive

Larval genotypes	Frequency	Emerging adults			Productivity
		DD	HD	HH	
DD,DD	p^4	2(1-E)	0	0	2(1-E)
DD,HD	$4p^3(1-p)$	(1-W)	W	0	1
DD,HH	$2p^2(1-p)^2$	(1-W)	0	W	1
HD,HD	$4p^2(1-p)^2$	0	1	0	1
HD,HH	$4p(1-p)^3$	0	½	½	1
HH,HH	$(1-p)^4$	0	0	1	1

2b. **Avoid** allele D is dominant

Larval genotypes	Frequency	Emerging adults			Productivity
		DD	HD	HH	
DD,DD	p^4	2(1-E)	0	0	2(1-E)
DD,HD	$4p^3(1-p)$	(1-E)	(1-E)	0	2(1-E)
DD,HH	$2p^2(1-p)^2$	(1-W)	0	W	1
HD,HD	$4p^2(1-p)^2$	0	2(1-E)	0	2(1-E)
HD,HH	$4p(1-p)^3$	0	(1-W)	W	1
HH,HH	$(1-p)^4$	0	0	1	1

IGS concept is not possible without some idea of how trait groups contribute to a global population (deme). However, the IGS concept is implicit in the models below where the larvae *within an individual seed* constitute a trait group, and seed productivity is a key feature of the genetic models.

3.1. A GENETIC MODEL OF THE TWO LARVAE GAME

Table 2 shows the scheme for two larvae in a seed. The important feature is that seeds containing two **Avoid** larvae produce on average 2(1-E) adult beetles, while a seed containing one or two **Attack** larvae only ever contributes one beetle to the deme of adults; this differential group (seed) productivity enhances maintenance of **Avoid** in the deme, even though **Avoid** is at a disadvantage to **Attack** within a seed.

Assume that two alleles determine larval behaviour, the **Attack** allele (H) and the **Avoid** allele (D). Let the parent population frequencies be p of D and (1-p) of H. Also assume random mating and that selection occurs only in larvae, such that allele frequencies in eggs are as in parents. If **Avoid** is recessive, parental allele frequency p is the square root of genotypic frequency in larvae p^2. With three genotypes, six types of genotype interaction occur; if oviposition is random with respect to genotype, frequencies are as in Table 2. Number of survivors of each genotype is listed for each interaction, and the final column (seed productivity) shows that seeds containing **Avoid** larvae contribute more to the adult population, on average, provided E<0.5.

Table 3 shows frequencies of the three genotypes in emergent adults. Average numbers of the DD genotype emerging per seed are obtained by multiplying numbers in the DD survivors column by frequency of interaction between two larval genotypes and then summing across interactions to give N_{DD}. The same procedure is followed for HD and HH genotypes and numbers are summed to give mean numbers of all genotypes emerging per seed N_{TOT}. Allele frequencies after larval competition are given in Table 4 for both cases, D recessive and D dominant to H.

These two genetical models are equivalent to the two larvae game theory model analysed by Smith and Lessells (1985). The results of the more rigorous genetical approach are essentially the same as those of game theory; the only stable equilibria in the simple two larvae game are fixation of one or other allele, depending on the relative sizes of E and W and (in the unstable polymorphism case when E<(1-W)<½) initial allele frequencies. The feature that Table 2 lays bare is that, provided E < ½, the higher group (seed) productivity when both larvae are **Avoid** enhances the maintenance of **Avoid** in the deme, even though **Avoid** is at a disadvantage (W > ½), in the presence of **Attack** within a seed.

3.2. AVOIDING CONTACT IN A LARGE SEED

Smith and Lessells (1985) noted that, if there is a non-zero probability q of **Avoid** escaping contact with **Attack**, **Attack** may no longer be evolutionarily stable. Here I will modify the genetic models to incorporate q and examine the effect on the dynamics of D and H. I will assume that, when both **Avoid** and **Attack** are in a seed, with probabili-

TABLE 3. Genotype frequencies in the survivors of the competition between two larvae in a seed assuming only one survivor when **Avoid** meets **Attack**. The frequency of the **Avoid** allele D in the parents of the larvae is p.

3a. mean numbers emerging/seed when **Avoid** D is recessive

Genotype	Numbers emerging/seed
DD	$2(E-W)p^4 - 4(E-W)p^3 + 2(1-W)p^2$
HD	$4(W-E)p^4 - 12(W-E)p^3 + 4(3W-2E-1)p^2 + 4(1-W)p$
HH	$(1-2W)p^4 - 4(1-2W)p^3 + 2(3-5W)p^2 + 4(1-W)p + 1$
Total	$(1-2E)p^4 + 1$

3b. mean numbers emerging/seed when **Avoid** D is dominant

Genotype	Numbers emerging/seed
DD	$2(E-W)p^4 - 4(E-W)p^3 + 2(1-W)p^2$
HD	$4(W-E)p^4 - 12(W-E)p^3 + 4(3W-2E-1)p^2 + 4(1-W)p$
HH	$(1-2W)p^4 - 4(1-2W)p^3 + 2(3-5W)p^2 - 4(1-W)p + 1$
Total	$(1-2E)(p^4 - p^3 + 4p^2) + 1$

TABLE 4. Allele frequencies following larval competition in the two larvae game. The frequency of the **Avoid** allele D is p in the parents and p' in the survivors of larval competition.

4a. Only one survivor when **Avoid** meets **Attack** (q = 0)

D	Frequency p' of D in emerging adult
Recessive	$((1\text{-}2E)p^4 - (1\text{-}2W)p^3 + (1\text{-}2W)p^2 + p)/((1\text{-}2E)p^4 + 1)$
Dominant	$2((E\text{-}W)(p^3\text{-}2p^2) + (1\text{-}W)p)/((1\text{-}2E)(p^4\text{-}4p^3 + 4p^2) + 1)$

4b. Probability q that **Avoid** escapes contact with **Attack**

D	Frequency p' of D in emerging adults
Recessive	$\dfrac{((1\text{-}2E\text{-}2q(2\text{-}W))p^4 - (1\text{-}2W\text{-}2Q(1\text{-}W))p^3 + (1\text{-}2W(1\text{-}q)p^2 + p)}{((1\text{-}2E\text{-}2q)p^4 + 2qp^2 + 1)}$
Dominant	$\dfrac{2((E\text{-}W(1\text{-}q))(p^3\text{-}2p^2) + (1\text{-}W(1\text{-}q))p)}{((1\text{-}2E\text{-}2q)(p^4\text{-}4p^3) + 2(2\text{-}4E\text{-}5q)p^2 + 4qp + 1)}$

ty q the **Avoid** larva suffers neither risk of being killed nor cost of exploitation competition and the **Attack** larva is also guaranteed successful emergence. Inclusion of q alters numbers of emerging adults in those rows of Table 2a,b where **Avoid** meets **Attack** and increases productivity from 1 to 1 + q. Table 4b gives allele frequencies in emerging adults.

Unfortunately, these expressions do not lend themselves to simple interpretation. I have used computer simulation to demonstrate the changed dynamics and the patterns turn out to be complicated. In general, q > 0 helps maintain variation; a stable polymorphic equilibrium may appear, either in addition to or replacing an unstable equilibrium.

Assume that q and E will be inversely related across a range of seed sizes and types; that is, a large seed in which costs of exploitation competition are small is also likely to allow escape from contact with an aggressive **Attack** larva chewing through the centre of the seed. Let W = 0.6 and represent increasing seed size by increasing q from 0, through 0.1, 0.3 to 0.5 and correspondingly decreasing E according to E + q = 0.6. Whether **Avoid** is recessive or dominant, the outcome is fixation of **Attack** H for q = 0, E = 0.6 > 1-W (parameters which represent a small seed). As q increases and E decreases, the D allele is retained in the population and there is stable polymorphism when E = 0.3, q = 0.3 in both recessive and dominant cases. For q = 0.5, E = 0.1 (corresponding to a large seed), there is stable polymorphism with p close to 0.58 when **Avoid** is recessive and fixation of **Avoid** when p is dominant.

The effect of increasing q when W and E are fixed is shown in Fig. 1 where change in frequency of **Avoid** allele D is plotted against phenotypic frequency of **Avoid** (phenotype frequency of 0.75 corresponds to p=0.5 when **Avoid** is dominant). The stabilising effect of increasing q is a consequence of increased seed productivity. Smith and Lessells (1985) noted, by considering the effect on E alone, that selection for increasing

358

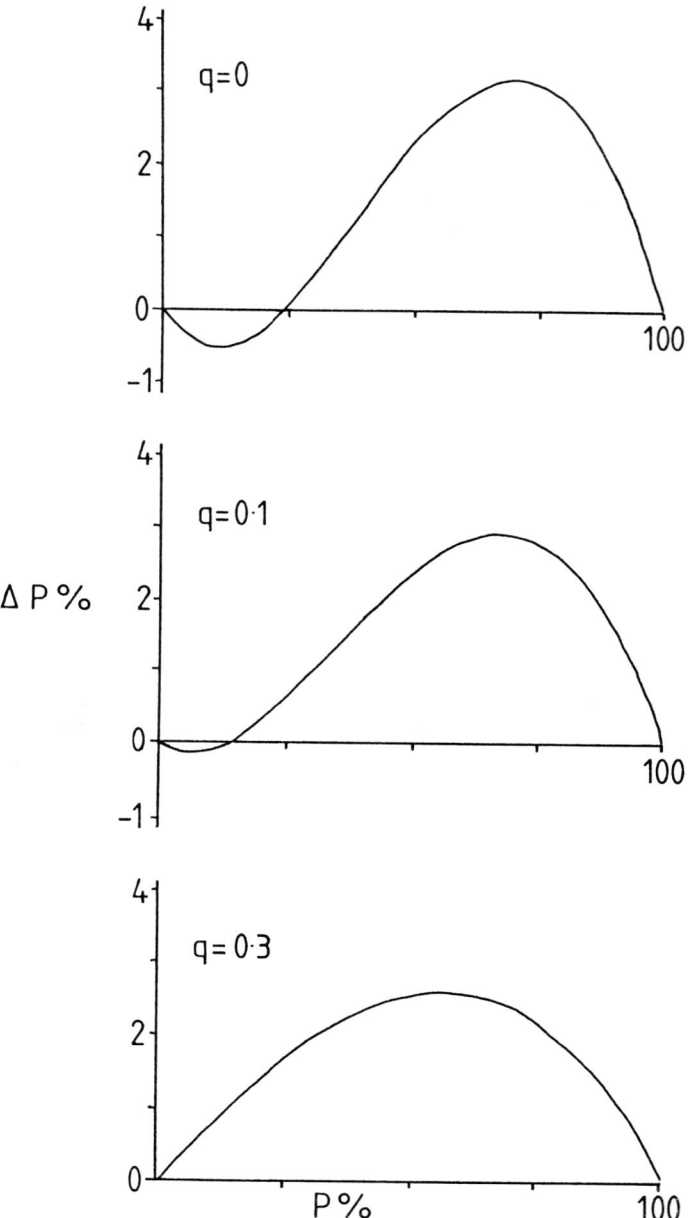

Figure 1. Change in frequency, P, of **Avoid** allele.

seed size may have favoured **Avoid**. The treatment here, considering q increasing as E decreases in larger seeds, adds further weight to our argument that **Attack** is more likely to be the evolved strategy in small seeds and **Avoid** in large seeds.

4. Discussion

The development of evolutionary optimisation theory and its application to insect behaviour (e.g. Charnov and Skinner, 1984; Parker and Courtney, 1984) have lead to fundamental changes in how we view insect life history. Functional explanations based on individual insect or individual gene as the unit of selection open up the possibility of understanding the evolution of interacting life-history traits by life-cycle modelling (e.g. Smith et al., 1987), quantitative genetical analysis and physiological manipulation (e.g. Moller et al., 1989a,b. Here, I have shown how some models of oviposition behaviour in *Callosobruchus* have been tested experimentally, and developed population genetics models of larval competition behaviour. These approaches are all reductionist, for example describing larval competition as one of two extremes deliberately ignores intermediate behaviours (e.g. a Yemen strain of *C. maculatus* which seems almost but not completely to show **Attack** behaviour; Credland et al., 1986). However, the very clear **Attack** behaviour demonstrated by *C. analis* (Umeya et al., 1975) is consistent with its association with the small-seeded mung bean *Phaseolus aureus* in S.E. Asia (Haines, 1989), and it may be no coincidence that the above-mentioned Yemen strain of *C. maculatus* was isolated on the small-seeded lentil *Lens culinaris*. Thus we are beginning to understand observed patterns of variation by reference to simple evolutionary models.

Seed size is clearly one important feature of the habitat of bruchids since intensity of larval competition determines both optimal larval strategy and optimal oviposition strategy of adult females (Smith and Lessells, 1985). However, it is important not to forget that we are talking about *legume* seeds. Legumes are characterised by the presence of different amounts and types of toxins in seeds, and few insects other than bruchids are able to feed on legume seeds. The ability successfully to detoxify different legumes may often be the main determinant of inter-and intra-specific differences in, for example, developmental rate (Giga and Smith, 1986) or body size (Credland et al., 1986), and these may be the primary traits which drive the evolution of other interacting life-cycle components. To take one example, a Yemen strain of *C. maculatus* was isolated from lentils on which it develops more successfully than other strains of the same species (Credland, this symposium); the Yemen strain has a low conversion efficiency (insect biomass produced in relation to seed biomass consumed), perhaps because much energy goes into detoxification, and development rate is also lower than in other strains but it emerges larger (Dick and Credland, 1984; Credland et al., 1986; Credland and Dick, 1987). Because they are both larger and consume relatively more seed, Yemen larvae are more likely than other strains to encounter larvae in a seed, even when reared on large-seeded cowpeas. Larval competition (and also adult oviposition behaviour) in the Yemen strain therefore resemble the extreme 'contest' type of *C. analis* more than the extreme 'scramble' type of a Brazilian (Campinas) strain of *C. maculatus* (Credland et al., 1986), but evolution of these behavioural traits in the Yemen strain may be in part a consequence of adaptation to physiological constraints imposed by toxins in the lentil host.

Acknowledgements. I am grateful to Rodger Mitchell and Richard Sibly for their comments on an earlier version of the manuscript, and to Chris Jagoe and Caroline Perry whose laboratory projects I have referred to.

References

Charnov, E. (1976) Optimal foraging: the marginal value theorem, Theor. Popul. Biol. 9, 129-136.
Charnov, E. and Skinner, S. W. (1984) Evolution of host selection and clutch-size in parasitoid wasps. Florida Entomol. 67, 5-21.

Credland, P. F. and Dick, K. M. (1987) Food consumption by larvae of three strains of *Callosobruchus maculatus* F. (Coleoptera: Bruchidae), J. stored Prod. Res. 23, 31-40.

Credland, P. F., Dick, K. M. and Wright, A. W. (1986) Relationships between larval density, adult size and egg production in cowpea seed beetle, *Callosobruchus maculatus*, Ecol. Entomol. 11, 41-50.

Dick, K. M. and Credland, P. F. (1984) Egg production and development of three strains of *Callosobruchus maculatus* F. (Coleoptera: Bruchidae), J. stored Prod. Res. 20, 221-227.

Giga, D. P. and Smith, R. H. (1985) Oviposition markers in *Callosobruchus maculatus* F. and *Callosobruchus rhodesianus* Pic. (Coleoptera: Bruchidae): asymmetry of interspecific responses, Agr. Ecosys. Environ. 12, 229-233.

Giga, D. P. and Smith, R. H. (1987) Egg production and development of *Callosobruchus rhodesianus* (Pic.) and *Callosobruchus maculatus* (F.) (Coleoptera: Bruchidae) on several commodities at two different temperatures, J. stored Prod. Res. 23, 9-15.

Haines, C. P. (1989) Observations on *Callosobruchus analis* (F.) in Indonesia, including a key to storage *Callosobruchus* spp. (Col., Bruchidae), J. stored Prod. Res. 25, 9-16.

Harvey, P. H. (1985) Intrademic group selection and the sex ratio, in R. M. Sibly and R. H. Smith (eds.) Behavioural Ecology: ecological consequences of adaptive behaviour, Blackwell, Oxford, pp. 59-73.

Jagoe, C. E. D. (1989) The effect of overcrowding on the oviposition behaviour of *Callosobruchus maculatus*. Unpublished BSc project, University of Reading.

Janarden, K. G. (1980) A stochastic model for the study of oviposition evolution of the pest *Callosobruchus maculatus* on mung beans, *Phaseolus aureus*, Math. Biosci. 50, 231-238.

Maynard Smith, J. (1982) Evolution and the Theory of Games, Cambridge University Press, Cambridge.

Mitchell, R. (1975) The evolution of oviposition tactics in the bean weevil, *Callosobruchus maculatus* (F.), Ecology 56, 696-702.

Moller, H., Smith, R. H. and Sibly, R. M. (1989a) Evolutionary demography of a bruchid beetle. I. Quantitative genetic analysis of the female life history, Functional Ecol. 3 (in press).

Moller, H., Smith, R. H. and Sibly, R. M. (1989b) Evolutionary demography of a bruchid beetle. II. Physiological manipulations. Functional Ecol. 3 (in press).

Parker, G. A. and Courtney, S. P. (1984) Models of clutch size in insect oviposition, Theor. Popul. Biol. 26, 27-48.

Perry, C. (1988) Optimal oviposition behaviour of *Callosobruchus maculatus*, Unpublished BSc project, University of Reading.

Skinner, S. W. (1985) Clutch size as an optimal foraging problem for insects, Behav. Ecol. Sociobiol. 17, 231-238.

Smith, R. H. (1987) Oviposition, competition and population dynamics in storage insects, in E. Donahaye and S. Navarro (eds.) Proc. fourth Intern. Working Conf. Stored Prod. Protection, September 21-26, 1986, pp. 426-433.

Smith, R. H. and Lessells, C. M. (1985) Oviposition, ovicide and larval competition in granivorous insects, in R. M. Sibly and R. H. Smith (eds.) Behavioural Ecology: ecological consequences of adaptive behaviour, Blackwell, Oxford, pp. 423-447.

Smith, R. H., Sibly, R. M. and Moller, H. (1987) Control of size and fecundity in *Pieris rapae*: towards a theory of butterfly life cycles, J. Anim. Ecol. 56, 341-350.

Umeya, K., Kato, T. and Kocha, T. (1975) Studies on the comparative ecology of bean weevils. VI. Intraspecific larva competition in *Callosobruchus analis* (F.), Jpn. J. appl. Entom. Zool. 19, 47-53.

COMPARISON OF THE DENSITY-DEPENDENT POPULATION PROCESS BETWEEN LABORATORY AND WILD STRAINS OF *CALLOSOBRUCHUS CHINENSIS* (L.)

M. SHIMADA
Department of Biology, College of Arts and Sciences,
University of Tokyo, Komaba, Meguro-ku, Tokyo 153,
Japan

ABSTRACT. The reproduction curves were determined for 6 strains of the azuki bean weevil, *Callosobruchus chinensis*. Various numbers of adult pairs from stocks were introduced into petri dishes with a constant amount of the azuki beans (*Vigna angularis*). Numbers of eggs deposited, eggs hatched and adult progeny emerged were counted.

Wild strains and experimentally selected strains showed loosely density-dependent control of oviposition at high parental densities. Consequently, hatchability of eggs declined more drastically at high densities and curves of hatched eggs to parental density became highly scramble in those strains. Larval competition process was similar in all the strains. Density-dependent population changes outside and inside the bean were simulated by a modified logistic-difference equation. The simulations followed the observed data closely. Cluster and principal component analyses based on 12 parameters in simulations were used to determine the similarities of strains. Changes of population parameters by density-dependent natural selection were discussed.

1. Introduction

Life history theories have explained within-species differences in ecological responses as indicators of adaptations to local habitats with different environmental stability and biotic interactions (MacArthur and Wilson, 1967; Gadgil & Solbrig, 1972; McNaughton, 1975; Stearns, 1976). Some laboratory studies have tested for changes of life history traits of experimental organisms under selection regimes in which resource renewal and population dynamical patterns were controlled (e.g., Luckinbill, 1978; Taylor and Condra, 1980). However, few studies on within-species variations and changes in density-dependent responses of a population have been made from the viewpoint of population ecology.

Recently, Mueller and Ayala (1981) and Bierbaum et al. (1989) found that *Drosophila* populations which experienced selection at different levels of population density showed opposite responses along a range of densities. Populations responding to different patterns of resource renewal and population dynamics should show different responses to a range of densities as a consequence of adaptation. Such variations in responses of a population to the density may be best analyzed in the competition curve within a life stage and the reproduction curve over an entire generation, which have been studied since classic population ecologists (Utida, 1941a; Nicholson, 1948).

In the present paper, I simulated competition curves in parental oviposition and larval development up to the adult in the azuki bean weevil, *Callosobruchus chinensis*

K. Fujii et al. (eds.), Bruchids and Legumes: Economics, Ecology and Coevolution, 361–371.
© 1990 *Kluwer Academic Publishers. Printed in the Netherlands.*

(L). Competition and reproduction curves simulated by the model were compared among strains; wild, experimentally selected and laboratory ones. The comparison reveals changes of population parameters by density-dependent natural selection.

2. Experiments

Six strains of the azuki bean weevil, *Callosobruchus chinensis* (L.) (Bruchidae) were used: one laboratory strain maintained for over 50 yr (jC), three wild strains (taC$_2$, mrC$_{88}$ and skC$_{88}$), two experimentally selected strains which originated from taC$_2$ and were maintained at the carrying capacity level for either 1 yr or 2 (Fujii and Shimada, 1980), termed as taC$_2$ K-1yr and taC$_2$ K-2yr, respectively. Their origin and profiles were shown in Table 1.

   Reproduction curves were determined for mrC$_{88}$ and skC$_{88}$ in a growth chamber at 30° C, 70% R.H. and 16L/8D photocycle. From 1 to 256 pairs in geometric series multiplied by 2 of *C. chinensis* were taken from stock cultures and introduced into a Petri dish (90mm in diameter and 15mm in depth) containing 5g (28 beans on average) of the azuki beans (*Vigna angularis* var. *dainagon*). Two or three replicates were established in each density except for 256 pairs (1 replicate). The numbers of eggs laid, eggs hatched and progeny emerged were counted. The same experiment was carried out by Fujii and Shimada (1980) for jC, taC$_2$, taC$_2$ K-1yr and taC$_2$ K-2yr. I used data of all six strains for simulations.

3. Models for Competition and Reproduction Curves

Adult *C. chinensis* lay eggs on the surface of beans. Both fecundity and hatchability of eggs decline with increasing adult density (Utida, 1941a). On the other hand, larval

Table 1. Origins and profiles of six strains of *Callosobruchus chinensis* in the present study.

Strain	Original Place	(year)	Profiles
jC	Kyoto	(1937)	A laboratory strain maintained at a middle density since Utida's studies.
mrC$_{88}$	Maruoka, Fukui-ken	(1988)	A wild strain originated from ca. 20 pairs taken in a field of azuki beans.
skC$_{88}$	Saiki, Tsukuba, Ibaraki-ken	(1988)	A wild strain originated from ca. 5 pairs taken in a field of azuki beans.
taC$_2$	Namiki, Tsukuba, Ibaraki-ken	(1978)	A wild strain originated from a few dozens of pairs emerged from azuki beans harvested in Tsukuba.
taC$_2$ K-1yr	-------		An experimental strain which originated from one pair of taC$_2$ and were maintained at the carrying capacity level for 1 year.
taC$_2$ K-2yr	-------		Ditto, for 2 years.

survival depends on the number of first instar larvae that bore into a bean (Utida, 1942). Therefore, the present study analyzed three population processes in *C. chinensis*; adults laying eggs, or producing hatched eggs outside the beans, and the first instar larvae surviving to newly emerged adults inside the beans.

The three processes were simulated according to Shimada (1989), which was based on a modified equation from a logistic-difference equation by Hassell (1975). Egg density (E) on the surface of beans laid by adults was formulated as,

$$E = f_0(N/2)[1 + (f_0/f_c - 1)N/K_{N1}]^{-b_{N1}} , \qquad (1)$$

where N was the adult density and $N/2$ was that of females, f_0 was fecundity per female in density-independent situation, K_{N1} was the adult density producing the maximal egg density (E_{max}), and b_{N1} was the degree of density-dependence in an adult population laying eggs. The degree of density-dependence is equal to 1 in the pure logistic-difference equation.

Hatchability of eggs was determined by the adult density, because mechanical injury by action of adults was considered to be the primary cause of egg mortality (Utida, 1941b). Hatched larval density (L) was also a function of adult density, and was formulated as,

$$L = r_0(N/2)[1 + (r_0/r_c - 1)N/K_{N2}]^{-b_{N2}}, \qquad (2)$$

where r_0 was the number of hatched eggs produced by a female in density-independent situation, K_{N2} was the adult density producing the maximal number of hatched eggs (L_{max}), and b_{N2} was the degree of density-dependence in an adult population producing hatched eggs. Eqs.(1) and (2) reveal the competition curves outside the beans.

The density of newly emerged adults (A), given as the first instar larvae that survive inside the beans, was formulated as

$$A = l_0L[1 + (l_0/l_c - 1)L/K_L]^{-b_L}, \qquad (3)$$

where l_0 was the survival rate during the larval and pupal stages in density-independent situation, K_L was the first instar larval density producing the maximal density of the newly emerged adults (A_{max}), and b_L was the degree of density-dependence in a population of the first instar larvae surviving to adults. Equation (3) reveals the competition curve inside the beans.

Since the reproduction curve is represented as the density of newly emerged adults (A) to the parental density (N) through one generation, I used Eqs.(2) and (3) for simulation of the reproduction curve, and used Eq.(1) only to simulate the competition curve for laying eggs.

4. Estimation of Parameters

Parameters in Eqs.(1), (2) and (3) were estimated from reproduction curve experiments.

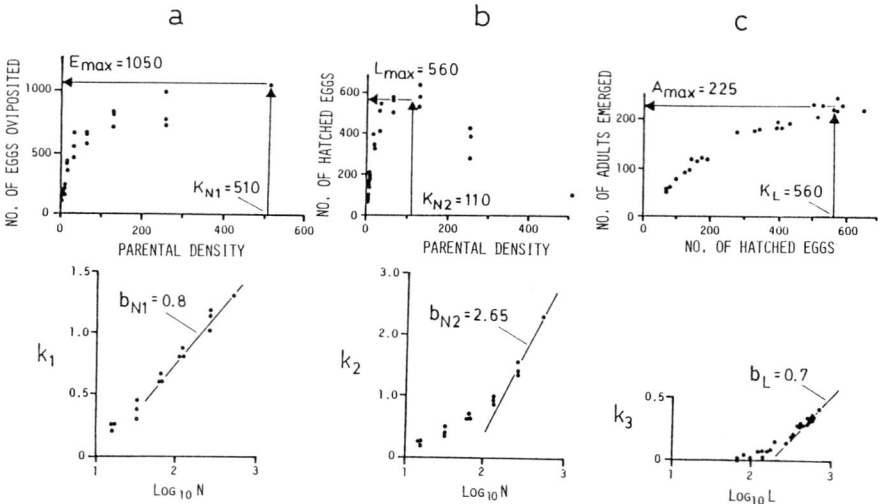

Figure 1. Estimation of parametric values for Eq.(1) (a), Eq.(2) (b) and Eq.(3) (c) in the case of jC. Estimations for other strains were the same as in jC. For detail, see text.

Figure 1 shows estimating methods of parametric values for the laboratory strain jC. The estimations for other strains were the same manner as in jC. Table 2 showed parametric values in all strains.

Fecundity per female in the density-independent situation (f_0) was 80, which was estimated by giving 1 pair of jC ca. 5g of azuki bean every day. The parental density (K_{N1}) producing the maximal density of eggs (E_{max}) was estimated to be 510, and E_{max}=1050 (upper panel of Fig. 1a). The parameter b_{N1} was estimated to be 0.8 as the gradient of the line $k_1 = \log(f_0N/2)$-log(observed no. of eggs) to $\log N$ (lower panel of Fig. 1a) (Varley and Gradwell, 1968). Fecundity per female (f_c) at the parental density K_{N1}, weighted by b_{N1}, was calculated to be 1.96 by substituting f_0, K_{N1}, E_{max} and b_{N1} into Eq.(1).

The number of hatched eggs (r_0) produced by a female in the density-independent situation was estimated to be 74 by counting hatched eggs in the preceding experiment for f_0. The observed number of hatched eggs was maximal at the middle density of parents, and the parental density (K_{N2}) producing the largest number of the hatched eggs (L_{max}) was estimated to be 110, and L_{max}=560 (upper panel of Fig. 1b). The parameter b_{N2} was estimated to be 2.65 as the gradient of the line $k_2 = \log(r_0N/2)$-log(observed no. of hatched eggs) to $\log N$ (lower panel of Fig. 1b). The number of hatched eggs (r_c) produced by a female at the parental density K_{N2}, weighted by b_{N2}, was calculated to be 35.0 by substituting r_0, K_{N2}, L_{max} and b_{N2} into Eq.(2).

Density-independent survival rate (l_0) from the first instar larvae (hatched eggs) to the newly emerged adults was determined to be 0.9 judging from the observed rate (0.87) in the 1-pair cultures where larval density per bean was low enough (ca. 2 to 3). The density of hatched eggs (K_L) producing the maximal number of newly emerged adults (A_{max}) was estimated to be 560, and A_{max}=225 (the upper panel of Fig. 1c). The parameter b_L was determined to be 0.7 as the gradient of the line $k_3 = \log(l_0L)$-log(observed no. of adults emerged) to $\log L$ (lower panel of Fig. 1c). The survival rate (l_c) at the density of hatched eggs K_L, weighted by b_L, was calculated to be 0.28 by substituting r_0, K_L, A_{max} and b_L into Eq.(3).

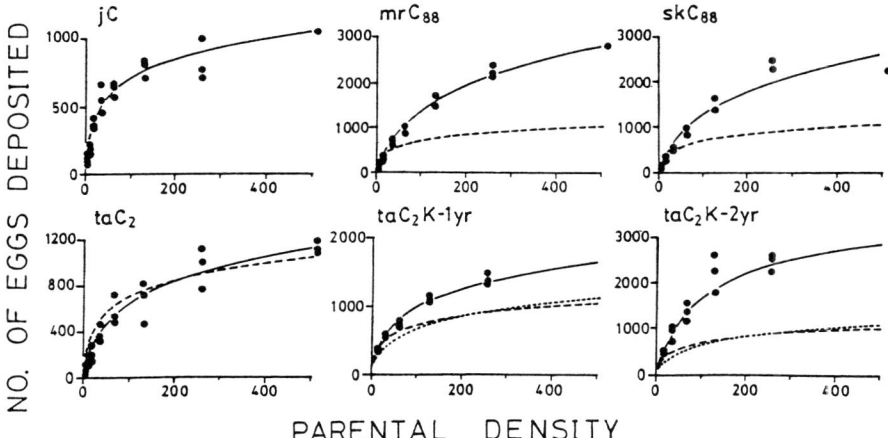

Figure 2. Simulations of competition curves for oviposition by parental insects. Plots showed data of each replicate. A dashed line shows the curve of jC, and a dotted line does that of taC$_2$.

5. Results

5.1. COMPETITION CURVES FOR OVIPOSITION BY PARENTS

Figure 2 shows how the number of eggs deposited changed with parental density in six strains. The laboratory strain jC showed a curve saturating at slightly more than 1000 eggs in spite of having the highest f_0 at the lower parental densities (Table 2). Two wild strains, mrC$_{88}$ and skC$_{88}$, controlled loosely the number of eggs deposited at middle and high parental densities, and their curves rose up to ca. 2500 eggs in relation to large E_{max} and small curvatures (b_{N1}). The strain taC$_2$ showed a similar curve to that of jC. However, curves of experimentally selected strains taC$_2$ K-1yr and K-2yr rose gradually in comparison with that of original taC$_2$, and became similar to wild strains mrC$_{88}$ and skC$_{88}$ in parallel with increase in E_{max}.

5.2. COMPETITION CURVES FOR PRODUCTION OF HATCHED LARVAE BY PARENTS

A competition curve for hatched larvae became a mountain-like curve with a long tail to the right in all strains (Fig. 3). The strain jC showed a higher curve than three wild

Table 2. Parametric values estimated from experimental data of each strain.

Strain	Eq.(1)				Eq.(2)				Eq.(3)			
	f_0	K_{N1}	E_{max}	b_{N1}	r_0	K_{N2}	L_{max}	b_{N2}	l_0	K_L	A_{max}	b_L
jC	80	510	1050	0.80	74	110	560	2.65	0.9	560	225	0.70
mrC$_{88}$	61	510	2850	0.65	55	100	430	2.8	0.88	430	205	0.90
skC$_{88}$	58	510	2650	0.65	53	75	430	2.9	0.93	400	205	0.85
taC$_2$	40	510	1130	0.74	38	45	300	3.0	0.95	350	203	0.53
K-1yr	60	510	1650	0.76	54	40	360	2.85	0.9	350	188	0.80
K-2yr	70	510	2900	0.90	63	100	760	3.2	0.9	700	225	0.85

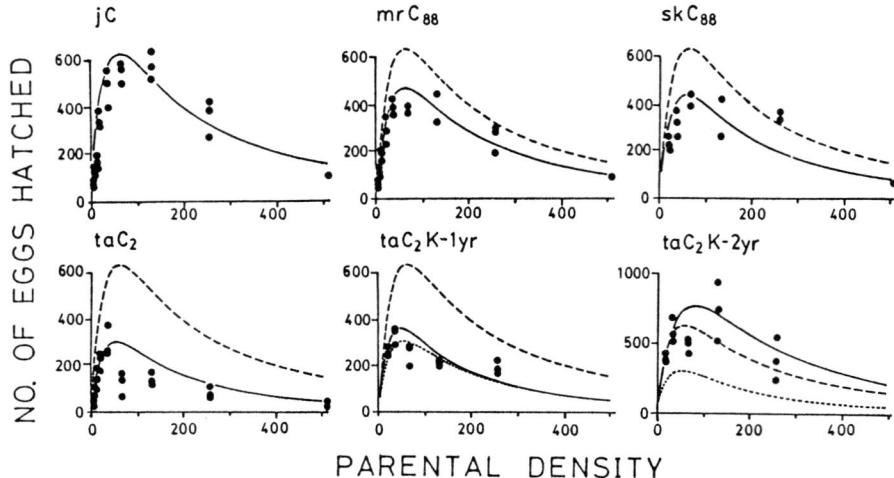

Figure 3. Simulations of competition curves for production of hatched larvae by parental insects. For plots, dashed and dotted lines, see legend of Figure 2.

strains, mrC$_{88}$, skC$_{88}$ and taC$_2$ in relation to larger L_{max} and K_{N2} and smaller b_{N2}, though jC showed the lowest curve for oviposition in Fig. 2. The reversal between jC and wild strains were dependent on hatchability of eggs at middle to high parental densities. Strain jC showed much higher hatchability of eggs at parental densities of 64 and 128 pairs than wild strains (Table 3).

The curve of an experimentally selected strain, taC$_2$ K-1yr, began to rise, and that of taC$_2$ K-2yr went over jC (Fig. 3). Hatchability of eggs at parental densities of 64 and 128 pairs was higher in taC$_2$ K-2yr than the original taC$_2$ (Table 3).

5.3. COMPETITION CURVES IN THE LARVAL STAGE

Larval competition curves did not show a large difference between jC and the three wild strains although those of wild strains became slightly higher (Fig. 4). Experimentally selected strains, taC$_2$ K-1yr and K-2yr, had curves progressively lower than the original taC$_2$, and that of taC$_2$ K-2yr was lower than that of jC.

Table 3. Hatchability (%) of eggs at parental densities of 64 and 128 pairs.

Parental Pairs		Strains					
		jC	mrC$_{88}$	skC$_{88}$	taC$_2$	K-1yr	K-2yr
64	observed	74.2	23.8	22.3	22.8	19.0	32.9
	simulated	69.5	24.5	24.5	31.3	23.4	38.4
128	observed	44.0	11.5	14.9	9.0	14.0	16.4
	simulated	34.9	10.6	10.2	13.2	9.5	19.4

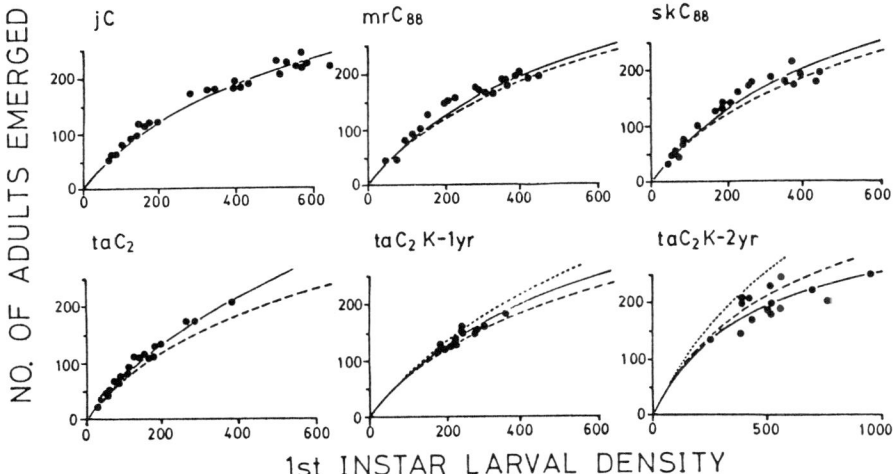

Figure 4. Simulations of competition curves in the larval stage. For plots, dashed and dotted lines, see legend of Figure 2.

5.4. REPRODUCTION CURVES

Equation (3) was combined with Eq. (2) to simulate a reproduction curve (parental density vs. progeny production) for each strain. The curves fit the observed reproduction curves as shown in Fig. 5. Strain jC had a higher reproduction curve than those of three wild strains, mrC_{88}, skC_{88} and taC_2, at all parental densities, which resulted from a higher production of hatched larvae in Fig. 3. The point at which the curve is intersect-

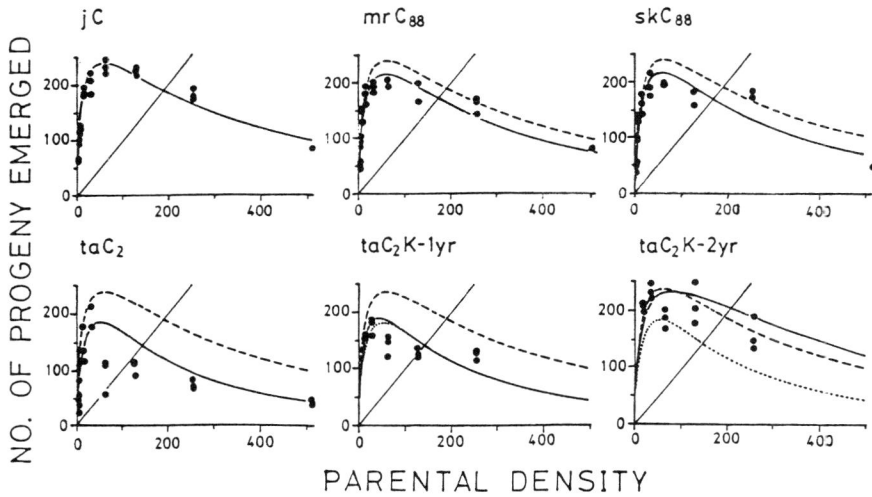

Figure 5. Simulations of reproduction curves. For plots, dashed and dotted lines, see legend of Figure 2.

368

ed by the line $A=N$ indicates the carrying capacity. The carrying capacity of jC was higher than those of mrC$_{88}$, skC$_{88}$ and taC$_2$.

Though a curve of the experimentally selected strain taC$_2$ K-1yr did not show a large change in the reproduction curve from that of taC$_2$, the rise of the curve of taC$_2$ K-2yr was largely in middle to high parental densities. The carrying capacity of taC$_2$ K-2yr was eventually higher than that of jC in parallel with increase in hatched larvae as shown in Fig. 3.

5.5. MULTIVARIATE ANALYSES FOR ORDINATION AND GROUPING OF STRAINS

I conducted principal component analysis for ordinating and cluster analysis for grouping strains with using 11 parameters except K_{N1} (constantly 510) in Table 2 and the carrying capacity in Fig. 5. Principal component analysis was carried out with variance-covariance matrix of parameters. Cluster analysis was based on group average method for calculating standardized square Euclidean distances.

An ordination map of principal component analysis was shown in Fig. 6a. The principal axis (Axis-1) was largely dependent on E_{max} and the subaxis (Axis-2) was related to L_{max}. Strain jC was located at the lowest position in Axis-1 and at the highest but one in Axis-2. Three wild strains, mrC$_{88}$, skC$_{88}$ and taC$_2$, were placed at lower positions in Axis-2, though they were widely ordinated along Axis-1. A wild strain, skC$_{88}$, was close to mrC$_{88}$ which came from geographically distant place rather than being close to taC$_2$ which came from the same locality. An experimentally selected strain taC$_2$ K-1yr was near its original taC$_2$, but taC$_2$ K-2yr located at upper-right corner of the plane in relation to high E_{max} and L_{max}.

These tendencies were seen in cluster analysis (Fig. 6b). It showed a cluster of taC$_2$ and taC$_2$ K-1yr and also a small cluster of mrC$_{88}$ and skC$_{88}$. Strains jC and taC$_2$ K-2yr were distant from it.

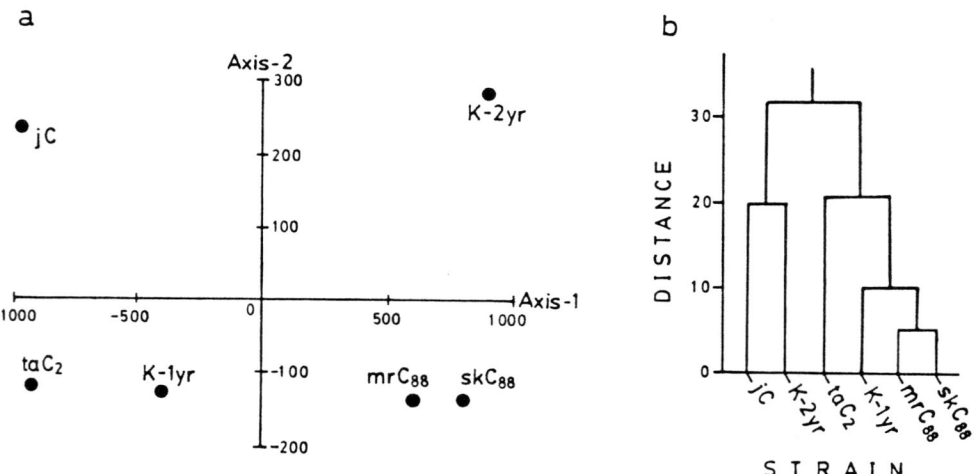

Figure 6. A map of principal component analysis for ordinating strains (a), and a dendrogram of cluster analysis for grouping strains by group average method (b). Ordination of the dendrogram is the standardized square Euclidean distance.

6. Discussion

6.1. DIFFERENCES IN OVIPOSITION AND EGG HATCHING PROCESSES OUTSIDE BEANS

As shown in the principal component analysis (Fig. 6a), the differences in density-dependent population processes among strains could be characterized mostly by the maximum number of eggs (E_{max}) deposited at high parental densities and the maximum number of hatched larvae (L_{max}) produced at middle parental densities. Larval development inside the bean was similar in the strains (Fig. 4).

The laboratory strain jC controlled the number of eggs deposited at high parental densities (Fig. 2). Wild strains, especially in mrC_{88} and skC_{88}, laid far more eggs than jC under crowded conditions, but most of which were unhatched (Fig. 2 and Table 3). Adults of wild strains move much more actively in a petri dish and are more likely to fly than the laboratory strain jC. In addition, their uncontrolled tendency to oviposit at high parental densities makes the competition for the surface of the bean quite severe. Those tendencies of adults should decrease hatchability of eggs laid on the bean under crowded conditions.

The uncontrolled tendency to oviposit under crowded conditions in wild strains may be the result of one of two mechanisms; (i) the oviposition marking pheromone (Ohshima et al. 1973) may have a different composition than that of jC, or (ii) responses of females to the oviposition marking pheromone may be different from jC. A chemical analysis for composition of the oviposition markers in both strains and bio-assaies designed in a 2 x 2 (markers x insects) contingency table are needed.

6.2. DENSITY-DEPENDENT NATURAL SELECTION ON ECOLOGICAL CHARACTERS

Pioneering studies on density-dependent natural selection by MacArthur (1962) and MacArthur and Wilson (1967) introduced a concept of density-dependent rate of population growth as a measure of fitness. They argued that a population experiencing high densities and strong competition for limiting though constantly renewed resources should evolve high intraspecific competitive ability and increase the carrying capacity. Later theoretical studies sometimes assumed a *priori* trade-off between the intrinsic rate of natural increase and the carrying capacity (e.g., Roughgarden, 1971), which led a rather stereotypic concept of r-K selection continuum hypothesis (Pianka, 1970).

However, the present results showed clearly that a density-independent rate of population growth per generation ($r_0 \times l_0$) increased as the carrying capacity increased instead of the trade-off between them in comparison with jC and wild strains. The jC strain differs from wild strains in having a larger body size, being less active, and strongly regulated oviposition under crowded conditions. These traits may contribute to a higher r and K because a larger female lays more eggs in *C. chinensis* and more energy can be allocated to reproduction if there is less activity. Such characteristics may have been selected under the density-dependent conditions in the laboratory for ca. 50 yr.

The strain taC_2 K-2yr also showed an increase in r and K over the original taC_2, but the high carrying capacity of taC_2 K-2yr was attained due to the increasing number of eggs deposited with a slight increase in hatchability at high parental densities (Fig. 2 and Table 3). A variety of combinations of traits may increase r and K together.

In comparison of taC_2 with taC_2 K-1yr and K-2yr, increase in E_{max} occurred during the first 1 yr in taC_2 K-1yr, then E_{max} increased far more in parallel with increase in L_{max} for the next 1 year, as seen in Figs. 2 and 3. Since increase in L_{max} results from increase in E_{max} and improved hatchability at middle and high parental densities, the

370

density-dependent natural selection should make an experimentally selected strain introduced from the field to first increase along Axis-1 in Fig. 6a, then to increase simultaneously in both Axis-1 and Axis-2. Eventually, the fecundity should become controlled
under crowded conditions and E_{max} may decrease while maintaining high L_{max} by improving hatchability at middle and high parental densities, judging from the results of the
strain jC. However, maintenance of a population at the carrying capacity level for 2 yr
did not result in $taC_2 K$-2yr gaining a tendency to control oviposition under crowded
conditions. More time may be necessary to shift the position of $taC_2 K$-2yr to near jC in
the map of principal component analysis in Fig. 6a.

The present study shows that the currently existing theories of density-dependent
natural selection may be too simple to predict the adaptations of a population of C.
chinensis competing for limiting resource. We should pay more attention to the parameters that change during the adaptation of bruchids for exploitation of stored beans.

Acknowledgements. I am grateful to Prof. K. Fujii of University of Tsukuba for his insightful discus-
sions throughout the present study. I also express my thanks to Prof. R.D. Mitchell of Ohio State
University for his valuable comments and grammatical improvements to an earlier manuscript and to
K. Kato of our laboratory for supporting multivariate analyses. The present study is supported by
Grant-in-Aid for Encouragement of Young Scientists from Ministry of Education, Science and Culture
(63740359).

REFERENCES

Bierbaum, T. J., Mueller, L. D., and Ayala, F. J. (1989) Density-dependent evolution of life-history
traits in *Drosophila melanogaster*, Evolution 43, 382-392.
Fujii, K. and Shimada, M. (1980) Process of domestication vs. theory of life history strategy, Abstract
XVI Intern. Congr. Entomol. Kyoto, 304.
Gadgil, M. and Solbrig, O. T. (1972) The concept of r- and K-selection: Evidence from wild flowers
and some theoretical considerations, Amer. Natur. 106, 14-31.
Hassell, M. P. (1975) Density-dependence in single species populations, J. Anim. Ecol. 44, 283-295.
Luckinbill, L. (1978) r- and K-selection in experimental population of *Escherichia coli*, Science 202,
1201-1203.
MacArthur, R. H. (1962) Some generalized theorems of natural selection, Proc. Nat. Acad. Sci. USA.
48, 1893-1897.
MacArthur, R. H. and Wilson, E. O. (1967) The Theory of Island Biogeography, Princeton Univ.
Press, Princeton, N.J.
McNaughton, S. J. (1975) r- and K-selection in *Typha*, Amer. Natur. 109, 251-261.
Mueller, L. D. and Ayala, F. J. (1981) Trade-off between r-selection and K-selection in *Drosophila*
populations, Proc. Nat. Acad. Sci. USA. 84, 1974-1977.
Nicholson, A. J. (1948) Competition for food amongst *Lucilia cuprina* larvae, Proc. VIII Intern. Congr.
Entomol. Stockholm 277.
Ohshima, K., Honda, H., and Yamamoto, I. (1973) Isolation of oviposition marker from azuki bean
weevil, *Callosobruchus chinensis* (L.), Agr. Biol. Chem. 37, 2679-2680.
Pianka, E. R. (1970) On r- and K-selection, Amer. Natur. 104, 592-596.
Roughgarden, J. (1971) Density-dependent natural selection, Ecology 52, 453-468.
Shimada, M. (1989) Systems analysis of density-dependent population processes in the azuki bean
weevil, *Callosobruchus chinensis*, Ecol. Res. 4, 145-156.
Stearns, S. C. (1976) Life history tactics: A review of the ideas, Quart. Rev. Biol. 51, 3-47.
Taylor, C.E. and Condra, C.(1980) r- and K-selection in *Drosophila pseudoobscura*, Evolution 34,
1183-1193.
Utida, S. (1941a) Studies on experimental populations of the azuki bean weevil, *Callosobruchus*

chinensis (L.), I. The effect of population density on the progeny population, Mem. Coll. Agr. Kyoto Imp. Univ. 48, 1-30.

Utida, S. (1941b) Studies on experimental populations of the azuki bean weevil, *Callosobruchus chinensis* (L.), III. The effect of population density upon the mortalities of different stages of life cycle, Mem. Coll. Agr. Kyoto Imp. Univ. 49, 21-42.

Utida, S. (1942) Studies on experimental populations of the azuki bean weevil, *Callosobruchus chinensis* (L.), VII. Analysis of the density effect in the preimaginal stage, Mem. Coll. Agr. Kyoto Imp. Univ. 53, 19-31.

Varley, G. C. and Gradwell, G. R. (1968) Population models for the winter moth, Sympo. Roy. Entomol. Soc. London, 4, 132-142.

POPULATION PROCESSES AND DYNAMICS OF LABORATORY POPULATIONS OF *CALLOSOBRUCHUS* SPP.

THOMAS S. BELLOWS, JR.
Department of Entomology, University of California,
Riverside, California 92521 U.S.A.

ABSTRACT. The results of single- and mixed- species studies are reviewed and the processes principally significant in forming the dynamics of these populations are discussed. The first of these processes is adult competition for oviposition sites, which is mediated by both exploitative and interference competition, and occurs both intraspecifically and interspecifically. Larval competition, apparently exploitative in nature, can produce very different outcomes in different populations, varying from nearly contest competition to scramble competition. The combinations of these processes result in different dynamics in different populations, varying from stable populations to those exhibiting cyclic behavior.

Interspecific competition involves similar mechanisms, and always results in the exclusion of one or the other of the species involved. Interactions with parasitoids involve additional mechanisms, including search behavior of the parasitoids, selection of specific ages of hosts for oviposition, and death of hosts due to host feeding. The processes, combined with intraspecific competition, can lead to extraordinary dynamical behavior of host-parasitoid systems.

1. Introduction

Laboratory populations of *Callosobruchus* spp., particularly *C. chinensis* (L.) and *C. maculatus* (F.), have served as model systems for the investigation of population dynamics for over half a century. In that time we have learned much about the biology of these species as they have served to advance our knowledge of the biological processes and dynamical behavior of single species systems, interspecific competitive systems, and parasitoid host systems.

In this paper I will review the vital rates and parameters known for *Callosobruchus*, and then discuss the density dependence mechanisms which operate in laboratory populations. The consequences of these processes in terms of population dynamics will be discussed for both single-species and multiple-species systems. I will conclude with a brief discussion of the general features of these laboratory systems and the role they have played in the development of population ecology.

2. Reproduction and Survival Parameters

Although adults of *Callosobruchus* spp. are not actively phytophagous, they will consume sugar solutions, honey, and feed on fungus. This increases their longevity and realized fecundity substantially (e.g., Shinoda and Yoshida, 1987). This may have a major impact on their ability to survive and reproduce successfully in the field, but laboratory studies

K. Fujii et al. (eds.), Bruchids and Legumes: Economics, Ecology and Coevolution, 373–383.
© 1990 *Kluwer Academic Publishers. Printed in the Netherlands.*

generally have not provided food resources for adults. This limits the fecundity of adults to the energy reserves stored during larval feeding. The fecundity of the two most widely employed species, C. chinensis and C. maculatus, has been reported by several workers for many different strains of both species. The recent work of Credland (1986) indicates that the fecundity reported in earlier studies may have been limited by the supply of seeds or the number of adults employed in those studies. Nonetheless, a survey of reported values (Table 1) indicates that fertility for most strains lies between 40 - 100 eggs per female. In two comparative studies between the two species, both Fujii (1968) and Bellows (1982b) report generally greater fecundities for C. maculatus than for C. chinensis.

Preimaginal survival is generally quite high in these species in the absence of density-dependent effects. Egg hatch percentages have ranged from 76.3 - 98.9%, and larval to adult survival has ranged from 55.6 - 98.8% (Table 1) in suitable hosts. Reported net reproductive rates have varied from 15.2 - 38.9 in C. chinensis and from 14.9 - 49.9 in C. maculatus.

Table 1. Vital rates from the laboratory for *Callosobruchus chinensis* and *C. maculatus*

Reference	Strain	Temp (°C)	Realized fertility	Egg Hatch (%)	Survival to Emergence(%)	Developmental Period(days)	Sex ratio (% fem.)	R_0
				(a) *C. chinensis*				
Utida (1941b)		25	76.3	88.8	98.8	34.1	.503	33.671
(host: *Phaseolus*		30	88.7	91.3	96.1	-	.5[a]	38.912
angularis)								
Fujii (1968)	jC	30	60.3	76.3	62.8	23.7	.540	15.603
(host: *P.*	kC	30	49.4	92.1	81.1	23.7	.504	18.597
angularis)	iC	30	53.3	85.6	70.6	23.3	.472	15.204
	nC	30	49.5	90.7	76.6	24.1	.506	17.402
Bellows (1982b)		30	36.7	98.9	94.0	22.0	.504	17.196
(host: *Vigna unguiculata*)								
				(b) *C. maculatus*				
Fujii (1968)	tQ	30	57.8	91.9	79.8	25.6	.505	21.406
(host: *P.*	bQ	30	64.4	92.1	69.6	26.2	.484	19.980
angularis)	aQ	30	81.0	92.8	55.6	27.2	.490	20.479
	cQ	30	67.4	86.4	76.1	27.3	.540	23.931
Bellows (1982b)		30	40.6	95.5	84.4	23.5	.455	14.890
(host: *V. unguiculata*)								
Credland (1986)								
(host: *V.*	Campinas	27	103.8	98.5	97.7	-	.5[a]	49.946
unguiculata)	Yemen	27	80.5	96.1	76.2	-	.5[a]	29.474
	IITA	27	100.0	97.9	93.7	-	.5[a]	45.866

[a] Sex ratio not given in original report; 0.5 used here to calculate R_0

3. Processes in Intraspecific Competition

3.1. COMPETITION IN THE ADULT STAGE

It is widely reported that the realized fecundity of *Callosobruchus chinensis* and *C. maculatus* is affected by amount of resource available for oviposition. Isolated pairs of adults have fecundity reduced below their maximum when offered few beans (e.g., Dick and Credland, 1984; Credland, 1986), reflecting a self-limiting or constrained oviposition behavior in the presence of limited resources. Whether this behavior is an adaptation to limit larval competition has not been well demonstrated, as the behavior is altered by changes in host seed offered (Wasserman, 1986), and oviposition per bean in many strains can exceed likely larval survival, often by a substantial amount. The degree to which oviposition is constrained varies substantially among strains (e.g., Credland, 1986), so that some strains reach maximum fecundity when provided with a few seeds, but others only when sufficient numbers of seeds are available that each seed bears only a single egg. The mechanisms underlying this behavior include recognition of the presence of eggs on the surface of the host seed (Messina and Renwick, 1985), as well as chemical stimuli that reduce oviposition. These chemical stimuli are the result of adult activity on the host seeds, and may be associated with oviposition (Messina and Renwick, 1985) or with other non-reproductive activity (Sakai et al., 1986).

The natural extension of this individual behavior to populations is a decline in fecundity as populations increase in density. This phenomenon has been reported widely (e.g., Utida, 1941a; Bellows, 1982a,b) and has at least two component behaviors. The first is limiting of fecundity by females caused by the prior presence of adults or eggs on the seeds (Bellows, 1982b), probably through the same chemical stimuli or egg assessment mechanism which mediates constrained oviposition by single pairs. Fecundity is further reduced as a direct result of adult crowding (Bellows, 1982b).

A further process significant in the adult stage is the interaction between adult density and egg survival. As pointed out by Utida (1941c) and others, egg survival is inversely related to adult density. Eggs are apparently adversely affected by the continued activity of adults on seeds, possibly through simple mechanical damage to the eggs, or perhaps through environmental contamination by the chemicals which mediate reduced fecundity, which are ovicidal at high concentrations (Oshima, Honda and Yamamoto, 1973).

The outcome of these combined processes in populations of *Callosobruchus* spp. is that both realized fecundity and egg survival decline as adult density increases. The severity of these processes varies with species and strain (Fig. 1), and also with the experimental arena employed. In arenas large relative to the number of seeds used, egg survival apparently is less affected than in arenas where the adults are confined in close proximity to the seeds (c.f., Utida, 1941c; Fujii, 1968, Bellows, 1982a). Depending on the severity of these affects, this density-dependence can contribute to stable or oscillatory population dynamics, as discussed below.

3.2. COMPETITION IN THE LARVAL STAGES

Competition in the larval stages appears largely exploitative in nature, with consumption of common resources the principle factor limiting the number of individuals surviving in beans. There is no record of intraspecific predation, or cannibalism, among larvae within a bean. In some strains, the maximum number of individuals surviving to adult in a single bean appears limited only by the total amount of resource or size of the bean, but in other strains some other behavioral factor appears to limit the number of survivors to numbers far below the potential productivity of the bean.

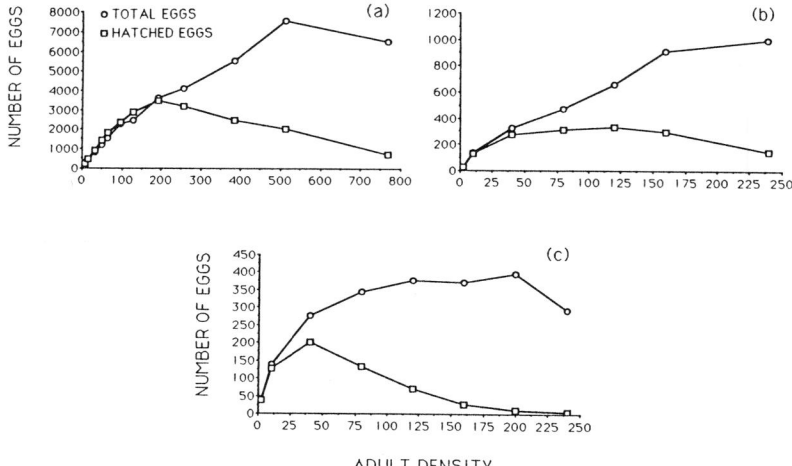

Figure 1.   Relationship between adult density and number of total (○) or hatched (□) eggs for (a) *Callosobruchus chinensis* (Utida, 1941a), (b) *C. chinensis* (Bellows, 1982a), and (c) *C. maculatus* (Bellows, 1982a).

The result of exploitative competition in many laboratory insects is scramble competition, with overcompensation resulting in few or no survivors at high initial densities. Such a result has been reported for *C. maculatus* (Fujii, 1968), who employed an arena in which little egg mortality occurred and the number of hatched eggs per bean was high.   A similar observation has been reported by M. P. Hassell (personal communica-

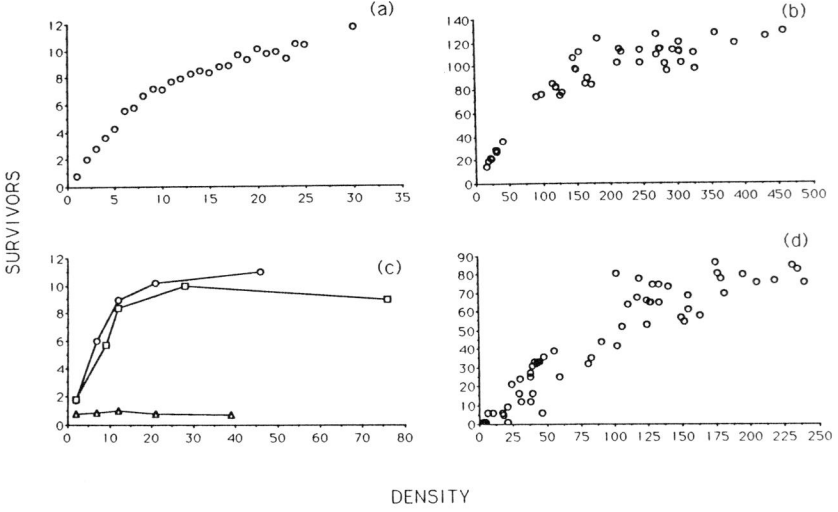

Figure 2.   Relationship between hatched eggs density and number of adult survivors for *Callosobruchus chinensis* [(a) and (b)] and *C. maculatus* [(c) and (d)]. (a) and (c) are densities per bean, (b) and (d) are densities per 3.7 g cowpea. (a) Utida (1942). (b) and (d) Bellows(1982a). (c) Dick and Credland (1984).

tion). More generally, the result of larval competition appears only slightly overcompensating in both *C. chinensis* and *C. maculatus* (e.g., Utida, 1942b; Bellows, 1982a,b) (Fig. 2). This result was found to be general in all four larval instars (Bellows, 1982b). The severity of this competition can vary with the host seed variety (Giga and Smith, 1981).

Some workers have reported decreasing adult weights (Utida, 1941a, Bellows, 1982a) and, in one study, decreasing fertility (Credland et al., 1986) with increasing larval density.

4. Single-Species Population Dynamics

4.1. RESULTS

The dynamics of laboratory populations of *Callosobruchus* spp. (e.g., Utida, 1941d, 1967; Fujii, 1968; Bellows 1982b) fall generally within the expected dynamics of discrete-generation systems, but span a range of behaviors from nearly monotonically-damped

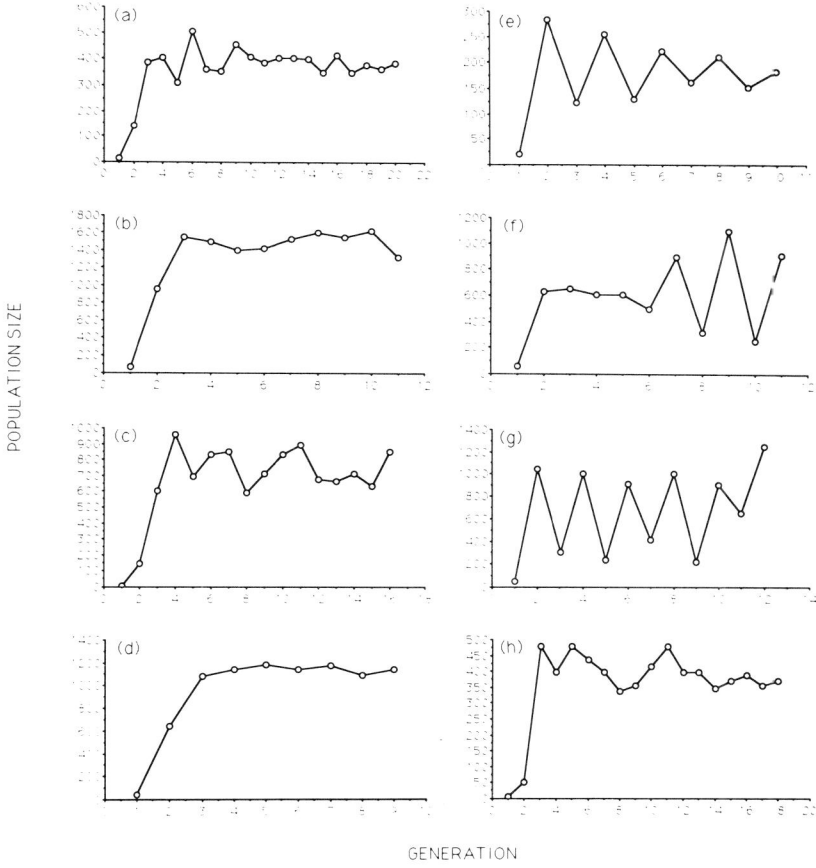

Figure 3. Selected examples of long-term single-species culture for *Callosobruchus chinensis* [(a)-(d)] and *C. maculatus* [(e)-(h)]. (a), (e) Utida (1967). (b), (d), (f), (g) Fujii (1968). (c) M. P. Hassell, unpublished. (h) Bellows (1982b).

populations to those which exhibit stable cycles (Fig. 3). Generally, *C. maculatus* has been reported more widely as showing cyclic behavior, while *C. chinensis* has shown more stable behavior.

4.2. A GENERAL TWO AGE-CLASS MODEL

Because the life cycle and competition of *Callosobruchus* spp. divide naturally into two distinct phases — adults and preimaginal stages — it is possible to construct a general model describing the outcomes of the competitive processes and the resulting population dynamics. One such framework is a two-stage reproduction curve (May et al., 1974) (Fig. 4). In such a reproduction curve, the effect of adult density on the number of hatched eggs is depicted by one curve. The effect of hatched egg density on larval survival is depicted on the companion curve, completing the cycle from adults in one generation to adults in the next. In either stage competition can be depicted by a curve which describes either no density dependence (e.g., line A in Fig 4), monotonically increasing or compensatory density dependence (lines B and D) or overcompensating density dependence (lines C and E). The appropriate selection of combinations of curves can model reproduction from most of the reported studies, from the case of little

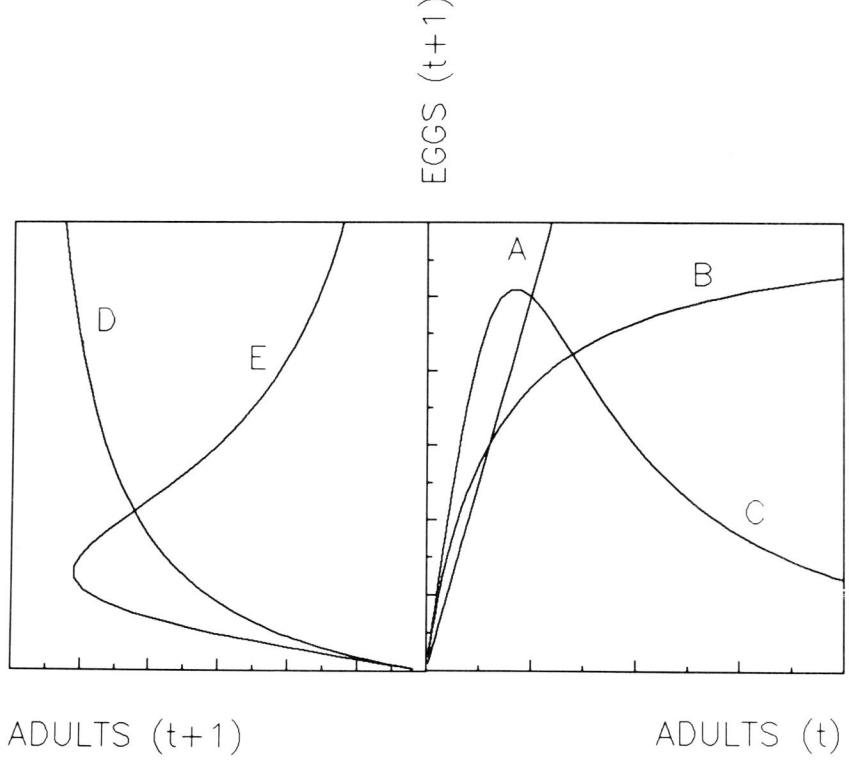

Figure 4. A two-stage reproduction curve. Adults in generation t give rise to eggs in generation t + 1 which in turn yield adults in generation t + 1. Different relationships, such as linear (line A), compensatory (B, D) or overcompensatory (C, E) may apply in either stage. Different combinations of relationships yield different population dynamics.

density dependence in the adult stage combined with scramble competition in the larval stage (e.g., Fujii, 1968) to the case of compensatory or overcompensatory density dependence in both stages (e.g., Bellows, 1982a).

A more formal presentation of the two-stage reproduction curve is possible using the density dependent model of Maynard Smith and Slatkin (1973), so that

$$E_{t+1} = rfA_t [1 + (aA_t)^b]^{-1}, \qquad\qquad (1)$$

$$A_{t+1} = sE_{t+1} [1 + (cE_{t+1})^d]^{-1}, \qquad\qquad (2)$$

where A and E are the adult and hatched egg population sizes, t and $t+1$ are generation indices, r is the proportion of females in the population, f is the number of hatched eggs laid by an isolated pair of adults, s is the proportion of larvae surviving in an uncrowded environment, a and c are scaling constants reflecting the size of the arena, and b and d determine the severity of density dependence in the two stages. As b (or d) increases from 0, competition increases from none through compensating (at $b = 1$) to overcompensating as b increases beyond 1. Typical curves depicted by these relationships are shown in Fig. 4.

The general dynamical features of *Callosobruchus* spp. populations can be closely mimicked by suitable combinations of parameters in this two age class system. For example, the cyclic populations of *C. maculatus* described by Fujii (1968) can be paralleled by a system with no density dependent reduction in the number of hatched eggs combined with overcompensating density dependence in the larval stage (Fig. 5a). More stable populations characterized by slightly overcompensating density dependence in both the adult and larval stages (e.g., Utida, 1941a; Bellows, 1982a) and the resulting dynamics (Utida, 1967; Fujii, 1968; Bellows 1982b) are similarly well characterized by the appropriate choice of density dependence parameters (Figs. 5b,c).

In general, the dynamics of these two-stage systems is characterized by the joint relationships among the parameter rfs, the effective fertility at low densities, and the values for b and d, the parameters describing density dependence. If either b or d takes a value close to 1, describing compensatory competition, then this feature will dominate the dynamics of the system and the populations will behave in a stable fashion. If either parameter takes a value near 0 (as is the case for no density dependent reduction in fecundity, line A of Fig. 4), then the population will show behavior characterized by the other parameter, either compensatory for values near 1 or overcompensatory for values greater than 1. Hence, the absence of adult competition in Fujii's results for *C. maculatus* leave the system dominated by the scramble competition he observed in the larvae (and hence subject to oscillatory behavior, e.g., Fig. 5a), whereas the scramble competition in the adult stage reported by Bellows (1982a) was compensated for by compensatory competition in the larval stage (resulting in more stable behavior, e.g., Fig. 5c).

5. Two-Species Competitive Systems

Competition between *Callosobruchus chinensis* and *C. maculatus* populations has been the subject of several studies (e.g., Fujii, 1968, 1969). The mechanisms involved, however, have been investigated in less detail than in single-species systems.

Some of the general processes occurring in single-species populations undoubtedly occur in mixed populations. Reductions in fecundity in relation to the presence of interspecific egg densities has been recorded (Bellows and Hassell, 1984). This is likely due to the same behaviors which limit oviposition in single-species experiments, vis.

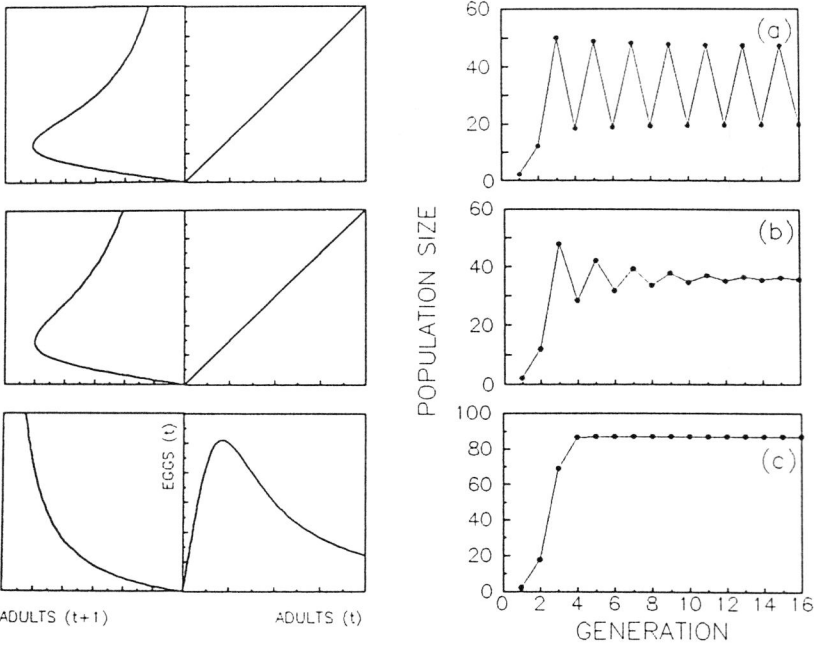

Figure 5. Simulated two-stage reproduction curves (equation (1)) and their resulting dynamical behavior. (a) No density dependence during oviposition, strongly overcompensatory (scramble) competition in the larval stage yielding cyclic behavior (cf. Fig. 3g). (b) No density dependence during oviposition, overcompensatory (scramble) competition in the larval stage yielding damped oscillations (cf. Fig. 3e). (c) Strongly overcompensatory density dependence during oviposition, compensatory (contest) density dependence in the larval stage yielding a stable population (c.f. Fig. 3b, d; see also Figs. 1 and 2).

recognition of eggs present on beans and chemical substances (Messina and Renwick, 1985); the chemical substances have been shown to be active interspecifically (Sakai et al., 1986). Additional reduction in fertility due to interspecific adult crowding has also been documented (Bellows and Hassell, 1984), as have interspecific effects on egg mortality. The net result of these processes is additional, interspecific density dependent reductions in hatched egg numbers (Bellows and Hassell, 1984) which, in at least one case (Fujii, 1970) may play a major role in determining the outcome of interspecific competition.

Similarly, the mechanisms of larval competition important in single-species culture are significant in mixed culture. Exploitation of resources appears to be the principal mechanism of competition, and interspecific competition generally results in the survival of the larger larva. The severity of the result may depend upon the strains involved, but in at least one case the effects between species were similar, although the larger *C. maculatus* had a slight competitive advantage (Bellows and Hassell, 1984).

The results of long-term mixed culture of these two species has provided varied results, always resulting in the extinction of one species, but not always the same one. Coexistence of the two species in a closed laboratory environment has not been demonstrated. *C. maculatus* appears to be generally the better competitor at the level of simple biological comparisons, being larger in size and generally outcompeting similar stages of *C. chinensis*, and several studies have demonstrated its ability to exclude *C. chinensis* in mixed culture. Other studies, however, have demonstrated that *C. chinensis* can exclude *C. maculatus*, even in cases where *C. maculatus* appears competitively superior. This may be due to inherent differences in the intraspecific responses to density, providing some strains of *C. chinensis* superiority over more inhibited strains of *C. maculatus* (e.g. Fujii, 1968, 1970). In other studies where *C. maculatus* appears competitively superior, *C. chinensis* manages to exclude *C. maculatus* via its faster development, securing a competitive advantage in every new lot of beans provided by occupying them before *C. maculatus* and thus being in a later (and competitively superior) stage when *C. maculatus* eggs hatch. A similar result has been shown for *C. analis* and *C. chinensis*. The larger *C. analis* appears inherently superior to *C. chinensis* in simple laboratory experiments, but *C. chinensis* can force the competitive exclusion of *C. analis* in long-term mixed culture.

Because of the importance of the developmental period in the outcome of these competitive systems, such systems generally require models of significantly more detail than the simple analytical model presented above. A detailed model has been developed for one such system (Bellows and Hassell, 1984) which successfully described the dynamics of mixed cultures of those strains.

6. Host-Parasitoid Systems

Callosobruchus spp. and their parasitoids have been the object of studies at several levels, including basic biology and behavior (e.g., Bellows 1983a, b; Heong 1985), population studies (e.g., Utida, 1950, 1957a, b), and population modeling (e.g., Bellows and Hassell, 1986). The principle species employed have been the braconic *Heterospilus prosopidis* Viereck and the pteromalid *Anisopteromalis calandrae* (Howard), usually with *C. chinensis* as the experimental host.

These systems are extraordinary in their richness of biological processes and interactions between the species. In addition to the single-species density dependent processes acting in the host population, there are additional intraspecific density dependent processes acting in the parasitoid population. Interactions between the two populations include a number of critical age-specific interactions, such as host selection, sex ratio allocation, and synchrony (or asynchrony) of the populations. Of these processes, certainly competition within populations and search efficiencies of the parasitoid play an important role in the dynamics of such systems. Equally important are the issues of sex-ratio allocation among hosts of different ages (and therefore sizes) by the parasitoid and asynchrony between host and parasitoid populations, a consequence of different developmental periods. The consequences of some of these effects have been explored by Bellows and Hassell (1988); further research into the critical processes and dynamics of such systems will likely increase our understanding of the relative significance of these various processes in shaping the dynamics of both natural and laboratory populations.

7. Discussion

Populations of *Callosobruchus* spp. maintained in the laboratory have provided excellent model systems for the study of intraspecific and interspecific processes responsible for shaping the dynamical behavior of biological systems. Their dynamical complexity and diversity of biological relationships provides excellent opportunities for examining all phases of population biology, from partially deterministic intraspecific density-dependent

382

mechanisms through interspecific competition and parasitism. Recently, the genetics of these species has received increasing attention, and the study of population genetics and selection may add further insights into processes affecting these species in both field and laboratory settings.

The advancement of population ecology depends on the study of tractable populations. In this sense, laboratory insect populations are some of the best suited systems for studying biological processes and their influence on population dynamics. *Callosobruchus* spp. provide one of the few systems where intraspecific, interspecific and intertrophic level interactions all can be studied, providing unique opportunities for exploring the dynamics of complex biological phenomena in systems where each of the underlying processes may be analyzed experimentally. It is largely because of this special attribute that they have been able to contribute so significantly to the development of population biology. Future exploitation of this feature of these laboratory populations will continue to be of major value in adding to our knowledge of insect population biology.

REFERENCES

Bellows, T. S. Jr. (1981) The descriptive properties of some models for density dependence, J. Anim. Ecol. 50, 139-156.

Bellows, T. S. Jr. (1982a) Analytical models for laboratory population of *Callosobruchus chinensis* and *C. maculatus* Coleoptera, Bruchidae), J. Anim. Ecol. 51, 263-287.

Bellows, T. S. Jr. (1982b) Simulation models for laboratory population of *Callosobruchus chinensis* and *C. maculatus*, J. Anim. Ecol. 51, 597-623.

Bellows, T. S. Jr. (1985a) Effects of host age and host availability on developmental period, adult size, sex ratio, longevity and fecundity in *Lariophagus distinguendus* Förster (Hymenoptera: Pteromalidae), Res. Popul. Ecol. 27, 55-64.

Bellows, T. S. Jr. (1985b) Effects of host and parasitoid age on search behavior and oviposition rates in *Lariophagus distinguendus* Förster (Hymenoptera: Pteromalidae), Res. Popul. Ecol. 27, 65-76.

Bellows, T. S. Jr. and Hassell, M. P. (1984) Models for interspecific competition in laboratory populations of *Callosobruchus* spp., J. Anim. Ecol. 53, 831-848.

Bellows, T. S. Jr. and Hassell, M. P. (1988) The dynamics of age-structured host-parasitoid systems, J. Anim. Ecol. 57, 259-268.

Credland, P. F. (1986) Effect of host availability on reproductive performance in *Callosobruchus maculatus* (F.) (Coleoptera: Bruchidae), J. stored Prod. Res. 22, 49-54.

Credland, P. F., Dick, K. M., and Wright, A. W. (1986) Relationships between larval density, adult size and egg production in the cowpea seed beetle, *Callosobruchus maculatus*, Ecol. Entomol. 11, 41-50.

Dick, K. M. and Credland, P. F. (1984) Egg production and development of three strain of *Callosobruchus maculatus* (F.) (Coleoptera: Bruchidae), J. stored Prod. Res. 20, 221-227.

Fujii, K. (1968) Studies on interspecies competition between the azuki bean weevil and the southern cowpea weevil. III. Some characteristics of strains of two species, Res. Popul. Ecol. 10, 87-98.

Fujii, K. (1969) Studies on interspecies competition between the azuki bean weevil and the southern cowpea weevil. IV. Competition between strains, Res. Popul. Ecol. 11, 84-91.

Fujii, K. (1970) Studies on interspecies competition between the azuki bean weevil and the southern cowpea weevil. V. The role of adult behavior in competition, Res. Popul. Ecol. 12, 233-242.

Giga, D. P. and Smith, R. H. (1981) Varietal resistance and intraspecific competition in the cowpea weevils *Callosobruchus maculatus* and *C. chinensis* (Coleoptera: Bruchidae), J. Appl. Ecol. 18, 755-761.

Heong, K. L. (1981) Searching preferences of the parasitoid, *Anisopteromalus calandrae* (Howard) for different stages of the host, *Callosobruchus maculatus* (F.) in the laboratory, Res. Popul.

Ecol. 23, 177-191.

May, R. M., Conway, G. R., Hassell, M. P., and Southwood, T. R. E. (1974) Time delays, Density dependence, and single-species oscillations, J. Anim. Ecol. 43, 747-770.

Maynard Smith, J. and Slatkin, M. (1973) The stability of predator-prey systems, Ecology 54, 384-391.

Messina, F. J. and Renwick, J. A. A. (1985) Mechanism of egg recognition by the cowpea weevil *Callosobruchus maculatus*, Entomol. Exp. Appl. 37, 241-245.

Oshima, K., Honda, H., and Yamamoto, I. (1973) Isolation of an oviposition marker from azuki bean weevil, *Callosobruchus chinensis* (L.), Agr. Biol. Chem. 37, 2679-2680.

Sakai, A., Honda, H., Oshima, K., and Yamamoto, I. (1986) Oviposition marking pheromone of two bean weevils, *Callosobruchus chinensis* and *Callosobruchus maculatus*, J. Pesticide Sci. 11, 163-168.

Shinoda, K. and Yoshida, T. (1987) Effect of fungal feeding on longevity and fecundity of the azuki bean weevil, *Callosobruchus chinensis* (L.) (Coleoptera: Bruchidae), in the azuki bean field, Appl. Entomol. Zool. 22, 465-473.

Utida, S. (1941a) Studies on experimental population of the azuki bean weevil, *Callosobruchus chinensis* (L.) I. The effect of population density on the progeny populations, Mem. College Agr. Kyoto Univ. 48, 1-30.

Utida, S. (1941b) Studies on experimental population of the azuki bean weevil, *Callosobruchus chinensis* (L.) III. The effect of population density upon the mortalities of different stages of life cycle, Mem. College Agr. Kyoto Univ. 49, 21-42.

Utida, S. (1941c) Studies on experimental population of the azuki bean weevil, *Callosobruchus chinensis* (L.) IV. Analysis of density effect with respect to fecundity and fertility of eggs, Mem. College Agr. Kyoto Univ. 51, 1-26.

Utida, S. (1941d) Studies on experimental population of the azuki bean weevil, *Callosobruchus chinensis* (L.) V. Trend of population density at the equilibrium position, Mem. College Agr. Kyoto Univ. 51, 27-34.

Utida, S. (1942) Studies on experimental population of the azuki bean weevil, *Callosobruchus chinensis* (L.) VII. Analysis of the density effect in the preimaginal stage, Mem. College Agr. Kyoto Univ. 53, 19-31.

Utida, S. (1950) On the equilibrium state of the interacting population of an insect and its parasite, Ecology 31, 165-175.

Utida, S. (1957a) Cyclic fluctuation of population density intrinsic to the host-parasite system, Ecology 38, 442-449.

Utida, S. (1957b) Population fluctuation, an experimental and theoretical approach, Cold Spring Harbor Symp. Quantitative Biol. 22, 149-151.

Utida, S. (1967) Damped oscillation of population density at equilibrium, Res. Popul. Ecol. 9, 1-9.

Wasserman, S. S. (1986) Genetic variation in adaptation to foodplants among populations of the southern cowpea weevil, *Callosobruchus maculatus*: Evolution of oviposition preference, Entomol. Exp. Appl. 42, 201-212.

SIMULATION OF PROFESSOR UTIDA'S CLASSIC EXPERIMENT ON AN INTERACTION BETWEEN THE AZUKI BEAN WEEVIL AND ITS PARASITIC WASP

T. ROYAMA

Forestry Canada—Maritimes Region, P. O. Box 4000,
Fredericton, New Brunswick E3B 5P7, Canada

ABSTRACT. This paper attempts to further analyze and synthesize Professor Utida's classic experiment with the azuki bean weevil, *Callosobruchus chinensis*, and the braconid wasp, *Heterospilus prosopidis* (Viereck). First, the density-dependent structure of the host-parasitoid interaction system is examined by the reproduction surfaces of both species. To facilitate the examination, the concept of conditional reproduction curves is introduced. A simple model is then constructed to represent the reproduction surfaces. The fluctuation in the wasp sex ratio is shown to be a major perturbation factor that causes the deviations of data points from the wasp reproduction surface. Finally, a simulation of the system dynamics is attempted, using the estimated density-dependent structure of the system and the hypothesis of perturbation by the variation in the wasp sex ratio. The dynamic pattern thus simulated closely resembles the observed pattern.

1. Introduction

Figure 1 illustrates the population fluctuations which Professor Utida (1956, 1957a, b) found experimentally in the laboratory, using *C. chinensis* as host and *H. prosopidis* as parasitoid. The graphs are routinely cited in ecology text books as a typical example of a host-parasitoid interaction system. The aim of this paper is to show that much of the observed pattern can be reproduced by a simple model built on a few basic principles underlying a host-parasitoid interaction.

2. Life Cycles and Experimental Setup

The parsitic wasp, *H. prosopidis*, is native to the southern United States and Mexico and is not a natural parasitoid of *C. chinensis*, although it readily parasitizes this adopted host in the laboratory. A female wasp attacks a late third to fourth instar larva or pupa inside a bean. The female controls the sex of its offspring by either fertilizing the eggs as they pass the oviduct (to produce female offspring), or not fertilizing them (to produce male offspring). Female offspring tend to issue from larger hosts and male offspring from smaller ones (Jones, 1982; see also Fujii and Khin Mar Wai, in this volume).

Utida kept experimental populations in petri dishes (8.5 cm across and 1.8 cm deep) containing dry azuki beans (*Vigna angularis*) in total darkness at a constant 30°C and 75% R.H. Under these conditions, the host completed its life cycle in about 20 days, while the parasitoid developed in half as many days. In order to maintain the host-parasitoid system without interruption, Utida used two series of host populations with the phase shift of one life cycle, i.e., 10 days, so that the single series of the wasp popu-

K. Fujii et al. (eds.), Bruchids and Legumes: Economics, Ecology and Coevolution, 385–394.
© 1990 *Kluwer Academic Publishers. Printed in the Netherlands.*

386

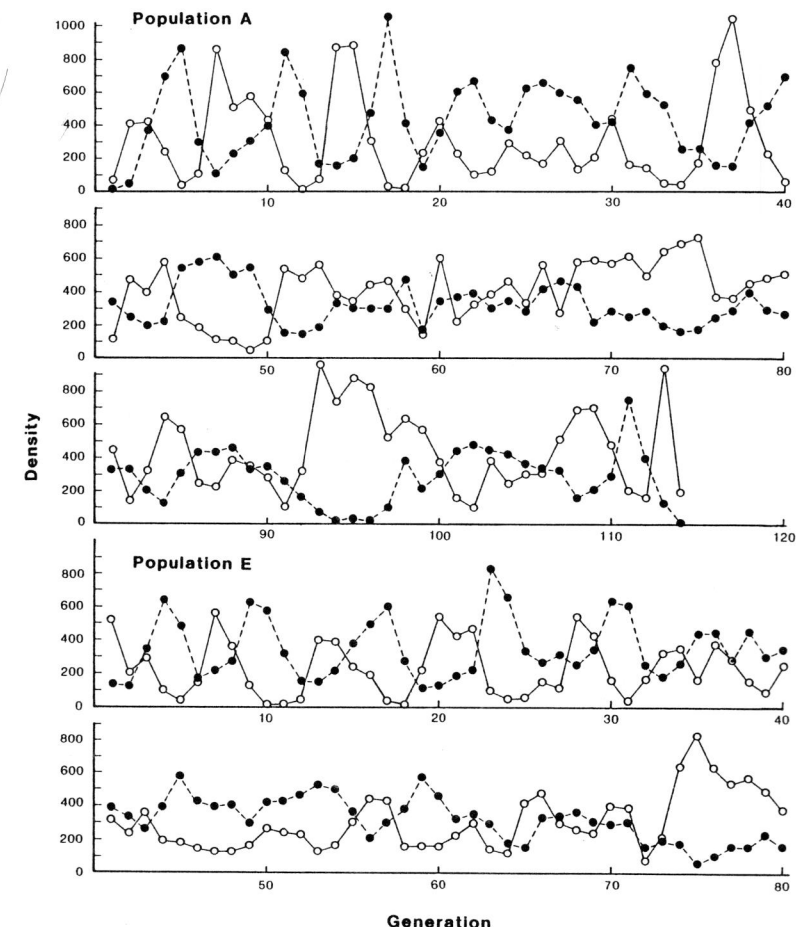

Figure 1. Fluctuations in the experimental populations of *C. chinensis* as host (open circles) and *H. prosopidis* as parasitoid (solid circles) by Utida (1956, 1957a, and b).

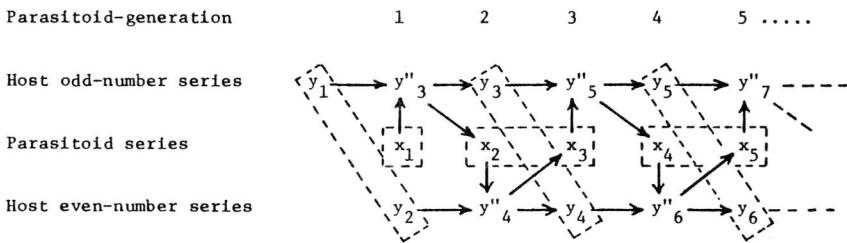

Figure 2. Experimental scheme by Utida (personal communication). The number of host adults, susceptible hosts, and parasitoid adults are denoted by y, y″, and x, respectively.

lation could attack the two host series alternately (Fig. 2).

Figure 3 shows the population fluctuations in the experimental setup, in which each data point is plotted against a parasitoid generation. When two successive data points in both species (enclosed in a dashed rectangle shown in Fig. 2) are combined, we get the graphs shown in Fig. 1.

3. Principles of Analysis

In general, a host-parasitoid interaction system has a density-dependen: structure, i.e., the rate of change in the density of each species from one generation to the next is dependent on the densities of both species. The density-dependent structure of one species (e.g., parasitoid) can be represented by a three-dimensional graph in which the rate of change in the parasitoid density is plotted against its own density, as well as against host density. Let us call such a graph a reproduction surface for the parasiatoid.

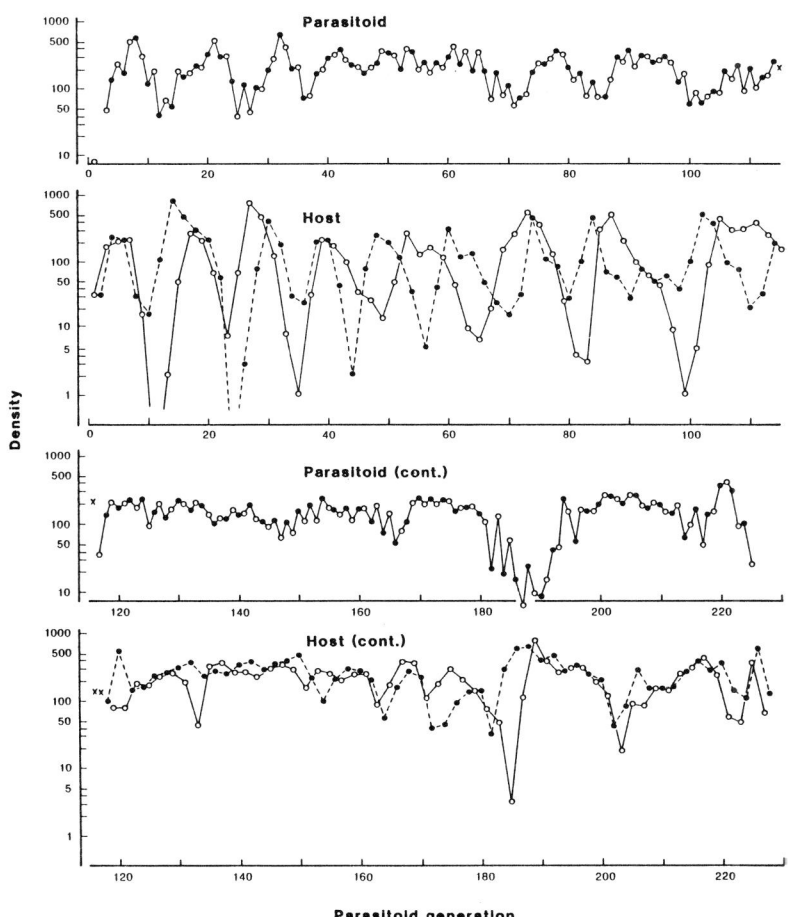

Figure 3.   The original, unpublished data for Population A in Fig. 1 by Utida under the experimental scheme of Fig. 2 (courtesy of Professor Utida).

388

If we plot the rate of change in the host density against its own density, as well as against the parasitoid density, we get a reproduction surface for the host.

Because three-dimensional graphs are awkward to draw and view, each surface can be sliced into its profiles to give the usual two-dimensional graphs. For example, we can plot the rate of change in the parasitoid density against its own density while the host density is fixed at an arbitrary value. This forms a profile of the reproduction surface for the parasitoid; we can slice the surface into as many profiles as we wish. Similarly, we can plot the rate of change in the parasitoid density against the host density while the initial density of the parasitoid is fixed at as many arbitrary values. These are profiles perpendicular to the first set of profiles. We can slice the host reproduction surface in the same way. I shall call each profile a conditional reproduction curve.

Figure 4 illustrates typical shapes often assumed by the four conditional reproduction curves. In particular, curves a and d represent the self-regulation of parasitoid and host, respectively. At a given density of the other species, the reproductive rate of one species generally declines as its density increases. Curve c shows the typical shape of a functional response of the parasitoid. As is well known, the parasitoid reproductive rate tends to increase as the host becomes more numerous because the parasitoid can attack more host individuals, although there is an upper limit to the number of hosts a female wasp can exploit. Curve b is, naturally, inversely proportional to parasitism, which increases as the parasitoid density increases. Needless to say, quantitative details of the shape in each conditional curve would depend on the value of the fixed density.

4. Data Analysis

Figure 5 shows examples of realized conditional reproductive rates (data points) in the experiment shown in Fig. 3, regressed on density of the parent generation in each species. The densities on which the rates are conditional are conveniently chosen in the interval from 100 to 200. These graphs, corresponding to those in Fig. 4, are now scaled

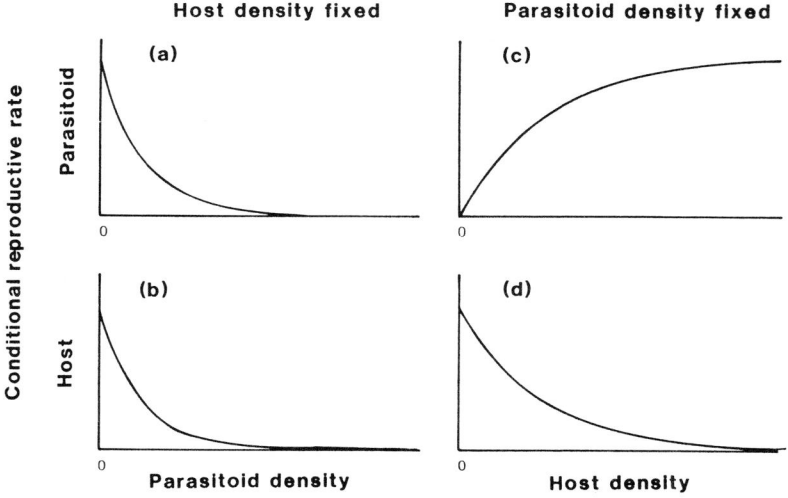

Figure 4. Typical conditional reproduction curves of a host-parasitoid interaction system, indicating self regulation ((a) and (d)), parasitoid functional response (c), and host responses to parasitoids (b).

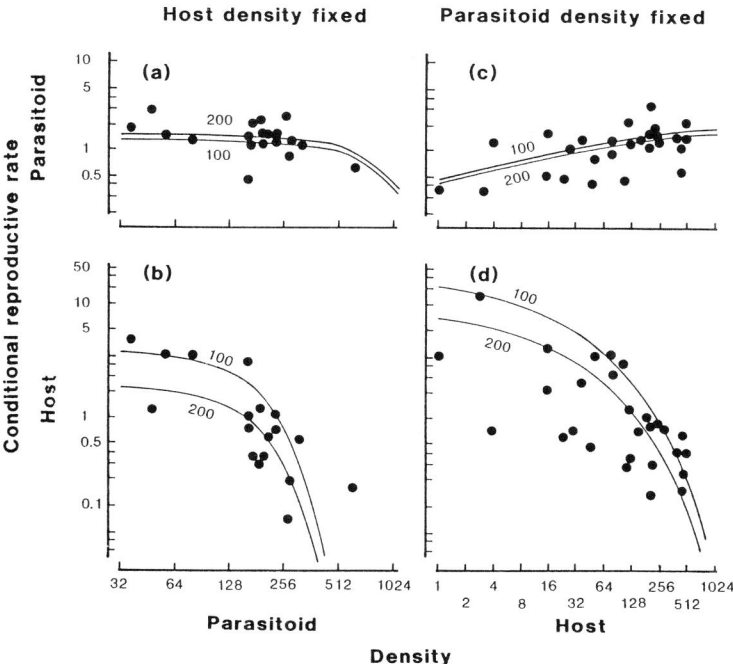

Figure 5. The reproductive rates of both species realized in the series in Fig. 3 but conditional on the fixed densities falling in the interval (100, 200). A pair of curves in each graph is the expected conditional reproduction curves of a simple model.

in logarithms. Fitted in each graph are two conditional reproduction curves - for densities fixed at 100 and 200 - expected from the simplest of a few models that I have formulated for the analysis of Utida's experimental system.

Given initial densities of both species, we can calculate their densities in the next generation from the model reproduction surfaces which give the rates of change in both species between the two generations. Recursive applications of the same operation to the model would successively generate series of populations in both species (Fig. 6). However, the above model, as it stands, ignores the factors which must have perturbed the density-dependent structure. Even though the experiment was carried out in a constant physical environment, a number of things fluctuated at random and perturbed the density-dependent structure. Thus, the deviations of data points in Fig. 5 from the expected curves reflect such perturbations. A major perturbation results from the variation in sex ratio of the wasps shown in Fig. 7 in which the wasp generations alternately attacking the two host series are distinguished from one another by open and solid circles. It appears that the two series of sex ratios fluctuate somewhat cyclically, which can by confirmed by calculating autocorrelations in the series. But why did the sex ratios fluctuate cyclically?

As already mentioned, female wasps tend to lay unfertilized (male) eggs on small hosts and fertilized (female) eggs on large hosts. In the experiment, the average age of susceptible hosts at the time wasps emerge as adults could change from generation to generation. This is because the length of the life cycle of each species was determined entirely by the constant temperature and humidity conditions, which could result in a

390

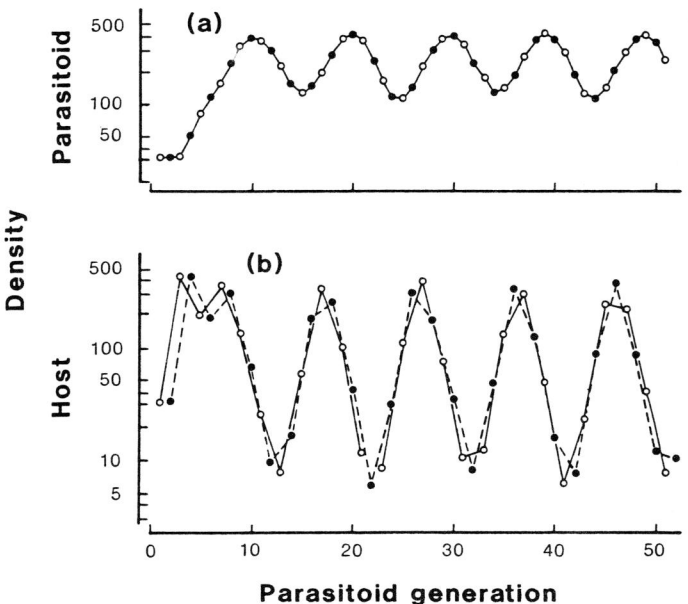

Figure 6. Deterministic series of parasitoid (graph a) and host (graph b) generated by the reproduction surfaces of the simple model.

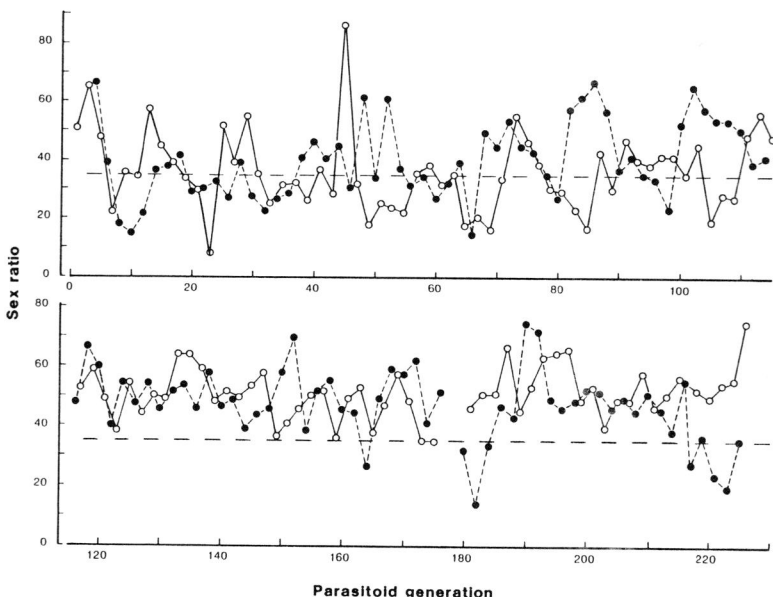

Figure 7. The observed variation in the wasp sex ratio (% females). A dashed horizontal line is the mean percentage up to the 110th generation.

small shift from generation to generation in the average age of the hosts at the time most wasps emerge. Consequently, in some generations, the searching wasps may encounter a larger proportion of comparatively young (hence, small) hosts and tend to produce more male wasps and, in other generations, more females are produced because hosts are, on average, larger. Thus, the sex ratio tends to vary cyclically. The autocorrelation analysis mentioned above shows an average cycle length of about 9 generations in each series, and the two series oscillate alternately with the phase shift of one half cycle.

Figure 7 also reveals that the average sex ratio for the early half of the experiment (up to the 110th generation) is about 35% female, but increases to about 50% for the latter half. This means, probably, that the phase shift in the life cycle synchrony between the two species has a much longer cycle, in addition to the 9-generation cycle. Sex ratio in the host population also fluctuates but does so in a nearly random fashion and steadily around an average of 45% females.

Notice now that the population fluctuations in Fig. 3 change their pattern coincidentally with the change in the average proportion of female wasps in about the 100 and 110th generations. The change in the pattern of fluctuation is particularly noticeable in the host population series. The fluctuations are strongly damped out after the 110th generation.

The coincidence between the population and sex ratio fluctuations appears to be associated with wasp reproductive capacity. Utida shows that the average number of female wasps was more or less the same before and after the 110th generation. However, the average proportion of female wasps was lower before the 110th generation because of an excessive number of male wasps. This means that the average number of offspring (both male and female) per female, was somehow higher before, than after, the 110th generation. In other words, the average reproductive output of a female wasp was inversely correlated with the wasp sex ratio (the proportion of female wasps).

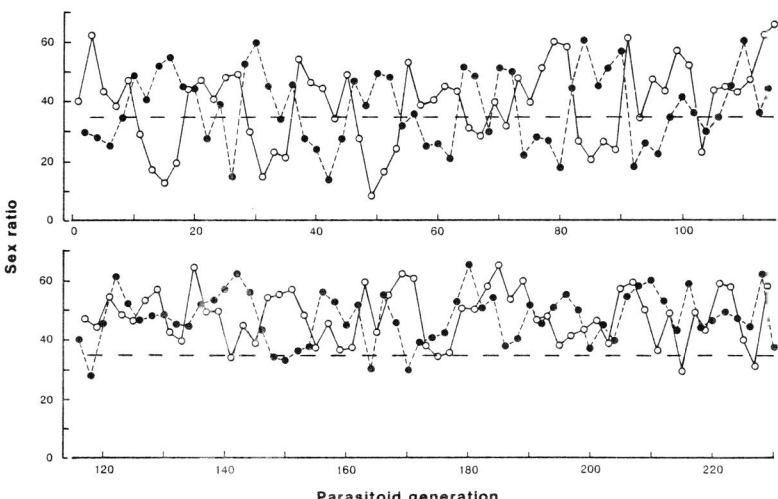

Figure 8. Simulation of the variation in the wasp sex ratio compared to the observed variation in Fig. 7. A dashed horizontal line is the same as one in Fig. 7.

392

5. Simulation

In Fig. 8, two series of sex ratios in the wasp population are generated to simulate the observed sex ratio fluctuations in Fig. 7. Each series was generated using a square periodic function superimposed by uncorrelated random numbers. The quantitative properties of the simulated series are such that their periodicity, phase shift, average sex ratios, and variances are the same as those in the observed series of Fig. 7.

As already mentioned, an average reproductive rate of a female wasp is inversely correlated with the wasp sex ratio. This property can be incorporated into the model by assuming that a realized reproductive rate tends to deviate from the expected rate in inverse proportion to the simulated sex ratio. A result of the simulation in Fig. 9 closely resembles the observed pattern in Fig. 3. The simple model thus explains much of the observed pattern.

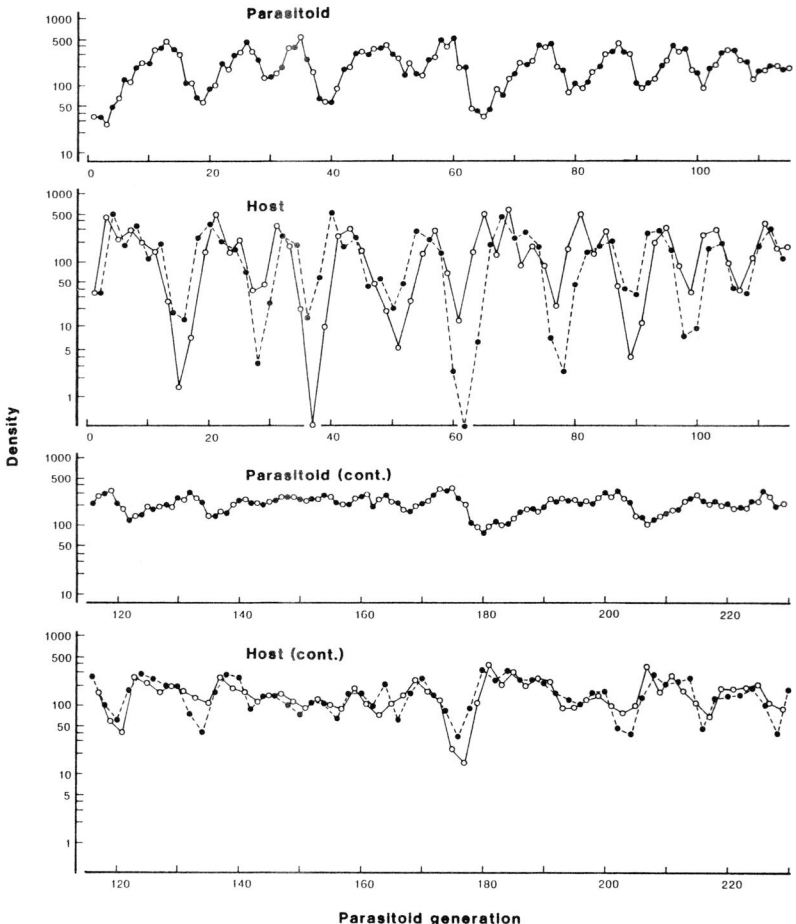

Figure 9. A simulation of the host-parasitoid series by incorporating the effect of sex-ratio variation in Fig. 8 in the deterministic population series of Fig. 6.

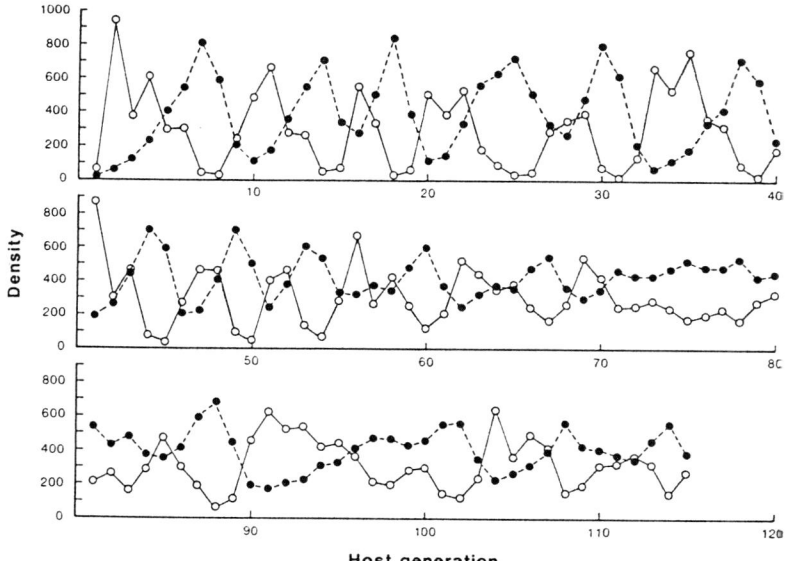

Figure 10. The same as Fig. 9 but the densities in the adjacent odd- and even-number generations are combined in both species after the scheme of Fig. 2 to make the simulation directly comparable with the observed series in Fig. 1.

Some obvious differences remain, however. Notably, the simulated wasp series is much smoother than the observed series which tends to exhibit a saw-tooth fluctuation. The reason for the difference is that the two observed host series tend to be more widely separated from one another than do those in the simulation. The two series are so separated from one another that the host populations in adjacent wasp generations tend to alternate between low and high densities. Consequently, the wasp population, which attacks low and high host densities alternately, tends to oscillate like saw-teeth. This tendency is due to a nonlinear interaction between the host and parasitoid populations. The present simulation model is too simplistic to reproduce the nonlinear property of the observed dynamics.

One way to eliminate the effect of the above nonlinear property of the experimental system is to combine data points in two successive generations as in Fig. 1. The results are shown in Fig. 10, which is almost indistinguishable from the observed dynamics.

Acknowledgements.   I am grateful to Professor Utida for most generously providing me with his original numerical data and for frequent discussions and valuable advice. I also thank Dr. M. Shimada of Tokyo University for critically reading my early manuscript.

REFERENCES

Fujii, K. and Khin Mar Wai (1989) Sex-ratio determination in three wasp species ectoparasitic on bean weevil larvae (this volume).

Jones, W. T. (1982) Sex ratio and host size in a parasitoid wasp, Behav. Ecol. Sociobiol. 10, 207-210.

Utida, S. (1956) Long-term fluctuation of population in the system of host-parasite interaction, Res. Popul. Ecol. 3, 52-59 (in Japanese with English summary).

Utida, S. (1957a) Cyclic fluctuation of population density intrinsic to the host-parasite system, Ecology

394

38, 442-449.

Utida, S. (1957b) Population fluctuation, an experimental and theoretical approach, Cold Spring Harbor Symp. Quantitative Biol. 22, 139-151.

A LIST OF THE LEGUMES TREATED IN THE BOOK AND IN THE SYMPOSIUM (ISBL-II)

Scientific Name [= *Synonym*]: Common Name

Aeschynomene indica L.
Albizia julibrissin Durazz.
Albizia procera (Roxb.) Benth.
Amphicarpaea edgeworthii Benth. var. japonica Oliver
Arachis hypogaea L.: peanut, groundnut
Bauhinia rufescens Lamarck
Caesalpinia sepiaria Roxb. var. japonica (Sieb. & Zucc.) Makino
Cajanus cajan (L.) Millsp. [= *C. indicus*]: pigeon pea
Cajanus indicus Spreng. —>Cajanus cajan
Calopogonium caeruleum (Benth.) Hemsl.
Calopogonium mucunoides Desv.: calopo
Canavalia braziliensis Mart. ex Benth.
Canavalia ensiformis (L.) DC.: jack bean
Canavalia gladiata (Jacq.) DC.: sword bean
Castanospermum australe A. Cunn.
Centrosema plumieri (Turp. ex Pers.) Benth.
Centrosema pubescens Benth.: butterfly pea, centro
Cercidium floridum Benth.
Cicer arietinum L.: chickpea, gram, garbanzo
Dioclea megacarpa Rolfe
Dolichos lablab L. —>Lablab purpureus
Dunbaria villosa (Thunb.) Makino
Enterobium cyclocarpum (Jacq.) Griseb.
Galactia striata (Jacq.) Urban
Gleditsia japonica Miquel
Glycine max (L.) Merr.: soybean
Glycine soja (L.) Sieb. & Zucc.
Indigofera pseudotinctoria Matsum.
Indigofera suffruticosa Mill.
Lablab niger Medic. —>Lablab purpureus
Lablab purpureus (L.) Sweet [= *L. niger, Dolichos lablab*]: hyacinth bean, lablab bean, bonavist
Lathyrus latifolius L.: sweet pea
Lathyrus maritimus (L.) Bigel.: beach pea, sea pea
Lathyrus sativus L.: grass pea
Lens culinaris Medic.[= *L. esculenta*]: lentil
Lens esculenta Moench —>Lens culinaris
Lespedeza bicolor Turcz.
Leucaena leucocephala (Lamarck) De Wit: leucaena
Lonchocarpus salvadorensis Pittier
Millettia japonica (Sieb. & Zucc.) A. Gray
Mimosa pudica L.: sensitive plant
Phaseolus angularis (Willd.) Wight —> Vigna angularis
Phaseolus aureus Roxb. —> Vigna radiata
Phaseolus calcaratus Roxb. —> Vigna umbellata
Phaseolus coccineus L.: scarlet runner bean
Phaseolus lunatus L.: lima bean, sugar bean

(continued)

K. Fujii et al. (eds.), Bruchids and Legumes: Economics, Ecology and Coevolution, 395–396.
© 1990 *Kluwer Academic Publishers. Printed in the Netherlands.*

Phaseolus mungo L. —> Vigna mungo
Phaseolus radiatus L. —> Vigna radiata
Phaeolus sinensis (L.) Savi & Hassk. —> Vigna unguiculata
Phaseolus vulgaris L.: kidney bean, common bean, bean, French bean, haricot bean
Pisum sativum L.: pea, common pea, field pea, garden pea
Psophocarpus tetragonolobus (L.) DC.: winged bean
Pueraria lobata (Willd.) Ohwi: kudzu
Sesbania emerus (Aubl.) Britton & Wilson
Sesbania grandiflora (L.) Pers.
Sophora flavescens Aiton
Tamarindus indica L.: tamarind
Vicia cracca L.: bird's tare, tuffed vetch
Vicia faba L.: broad bean
Vicia sepium L.
Vigna aconitifolia (Jacq.) Maréchal: moth bean
Vigna adenantha (G. F. Meyer) Maréchal, Mascherpa & Stainier
Vigna angularis (Willd.) Ohwi & Ohashi [= *Phaseolus angularis, Azukia angularis*] : adzuki bean, azuki bean, small red bean
Vigna angularis (Willd.) Ohwi & Ohashi var. nipponensis (Ohwi) Ohwi & Ohashi [= *V. trilobata* sensu auctt. Jap., *Azukia angularis* var. *nipponensis*]
Vigna glabrescens Maréchal, Mascherpa & Stainier
Vigna khandalensis (Santapau) Raghavan & Wadhwa
Vigna kirkii (Baker) Gillett
Vigna luteola (Jacq.) Benth.
Vigna marina (Burm.) Merr.
Vigna mungo (L.) Hepper [= *Phaseolus mungo*]: black gram, black matpe, urd bean
Vigna mungo (L.) Hepper var. *silvestris* Lukoki, Maréchal & Otoul
Vigna nakashimae (Ohwi) Ohwi & Ohashi
Vigna nuda N. E. Br.
Vigna oblongifolia (Benth.) Verdc.
Vigna racemosa (G. Don.) Hutch. & Dalz.
Vigna radiata (L.) Wilczek [= *Phaseolus radiatus, P. aureus*]: green gram, mung bean
Vigna radiata (L.) Wilczek var. sublobata (Roxb.) Verdc. = Vigna sublobata
Vigna reflexo-pilosa Hayata
Vigna riukiuensis (Ohwi) Ohwi & Ohashi
Vigna sesquipedalis (L.) Fruwirth —> Vigna unguiculata var. sesquipedalis
Vigna sinensis (L.) Savi & Hassk. —> Vigna unguiculata
Vigna stipulacea (Lamarck) Tateishi
Vigna sublobata (Roxb.) Babu & Sharma = V. radiata var. sublobata
Vigna umbellata (Thunb.) Ohwi & Ohashi [= *Phaseolus calcaratus*]: rice bean, bamboo bean, peyin bean
Vigna unguiculata (L.) Walp.[= *V. sinensis, Phaeolus sinensis*]: cowpea, black eye (for special variety)
Vigna unguiculata (L.) Walp. var. sesquipedalis (L.) Ohashi [= *V. sesquipedalis*]: yard long bean, string bean, asparagus bean
Vigna vexillata (L.) A. Rich.
Vigna vexillata (L.) A. Rich. var. tsushimensis Matsum.
Voandzeia subterranea (L.) Thou.: bambara groundnut

Prepared by Yoichi Tateishi[1] and Naoshi Watanabe[2]
[1] Biological Institute, Faculty of Science, Tohoku University, Sendai 980, Japan
[2] Kobe Plant Protection Station, 1-1 Hatobacho, Chuoku, Kobe 650, Japan

Index